FLUID DYNAMICS FOR PHYSICISTS

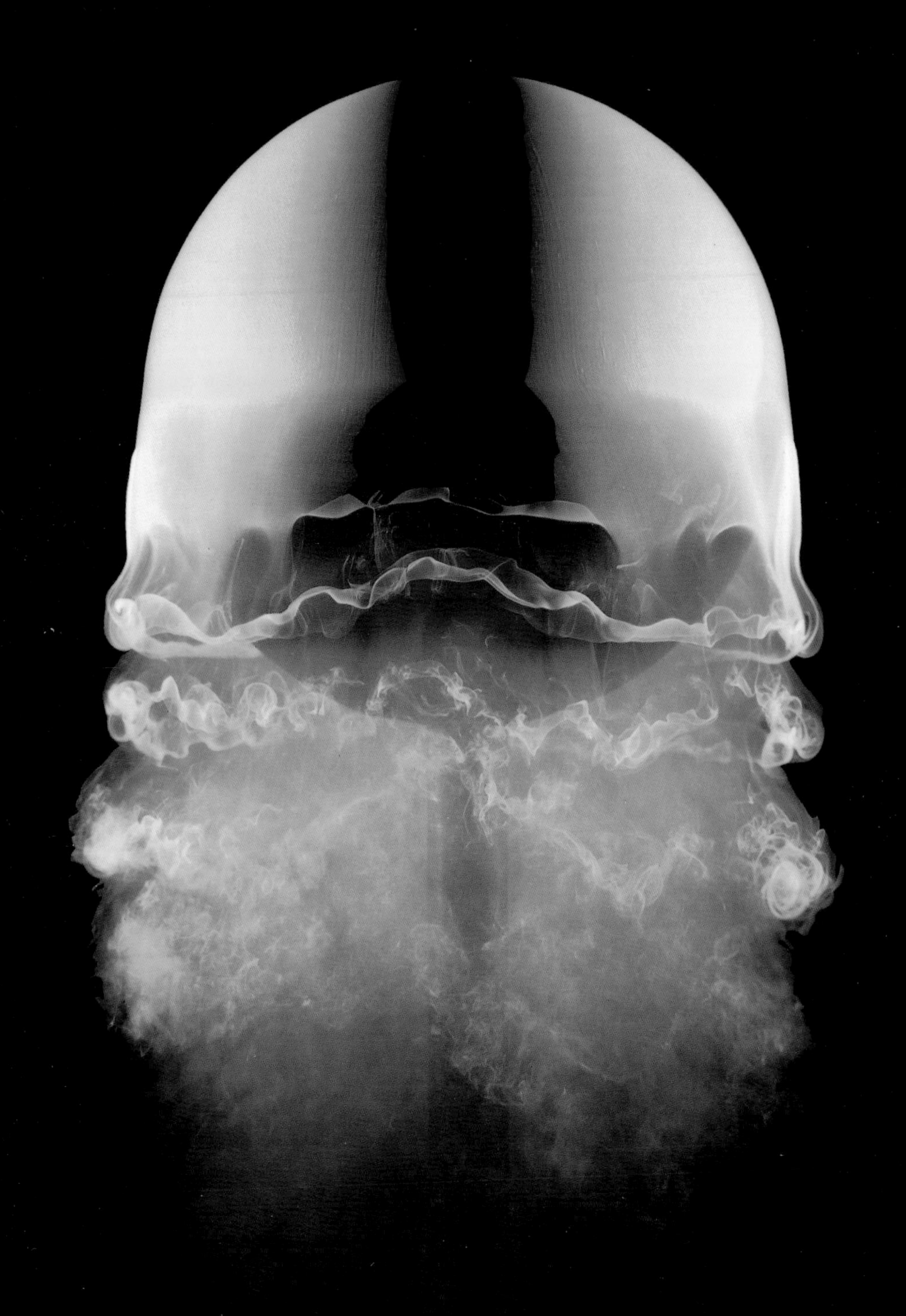

T. E. Faber

It is over 350 years since Toricelli discovered the law obeyed by fountains, yet fluid dynamics remains an active and important branch of physics. Modern examples of the extraordinarily varied and intriguing phenomena with which it deals include solitons, chaotic behaviour in convecting systems and non-Newtonian effects in polymer solutions. This book provides an accessible and comprehensive account of the subject, emphasising throughout the fundamental physical principles, and stressing the connections with other branches of physics.

Beginning with a gentle introduction, the book goes on to cover: Bernoulli's theorem and its implications; compressible flow and shock waves; potential flow; surface waves and ship waves; viscosity and highly viscous flow; vorticity dynamics and boundary layers; thermal convection and instabilities; turbulence and turbulent shear flow; the behaviour of non-Newtonian fluids, including nematic liquid crystals and liquid helium; and the propagation and attenuation of sound in gases.

Undergraduate or graduate students in physics or engineering who are taking courses in fluid dynamics will find this book invaluable, but it will also be of great interest to anyone who wants to find out more about this fascinating subject.

FLUID DYNAMICS FOR PHYSICISTS

FLUID DYNAMICS FOR PHYSICISTS

T. E. FABER

Cavendish Laboratory
and
Corpus Christi College,
Cambridge

Published by the Press Syndicate of the University of Cambridge
The Pitt Building, Trumpington Street, Cambridge CB2 1RP
40 West 20th Street, New York, NY 10011-4211, USA
10 Stamford Road, Oakleigh, Melbourne 3166, Australia

First published 1995

Printed in Great Britain at the University Press, Cambridge

A catalogue record for this book is available from the British Library

Library of Congress cataloguing in publication data
Faber, T. E.
Fluid dynamics for physicists/T. E. Faber. p. cm.
Includes index.
ISBN 0 521 41943 3 – ISBN 0 521 42969 2 (pbk.)
1. Fluid dynamics. I. Title.
QC151.F33 1995
532'.05–dc20 94-19245 CIP

ISBN 0 521 41943 3 hardback
ISBN 0 521 42969 2 paperback

PN

For
Liesbeth, Sophie and Benjamin

Contents

Preface

Every physicist should know some fluid dynamics, and every university physics department should include the subject in its core curriculum. Those propositions can readily be justified by pointing out the usefulness of the subject – its relevance to diverse areas of contemporary research and to a vast range of problems of practical importance. What counts as much for me, however, is that most of the students I have known at Cambridge have enjoyed their limited exposure to it. The notion that the only way to arouse the enthusiasm of physicists is to teach them about quarks and black holes is in my view a myth.

I hope that this book will slightly increase the chance that future generations of physicists will be taught the subject systematically, in a way that I and my contemporaries were not. However, since newer branches of physics may continue to displace it, I have tried to write something that may be read for pleasure as well as for instruction by physicists of any age and at almost any level of sophistication who want to learn fluid dynamics for themselves. They do, of course, have many books to choose from already, but most of them were written for mathematicians or engineers. Students of all three disciplines – mathematics, physics and engineering – speak the same language and have many objectives in common, but they differ in their approach to new problems because their intuition has been honed in different ways, and they also tend to differ in what they find interesting. Books written for people who are studying or have studied mathematics may put off a reader trained as a physicist because they presuppose skills which he does not possess, or because they seem to him to lay undue emphasis on problems for which analytical solutions happen to be available, or simply because they are too narrow in their scope. Books written for engineers are likely to put him off for other reasons: perhaps because the treatment seems imprecise in places, or perhaps because it is so compendious that the wood cannot be seen for the trees. I do not wish to seem dismissive of books in these categories. I have learnt most of what I know about fluid dynamics by reading them, and in particular by reading (as every serious student of the subject should) George Batchelor's *An Introduction to Fluid Dynamics*, Cambridge University Press, 1967, a classic text which has no rival in its class. None of them is wholly suited, however, for the readership I aim to address.

I know of three books on fluid dynamics which were written by physicists for physicists, each of them excellent in its own way. Ludwig Prandtl's *Essentials of Fluid Dynamics*, Blackie, London and Glasgow, 1952, an English translation of his *Führer durch die Strömungslehre*, was written by one of the great pioneers of the subject; it is coloured by his exceptional insight and enlivened by intriguing tidbits of information concerning an exceptionally wide range of topics. However, it is too unsystematic to form a satisfactory textbook, besides being out of date in some respects and also out of print. Landau and Lifshitz's *Fluid Dynamics*, Pergamon Press, Oxford, 2nd edition 1987, is widely regarded as one of the more successful volumes in a famous series which covers most branches of theoretical physics, but it is too advanced to serve the needs of undergraduate students. D. J. Tritton's *Physical Fluid Dynamics*, Clarendon Press, Oxford, 2nd edition 1988, provides an individual and very readable account, informed by personal experience and supplemented by extensive references, of developments in the subject during the last few decades, but it is perhaps better suited to readers who have already learnt some fluid dynamics and wish to have their perspectives enlarged than to absolute beginners.

My qualifications for writing a book to stand comparison with those just mentioned are slender. Where the authors of those books are all as it were professionals, I am merely an enthusiastic amateur: I was not taught the subject properly as a student myself; the level at which I have since taught it to others has been superficial; and although my research interests have at different times embraced liquid metals and liquid crystals I have had little to do with the flow properties of these substances. Still, I have tried to make up for lack of experience in the field by reading widely – the sources on which I have principally relied are indicated in the sections headed 'Further reading' which are to be found at the end of each chapter – and by thinking hard about arguments which I have found obscure or otherwise unsatisfactory. Substantial parts of the text are original, in presentation if not in content, and have something useful to offer, or so I believe, to others besides those for whom it is primarily intended. Dr Johnson compared a woman's preaching with a dog's walking on its hinder legs in the withering phrase: 'It is not well done; but you are surprised to find it done at all'. I have detected surprise-to-find-it-done-at-all in several colleagues who have learnt of this book before its publication, and I expect it to be greeted with a good deal of such surprise by the professional sorority. But without evidence that it would be judged overall in terms less dismissive than Dr Johnson's the book would never have been completed or found a publisher.

Incidentally, my views on matters of gender are not as prejudiced as those of Dr Johnson. I have used the word 'sorority' in the previous paragraph to demonstrate my impartiality, and because it is shorter than 'fraternity'. It is on the same principle that I refer to the reader in several places later on as 'he' rather than 'she'. That pronoun is to be read, of course, as an abbreviation for 'he or she'.

The hardest things in writing any book are to decide where to start and when to stop. I have started from an assumption that my readers have no previous experience of continuum mechanics and therefore need to be introduced to the concept of stress; I effect the introduction in the gentlest possible way, without plunging into tensor notation. I also provide introductions to peripheral topics such as dimensional analysis and group velocity; most readers will have met these before in other contexts, no doubt, but perhaps only at an elementary level. On the other hand, I assume in places a superficial familiarity with the kinetic theory of gases and with electromagnetism. More significantly, I do not provide line-by-line derivations for every mathematical result, since I expect readers to be able to effect the necessary manipulations and integrations for themselves, or else to take the results on trust. I use partial differentials and vector notation without explanation, and I expect readers to recognise a number of standard identities in vector algebra which are collected for reference at the beginning of the book (where a list of symbols is also to be found). To claim that the book is genuinely accessible 'at almost any level' would obviously be an exaggeration, but anyone who has survived at least one year of undergraduate physics should be able to make some headway through it.

So much for starting; what about stopping? My general aim has been to provide readers with a solid understanding of each of the topics discussed but to stop short of technical details that I myself find unexciting. As regards the choice of topics I became increasingly ambitious as the writing progressed, so that what was to have been a fairly slim book, mainly devoted to the laminar flow of fluids that can be treated as incompressible and Newtonian, is now distinctly plump. It has put on weight in chapter 3, which concerns compressible flow and shock fronts, in chapters 7, 8, 9 and 10, which respectively concern vorticity and boundary layers, instabilities and thermal convection, turbulence, and non-Newtonian fluids (including liquid crystals and liquid helium), and in an appendix which deals with sound propagation and the attenuation of sound in particular. These chapters, together with chapter 5 on surface waves, make the book more comprehensive than most others and will, I hope, enhance its usefulness. There are of course significant omissions, but most of the interesting topics I am conscious of having neglected, such as the Taylor–Proudman column in rotating fluids, are dealt with by Tritton with an authority which I could not pretend to match.

I am grateful to Mike Cross, who put at my disposal for a few pleasant months during the summer of 1990 a quiet room at CalTech, a word processor, and some empty book shelves previously used by Professor Feynman, all of which helped to get me started. First drafts of chapters 1–3 and of the appendix were subsequently read by Brian Pippard; he drew my attention to the formula which appears in the appendix as (A.25), besides making numerous useful suggestions concerning points of detail. A first draft of chapter 10 was read by Adrian Rennie; his constructively critical remarks led to a thorough revision. Others who have enlightened me on isolated matters include Robin Ball, David Maull, Howell Peregrine, Alan Townsend and Mark Warner. I am grateful to all of these for their help and encouragement; to the copy editor, Jo Clegg, who subjected the whole text to a more thorough scrutiny than it will ever receive again and whose comments were extremely helpful; to Colin Yeomans, who redrew all the diagrams for me; and to several anonymous students who, by asking searching questions about fluid dynamics, have forced me to think about it more deeply.

Books on fluid dynamics need to be illustrated not only by diagrams but also by photographs, and perhaps my greatest debt of gratitude is to those who took the beautiful photographs which are reproduced below. I have made strenuous efforts to contact these people individually, besides seeking permission where appropriate from journals in which some of the photographs were first published, and everyone I have succeeded in contacting has been helpful. Many of them have gone to considerable trouble to supply me with prints of their work. Their cooperation has encouraged me to believe that the few who have not answered my letters are either no longer active or have moved without leaving forwarding addresses behind them, and that they too will not object to their photographs being reused.

The sources of photographs are of course acknowledged individually, in the figure captions. I should also acknowledge, however, that but for Professor Milton Van Dyke's marvellous collection, *An Album of Fluid Motion*, Parabolic Press, Stanford, 1982, I would never have become aware of several of them. The best way I can express my gratitude to him, not only for putting this collection together in the first place but for letting me copy one or two items from it and for supplying me so freely with addresses, is to urge every reader to give his book the close attention it deserves. Splendid photographs are also to be found in a collection edited by the Japanese Society of Mechanical Engineers, published in an English version by Pergamon Press in 1988 under the title of *Visualized Flow*.

Tom Faber

Mathematical conventions

Coordinate systems

Where cartesian coordinates are employed, the axes are normally distinguished by suffices thus, (x_1, x_2, x_3), using the right-handed labelling convention. The vector $\boldsymbol{x}$ then represents position, and integration over $\mathrm{d}\boldsymbol{x}$ implies three-dimensional integration over $\mathrm{d}x_1$, $\mathrm{d}x_2$ and $\mathrm{d}x_3$. Components of velocity $\boldsymbol{u}$ along these axes are labelled (u_1, u_2, u_3), and components of other vectors are distinguished in the same way. In contexts where the flow is essentially one-dimensional, however, suffices are normally suppressed; the relevant axis is labelled x and velocity along it is labelled u. Elsewhere, coordinates (x, y, z) are sometimes used when the Earth's gravitational field is important, in which case velocity components are labelled (u_x, u_y, u_z); the z axis is always vertical, and gz represents gravitational potential throughout.

Spherical polar coordinates (R, θ, ϕ) are employed for problems to do with flow past spheres and suchlike. In these problems the flow is always axially symmetric, so the azimuthal angle is irrelevant and does not appear; the fact that ϕ is also conventionally used to denote flow potential should not, therefore, lead to confusion. Components of velocity in the directions of increasing R (for constant θ) and increasing θ (for constant R) are denoted (u_R, u_θ). Where it is desirable to use cartesian and spherical polar coordinates simultaneously, the symmetry axis, i.e. the axis along which $\theta = 0$, is taken as the x_1 direction.

For certain two-dimensional problems, e.g. for the problems of circulation round a line vortex or flow past a long cylinder with its axis perpendicular to the general direction of flow, cylindrical coordinates (r, θ, x_3) are appropriate. In such cases the azimuthal angle *is* important, and to prevent confusion with flow potential it has to be labelled θ rather than ϕ. The direction for which $\theta = 0$, which in the case of flow past a transverse cylinder corresponds to the general direction of flow, is equated to the x_1 direction, as in the spherical polar case.

Vectors

Vectors are represented by symbols in bold type. The differential operator $\boldsymbol{\nabla}$ is employed, rather than 'grad', 'div' and 'curl'. In cartesian coordinates this has components $(\partial/\partial x_1, \partial/\partial x_2, \partial/\partial x_3)$. Thus $\boldsymbol{\nabla}\phi$, where ϕ is a scalar quantity, is the vector otherwise denoted by grad ϕ. The scalar product $\boldsymbol{\nabla}\cdot\boldsymbol{u}$, where $\boldsymbol{u}$ is a vector, is the scalar otherwise denoted by div $\boldsymbol{u}$, i.e.

$$\boldsymbol{\nabla}\cdot\boldsymbol{u} = \frac{\partial u_1}{\partial x_1} + \frac{\partial u_2}{\partial x_2} + \frac{\partial u_3}{\partial x_3}.$$

The vector product $\boldsymbol{\nabla}\wedge\boldsymbol{u}$ corresponds to curl $\boldsymbol{u}$, i.e. (with right-handed axes)

$$(\boldsymbol{\nabla}\wedge\boldsymbol{u})_1 = \frac{\partial u_3}{\partial x_2} - \frac{\partial u_2}{\partial x_3}, \quad (\boldsymbol{\nabla}\wedge\boldsymbol{u})_2 = \frac{\partial u_1}{\partial x_3} - \frac{\partial u_3}{\partial x_1}, \quad (\boldsymbol{\nabla}\wedge\boldsymbol{u})_3 = \frac{\partial u_2}{\partial x_1} - \frac{\partial u_1}{\partial x_2}.$$

The scalar product of $\boldsymbol{\nabla}$ with itself, i.e. $\boldsymbol{\nabla}\cdot\boldsymbol{\nabla}$ or ∇^2, is the *Laplacian* operator,

$$\nabla^2 = \frac{\partial^2}{\partial x_1^2} + \frac{\partial^2}{\partial x_2^2} + \frac{\partial^2}{\partial x_3^2}.$$

When ∇^2 operates on a scalar quantity the result is a scalar. When it operates on a vector, however, the result is a vector. Thus $\nabla^2\boldsymbol{u}$ represents a vector such that

$$(\nabla^2\boldsymbol{u})_1 = \nabla^2 u_1, \quad (\nabla^2\boldsymbol{u})_2 = \nabla^2 u_2, \quad (\nabla^2\boldsymbol{u})_3 = \nabla^2 u_3.$$

Familiarity is assumed with the following aspects of vector analysis.

(i) *Gauss's divergence theorem*, which states that for any vector $\boldsymbol{u}$ we have

$$\int_\Sigma \boldsymbol{u}\cdot \mathrm{d}\boldsymbol{\Sigma} = \int_\mathrm{V} (\boldsymbol{\nabla}\cdot\boldsymbol{u})\mathrm{d}\boldsymbol{x},$$

the integral on the right being taken over all values of $\boldsymbol{x}$ lying within some volume V, and the integral on the left being taken over the whole of the surface Σ which encloses V, $\mathrm{d}\boldsymbol{\Sigma}$ being an element of area of this surface, represented by a vector pointing outwards.

(ii) *Stokes's theorem*, which states that

$$\oint_\mathrm{C} \boldsymbol{u}\cdot \mathrm{d}\boldsymbol{l} = \int_\Sigma (\boldsymbol{\nabla}\wedge\boldsymbol{u})\cdot \mathrm{d}\boldsymbol{\Sigma}.$$

Here the integral on the right is over any connected surface Σ which is bounded by the continuous curve C. On the left, $\mathrm{d}\boldsymbol{l}$ is an element of length along this curve, and the integral is over one complete circuit. If the surface and its bounding curve are viewed in a direction such that $\mathrm{d}\boldsymbol{\Sigma}$ points towards the observer rather than away from him, then $\mathrm{d}\boldsymbol{l}$ points round the curve in an anticlockwise direction.

(iii) *The triple product rules*,

$$\boldsymbol{a} \wedge (\boldsymbol{b} \wedge \boldsymbol{c}) = (\boldsymbol{a}\cdot\boldsymbol{c})\boldsymbol{b} - (\boldsymbol{a}\cdot\boldsymbol{b})\boldsymbol{c},$$

and

$$\boldsymbol{\nabla} \wedge (\boldsymbol{\nabla} \wedge \boldsymbol{u}) = \boldsymbol{\nabla}(\boldsymbol{\nabla}\cdot\boldsymbol{u}) - \nabla^2\boldsymbol{u}.$$

(iv) The fact that in circular polar coordinates, i.e. in cylindrical coordinates in circumstances such that there is no variation with x_3,

$$\boldsymbol{\nabla}\cdot\boldsymbol{u} = \frac{1}{r}\frac{\partial(ru_r)}{\partial r} + \frac{1}{r}\frac{\partial u_\theta}{\partial\theta},$$

and

$$\nabla^2\phi = \frac{1}{r}\frac{\partial}{\partial r}\left(r\frac{\partial\phi}{\partial r}\right) + \frac{1}{r^2}\frac{\partial^2\phi}{\partial\theta^2}.$$

(v) The equivalent results in spherical polar coordinates, in circumstances such that there is no variation with azimuthal angle, namely

$$\boldsymbol{\nabla}\cdot\boldsymbol{u} = \frac{1}{R^2}\frac{\partial(R^2u_R)}{\partial R} + \frac{1}{R\sin\theta}\frac{\partial(\sin\theta\, u_\theta)}{\partial\theta},$$

$$\nabla^2\phi = \frac{1}{R^2}\frac{\partial}{\partial R}\left(R^2\frac{\partial\phi}{\partial R}\right) + \frac{1}{R^2\sin\theta}\frac{\partial}{\partial\theta}\left(\sin\theta\frac{\partial\phi}{\partial\theta}\right).$$

Tensors

Conventional suffix notation is used to denote tensors. Only in one section, however, are they encountered in the raw. Elsewhere they are disguised, the normal and shear components of the stress tensor, for example, receiving separate discussion. In the one section devoted entirely to tensors, familiarity is assumed with the convention

$$\delta_{ii} = 1, \quad \delta_{ij} = 0 \ (i \neq j),$$

but the summation convention for repeated dummy suffices is spelt out.

Complex representation of oscillatory quantities

The function $\cos\omega t$ is represented by $e^{-i\omega t}$ rather than by $e^{+i\omega t}$.

Brackets

Square brackets [] are used to isolate cross references to equations, figures or sections which are relevant to the material under discussion. Curly brackets {} are used to enclose the arguments of a function, as in $f\{\boldsymbol{x}, \mathrm{t}\}$. In addition, both these types of bracket are used where necessary in conjunction with round brackets, nestling thus: $[\{()\}]$. Angle brackets $\langle\rangle$ are used to indicate spatial averages, or ensemble averages. Averages over time are indicated by an overbar, however, and where time averages and spatial or ensemble averages amount to the same thing the overbar notation is normally preferred.

Symbols

Roman

e	exponential constant
i	square root of -1
ln	operator denoting natural logarithm
S, S′	inertial frames of reference

Italic

a	radius; semi-major axis of ellipse; coefficient
A	area (especially cross-sectional area); coefficient
b	radius; semi-minor axis of ellipse; coefficient
$\boldsymbol{B}$, B	magnetic field (induction); coefficient
c	phase velocity of waves
c_{g}	group velocity of waves
c_{mol}	molecular velocity
c_{s}	velocity of sound
c_p, c_V	specific heat per unit mass at constant pressure or volume
C	velocity of tidal bore or shock front
d	diameter; thickness or depth; side of cube
D	breadth (i.e. $d < D < L$ for a rectangular body); chord of aerofoil; diffusion coefficient
e	intrinsic internal energy per unit mass; elementary charge
E	energy; energy per unit mass and per unit range of k in turbulence
$\boldsymbol{f}$, f	force per unit mass
$\boldsymbol{F}$, F	force
F_{D}, F_{L}	drag force, lift force
$f\{\ldots\}$, $F\{\ldots\}$, $g\{\ldots\}$ etc.	functions of stated arguments
$\boldsymbol{g}$, g	gravitational acceleration; torque per unit length or per unit volume

G	shear modulus
$\boldsymbol{G}$, G	torque
h	intrinsic enthalpy per unit mass
$\boldsymbol{H}$, H	heat flux per unit area and time; magnetic field (intensity)
$\boldsymbol{i}$, i	electric current
$\boldsymbol{I}$, I	magnetisation
$\boldsymbol{k}$, k	wavevector ($2\pi/\lambda$)
k_{B}	Boltzmann constant
K	circulation round closed loop embedded in fluid
$\boldsymbol{K}$, K	strength of line vortex
$K_{1,2,3}$	splay, twist, bend coefficients for nematic liquid crystal
$\boldsymbol{l}$, l	arc length along continuous curve
L	length; scale length
L_{i}	inlet length (of pipe)
m	mass; magnetic pole strength; $u_{\mathrm{o}}\delta$ in a turbulent layer
m_{mol}	molecular mass
M	magnetic dipole strength; $u_{\mathrm{o}}^2\delta$ in a turbulent layer
n	exponent
p	mean pressure, i.e. $\frac{1}{3}(p_1 + p_2 + p_3)$
p^*	excess (mean) pressure
$p_{1,2,3}$	normal components of compressive stress
p_{A}	atmospheric pressure
q	discharge rate per unit width; source strength per unit area; heat loss per unit length
Q	volume discharge rate; source strength; heat loss
r	radius of curvature
s	entropy per unit mass; imaginary part of complex ω
$s_{1,2,3}$	components of shear stress
s_T	Soret coefficient
T	temperature; kinetic energy
$\boldsymbol{u}$, u	velocity of fluid
$\boldsymbol{U}$, U	velocity of solid object relative to fluid, or *vice versa*
V	volume
Z	wave height between crest and trough

Script

ℓ	molecular mean free path; mixing length

Sans serif (used only for dimensionless quantities)

A	constant
B,C,D	'constants' (with scope for variation)

c	mass fraction
C	contraction coefficient
C_D, C_L	drag coefficient, lift coefficient
f	fraction less than unity
Gr	Grashof Number
H	aspect ratio
K	von Kármán constant
M	Mach Number
n	depolarising factor for ellipsoid
Nu	Nusselt Number
P	roll parameter
Pr	Prandtl Number
r	reduced density difference, i.e. $(\rho - \rho')/(\rho + \rho')$
Ra	Rayleigh Number
Re	Reynolds Number
s	condensation, i.e. $(\rho - \rho_o)/\rho_o$
T	Taylor Number

Greek

α	volume expansion coefficient; angle; attenuation coefficient
β	compressibility; angle
γ	specific heat ratio, c_p/c_V
Γ	inverse of relaxation time
δ	boundary layer thickness; thickness of turbulent shear layer; angle
ε	ideal Ekman layer thickness
ε_o	energy dissipation per unit mass in turbulence
ζ	rate of shear; displacement, especially in vertical direction
ζ_{ij}	symmetric part of rate of deformation tensor
η	shear viscosity
η_b	bulk viscosity
θ	excess temperature; angle
κ	thermal conductivity
λ	wavelength
μ	Mach angle
μ_o	permittivity of free space
ν	kinetic viscosity, η/ρ
$\boldsymbol{\xi}$, ξ	displacement
ρ	density

σ	surface tension; strength of shock front
$\mathrm{d}\boldsymbol{\Sigma}$	element of area of surface Σ
τ	relaxation time; enhancement of temperature due to convection
τ_{ij}	stress tensor
ϕ	flow potential
Φ	potential energy
χ	magnetic susceptibility; thermal diffusivity; phase shift
ψ	phase
$\boldsymbol{\omega}$, ω	angular velocity
ω_{ij}	antisymmetric part of rate of deformation tensor
Ω	solid angle
$\boldsymbol{\Omega}$, Ω	vorticity

1

A bird's eye view

1.1 Introduction

Physics is a tree with many branches, and fluid dynamics is one of the older and sturdier ones. It began to form in the eighteenth century, when Euler and Daniel Bernoulli set out to apply the principles which Newton had enunciated for systems composed of discrete particles to liquids which are virtually continuous, and it has been in active growth ever since. Nowadays it is partially obscured from view by branches of more recent origin, such as relativity, atomic physics and quantum mechanics, and students of physics pay rather little attention to it. This is a pity, for several reasons. Firstly, because of the engineering applications of the subject, which are many and various: the design of aeroplanes and boats and automobiles, and indeed of any structure intended to move through fluid or propel fluid or simply to withstand the forces exerted by fluid, depends in a critical way upon the principles of the subject. Secondly, because fluid dynamics has important applications in other branches of physics and indeed in other realms of science, including astronomy, meteorology, oceanography, zoology and physiology: dripping taps, solitary waves on canals, vortices in liquid helium, seismic oscillations of the Sun, the Great Red Spot on Jupiter, small organisms that swim, the circulation of the blood – these are just a few of the very varied topics involving fluid dynamics which have been occupying research scientists and mathematicians of international reputation over the past few decades. Thirdly, because most other subjects in the physics curriculum are almost exclusively concerned with *linear* processes, whereas fluid dynamics leads one into the *non-linear* domain. And lastly, because there are so many curious and beautiful natural phenomena, visible every day in the world about us, which a physicist with no knowledge of fluid mechanics is unable to appreciate to the full.

The Newtonian principles of conservation of momentum and conservation of energy which lie at the core of fluid dynamics are straightforward. Nevertheless, the subject is not an easy one; if it were, it would not attract researchers in the way that it still does, three centuries after the time of Newton, two centuries after Euler and Bernoulli, and one century after other giants such as Stokes, William Thomson (Lord Kelvin), and Rayleigh. The principal aims of this introductory chapter are humdrum enough: to refresh the reader's memory concerning a few concepts and results that students of physics tend to meet quite early in their education, to relate these to one another and to establish terminology and notation. It has an important secondary aim, however: to ease the reader's path through the rest of the book by explaining, in outline only, why fluids can often – though certainly not always – be treated as though they were incompressible and non-viscous. That secondary aim cannot be achieved without delving quite deeply in places, but the mathematics is kept as simple as possible. Gaps in the argument, of which there are many, are filled by the more systematic treatment which later chapters provide.

Abstract principles are easier to grasp when they are illustrated by concrete examples, so the chapter is built around a typical fluid dynamics problem – particularly typical in that it has more subtleties in it than one might imagine at first sight – concerning flow through a syringe. The problem is posed in §1.4, after some substantial but necessary preliminaries have been dealt with in §§1.2 and 1.3. Section 1.5 concerns the application to it of dimensional analysis, and introduces the important general concept of *dynamical similarity*. Sections 1.6–1.10 concern the flow pattern in the barrel of the syringe and the associated pressure drop along the barrel; these are determined almost exclusively by the inertia of the fluid, which means that *Bernoulli's theorem* is applicable. Section 1.11 concerns the boundary conditions at a fluid–solid interface, while §1.12 introduces, with illustrative examples, the concept of a *boundary layer*. Section 1.13 concerns *Poiseuille's law*, which might naively be expected to describe the pressure drop along the needle of the syringe, where viscosity can no longer be ignored, and the problem posed in §1.4 is finally answered in §1.14. But naive expectations are often confounded in fluid mechanics by instabilities inherent in the flow which lead to *turbulence*; this is the message of §1.15. The final section is a summary, not only of this chapter but of the contents of the rest of the book.

1.2 What is a fluid?

Imagine that you are holding a brick in front of you between the palms of your hands, and that you move, or try to move, your right hand away from you and your left hand towards you. The forces which you are exerting with your hands are transmitted through the brick from one layer to the next, every layer experiencing the same 'away' force on its right side and an equal and opposite 'towards' force on its left side [fig. 1.1]. When a material object is treated in this way there is said to be *shear stress* within it.

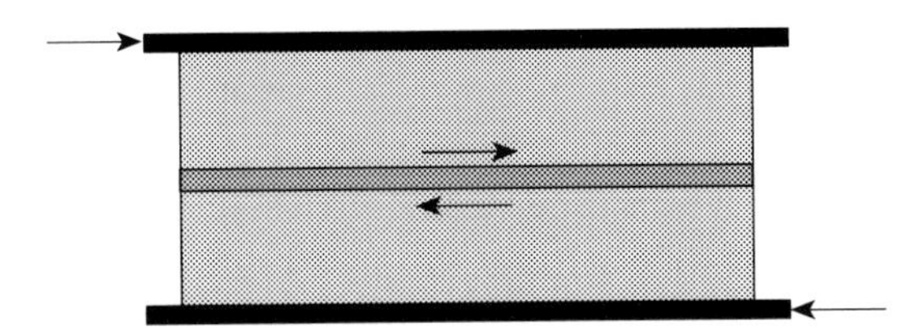

Figure 1.1 A brick between two plates which are being forced in opposite directions. Shear stresses are generated within the brick and every layer of it suffers in the same way.

Normal bricks are made of solid materials which possess the property of *rigidity*. This means that although they tend to shatter when subjected to very large stresses they can withstand moderate shear stress for an indefinite period. When stress is first applied they deform slightly but the deformation is not permanent; they spring back to their original shape when the stress is relieved. Plastic materials such as modelling clay or glaziers' putty also have some rigidity. The shear stress at which they yield is relatively small, however, and once this critical limit is exceeded a plastic material deforms continuously and irreversibly; it does not recover its shape when the stress is relieved. A true fluid is by definition a material with no rigidity at all. Subjected to shear stress, no matter how small this stress may be, a true fluid is bound to flow. An obvious corollary is that within a true fluid which is in mechanical equilibrium the shear stress vanishes everywhere.

The fluids with which all of us are most familiar are water and air. Water is a liquid, while air is a gas, but that distinction is less important in fluid dynamics than might be imagined. The molecules of which liquids are composed are in intimate contact with their immediate neighbours, and their migration from one place to another is the cumulative result of a succession of very small rearrangements. In gases, on the other hand, the molecules are normally well separated, which means they move in straight lines between one collision and the next, with a *mean free path* which is much larger than the molecular diameter. On a molecular scale, therefore, the processes whereby liquids and gases respond to shear stress look very different. In fluid dynamics, however, we forget about molecules and treat the fluid, whether it be liquid or gas, as an effectively continuous medium characterised by a relatively small number of bulk properties, of which *density* (ρ), *compressibility* (β) and *viscosity* (η) are the most obviously significant. Under normal conditions of temperature and pressure the density of water exceeds that of air by a factor of about 800, the compressibility of air exceeds that of water by a factor of about 16 000, and the viscosity of water exceeds that of air by a factor of about 55. These differences may appear considerable, but they are differences of degree only. For a fluid dynamicist, the only essential distinction between liquids and gases is that liquids *cohere*; a liquid specimen can sit in a beaker with a free surface where it meets only its own vapour, or it can exist as a drop with its whole surface free, whereas a gaseous specimen normally disperses unless it is completely surrounded by an impermeable container.

Most of the practical applications of fluid dynamics concern water or air, or fluids that closely resemble one of these. We are therefore justified in approaching the subject with three assumptions in mind, (a) that the properties of the fluid we are to discuss are *isotropic*, (b) that it is *Newtonian* – a Newtonian fluid being one which obeys the linear relations which Newton first postulated between shear stress and rate of deformation – and (c) that its behaviour is governed by the laws of *classical* as opposed to quantum physics. Water and air are undoubtedly isotropic, Newtonian and classical, and so are many other fluids of common experience. A reader with a digital watch with a liquid crystal display, however, will be carrying some anisotropic fluid on his wrist, anyone who has tried to whip egg whites has encountered a fluid which is non-Newtonian, and liquid helium is essentially non-classical. Such complex fluids display fascinating phenomena peculiar to themselves, and we shall take a brief look at them in the final chapter.

1.3 Some facts about stress

At each point in a continuous medium, whether it be solid or fluid, we need *six* numbers, each of them representing a component of force per unit area, to define the local stress completely.

To understand why this is so, consider an infinitesimal cubic element of the medium which has the point of interest at its centre and whose square faces, of side d, lie perpendicular to cartesian axes labelled x_1, x_2 and x_3. This element is shown in cross-section in fig. 1.2. The arrows labelled p_1 and p_2 in the figure indicate forces, of magnitude d^2p_1 and d^2p_2 respectively, exerted on the cube by two *normal stress components*, or *pressures*, which act on the planes AB and BC. The arrows labelled s_{21} and s_{12} indicate forces of magnitude d^2s_{21} and d^2s_{12} exerted by *shear stress components* acting on the same planes; here the first suffix indicates the direction of the force and the second the direction of the normal to the plane under consideration. Unlabelled arrows indicate balancing forces which are exerted on the opposite faces CD and DA. The balance must be exact in the limit $d \rightarrow 0$, whether or not the medium is in mechanical equilibrium: any difference between say p_1 on one side and p_1 on the other would result in a force proportional to d^2 acting on a mass proportional to d^3; the infinite accelerations which would result would eliminate the difference in an infinitesimal time. Stress components such as p_1 may well vary between two points in a medium which are separated by a distance which is *not* infinitesimal, but we may defer consideration of the effects of such variation until §1.7 below.

A similar argument shows that s_{21} and s_{12} must always be equal to one another, whether or not the medium is in mechanical equilibrium. If they were not equal

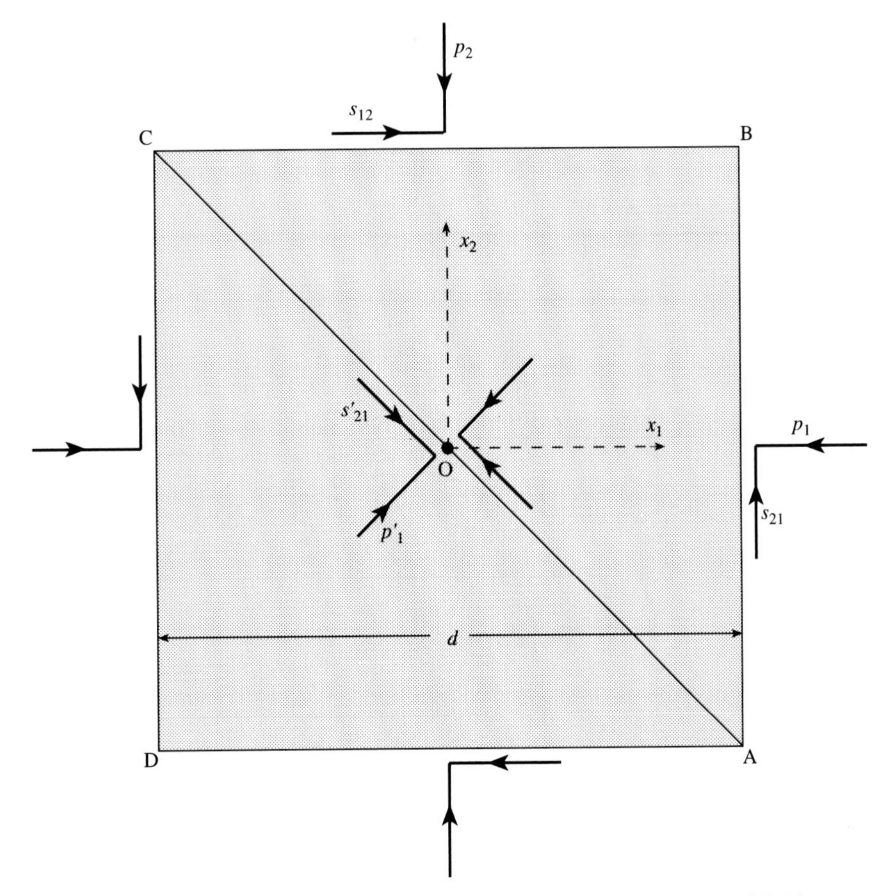

Figure 1.2 Normal and shear stresses acting on a cubic element of fluid.

the cube would experience a couple about the x_3 axis, anticlockwise in the diagram and therefore positive if the coordinate system is right-handed, of magnitude $d^3(s_{21} - s_{12})$, and the resultant angular acceleration would be infinite because the moment of inertia of the cube is proportional to d^5. The fact that shear stress is always symmetric in this way is one of the cardinal principles of elasticity theory as well as of fluid mechanics.[1] Since s_{21} and s_{12} are the same we may denote them by the single symbol s_3, where the suffix now refers to the axis about which the shear acts. Likewise we may denote s_{32} and s_{23} (not represented in the figure) by the symbol s_1, and s_{13} and s_{31} by s_2. In this notation the six numbers which are needed to describe the stress at a point completely are the magnitudes of p_1, p_2, p_3, s_1, s_2 and s_3.

[1] The principle fails, of course, if the medium is subject to an 'external' torque per unit volume for which shear stress is not responsible. Situations in which it does not apply are discussed in §§10.4 and 10.5.

The description is complete, given these six numbers, because the stresses acting across any plane through O which lies at an arbitrary angle to the axes of our cartesian reference frame are fixed uniquely by them. Consider, for example, the plane AC in fig. 1.2, which divides the cube into two rectangular prisms, ABC and ADC. Because the forces acting on either prism must balance exactly in the limit $d \to 0$, for the reason outlined above, it follows by resolving forces along DB and AC that if p_1' and s_{21}' $(= s_3')$ are the normal and shear stresses acting across AC then

$$2p_1' = p_1 + p_2 - 2s_3, \tag{1.1}$$

$$2s_3' = p_1 - p_2. \tag{1.2}$$

These equations describe, in effect, how the stress components *transform* if one chooses a new reference frame S′, rotated with respect to the original frame S by 45° about the x_3 axis. Since it can readily be shown by considering the balance of forces acting on the prisms BAD or BCD that

$$2p_2' = p_1 + p_2 + 2s_3, \tag{1.3}$$

it follows that

$$p_1' + p_2' = p_1 + p_2.$$

This is one aspect of a more general result, that the *mean pressure* p defined by

$$p = \frac{1}{3}(p_1 + p_2 + p_3) \tag{1.4}$$

is invariant under any rotation of the reference frame.

In a fluid which is in mechanical equilibrium, shear stress components such as s_3 and s_3' necessarily vanish [see §1.2]. It follows immediately from (1.1) and (1.2) that in such a fluid

$$p_1' = p_1 = p_2, \tag{1.5}$$

and it requires little imagination to extend the argument so as to prove that all three normal components of stress are equal to one another and unaffected by rotation of the reference frame. This result, that stress in a fluid in mechanical equilibrium is a scalar quantity, completely described by an invariant isotropic pressure p, is known as *Pascal's theorem*. It is the cornerstone of hydrostatics, the elementary subject that has to do with liquids – water in particular – at rest.

1.4 Statement of the syringe problem: Mach and Reynolds Numbers

We are now to imagine a syringe of the type used for medical injections, filled with a liquid resembling water. The barrel has a cylindrical bore of uniform cross-section, whose length is significantly larger than its diameter; at one end of the barrel there is a piston of area A, and at the other a small circular orifice. This orifice leads to a needle of length L and internal radius a. To simplify the discussion we shall initially assume that the internal surfaces of the barrel are ideally smooth and exert no frictional shear stresses on liquid that flows over them, but the internal surfaces of the needle will be treated more realistically, as rough. What is the relation between the *discharge rate* Q of the syringe, defined as the volume of liquid which emerges through the needle per unit time, and the force F applied to the piston?

When one first applies force, the piston and the liquid immediately in contact with it are bound to accelerate. Pretty soon, however, if the force is kept constant, they reach a steady state in which their speed is constant, and it is only this steady state which we shall attempt to analyse below. Now liquid which is moving steadily in an inertial frame of reference – the frame in which the syringe is stationary – is at rest in another inertial frame, and there is no doubt that Pascal's theorem applies to it. Hence immediately in front of the piston there exists an isotropic pressure p_{P}, say, which is higher than the atmospheric pressure p_{A}. It is related to the force F applied to the piston by the equation

$$F = (p_{\mathrm{P}} - p_{\mathrm{A}})A. \tag{1.6}$$

The problem therefore can be rephrased. We may study the relation between Q and $(p_{\mathrm{P}} - p_{\mathrm{A}})$ rather than F; $(p_{\mathrm{P}} - p_{\mathrm{A}})$ will be called the *excess pressure* at the piston and denoted by p^*.

It will be helpful in what follows to have some numbers in mind. A typical syringe has a capacity of 3 ml and a needle for which a is say 0·2 mm and L is 3 cm. A full load of water can readily be discharged through such a syringe in 10 s. Hence Q may be about 0·3 ml s^{-1}, which corresponds to a mean velocity along the needle of

$$U = \frac{Q}{\pi a^2} \approx 2{\cdot}5 \text{ m s}^{-1},$$

enough to carry the emerging jet to a height of 30 cm. Numbers such as those acquire significance only when compared with something else. As we shall discover, the ratio between U and the velocity of sound in water is significant; it constitutes what is called the *Mach Number* for this problem, which will be denoted here by M. The velocity of sound turns out to be equal to $\sqrt{1/\beta\rho}$, so we have

$$M = U\sqrt{\beta\rho}. \tag{1.7}$$

Another significant dimensionless quantity in which U features is the *Reynolds Number*,

$$Re = \frac{2\rho a U}{\eta}. \tag{1.8}$$

Now the velocity of sound in water is about 1·5 km s^{-1}, while the ratio η/ρ, which is referred to as the *kinematic viscosity* and often denoted by the symbol ν, is about 10^{-6} m^2 s^{-1} at normal temperatures. When the mean velocity in the needle is 2·5 m s^{-1}, therefore, the Mach Number is about 0·002 and the Reynolds Number is 1000.

Mach Numbers and Reynolds Numbers recur repeatedly in the arguments of fluid mechanics. Their definitions vary according to the problem in hand, but M is always the ratio of a velocity to the velocity of sound, while Re is always the product of a velocity and some characteristic length, divided by kinematic viscosity. Sometimes they are treated as quantities which vary from point to point as the fluid velocity varies. More often, however, they describe a single velocity which is obviously of critical importance and which serves to scale velocities throughout the field of flow; it might be, as here, the mean flow velocity along a pipe, or it might be the velocity of a moving object which is displacing otherwise stationary fluid. The factor 2 in (1.8) is a matter of convention and of no particular significance; it is conventional in discussions of pipe flow to use the diameter of the pipe rather than its radius as the characteristic length.

Some of the work which is done when the piston of a syringe is pushed in may be dissipated as heat in the liquid, rather than transformed into kinetic energy of the emergent jet. Is the resultant temperature rise likely to be significant or not? To answer this question, let us consider what would happen if *all* the work were dissipated internally as heat. In that case the temperature rise ΔT would evidently be given by

$$p^*Q = c_p\rho Q\Delta T, \tag{1.9}$$

where c_p is the specific heat of the liquid per unit mass. Now the result we finally obtain, in §1.14 below, is that the excess pressure needed to discharge water at a rate of about 0·3 ml s^{-1} is about 2×10^4 N m^{-2}. Insertion of this result into (1.9), with $c_p\rho \approx 4$ J ml^{-1}, gives $\Delta T \approx 5$ mK. Since this is clearly an upper limit to the temperature rise which occurs in practice, the heating can surely be neglected. This is just as well, because if it were not negligible we might have to worry about the variation of ν from place to place. The viscosity of water, far more than its density, is sensitive to temperature changes: it decreases by about 2% for every 1 K temperature rise. Nevertheless, we can treat ν as uniform.

1.5 Dimensional analysis and dynamical similarity

Before we embark on a detailed examination of the syringe problem it is worth asking what dimensional arguments can tell us about it. They turn out to provide, as is often the case in fluid dynamics, a useful framework within which later results can be set.

The principle at the heart of dimensional analysis is that any equation which expresses a scientific law must be satisfied in all possible systems of units. What differentiates one system from another is, of course, the choice of units for the so-called *primary* quantities, which are measured by comparison with essentially arbitrary standards such as the platinum cylinder whose mass constitutes the imperial standard pound. In the context of fluid dynamics only four primary quantities concern us: *mass*, *length*, *time* and *temperature*. All the other quantities to be met with are *secondary*, their units being fixed in terms of the primary units by the way in which they are defined and measured. A straightforward example of a secondary quantity is *velocity*, which is defined as a ratio between distance covered and time elapsed, and which can in principle be measured by measuring these two ingredients separately and comparing them with one's chosen standards of length and time. It is clear that if the primary unit of length is doubled in size then the secondary unit of velocity is doubled also, while if the primary unit of time is doubled the unit of velocity is halved. This dependence of the secondary unit on the primary units is conventionally expressed by saying that velocity has *dimensions* $[L][T]^{-1}$. In that language, acceleration has dimensions $[L][T]^{-2}$, force (which can be defined as the product of a mass and the acceleration which it produces when acting on that mass) has dimensions $[M][L][T]^{-2}$, energy (which is a product of force and distance) has dimensions $[M][L]^2[T]^{-2}$, specific heat (which yields an energy when multiplied by a temperature) has dimensions $[M][L]^2[T]^{-2}[\Theta]^{-1}$, and so on. In general, whenever two physical quantities with dimensions $[M]^a[L]^b[T]^c[\Theta]^d$ and $[M]^{a'}[L]^{b'}[T]^{c'}[\Theta]^{d'}$ are multiplied together, the result has dimensions $[M]^{a+a'}[L]^{b+b'}[T]^{c+c'}[\Theta]^{d+d'}$, and this rule enables the dimensions of complex combinations of physical variables to be established without difficulty. Now if an equation representing some scientific law is to work equally well in all systems, its two sides must scale in the same way when the system of units is changed; *their dimensions must therefore be the same*.

All physics students learn at an early stage in their training how to apply that principle in solving simple problems where the answer is well defined. How does the time of swing T of a simple pendulum depend upon the length L of the string, the mass m of the bob and the gravitational acceleration g? We are taught to postulate an answer in simple power law form,

$$T \propto m^a L^b g^c,$$

such that the left-hand side has dimensions $[T]$ while the right-hand side has dimensions $[M]^a[L]^{b+c}[T]^{-2c}$. For the dimensions on both sides to be the same we must have $a = 0$, $b + c = 0$, $2c = -1$, and the answer required is therefore

$$T \propto \sqrt{\frac{L}{g}};$$

only the dimensionless constant of proportionality remains undetermined.

However, that simple approach to dimensional analysis works only when the number of variables is small, in which case the problem is normally a trivial one, readily soluble by more detailed arguments which yield the constant of proportionality in the desired equation as well as the exponents. Suppose that the pendulum is a compound one, where the bob needs to be characterised by its radius of gyration k as well as its mass. What does the formula for its period look like then? In fact it becomes

$$T = 2\pi\sqrt{\frac{L^2 + k^2}{Lg}},$$

but we cannot deduce even the form of this by dimensional analysis. Even if it were a simple power law, which of course it is not, it would contain four unknown exponents, and in circumstances where the fourth primary quantity (temperature) plays no part the elementary procedure outlined above allows us to determine only three.

To make progress in such cases it is better to start from the principle that *all equations representing scientific laws must be expressible in such a way that both sides are dimensionless*. This is a trivial extension of the principle stated above; if X and Y possess the same dimensions, the equation $X = Y$ can always be recast in dimensionless form as $X/Y = 1$. Having identified the physical variables which are relevant in a problem, look first for ways of combining them into dimensionless groups, A, B, C etc. These groups need to be be independent of one another; if A and B are dimensionless then so are AB, AB^2, A/B etc., but none of these is independent of A and B and they should not be counted separately, though any one of them may be included *instead* of A or B if this seems advantageous. Choose the groups in such a way that the 'unknown' of interest – e.g. the period T in the pendulum problem – occurs in only one of them, say in A. Then the law connecting the variables must take the form

$$A = f\{B, C, \ldots\};$$

instead of an undetermined constant of proportionality it contains an undetermined function f, not normally expressible in simple power law form. In the case of a compound pendulum, it is easy to see that gT^2/L and k/L are dimensionless groups and that no other independent dimensionless groups can be formed. What dimensional analysis tells us about a compound pendulum, therefore, is that

$$\frac{gT^2}{L} = f\left\{\frac{k}{L}\right\}.$$

In the syringe problem, we are seeking a formula to relate an excess pressure p^* to a discharge rate Q and say five other variables whose values specify the dimensions of the syringe and the properties of the liquid, namely a, L, ρ, η and β.

The list of variables could obviously be made longer, for example by inclusion of the radius and length of the barrel, or of the surface tension of the liquid, or of the gravitational acceleration g. All of these could be relevant in certain circumstances, but they seem likely to be of secondary importance, so let us leave them out of account for the time being. The excess pressure has dimensions $[M][L]^{-1}[T]^{-2}$, and so too does the combination of relevant variables $\rho Q^2/a^4$. Hence $p^*a^4/\rho Q^2$ is dimensionless, as are the Mach and Reynolds Numbers, M and Re, and the ratio a/L. Since it can be shown that no other independent dimensionless groups exist, the formula must be expressible as

$$\frac{p^*a^4}{\rho Q^2} = f\left\{M,\ Re,\ \frac{a}{L}\right\}. \tag{1.10}$$

In so far as it includes the ratio a/L, (1.10) acknowledges that the performance of a syringe may depend upon its *shape* as well as its *size*, but the variations in shape which can be allowed for by varying a/L are naturally very limited. To take account of all possible variations in shape – the cross-section of the needle might, for example, be elliptical or rectangular rather than circular – the formula would have to include an infinite set of dimensionless length ratios, in which case it would be of no conceivable use.

In the absence of an exact theory, the nature of the undetermined function $f\{B, C, \ldots\}$ in an expression generated by dimensional analysis has to be established by systematic experimentation, or perhaps by computer simulation. Experiments may reveal, of course, that the list of variables fed into the analysis at the start was not long enough. Thus if measurements on two syringes having the same a/L ratio, filled with different fluids but operating under conditions such that M and Re were also the same for both, failed to yield the same values for $p^*a^4/\rho Q^2$, one would need to reconsider the secondary variables which were set aside above. In the present example, however, we probably started with too many variables rather than too few. Although compressibility is certainly relevant in principle, common sense suggests that the sort of pressure we can generate in a syringe with finger and thumb is unlikely to change the density of water to any significant degree. In that case, we should be able to omit β from our list of relevant variables and treat the density as constant; omission of β means, of course, omission of M from (1.10). This suggestion is put on a sounder footing in §1.8 below.

Two systems enclosing or enclosed by fluid which are identical in shape but which differ in scale are said to be *dynamically similar* when all the dimensionless groups of relevant variables take the same values in both of them. Thus two syringes which are scaled versions of one another, one containing water and the other containing air perhaps, should, if M is indeed irrelevant, be dynamically similar when Re is the same for both. Under these conditions $p^*a^4/\rho Q^2$ is evidently the same for both, but one can go further than that and say that the flow patterns within the syringes are perfectly scaled versions of one another, as long as both patterns are steady. A formal proof of this statement will not be given here but is straightforward enough; it involves transformation of the equations of motion for fluid flow into a form where all the variables appear in *reduced*, dimensionless, form, and subsequent discussion of the *uniqueness theorems* which dictate that for any given set of boundary conditions there is only one possible solution. The principle of dynamical similarity is of enormous practical importance. It is extensively relied upon by those who use wind tunnels and wavetanks to test models of structures which, in their final form, will be much larger.

1.6 Streaklines, streamlines, pathlines and lines of flow

It is often helpful to be able to see what is going on in a complicated flow process, and one way to make things visible is to release some sort of dye continuously into the fluid stream at a number of fixed points. The dye spreads out from each point into a continuous streak, and the instantaneous configuration of the streaks is shown up by a flash of light and photographed. This book contains photographs which have been taken in this way, and the lines which they exhibit are known as *streaklines*. The book is also illustrated by short time-exposure photographs of fluids which are carrying small solid particles in suspension; each particle scatters light and produces a short trace on the photograph, the length and orientation of which indicate the magnitude and direction of the instantaneous fluid velocity $\boldsymbol{u}$ at that point. The lines which the viewer of such a photograph forms in his imagination, by joining adjacent traces end to end, are known as *streamlines*. If one were to use a long time-exposure instead, and a fluid carrying only a small number of particles in suspension, the long trace left by each of the particles would be its *pathline*. A streakline is the locus of fluid elements – or fluid 'particles', as we often say – which have all originated from the same point; a streamline is a continuous curve whose tangent coincides in direction with $\boldsymbol{u}$; a pathline is the trajectory of a single fluid particle.

Now flow patterns may be *steady*, in the sense that the value of $\boldsymbol{u}$ associated with each point in space does not change as time passes, or they may be *unsteady*; in one case the partial derivative $\partial \boldsymbol{u}/\partial t$ is zero everywhere and in the other it is not. In cases of unsteady flow the streaklines, streamlines and pathlines must be thought of as continuously changing, and there is no reason why they should coincide; indeed, given a streakline photograph of unsteady flow one may need considerable imagination to visualise what a streamline photograph taken at the same instant would look like, and *vice versa*. Very many applications of fluid dynamics concern steady flow, however, and since streaklines, streamlines and pathlines do not alter as time passes in cases of steady flow there can be no distinction between them.

In this book (though not elsewhere), the words *streakline* and *streamline* are reserved for application to unsteady flow. Wherever the flow pattern under

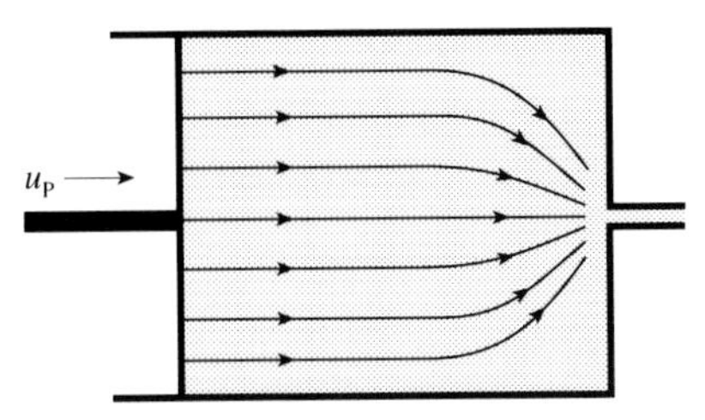

Figure 1.3 Lines of flow in the barrel of an idealised syringe; the piston is being driven inwards with velocity u_p.

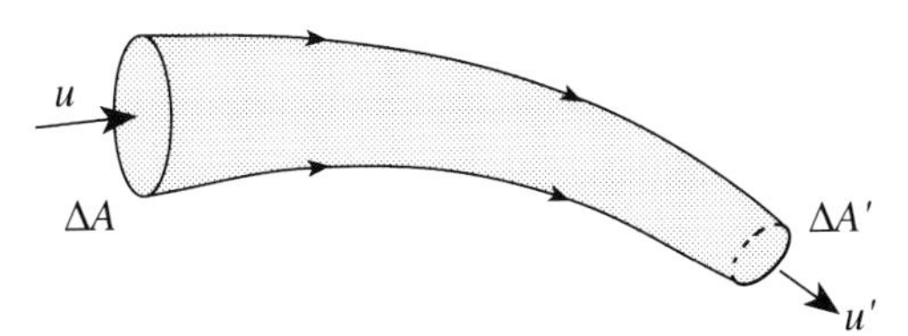

Figure 1.4 Steady convergent flow through a stationary flow tube, with $u\Delta A = u'\Delta A'$.

discussion is a steady one, such that streaklines and streamlines necessarily coincide, they are referred to as *flowlines*, or *lines of flow* instead. This appellation may help to emphasise the resemblance between lines of flow in fluid dynamics and lines of force in electromagnetism.

The flowline pattern in the smooth barrel of the syringe, once steady flow has been established, must be something like what is drawn in fig. 1.3; the lines there are horizontal at the left of the diagram, where they meet the face of the piston at right angles, but they have to converge on the orifice on the right. In doing so, they get closer to one another. Now consider a thin tube of liquid the sides of which are defined by fixed lines of flow [see fig. 1.4]. The mass of liquid moving through this tube per unit time must be the same at all points along its length: no liquid moves in or out across the sides of the tube, and if liquid were to start piling up somewhere then the tube would have to expand at that point, in which case the pattern would not be a steady one. Hence the product $\rho u\Delta A$, where ΔA is the cross-sectional area of the tube, must be uniform along it, and if changes of density can indeed be ignored then $u\Delta A$ must be uniform. Where lines of flow draw together, therefore, the liquid speeds up, and *vice versa*. What the diagram in fig. 1.3 tells us is something obvious enough but nevertheless quite liable to be overlooked, that much of the momentum with which the liquid ultimately emerges from the syringe is imparted to it before it ever enters the needle, rather than while it is travelling through the needle.

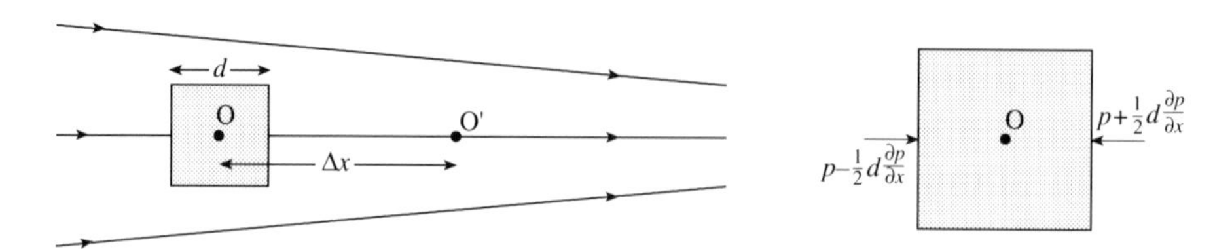

Figure 1.5 Fluid element accelerating along a line of flow. The enlarged view on the right shows the stress components which, because the pressure gradient is negative, cause it to accelerate.

1.7 Bernoulli's theorem

We must now consider in more detail the acceleration of liquid particles as they approach the entry to the needle. Consider, in particular, the small (but not necessarily infinitesimal) cube of liquid which is drawn in cross-section as a square of side d in fig. 1.5. Instantaneously, its centre of mass at O is moving along the straight central line of flow in the figure with velocity u_{O}, which means that after a time Δt it will have advanced through a distance $\Delta x = u_{\mathrm{O}}\Delta t$ to the point O′. Here its velocity is $u_{\mathrm{O}'}$, which is greater than u_{O} because the lines of flow are closer together near O′. In fact

$$u_{\mathrm{O}'} = u_{\mathrm{O}} + \left(\frac{\partial u}{\mathrm{d}x}\right)_{\mathrm{O}} \Delta x = u_{\mathrm{O}} + \left(\frac{\partial u}{\partial x}\right)_{\mathrm{O}} u_{\mathrm{O}}\Delta t.$$

It follows that in this essentially one-dimensional situation the acceleration of a fluid particle at O is

$$\lim_{\Delta t \to 0}\left(\frac{u_{\mathrm{O}'} - u_{\mathrm{O}}}{\Delta t}\right) = \left(\frac{\partial u}{\partial x}\right)_{\mathrm{O}} u_{\mathrm{O}} = \frac{\partial}{\partial x}\left(\frac{1}{2}u^2\right)_{\mathrm{O}}. \tag{1.11}$$

Equation (1.11) is no surprise, because in particle dynamics acceleration is frequently expressed in a similar way. In one-dimensional particle dynamics, however, we use u to represent the velocity of a single particle, which means that it is a function of a single variable – either of displacement x or of time t, but not of x and t separately. The acceleration of a particle is then

$$\dot{u} = \frac{\mathrm{d}u}{\mathrm{d}t} = \frac{\mathrm{d}u}{\mathrm{d}x}\frac{\mathrm{d}x}{\mathrm{d}t} = \frac{\mathrm{d}u}{\mathrm{d}x}u = \frac{\mathrm{d}}{\mathrm{d}x}\left(\frac{1}{2}u^2\right),$$

which involves full differentiation rather than partial differentiation as in (1.11). The distinctive feature of fluid dynamics in one dimension is that we use u to represent the velocity field of the whole sample, so that it is in general a function of position *and* time, i.e. $u = u\{x, t\}$. In these circumstances we must *not* represent acceleration by $\dot{u}$. If $\dot{u}$ means anything at all it means $\partial u/\partial t$, i.e. the rate of change of fluid velocity at a fixed point in space. That is quite different from acceleration, which is the rate of change of velocity of a fluid particle which is moving from point to point. In steady flow $\partial u/\partial t$ is everywhere zero, as we noted in the previous section, but the acceleration described by (1.11) is clearly not zero.

The above remarks can be extended readily enough to situations where more than one dimension is relevant because the lines of flow are locally curved. In general [§2.4] the acceleration of a fluid in steady flow is the vector

$$\frac{\partial \boldsymbol{u}}{\partial x_1} u_1 + \frac{\partial \boldsymbol{u}}{\partial x_2} u_2 + \frac{\partial \boldsymbol{u}}{\partial x_3} u_3.$$

This turns out to have a transverse component perpendicular to $\boldsymbol{u}$, the familiar *centripetal acceleration*, but it is easy to see by choosing axes such that u_2 and u_3 are both equal to zero at the point of interest that its longitudinal component is

$$\frac{\partial}{\partial x_1}\left(\frac{1}{2} u_1^2\right) = \frac{\partial}{\partial x_1}\left\{\frac{1}{2}(u_1^2 + u_2^2 + u_3^2)\right\} = \frac{\partial}{\partial l}\left(\frac{1}{2} u^2\right), \tag{1.12}$$

where differentiation with respect to l means differentiation in the direction of motion, i.e. along a line of flow.

Now let us make a drastic assumption, to which we return in §1.9, that throughout the barrel of the syringe, even close to the needle where the liquid is accelerating fast, the shear stresses are everywhere negligible. In that case the pressure is everywhere isotropic, but it may – indeed must – vary with position. The cube of fluid which is sketched in fig. 1.5 must experience a higher pressure on its left-hand face than on its right-hand face in order to accelerate. When d is small these two pressures may be expanded as $\{p - \frac{1}{2}d(\partial p/\partial x)\}_{\mathrm{O}}$ and $\{p + \frac{1}{2}d(\partial p/\partial x)\}_{\mathrm{O}}$ respectively. The difference between them is responsible for a force on the cube of magnitude $-d^3(\partial p/\partial x)$. Equating force to the product of mass and acceleration, we have

$$-d^3 \frac{\partial p}{\partial x} = \rho d^3 \frac{\partial}{\partial x}\left(\frac{1}{2} u^2\right)$$

at O, and in more general cases of steady flow, such that the lines of flow are curved, we likewise have

$$-\frac{\partial p}{\partial l} = \rho \frac{\partial}{\partial l}\left(\frac{1}{2} u^2\right).$$

If the compressibility is negligible so that ρ does not vary with position, this may be integrated along any line of flow to show that

$$\frac{p}{\rho} + \frac{1}{2} u^2 = \text{constant}. \tag{1.13}$$

This famous theorem concerning the steady flow of fluids in circumstances where compressibility and shear stresses may be ignored is attributed to Bernoulli, though Euler seems to have derived it before him.

The 'constant' on the right-hand side of (1.13) is not necessarily the same on adjacent lines of flow, but in the syringe problem it must be so. We can tell this by tracing each line of flow in fig. 1.3 back to its point of origin on the face of the piston. There the velocity is certainly the same on each flowline because it is the velocity of the piston. As for the pressure at the piston, that must be the same on each flowline too. Were the pressure to vary across the face of the piston the particles of liquid adjacent to the face would experience forces at right angles to their direction of motion which would endow them with transverse acceleration, and the flowlines would be curved in consequence. It is a useful point to bear in mind that straight flowlines always imply the absence of a transverse pressure gradient.

According to (1.13), the pressure in a fluid stream is smaller in constrictions, where the fluid is obliged to travel more swiftly, than it is elsewhere. When people are told this for the first time they often express disbelief. Surely, they say, a constriction would be expected to squeeze the fluid and *increase* its pressure? The fact is, however, that any increase of pressure for which the constriction is responsible is felt upstream of it; unless the pressure falls before the constriction itself is reached there is no pressure gradient to produce the acceleration which is needed to carry the fluid through it. The prediction is, of course, amply confirmed by experiment. Indirect but nevertheless dramatic confirmation is provided by fig. 1.6.

1.8 When is compressibility negligible?

Bernoulli's theorem tells us that if the velocity of the liquid in contact with the piston of the syringe is u_P and the velocity at the orifice is comparable with the mean velocity inside the needle $U(\gg u_P)$, then the pressure drop between these two regions is

$$\Delta p \approx \frac{1}{2}\rho(U^2 - u_P^2) \approx \frac{1}{2}\rho U^2.$$

It follows from the definition of compressibility and from (1.7) that the difference of density is such that

$$\frac{\Delta\rho}{\rho} = \beta\Delta p \approx \frac{1}{2}\beta\rho U^2 = \frac{1}{2}M^2.$$

This difference is surely small enough to be neglected when the Mach Number is only about 0·002.

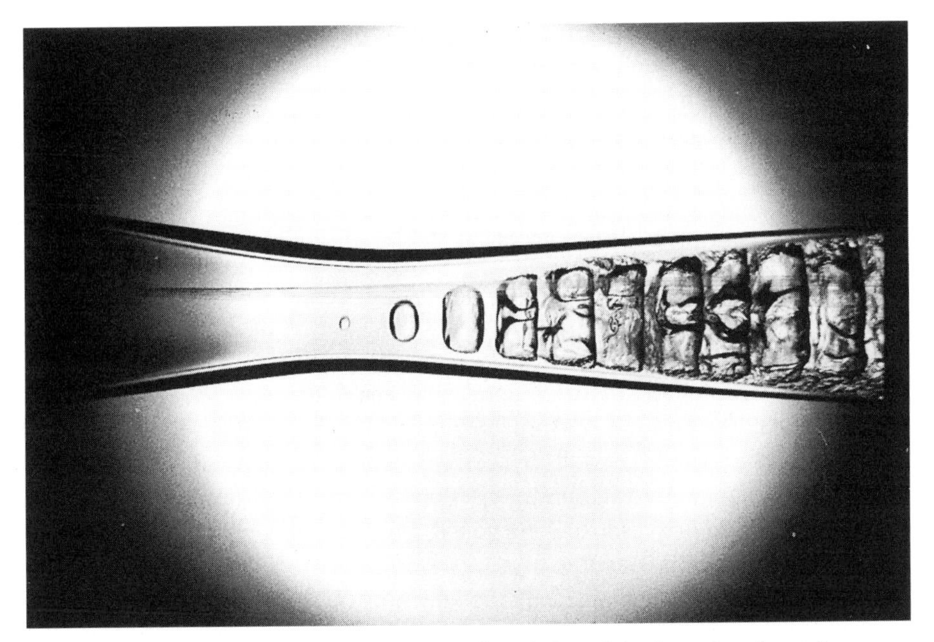

Figure 1.6 Superheated water flowing from left to right through a glass tube; the pressure drop at the constriction enables a regular succession of steam-filled bubbles to nucleate there, which grow in size as they are swept downstream.
[Photograph by E. Klein, by courtesy of the director of the Hermann-Föttinger-Institut, Berlin, Germany.]

It is one of the general principles of fluid dynamics that *in circumstances where Bernoulli's theorem is applicable in the form of* (1.13) the compressibility of the fluid can always be ignored, even if the fluid is a gas, provided that the velocity, measured in a frame of reference such that at some point within the region of interest u is zero, is less than the velocity of sound at all other points by a factor of five or so. The subject is frequently dealt with under two headings, therefore: *incompressible flow*, meaning flow that is, by a significant margin, subsonic, and *compressible flow*, meaning flow at speeds which approach or exceed the velocity of sound. Compressible flow is discussed in chapter 3, and in the appendix. The rest of this book is concerned almost exclusively with incompressible flow.

The clause in italics at the start of the previous paragraph requires emphasis because the condition $M \ll 1$, though necessary, is not always sufficient to justify neglect of compressibility. Compressibility cannot be neglected in problems to do with circulation of the atmosphere, where the air moves at speeds very much less than that of sound but explores a range of heights over which its pressure varies appreciably. It cannot be neglected in problems to do with the flow of gases through narrow channels, where an appreciable pressure difference may exist between the inlet and outlet regions. And it obviously cannot be neglected in acoustic problems, though in a sound wave of small amplitude the velocity of the fluid, as distinct from the velocity of the wave, may be very small indeed. These situations are, of course, discussed in later chapters.

1.9 When are shear stresses negligible?

A plausible first answer to the question posed above is that shear stresses are negligible when they are small compared with that part of the normal stress which varies from place to place and is thereby responsible for accelerating the fluid, which, when Bernoulli's theorem applies, is just $\frac{1}{2}\rho u^2$. Now shear stresses are proportional to the product of η and gradients of the fluid velocity, so they should be of order $\eta u/L$, where L is a distance over which, in the particular flow problem of interest, u changes by a significant fraction of itself. If this distance is chosen as the characteristic length used to define the Reynolds Number for the problem, i.e. if $Re = \rho L u/\eta$, then the condition $\eta u/L \ll \frac{1}{2}\rho u^2$ corresponds to $Re \gg 1$.

That answer turns out to need qualification in several respects, and two qualifications are important in the context of the present chapter.

(i) Although normally correct where the fluid is accelerating or decelerating, it is certainly not correct where the fluid is moving without acceleration through a confined channel of some sort. The lack of acceleration in such a situation means that forces in one direction due to gradients of the mean pressure must be exactly balanced by forces in the other direction due to viscous shear stresses. In that case the viscous shear stresses *must* be significant whatever the magnitude of Re. Thus although Bernoulli's theorem may apply in the barrel of the syringe it certainly cannot apply within the needle.

(ii) Where effectively incompressible liquid in the barrel of a syringe is accelerating towards the orifice, and in other comparable situations where lines of flow converge or diverge under conditions such that $M \ll 1$, *Bernoulli's theorem applies exactly at all values of the Reynolds Number and not just as an approximation which becomes reliable when* $Re \gg 1$. It applies because the viscous shear stresses contribute in two ways to the force which accelerates the liquid, and in the special circumstances expected to apply near the orifice the two contributions exactly cancel one another.

The second of those qualifications is a subtle one which deserves illustration. Let us suppose that near the orifice the lines of flow are straight and are all directed to a common centre O, as in fig. 1.7. Let us suppose that in spherical polar coordinates (R, θ, ϕ), where $R = 0$ at the centre of the orifice in the figure and where the $\theta = 0$ axis is horizontal there, the fluid velocity $\boldsymbol{u}$ has components

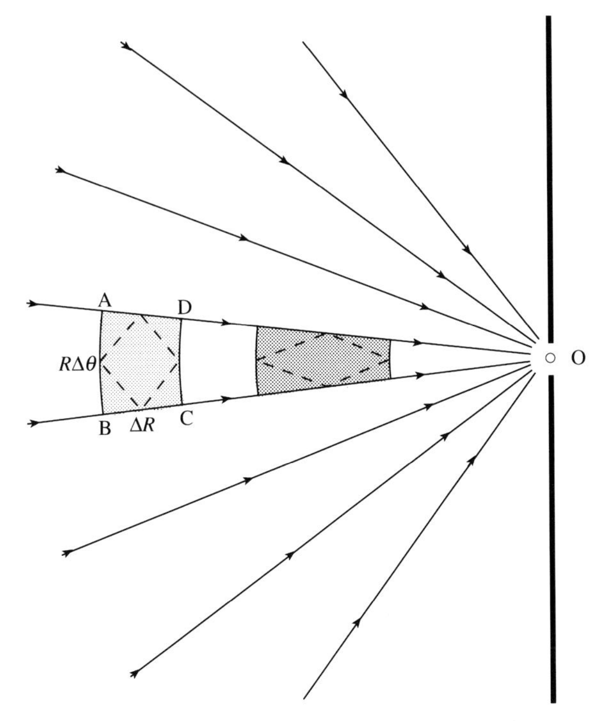

Figure 1.7 Lines of flow converging to a point.

$(-u, 0, 0)$, that u depends only on radius R, and that it is everywhere small compared with the velocity of sound. Evidently the volume of fluid flowing inwards per unit time across any hemispherical surface of radius R is

$$Q = 2\pi R^2 u,$$

and since Q has to be the same for all R if the fluid is effectively incompressible we know that

$$\frac{\mathrm{d}u}{\mathrm{d}R} = -\frac{Q}{\pi R^3}. \tag{1.14}$$

Rather than consider a fluid element of cubic shape, as in fig. 1.5, consider an element which forms part of a spherical shell of radius R and thickness ΔR, bounded by a circular cone of flowlines which enclose a small solid angle $\Delta\Omega$, as suggested in cross-section by the lightly shaded area in fig. 1.7. An element of this shape experiences no shear stresses on any of its faces, because shear stresses result from friction where one layer of fluid slides over another and no such relative motion occurs, either on the spherical caps labelled AB and CD in the figure, or on the side surface which includes AD and BC. As the element moves inwards, however, its length ΔR increases while its width gets correspondingly less: compare the lightly shaded area with the heavily shaded one, which shows where the element has got to a moment later. As a result, the inscribed square which is indicated by a broken line in the figure deforms into a rhombus. The deformation reveals that a shear stress, say s', exists on planes which are at 45° to the lines of flow, and hence that Pascal's theorem does not apply. To cause this sort of deformation, in fact, the radial pressure p_R which acts on AB and CD must be less than the pressure p_θ which acts on AD and BC.

If the fluid is a Newtonian one, then in the situation depicted in fig. 1.7 we have [(1.2)]

$$p_R = p - \frac{4}{3}s', \quad p_\theta = p + \frac{2}{3}s', \tag{1.15}$$

with

$$s' = -\frac{3}{2}\eta\frac{\mathrm{d}u}{\mathrm{d}R} = \frac{3\eta Q}{2\pi R^3}. \tag{1.16}$$

But the net force accelerating the element inwards is

$$\Delta\Omega\Delta R\left\{\frac{\mathrm{d}(R^2 p_R)}{\mathrm{d}R} - 2Rp_\theta\right\} = R^2\Delta\Omega\Delta R\left(\frac{dp}{\mathrm{d}R} - \frac{4}{3}\frac{\mathrm{d}s'}{\mathrm{d}R} - 4\frac{s'}{R}\right), \tag{1.17}$$

and if (1.16) is used in evaluating (1.17) it will be found that the terms involving s' cancel. Thus the accelerating force per unit volume of the element is exactly what it would be in the absence of shear stresses, namely $\mathrm{d}p/\mathrm{d}R$, and the proof of Bernoulli's theorem then proceeds just as in §1.7. We can deduce without any approximation that if the pressure at large R, where u is negligible, is say p_P, then elsewhere we have

$$p = p_\mathrm{P} - \frac{1}{2}\rho u^2 = p_\mathrm{P} - \frac{\rho Q^2}{8\pi^2 R^4}. \tag{1.18}$$

However, it is one thing to say that Bernoulli's theorem is exactly valid and another to say that viscous shear stresses can therefore be ignored. The theorem refers only to the mean pressure, and there are situations in which what really matters is one of the components p_1, p_2 or p_3, which individually may differ from p. In the syringe problem, for example, the force which drives liquid through the needle against the viscous retarding forces which arise there presumably depends upon the longitudinal component of normal stress, i.e. upon the magnitude of p_R where R is something like the needle radius a, rather than on the magnitude of p_θ or p there. To be able to rely on Bernoulli's theorem to estimate this force, we need to know that the shear component s' is relatively small – small, that is, compared with the pressure drop of $\frac{1}{2}\rho u^2$ which Bernoulli's theorem describes. The condition for shear stresses to be entirely negligible is therefore

$$s'_{R=a} = \frac{3\eta Q}{2\pi a^3} \ll \frac{\rho Q^2}{8\pi^2 a^4},$$

or [see (1.8) for the definition of the Reynolds Number Re]

$$\mathsf{Re} = \frac{2\rho Q}{\eta\pi a} \gg 24. \tag{1.19}$$

Despite the cancellation from (1.17) of the terms involving s', therefore, we come back in the end to where we started: it is only when $\mathsf{Re} \gg 1$ that viscous shear stresses in the barrel have no significant effect whatever. Needless to say, the condition which (1.19) represents is adequately satisfied when Re is about 1000.

It is customary to regard Bernoulli's theorem as a statement of the law of conservation of energy as applied to incompressible flow, and for an idealised non-viscous fluid that is just what it is. For a real fluid with viscosity, however, the content of the theorem is a bit more subtle than that, because in the sort of convergent flow which we have just considered – and equally in divergent flow – kinetic energy is continuously dissipated as heat. In the situation to which fig. 1.7 refers, the rate at which work is being done on the shaded element of fluid per unit volume is

$$\frac{1}{R^2}\left\{\frac{\mathrm{d}(R^2 u p_R)}{\mathrm{d}R}\right\} = u\,\frac{\mathrm{d}p}{\mathrm{d}R} - \frac{4}{3}\,u\,\frac{\mathrm{d}s'}{\mathrm{d}R}. \tag{1.20}$$

In telling us, as it does, that the first term on the right-hand side of (1.20) can be equated to the rate of increase of kinetic energy of the element, Bernoulli's theorem is telling us that the second term provides all the energy that is dissipated. It supplies exactly this amount of energy whether the Reynolds Number is large or small; the magnitude of Re is relevant only in so far as it determines the degree of anisotropy in the pressure.

1.10 Potential flow

The convergent flow pattern considered at some length in the previous section is perhaps the simplest non-trivial example of a very important class of steady incompressible flow patterns for which it is universally true that viscous stresses contribute nothing to the acceleration of the fluid, that mean pressure is therefore described by Bernoulli's theorem, and, incidentally, that the 'constant' in Bernoulli's theorem is the same on all lines of flow. *What members of this class have in common is that the vector quantity* $\boldsymbol{\nabla} \wedge \boldsymbol{u}$ *is zero everywhere.*

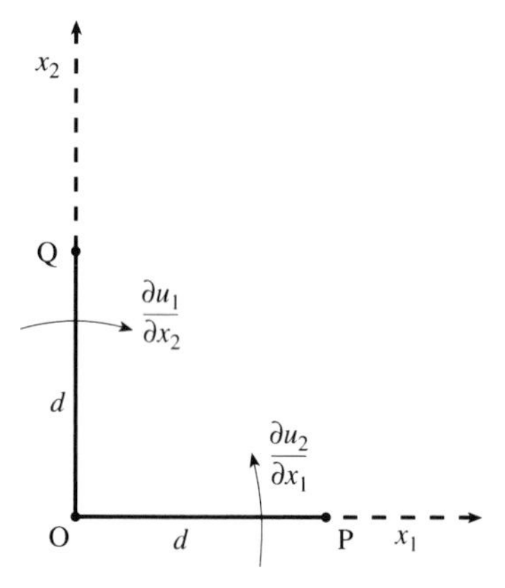

Figure 1.8 Angular velocities of two short lines embedded in moving fluid.

This quantity is called the vorticity of the fluid and is represented by the symbol $\boldsymbol{\Omega}$. That the name is appropriate can be seen by reference to fig. 1.8. The two short straight lines of length d in this figure, OP and OQ, are to be regarded as embedded in moving fluid. The velocity of P relative to that of O is

$$\boldsymbol{u}_{\mathrm{P}} - \boldsymbol{u}_{\mathrm{O}} \approx \frac{\partial \boldsymbol{u}}{\partial x_1} d,$$

and the component of this relative velocity in the x_2 direction implies that the line OP is instantaneously rotating anticlockwise in the plane of the diagram with angular velocity $\partial u_2/\partial x_1$. Similarly, OQ is instantaneously rotating anticlockwise with angular velocity $-\partial u_1/\partial x_2$. The quantity

$$\frac{1}{2}\left(\frac{\partial u_2}{\partial x_1} - \frac{\partial u_1}{\partial x_2}\right),$$

which is of course the component along x_3 of $\frac{1}{2}\boldsymbol{\Omega}$, can therefore be thought of as describing the *mean* angular velocity of the fluid near O about the x_3 axis. The other components of $\frac{1}{2}\boldsymbol{\Omega}$ are open to similar interpretations. Because vorticity is associated in this way with local rotation of the fluid, flow which is everywhere vorticity-free is often referred to as *irrotational flow*. This usage is avoided, however, in the present book.

Another name for vorticity-free flow is *potential flow*, the reason being that any vector field $\boldsymbol{V}\{\boldsymbol{x}\}$ which satisfies the condition $\boldsymbol{\nabla} \wedge \boldsymbol{V} = 0$ for all $\boldsymbol{x}$ can always be derived from a scalar potential $\phi\{\boldsymbol{x}\}$, using the relation

$$\boldsymbol{V} = \pm\boldsymbol{\nabla}\phi \tag{1.21}$$

(in which the sign is a trivial matter of convention). Static electric and magnetic fields in free space provide familiar examples, and the description of such fields in terms of scalar potentials is well known. Since electromagnetic fields in free space also satisfy the relation

$$\nabla \cdot \boldsymbol{V} = 0,$$

the electromagnetic scalar potentials satisfy Laplace's equation,

$$\nabla^2 \phi = 0. \tag{1.22}$$

The same is true of the potential used to describe vorticity-free flow, provided that the fluid's compressibility is negligible.

It seems very unlikely that the forces associated with gradients in the mean pressure p, as opposed to gradients of shear stress, can exert torques on small (but not necessarily infinitesimal) fluid elements and thereby change their angular momentum or mean angular velocity. Thus if fluid is vorticity-free at one instant of time, and if it is true that shear stresses are irrelevant to the motion of vorticity-free fluid, the fluid should surely remain vorticity-free at all later times. Kelvin's formal proof of this conjecture, which is valid whether the flow pattern is steady or not, is given in chapter 4. Now the liquid in the barrel of a syringe is stationary before one starts to push the piston, and at that stage it is clearly vorticity-free. Kelvin's theorem appears to tell us that it remains vorticity-free after it has been set into motion. Hence we can relate the lines of flow sketched in fig. 1.3 to a potential, and if we need to know exactly what they look like we can use one of the very many methods that exist – both analytical and numerical – for the solution of Laplace's equation subject to given boundary conditions; we can handle the problem just as we would handle an equivalent problem in electromagnetism, or for that matter in thermal conduction. The full answer involves Bessel functions, but the simple vorticity-free pattern sketched in fig. 1.7 is an adequate approximation for values of R which are small compared with the radius of the barrel and yet large compared with a.

Behind every application of potential theory to fluid dynamics lies an argument resembling the one just outlined, and the argument is always a bit suspect. It is suspect because Kelvin's proof pays inadequate attention to what happens at the boundaries of the vorticity-free fluid.

1.11 The no-slip boundary condition at fluid–solid interfaces

The idealisation of a solid which is perfectly smooth is so familiar from rigid body dynamics that it has been used without apology above. However, there is always *some* friction in practice between two solids that are sliding over one another, and friction is likewise inevitable between solids and fluids. Thus if two large plates in the planes $x_2 = \pm\frac{1}{2}d$, with fluid between them, are set into motion in the x_1 direction with velocities $\pm U$ the fluid is bound to respond. Presumably it adopts,

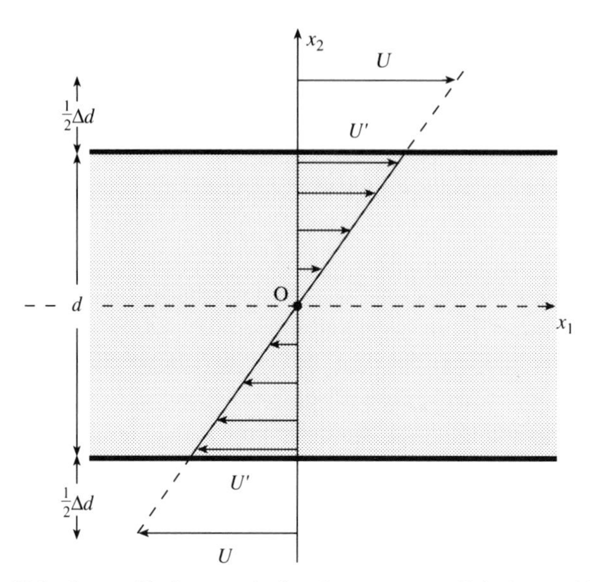

Figure 1.9 Velocity profile for steady flow between parallel plates which are moving at $\pm U$. The fluid velocity extrapolates to $\pm U$ at distances $\frac{1}{2}\Delta d$ beyond the plates' surfaces.

after whatever interval is needed for the establishment of steady flow, the velocity profile sketched in fig. 1.9, for which

$$u = 2U' \frac{x_2}{d},$$

with U' slightly less than U. How much less is it likely to be, or in other words how large is the distance marked as $\frac{1}{2}\Delta d$ in the figure, over which the fluid velocity extrapolates to that of the plates?

If the fluid is a gas to which the ideas of kinetic theory can be applied, a fairly precise answer to the latter question can be given. Let us assume the *accommodation coefficient* to be unity, i.e. that all molecules which hit the surface stick to it for long enough to reach thermal equilibrium before leaving the surface again; there is experimental evidence that this assumption is normally a sound one. Then the molecules of gas immediately adjacent to the surface of say the upper plate which are just leaving it have a mean velocity U in the x_1 direction, while the mean velocity of the molecules which are just about to hit the surface is the drift velocity of the gas at a distance of about one mean free path ℓ below the plate, i.e. $U'\{1 - (2\ell/d)\}$. Averaging over both groups of molecules we have

$$U' = \frac{1}{2}\left\{U + U'\left(1 - \frac{2\ell}{d}\right)\right\},$$

so that

$$\frac{1}{2}\Delta d = \frac{1}{2} d \frac{U - U'}{U'} = \ell. \tag{1.23}$$

This argument cannot be applied as it stands if the fluid in fig. 1.9 is a liquid, for the concept of a mean free path has little meaning in the liquid phase. It can be modified, however, to suggest that in this case $\frac{1}{2}\Delta d$ is unlikely to exceed the molecular diameter.

The mean free path in air at standard temperature and pressure is only about 10^{-7} m and the diameter of a water molecule is of course much less – say 3×10^{-10} m. These lengths are so small compared with the dimensions of the sort of apparatus that is normally used for experiments on fluids that one can safely ignore $\frac{1}{2}\Delta d$, and therefore ignore the difference between U' and U. So many predictions based upon the assumption that where fluids meet solids the two must be moving at the same velocity – that there can be no slip between them – have been verified experimentally, that this boundary condition can be taken for granted. It is liable to fail in the *Knudsen regime*, of course, i.e. in gases at low pressures where ℓ is no longer small, but Knudsen flow is a subject outside the domain of classical fluid dynamics.

1.12 Boundary layers

Let us now consider how the steady flow pattern represented by fig. 1.9 becomes established when the plates, initially stationary, are suddenly set into uniform motion. Presumably the upper plate drags with it the topmost layer of fluid; that layer then exerts shear stresses on the next layer down which cause it to accelerate; and the motion is gradually communicated in that way to layers which are deeper still. The velocity profile before the steady state is reached is likely to resemble the one sketched in fig. 1.10.

For a Newtonian fluid in this situation the shear stress s_3 which acts on any plane perpendicular to the x_2 axis is

$$s_3 = \eta \frac{\partial u_1}{\partial x_2}. \tag{1.24}$$

Hence each layer of fluid of thickness Δx_2 can be shown to experience a net force per unit area in the x_1 direction of magnitude

$$\frac{\partial s_3}{\partial x_2} \Delta x_2 = \eta \frac{\partial^2 u_1}{\partial x_2^2} \Delta x_2.$$

This is the force responsible for the acceleration which ultimately turns the flow pattern represented by fig. 1.10 into the steady state represented by fig. 1.9. Now the acceleration of the layer is just $\partial u_1/\partial t$ – the non-linear term expressed by (1.11)

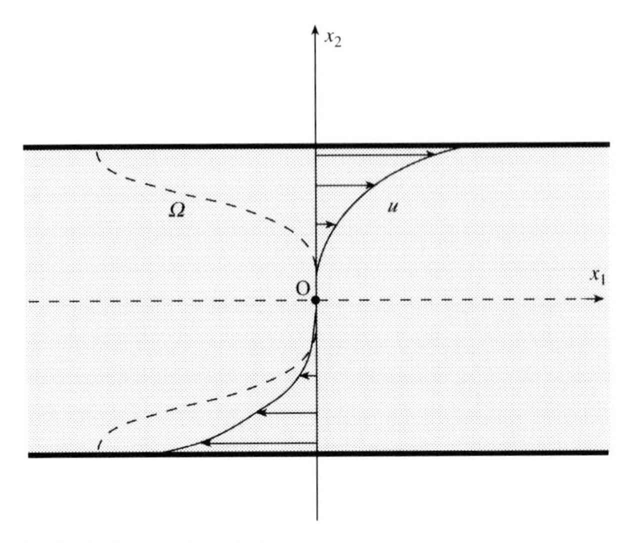

Figure 1.10 Velocity profile during acceleration towards the steady flow of fig. 1.9 (with Δd assumed to be negligible). A broken curve shows the corresponding profile for vorticity Ω.

is zero when u_1 does not vary with x_1 – so by equating the force per unit area on the layer to its mass per unit area times its acceleration we may arrive at the partial differential equation

$$\frac{\partial u_1}{\partial t} = \frac{\eta}{\rho} \frac{\partial^2 u_1}{\partial x_2^2}. \tag{1.25}$$

This is an example of the *one-dimensional diffusion equation*, and the solution which describes the velocity profile in fig. 1.10 is known; for small values of t, such that the fluid half way between the plates is still stationary, it can be expressed in terms of error functions. Suppose, however, that we are interested in the *vorticity* of the flow rather than in its velocity. This lies in the x_3 direction and is given by

$$\Omega_3 = (\nabla \wedge \boldsymbol{u})_3 = -\frac{\partial u_1}{\partial x_2};$$

its variation with x_2 is indicated by the broken curve in fig. 1.10. Now we have only to differentiate both sides of (1.25) with respect to x_2 to see that the vorticity obeys the similar diffusion equation

$$\frac{\partial \Omega_3}{\partial t} = \frac{\eta}{\rho} \frac{\partial^2 \Omega_3}{\partial x_2^2}, \tag{1.26}$$

and the solution which describes the broken curve at small values of t is likely to be rather more familiar to most readers than the solution to (1.25); it involves Gaussians rather than error functions and can be written as

$$\Omega_3 \approx -\frac{U}{\delta}\sqrt{\frac{\pi}{8}}\left[\exp\left\{-\frac{(x_2-\frac{1}{2}d)^2}{2\delta^2}\right\} + \exp\left\{-\frac{(x_2+\frac{1}{2}d)^2}{2\delta^2}\right\}\right], \tag{1.27}$$

with

$$\delta = \sqrt{\frac{2\eta t}{\rho}}. \tag{1.28}$$

Here, then, we have a situation where a slab of fluid which is stationary and therefore vorticity-free at $t = 0$, and which according to Kelvin's argument unthinkingly applied should remain vorticity-free at all later times, picks up vorticity at its boundaries. Once present in the fluid the vorticity disturbs the conditions required for Kelvin's argument to be valid, and on this account it is able to diffuse further, penetrating the fluid by diffusion in the way that a soluble dye released at its boundaries would do.

The idea first formulated by Prandtl, that where fluid undergoing potential flow moves over a solid surface it becomes contaminated by vorticity within a *boundary layer*, is vital to the understanding of fluid dynamics. The thickness of the boundary layer can often be estimated from (1.28), even where the flow pattern is much less simple than it is between two effectively infinite parallel plates. Consider, for example, steady flow past a flat strip of solid material which lies edge on to the direction of flow, as in fig. 1.11; it has an effectively infinite length perpendicular to the plane of that diagram but its breadth D is finite. In this situation there is a boundary layer attached to the strip with a thickness δ which increases with distance x from the leading edge, as shown. The vorticity present in this boundary layer gets swept downstream by the motion of the fluid, so that the inevitable wake behind the strip, where the fluid velocity is less than the incident velocity U, contains vorticity; extra vorticity to replace that which is lost downstream is continuously supplied to the boundary layer, however, by diffusion from the plate's leading edge. As is shown in §7.2, $\delta\{x\}$ may be estimated from (1.28) by using x/U for the time t.

Whether potential theory can be usefully applied or not often depends crucially on just how thick the boundary layer is, compared with the dimensions of the 'apparatus' under consideration. In the case of flow past an aircraft wing of breadth D, for example, the boundary layer thickness reaches a maximum value near the trailing edge of

$$\delta\{D\} \approx \sqrt{\frac{2\eta D}{U\rho}} = D\sqrt{\frac{2}{\mathit{Re}}}, \tag{1.29}$$

where

$$\mathit{Re} = \frac{\rho D U}{\eta}$$

is an appropriate Reynolds Number for this problem. If the Reynolds Number is of order unity, the boundary layer is so thick that the whole concept of such a layer

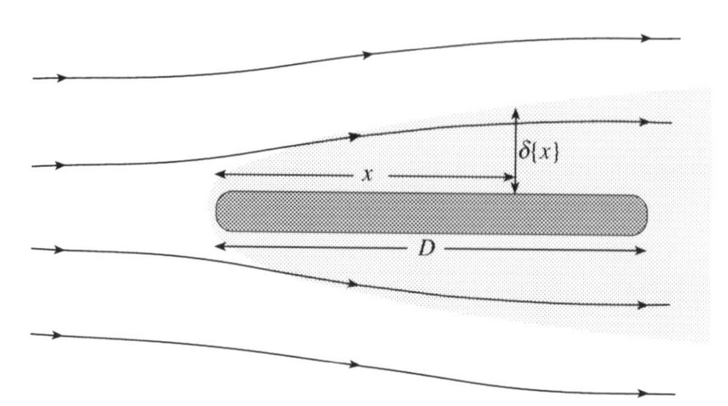

Figure 1.11 Lines of two-dimensional flow past a stationary plate, with vorticity present in boundary layers (shaded) of thickness δ.

becomes rather meaningless; the forces exerted on the wing are determined largely by the viscosity of the fluid, and potential theory and Bernoulli's theorem are of little relevance. When the Reynolds Number is large, however, the boundary layer is thin, and the pressure exerted on the wing is the pressure that exists in the vorticity-free fluid just outside this layer. This pressure, and therefore the lift force on the wing, can safely be calculated by using potential theory to map out the variation of u over the wing's surface and by subsequent application of Bernoulli's theorem; remember that when the Reynolds Number is large we expect the pressure to be effectively isotropic [§1.9], so we have no need to distinguish between the mean pressure given by that theorem and the pressure transverse to $\boldsymbol{u}$ which is felt by the wing. Just how large *is* Re, for an aircraft in flight at the relatively modest speed of 360 km per hour or 100 m s^{-1}, at which the Mach Number M is about 0·3 and the assumption that air is an incompressible fluid is not seriously misleading? Well, the kinematic viscosity η/ρ for air is about $1{\cdot}5 \times 10^{-5}$ m^2 s^{-1}, while for a typical aircraft wing D is say 3 m; these figures imply $Re \approx 2 \times 10^7$. The boundary layer should be no more than a millimetre or two in thickness, and its influence on the lift force is certainly negligible.

Some of the applications of potential theory which are of particular interest to physicists concern problems where the flow is oscillatory rather than steady. In such circumstances the boundary layer thickness normally corresponds to the attenuation length δ in the solutions that exist for (1.26) of the form

$$\Omega_3 \propto \exp\left(-\frac{x_2}{\delta}\right)\exp\left(\mathrm{i}\,\frac{x_2}{\delta}\right)\exp\left(-\mathrm{i}\omega t\right), \qquad (1.30)$$

which is given in terms of the angular frequency of oscillation ω by

$$\delta = \sqrt{\frac{2\eta}{\omega\rho}}. \qquad (1.31)$$

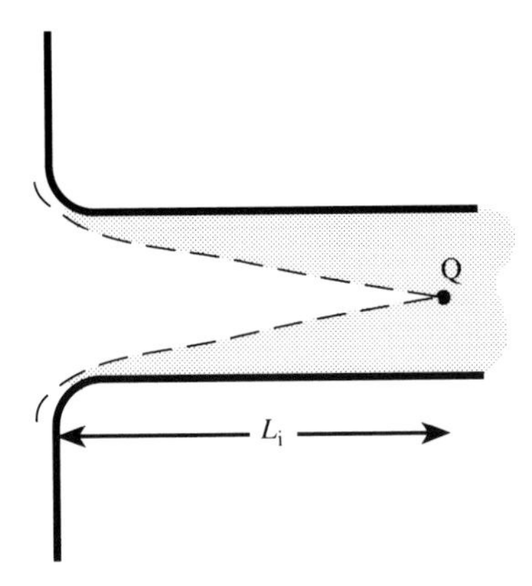

Figure 1.12 Boundary layer (shaded) near the inlet to a pipe.

Note that when the motion is oscillatory the boundary layer thickness and hence the total amount of vorticity present in the fluid do not continuously increase; there is cancellation between vorticity of positive sign that enters the fluid during one half-cycle of the motion and vorticity of negative sign that enters during the other half-cycle.

One illustrative example will suffice at this stage: the oscillatory flow patterns in deep water which manifest themselves as waves or ripples on the surface. One can find a potential which is a solution of Laplace's equation to describe this motion, but the motion involves shear and therefore dissipates energy which must come from somewhere. Examination of the solution shows that it presupposes shear stresses to exist at the free surface, which urge the wave crests forwards and the troughs backwards and which thereby supply the energy which is needed. Because these surface stresses are in practice absent, the vorticity-free motion which the solution describes is disturbed by diffusion of vorticity into the water from the crests and troughs. The depth to which it penetrates may be estimated from (1.31), and for ripples which have a wavelength λ of 1 cm, and for which ω is about 150 s^{-1}, this indicates a boundary layer thickness about 0·1 mm. This is surely small enough compared with λ for boundary layer effects to be neglected in the first instance. For ocean waves whose wavelength is several metres the ratio δ/λ is even less than 1%, and the effects of viscosity are very small indeed.

1.13 Poiseuille's law

When vorticity-free liquid enters the rough needle of the syringe it picks up vorticity, and a boundary layer forms inside the needle with a thickness that increases with distance from the needle's entry. A dwindling core of liquid which is virtually vorticity-free survives over what is called the *inlet length*, labelled L_i in fig. 1.12, but beyond that potential theory and Bernoulli's theorem are no longer

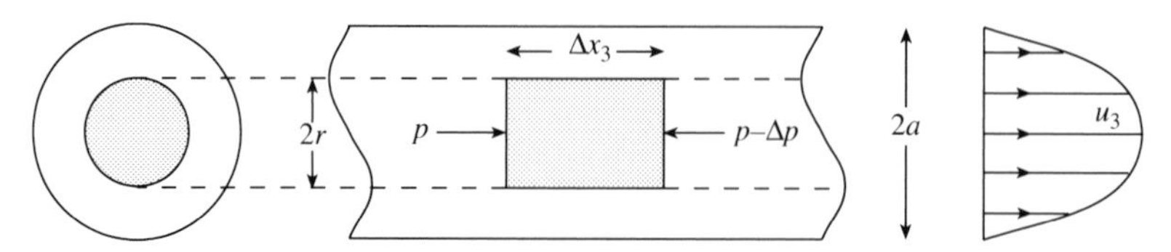

Figure 1.13 Two views of a cylindrical element of fluid in laminar motion along a pipe. The parabolic velocity profile is sketched on the right.

relevant. Over the rest of the length of the pipe it is the viscosity of the liquid which matters, rather than its inertia.

The longitudinal pressure gradient needed to force a volume Q of incompressible viscous fluid per unit time through a cylindrical pipe of radius a is described by *Poiseuille's law*, otherwise known as the *Hagen–Poiseuille law*; Poiseuille was a French physician who studied the flow of blood through arteries, while Hagen was a German hydraulic engineer. This result is based on the reasonable assumption that the fluid is in a steady state of what is called *laminar* motion parallel to the axis, with a longitudinal velocity u_3 which, in cylindrical polar coordinates (r, θ, x_3), is a function of r but not of θ. Consider the forces acting on a cylindrical column of fluid, of radius r and length Δx_3, over which the pressure drops by Δp $(= -\nabla_3 p \Delta x_3$, where $\nabla_3 p$, a negative quantity, is the pressure gradient $\mathrm{d}p/\mathrm{d}x_3)$. The column [fig. 1.13] experiences a force due to the pressure drop which drives it forwards, of magnitude $\pi r^2 \Delta p$. It also experiences a viscous retarding force which is the product of an area $2\pi r \Delta x_3$ and a shear stress which is equal, if the liquid is Newtonian, to $-\eta(\mathrm{d}u_3/\mathrm{d}r)$. Since the column is not accelerating the two forces must balance, so

$$\frac{\mathrm{d}u_3}{\mathrm{d}r} = -\frac{r\Delta p}{2\eta \Delta x_3} = \frac{r\nabla_3 p}{2\eta}.$$

By integrating over r, with the integration constant adjusted to ensure that the no-slip boundary condition is satisfied at $r = a$, one may show that

$$u_3 = -\frac{(a^2 - r^2)}{4\eta} \nabla_3 p. \tag{1.32}$$

The predicted velocity profile within the pipe is therefore parabolic in form, and the discharge rate is given by

$$Q = \int_0^a 2\pi u_3 r \mathrm{d}r = -\frac{\pi a^4}{8\eta} \nabla_3 p. \tag{1.33}$$

This law appears to be applicable in the needle of a syringe, except over the inlet length L_i.

How long is the inlet length likely to be? We may base a very rough estimate of it upon (1.28), by using L_i/U for t in that equation [cf. the discussion in §1.12

regarding the thickness of the boundary layer attached to a strip; note that U denotes the mean velocity of the fluid in the needle, as in §1.4] and by equating the value of δ which it yields to the needle's radius a. This procedure suggests

$$L_i \approx \frac{\rho U a^2}{2\eta} = 0{\cdot}25\ Re a,$$

where Re is the Reynolds Number defined by (1.8). A more careful though nevertheless approximate analysis shows that this rough estimate is too large, and that

$$L_i \approx 0{\cdot}06\ Re a \tag{1.34}$$

should be used instead. Thus when a is 0·2 mm and Re is 1000 the inlet length is about 12 mm – a considerable fraction of the whole length L of the needle.

1.14 The syringe problem answered

We are now at last in a position to suggest an answer to the problem posed in §1.4. Where the parabolic profile associated with Poiseuille flow is fully developed, the velocity on the axis of the needle, at $r = 0$, is twice the mean velocity U. Presumably the velocity of the vorticity-free core of fluid in the inlet zone reaches something like this value at the point labelled Q in fig. 1.12, in which case we may use Bernoulli's theorem to estimate that the pressure drop between the piston and Q is

$$p_P - p_Q \approx \frac{1}{2}\rho(u_Q^2 - u_P^2) \approx 2\rho U^2. \tag{1.35}$$

We may add to this the pressure drop over the length $(L - L_i)$ which Poiseuille's law implies, namely

$$p_Q - p_A \approx \frac{8\eta Q}{\pi a^4}(L - 0{\cdot}06\ Re a) = 16\rho U^2\left(\frac{L}{aRe} - 0{\cdot}06\right),$$

to obtain an expression for the excess pressure at the piston, $p^* = p_P - p_A$. This conforms to expectations based upon dimensional analysis alone, with the function f in (1.10) given approximately by

$$f\left\{Re, \frac{a}{L}\right\} \approx \frac{16}{\pi^2}\left(\frac{L}{aRe} + 0{\cdot}07\right). \tag{1.36}$$

To discharge water at a rate of 0·3 ml s^{-1} through a needle with $a = 0{\cdot}2$ mm and $L = 3$ cm [§1.4], we therefore need to establish an excess pressure of about 2×10^4 N m^{-2} (one fifth of normal atmospheric pressure). The corresponding force F on the piston may need to be about 2 N, i.e. the weight of 200 grams.

That answer is a reasonable one and correct in essence, though the derivation is weak in places. One obvious weakness is that the barrel has been treated as smooth; in reality it is just as rough as the needle, and our assumption that until the liquid enters the needle it remains completely uncontaminated by vorticity is surely fallacious on that account.[2] A less obvious but potentially more serious weakness lies in the assumption that within the needle the flow is laminar. This assumption is liable to fail, in fact, at discharge rates which are only slightly greater than those we have discussed [§1.15].

1.15 Instabilities and turbulence

Systems described by non-linear equations of motion are notoriously prone to instabilities. Suppose that a body of fluid in a state of steady flow described by a time-independent velocity field $\boldsymbol{u}\{\boldsymbol{x}\}$ is subjected to a small perturbation which changes this velocity field to say $\boldsymbol{u}' = \boldsymbol{u} + \boldsymbol{v}$, where $\boldsymbol{v}\{\boldsymbol{x}, t\}$ by itself describes some sort of weak, local, time-dependent circulation. Now imagine trying to predict what subsequently happens, by substituting $\boldsymbol{u}'$ into the relevant equations of motion. If these equations were linear, they would be found to contain terms involving $\boldsymbol{u}$ only, and terms involving $\boldsymbol{v}$ only. The former would cancel, because $\boldsymbol{u}$ is a solution of the equations. The latter would describe the evolution of $\boldsymbol{v}$, and except in special cases the perturbing circulation would be found to decay, due to the dissipation of kinetic energy as heat through the agency of viscosity. In practice, however, the equations of motion are *not* linear: the acceleration introduces terms which are of *second order* in $\boldsymbol{u}$ [(1.12)]. That being so, the substitution of $\boldsymbol{u}'$ generates cross terms proportional to both $\boldsymbol{u}$ and $\boldsymbol{v}$, which affect the evolution of the circulation by coupling it to the steady motion. In some circumstances the coupling enables the perturbation to draw energy from the steady motion at a rate which more than compensates for viscous dissipation. The velocity field $\boldsymbol{u}$ is then unstable, since even the tiniest perturbation, due perhaps to chance fluctuations on a molecular scale, is bound to grow at an exponential rate.

Flow instabilities are often very difficult to anticipate in the absence of clues provided by experimental observations. To take a single example, it has been generally supposed for over a century that regular waves of uniform amplitude on the surface of the ocean represent a state of motion which, in a frame of reference which moves with the waves, is steady apart from very small viscous dissipation effects associated with the surface boundary layer. It has recently been discovered, however, as a result of experiments with a wave-making machine in a giant tank, that when their amplitude is large – almost but not quite large enough to make them break – such waves are unstable to a perturbation which doubles the distance in the direction of propagation over which the disturbance repeats itself

[2] The relevance of upstream vorticity to the pressure drop at a constriction is examined in §2.10, where Bernoulli's theorem is applied to flow through a Venturi meter. See footnote on p. 53.

Figure 1.14 Waves travelling from left to right on the surface of water in a very large tank. They began as two-dimensional Stokes waves [§5.7] of frequency 1·2 Hz, but an unsuspected instability to which that mode of motion is prone at large amplitudes has altered their appearance. The wave crests are no longer straight, the motion is no longer two-dimensional and the distance in the direction of propagation over which the pattern repeats itself is twice the original wavelength.
[Photograph by Ming-Yang Su. [*J. Fluid Dyn.*, **124**, 73, 1982.]]

and which gives to each wave crest a scalloped appearance [fig. 1.14]. In this particular instance, growth of the initial perturbation leads to a new flow pattern which still has regularity in it and which may even be genuinely stable. The same can be said of the instability that leads ripples to appear on the surface of an otherwise still lake when the wind gets up. In a great many other cases, however, one instability leads to another and the end result is a *turbulent* state of motion which is not regular at all.

The laminar flow pattern in a pipe to which Poiseuille's law refers becomes turbulent when the Reynolds Number defined by (1.8) reaches a critical value which may be as low as 2300 if the pipe has what is called a *bluff entry*, i.e. if its cross-section is uniform and its ends are squared off. When $Re \approx 2300$, of course, the inlet length for a pipe with a bluff entry is about 70 times the pipe's diameter, according to (1.34). It seems that this is long enough for certain primary eddies to get out of hand before viscosity can quench them. These primary eddies seed smaller secondary eddies, which in turn seed tertiary eddies and so on, until turbulence ensues. Presumably there is a mean fluid velocity at each point which remains steady if Q is kept constant, and presumably this varies smoothly with r, though not necessarily in the parabolic fashion which (1.32) describes. There are large fluctuations about the mean, however, which one cannot hope to describe except in statistical terms, i.e. in terms of r.m.s. amplitudes, distribution functions, correlation times and correlation lengths.

Experiments show that above the critical Reynolds Number at which turbulence sets in the pressure gradient needed to drive fluid through a pipe increases more rapidly than Poiseuille's law suggests; it is roughly proportional in many cases to Q^2 rather than Q, though experimental results depend upon what the pipe is lined with and so forth. Hydraulic engineers who need to transfer fluid from place to place, and to lose as little pressure head as possible in the process, would obviously prefer to avoid turbulence if they could do so. If pipes with carefully flared entries rather than bluff ones are used, if sudden changes of cross-section are everywhere avoided, and if the pipe is as far as possible vibration-free, then discharge rates which correspond to $Re \approx 10^5$ may be achieved without it. That seems to be about the limit, however, and in engineering practice turbulence in pipe flow is the rule rather than the exception.

1.16 Summary

When Euler and Bernoulli laid the foundations of fluid dynamics they treated the simplest conceivable model, in which fluid density is constant, viscosity zero, and pressure therefore isotropic. What have we learnt so far about when and why this ideal model can, or cannot, be justified?

(i) In circumstances where the model is justified the fluid obeys Bernoulli's theorem, and we may use this to deduce how pressure, and hence density, varies from place to place in cases of steady flow. [§1.7]

(ii) The variations of density predicted by Bernoulli's theorem are insignificant provided that the variations of fluid velocity from place to place are small compared with the velocity of sound. In such circumstances, i.e. when a suitably defined Mach Number M is small compared with unity, the assumption that density is constant is normally legitimate. [§1.8]

(iii) Ideal fluid which is vorticity-free at one instant (when it is stationary, for example, or moving with uniform velocity) should remain vorticity-free. Its velocity can then be described at all times by a potential which obeys Laplace's equation, which makes the velocity distribution consistent with given boundary conditions quite easy to determine. [§1.10]

(iv) Shear stresses associated with viscosity, and the anisotropic components of normal stresses, should not affect the motion of a fluid, at any rate in its interior, provided that it is vorticity-free. That being so, the finite viscosity of real fluids need not invalidate the use of potential theory, or the use of Bernoulli's theorem to find how mean pressure varies from place to place. [§1.9]

(v) Nevertheless, a real fluid undergoing vorticity-free flow is not matched in every particular by the ideal model. The pressure within it is not always isotropic, and it tends to dissipate energy as heat. These distinguishing features are expected to become less significant as the flow velocity increases, i.e. when a suitably defined Reynolds Number Re is large compared with unity. [§1.9]

(vi) To replace the energy which is dissipated, a real fluid in a state of vorticity-free flow needs to experience external stresses at its boundaries which are not needed by an ideal fluid in the same state. When these external stresses are not supplied, as is normally the case, vorticity diffuses into the fluid and contaminates a boundary layer. [§1.12]

(vii) Boundary layers become very thin at large values of Re. Nevertheless, the differences of behaviour between real and ideal fluids remain significant even in this limit. Real fluids become unstable in different ways, and in particular they become turbulent; the fact that the onset of turbulence depends upon the magnitude of Re is enough to show that viscosity somehow plays a part in it. [§1.15]

That brief summary will do to be going on with, though it is incomplete in many respects. It pays no attention to the effects of gravity, which have indeed been largely ignored throughout this chapter. It does not refer to the subsidiary question, touched on in §1.4, of whether or not the temperature of moving fluid is likely to be uniform. It says virtually nothing about unsteady flow, though the ideal model can often be applied to that, particularly to oscillatory flow in which $\boldsymbol{u}\{\boldsymbol{x}, t\}$ varies with time in a sinusoidal fashion.

The effects of gravity are taken into consideration in chapter 2, where the ideal model and Bernoulli's theorem are used to explain in an elementary way a number of familiar phenomena to do with flowmeters, jets, windmills, tidal bores and suchlike.

In chapter 3, which is devoted to the compressible flow of gases, the assumption of constant density is relaxed. Many of the results obtained in this chapter are based upon a version of Bernoulli's theorem which is applicable even when $M > 1$ to ideal gases in steady flow. One prominent phenomenon which involves unsteady flow and which depends in an essential way upon compressibility is that of sound propagation. The treatment of sound propagation in chapter 3 itself is incomplete; a fuller treatment of the topic, which covers sound attenuation in gases, is given in the appendix. Considerable space is devoted, however, to the behaviour of shock fronts in gases, and their relevance to the forces which act on supersonic projectiles is explained.

We return to ideal incompressible fluids in chapter 4, which concerns potential theory, and in which Kelvin's proof that vorticity-free fluid remains vorticity-free is established. Emphasis is placed on analogies with electromagnetic phenomena which physicists may find intriguing and helpful; they help us to understand, for example, the behaviour of vortices and smoke rings. Chapter 5 concerns the application of potential theory to waves on deep water. Although the treatment is, of necessity in a book at this level, very largely restricted to waves of small amplitude, the chapter includes a brief account of solitary waves.

In chapter 6 the relations between shear stress and rate of deformation for a Newtonian fluid are formulated for the first time, and some results concerning shear stress which in the present chapter have been quoted without proof receive justification. The Navier–Stokes equation is derived. Viscometers are discussed in chapter 6, together with a variety of problems involving what is called *creeping flow*, where the effects of viscosity completely dominate the effects of fluid inertia.

Chapter 7, though entitled 'Vorticity', is largely concerned with boundary layers and with the phenomenon of *boundary layer separation*, to which Prandtl first drew attention in a seminal paper dated 1904. The drag forces which obstacles to flow experience at high values of the Reynolds Number, and the lift forces which make flight possible at speeds below the speed of sound, are profoundly affected by this phenomenon

A variety of instabilities, together with some aspects of thermal convection, are examined in chapter 8, and turbulent flow is the subject of chapter 9. Finally, the behaviour of complex fluids for which the Newtonian shear stress–strain rate relations are for one reason or another inadequate is surveyed in chapter 10.

Further reading

Most of the topics dealt with in this chapter recur later on, but we shall not come back to the problem of the inlet length in pipe flow [§1.14]. For more information about this, consult *Modern Developments in Fluid Dynamics*, Clarendon Press, Oxford, 1938, edited by S. Goldstein, Vol I, §VII.139.

2

The Euler fluid

2.1 The model

The title of this chapter refers to the idealised model discussed in chapter 1, on which Euler and Bernoulli based their contributions to fluid dynamics. An Euler fluid by definition has zero viscosity and zero compressibility. A fluid without viscosity cannot sustain shear stress, and the pressure p within it is therefore isotropic at all points. A fluid without compressibility has a density ρ which is unaffected by variations of p from place to place. The model need not exclude small variations of density due to thermal expansion if the temperature is non-uniform, but such variations are normally irrelevant except in so far as they may drive thermal convection currents in the fluid. Consideration of the topic of convection is deferred to chapter 8. For the time being we may regard temperature as something which has no influence on the flow behaviour of our model fluid and which may therefore be ignored.

Some of the conditions which need to be satisfied if the model is to match the behaviour of real fluids have been discussed in chapter 1. The reader may wish to refer back to that, and to the summary in §1.16 in particular.

2.2 The continuity condition

It is usually safe to assume that fluids remain continuous, and in that case the mass of fluid which occupies any volume V whose boundaries are fixed in space is just the integral over this volume of $\rho \mathrm{d}\boldsymbol{x}$, where $\mathrm{d}\boldsymbol{x}$ is a volume element. Then, by equating the rate at which mass is leaving this volume to the rate at which the mass within it is diminishing, we may arrive at the equation

$$\int_{\Sigma} \rho \boldsymbol{u} \cdot \mathrm{d}\boldsymbol{\Sigma} = \int_{\mathrm{V}} \boldsymbol{\nabla} \cdot (\rho \boldsymbol{u}) \mathrm{d}\boldsymbol{x} = -\int_{\mathrm{V}} \frac{\partial \rho}{\partial t} \mathrm{d}\boldsymbol{x},$$

where the left-hand integral is over the whole surface Σ that encloses V, $\mathrm{d}\boldsymbol{\Sigma}$ being an element of area of this surface, and where Gauss's divergence theorem has been used to transform this into the integral over V which follows it. Since the equation is true for all V, it follows that

$$\nabla \cdot (\rho \boldsymbol{u}) = \rho(\nabla \cdot \boldsymbol{u}) + (\boldsymbol{u} \cdot \nabla)\rho = -\frac{\partial \rho}{\partial t}. \tag{2.1}$$

This is the general *continuity condition* for fluid flow. For the Euler fluid, in which ρ does not vary with position or time, it evidently reduces to the simpler equation

$$\nabla \cdot \boldsymbol{u} = 0. \tag{2.2}$$

A breakdown of continuity is not inconceivable if the fluid is a liquid: where the pressure is low the liquid may *cavitate*, that is to say it may rupture to allow the formation of a vapour-filled bubble. This process is discussed below, in §2.11. If it occurs, needless to say, (2.1) holds only in the liquid phase outside the bubble.

2.3 Euler's equation

A small element of fluid subject to an isotropic pressure p which varies with position experiences a force per unit volume in the x_1 direction which is given by $-\partial p/\partial x_1$ [§1.7 and fig. 1.5]. It experiences forces in the x_2 and x_3 directions which are similarly given by $-\partial p/\partial x_2$ and $-\partial p/\partial x_3$ per unit volume, so in vector notation the force *per unit mass* due to pressure gradients alone is $-\nabla p/\rho$. The total force per unit mass $\boldsymbol{f}$ normally includes a term due to the weight of the element, which may be expressed as $-\nabla \phi_G$, where ϕ_G is the gravitational potential energy per unit mass; in terrestrial applications ϕ_G may usually be written as the product of a constant acceleration due to gravity, g, and a height z measured from some arbitrary zero, and we shall write it in that way here. Hence we arrive at *Euler's equation* (otherwise known as the *momentum equation* for an Euler fluid),

$$\boldsymbol{f} = -\frac{1}{\rho}\nabla p - \nabla(gz) = \frac{\mathrm{D}\boldsymbol{u}}{\mathrm{D}t}. \tag{2.3}$$

Here $\mathrm{D}\boldsymbol{u}/\mathrm{D}t$ is the local acceleration of the fluid, written in a notation which is unlikely to be familar to the reader but which is elucidated in §2.4.

The vector equations (2.3) and (2.2) are the basic equations of motion for the Euler fluid, which in principle determine completely, once initial conditions and boundary conditions have been stated, the spatial and temporal dependence of both pressure and velocity.

2.4 The operator D/Dt

It was pointed out in §1.7 that, since $\boldsymbol{u}$ is used in fluid dynamics to describe the whole velocity field rather than the velocity of a single particle, one cannot represent acceleration by a simple time derivative. The partial derivative with respect to time, $\partial \boldsymbol{u}/\partial t$, represents the rate of change of velocity at a point which is fixed in space and which is occupied by a succession of different fluid particles in turn. Acceleration is the rate of change of velocity of a single fluid particle which is on the move, and which therefore occupies a succession of different points.

Let us forget about fluid acceleration for a moment, and imagine that we are interested in some scalar quantity X which is a function of position as well as of time. If its value at some point $\boldsymbol{x}_o$ – a fixed point in reference frame S – is X_o at time t_o, what is its value at

$$\boldsymbol{x} = \boldsymbol{x}_o + \boldsymbol{v}\mathrm{d}t, \quad t = t_o + \mathrm{d}t,$$

in the limit when dt is an infinitesimal quantity? The answer is readily found by a Taylor expansion; it is $X_o + \mathrm{d}X$, where

$$\mathrm{d}X = \frac{\partial X}{\partial t}\,\mathrm{d}t + \frac{\partial X}{\partial x_1}\,v_1\mathrm{d}t + \frac{\partial X}{\partial x_2}\,v_2\mathrm{d}t + \frac{\partial X}{\partial x_3}\,v_3\mathrm{d}t. \tag{2.4}$$

Now in a frame S′ which is moving with respect to S at the velocity $\boldsymbol{v}$, these two points are coincident. Hence the derivative obtained by dividing (2.4) by dt throughout, namely

$$\frac{\mathrm{d}X}{\mathrm{d}t} = \frac{\partial X}{\partial t} + \frac{\partial X}{\partial x_1}\,v_1 + \frac{\partial X}{\partial x_2}\,v_2 + \frac{\partial X}{\partial x_3}\,v_3,$$

is also the partial differential of X with respect to t in the frame S′. In the particular case where $\boldsymbol{v}$ is the fluid velocity $\boldsymbol{u}$ at the point in question, S′ becomes what is called the *co-moving frame* – the frame in which the fluid appears, locally and instantaneously, to be at rest. We define DX/Dt to mean $\partial X/\partial t$ in the co-moving frame, so

$$\begin{aligned}\frac{\mathrm{D}X}{\mathrm{D}t} &= \frac{\partial X}{\partial t} + \frac{\partial X}{\partial x_1}\,u_1 + \frac{\partial X}{\partial x_2}\,u_2 + \frac{\partial X}{\partial x_3}\,u_3 \\ &= \frac{\partial X}{\partial t} + (\boldsymbol{u}\cdot\boldsymbol{\nabla})X. \end{aligned}\tag{2.5}$$

Another way to describe what is meant by DX/Dt is to say that it is the rate of change of X with time *following the fluid*.

We may note here for future reference that (2.5) allows us to express the general continuity condition in a form which is neater and more transparent than (2.1). The condition

$$\rho\boldsymbol{\nabla}\cdot\boldsymbol{u} = -\frac{\partial \rho}{\partial t} - (\boldsymbol{u}\cdot\boldsymbol{\nabla})\rho$$

is clearly equivalent to

$$\rho\boldsymbol{\nabla}\cdot\boldsymbol{u} = -\frac{\mathrm{D}\rho}{\mathrm{D}t}. \tag{2.6}$$

To revert now to the acceleration of the fluid, this is a vector quantity, the rate of change with time following the fluid of the vector velocity $\boldsymbol{u}$. Its three components are scalars, which may be separately expressed with the aid of (2.5) as

$$\frac{\mathrm{D}u_1}{\mathrm{D}t} = \frac{\partial u_1}{\partial t} + (\boldsymbol{u}\cdot\boldsymbol{\nabla})u_1 \quad \text{etc.}$$

and collectively expressed by the single vector equation

$$\frac{\mathrm{D}\boldsymbol{u}}{\mathrm{D}t} = \frac{\partial \boldsymbol{u}}{\partial t} + (\boldsymbol{u}\cdot\boldsymbol{\nabla})\boldsymbol{u}. \tag{2.7}$$

2.5 Transverse pressure gradients in steady flow

In the present chapter, and indeed throughout the rest of the book, we are largely concerned with situations where the flow is *steady*, i.e. where $\partial\boldsymbol{u}/\partial t$ is zero [§1.6]. In such situations the acceleration is just $(\boldsymbol{u}\cdot\boldsymbol{\nabla})\boldsymbol{u}$. If we choose our axes so that at the point of particular interest, say P, $\boldsymbol{u}$ lies along the x_1 direction, with $u_2 = u_3 = 0$, then the *longitudinal* component of acceleration is its component in the x_1 direction, and the magnitude of this in steady flow is

$$(\boldsymbol{u}\cdot\boldsymbol{\nabla})u_1 = u_1\frac{\partial u_1}{\partial x_1} = \frac{\partial(\frac{1}{2}u^2)}{\partial l}$$

as in (1.12). However, acceleration is liable to have transverse components as well; in steady flow they are $u_1(\partial u_2/\partial x_1)$ and $u_1(\partial u_3/\partial x_1)$ in the x_2 and x_3 directions respectively. We need consider only the relatively simple case in which one of these, say $u_1(\partial u_3/\partial x_1)$, happens to be zero. In that case u_3 is not only zero at P itself, it is also zero at nearby points on the line of flow through P; over small distances, therefore, this line of flow is confined to the (x_1, x_2) plane. Let O be its local centre of curvature and r_c = OP its local radius of curvature, as in fig. 2.1. Since $u_2 = 0$ at P itself, u_2 at the neighbouring point Q is given by

$$u_{2,\mathrm{Q}} \approx \left(\frac{\partial u_2}{\partial x_1}\right)_\mathrm{P} r_c \sin\theta + \left(\frac{\partial u_2}{\partial x_2}\right)_\mathrm{P} r_c(1-\cos\theta),$$

or

$$u_{2,\mathrm{Q}} \approx \left(\frac{\partial u_2}{\partial x_1}\right)_\mathrm{P} r_c\theta$$

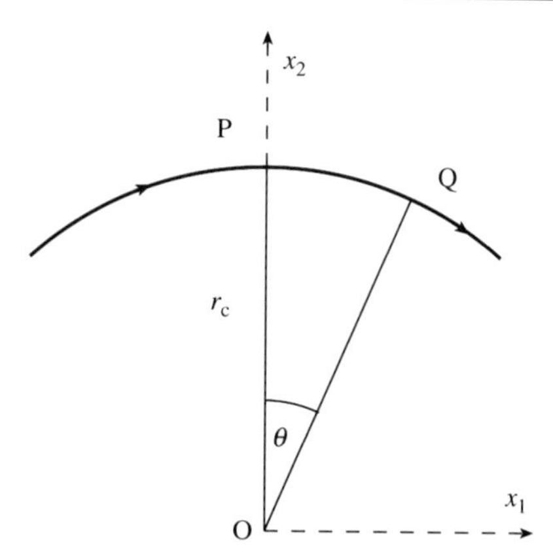

Figure 2.1 A curved line of flow on which the fluid accelerates towards the centre of curvature, O.

to first order in the angle θ. But

$$u_{2,Q} = u_Q \sin\theta \approx u_P\theta,$$

and by equating these two expressions for $u_{2,Q}$ we may deduce that

$$\left(\frac{\partial u_2}{\partial x_1}\right)_P = \left(\frac{u}{r_c}\right)_P,$$

and hence that

$$u_1\left(\frac{\partial u_2}{\partial x_1}\right)_P = \left(\frac{u^2}{r_c}\right)_P. \tag{2.8}$$

This result is very familiar in elementary particle dynamics. It describes the *centripetal acceleration* for a particle moving in two dimensions along a trajectory which has a radius of curvature r_c.

Where lines of flow are curved in a two-dimensional manner, Euler's equation evidently implies a transverse pressure gradient:[1] p must increase with distance from the local centre of curvature according to the equation

$$\frac{\partial p}{\partial r} = \frac{\rho u^2}{r_c}. \tag{2.9}$$

A transverse pressure gradient described by (2.9) exists, for example, in a bucketful of water which is set spinning about its vertical axis with a uniform

[1] When curved lines of flow lie in a vertical rather than a horizontal plane, there may be a transverse pressure gradient which has nothing to do with their curvature, namely a component of the vertical pressure gradient $\mathrm{d}p/\mathrm{d}z = -\rho g$ which is a feature of hydrostatic equilibrium. In such cases the right-hand side of (2.9) describes the transverse gradient of *excess pressure* p^* rather than p. The distinction between p and p^* is explained in §2.9.

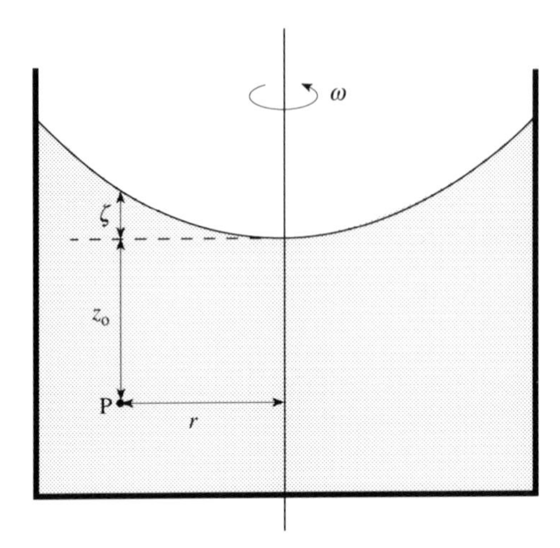

Figure 2.2 Water in a vessel rotating with uniform angular velocity. Its surface is a paraboloid.

angular velocity ω, and it manifests itself by the way in which the free surface of the water becomes curved. At the point P in fig. 2.2 the pressure must be sufficient to support the weight of the water above it, so

$$p = p_A + \rho g(z_o + \zeta),$$

where p_A is the atmospheric pressure while ζ is the local elevation of the surface at radius r from the axis. Hence the radial pressure gradient near P is $\rho g(\mathrm{d}\zeta/\mathrm{d}r)$, and by equating this to the pressure gradient which (2.9) describes we have

$$\frac{\mathrm{d}\zeta}{\mathrm{d}r} = \frac{u^2}{rg} = \frac{\omega^2 r}{g},$$

and therefore

$$\zeta = \frac{\omega^2 r^2}{2g}. \tag{2.10}$$

The surface is a paraboloid of revolution.

Note that it would be difficult to set a bucketful of ideal Euler fluid into the state of uniform rotation described above. With real liquids there is no such difficulty – one has only to rotate the bucket with uniform angular velocity and this angular velocity will in due course be imparted to all the fluid inside the bucket – but that is because real liquids are viscous. Once the liquid is rotating uniformly, however, the shear stresses necessarily vanish whatever the viscosity, as is shown in §6.9(ii). Thus the assumption we have made in deriving (2.9) and (2.10) that the viscosity of the fluid is zero is not misleading for real liquids in the context of this particular problem.

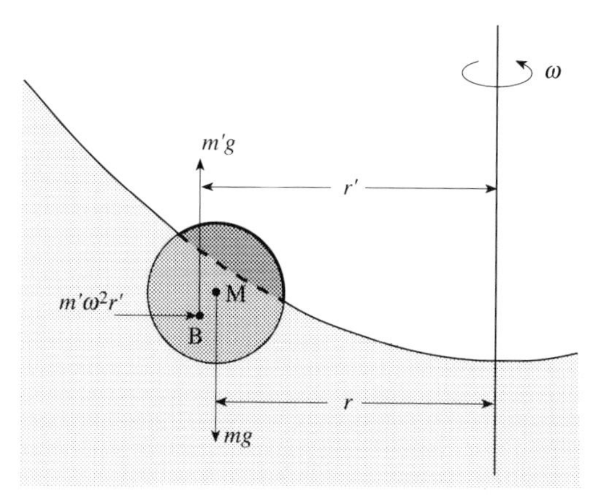

Figure 2.3 Forces acting on a uniform solid sphere floating in rotating water.

2.6 Floating objects

What happens when a solid object which is lighter than water floats in a bucketful of water which is uniformly rotating and which has the paraboloidal surface described by (2.10)?

Figure 2.3 shows a cross section through the surface and through a partly submerged solid sphere. The distributed forces which act upon the sphere may be treated as the three concentrated forces which are represented by arrows in the figure. The first arrow, which represents the weight of the sphere mg, acts downwards through its centre of mass M, which is envisaged as being at the centre of the sphere at a distance r from the axis of rotation. The other two forces represent the effect on the sphere of the pressure of the water and air in contact with it. These must be such as would preserve the system in equilibrium, were the sphere to be removed and the volume indicated by intermediate shading in the figure to be occupied by water instead. They therefore act through the centre of mass of this displaced water, a point which is called the *centre of buoyancy* and which is labelled B in the figure. The vertical component of the Archimedean upthrust through B is of course $m'g$, where m' is the mass of water displaced, while its horizontal component must be $m'\omega^2r'$, this being, if r' is the distance of B from the axis of rotation, the radial force required to maintain the centripetal acceleration of the displaced water. Now by moving vertically the sphere can readily reach a situation where the two vertical forces which act upon it are equal and opposite, and in this situation $m' = m$. Because r' is greater than r, however, the horizontal force $m'\omega^2r'$ is then greater than the force $m\omega^2r$ which is needed to keep M moving round a circle of fixed radius. Consequently, the sphere moves inwards and does not reach equilibrium until M lies on the axis of rotation, with $r = 0$.

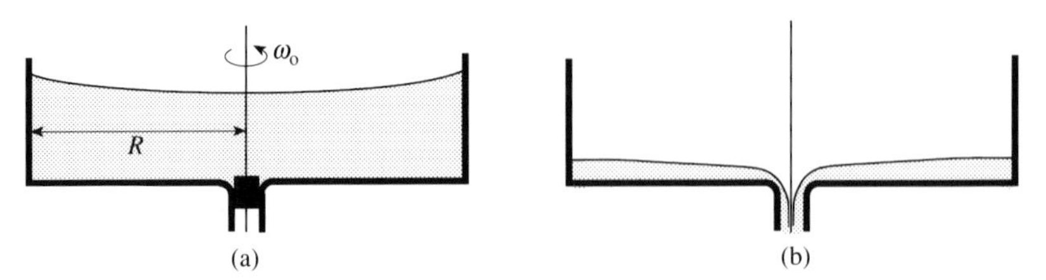

Figure 2.4 Origin of the plug-hole vortex.

What is true for floating spheres is true for floating objects of any shape if their density is uniform. But an object which is 'ballasted', as ships are ballasted to make them more stable, may have a centre of mass which is below its centre of buoyancy. Such an object, when floating in rotating liquid, would move away from the axis of rotation rather than towards it.

2.7 The plug-hole vortex

Figure 2.4(a) shows in cross-section a circular dish of radius R with a small hole at the centre of its base, initially plugged, containing Euler liquid which is rotating with uniform angular velocity ω_0 about a vertical axis through the hole. When the plug is removed the liquid starts to drain slowly away, and by the time that most of it has gone the situation is as shown in fig. 2.4(b). The particles of liquid which remain in the dish at this stage were presumably near the sides of the dish initially, at distances from the hole not much less than R. Since none of them, in the absence of viscous shear stresses, can have experienced any torque about the central axis while draining was in progress their angular momentum about this axis must be unchanged, which means that the angular velocity of the liquid in fig. 2.4(b) must vary with radius r according to the approximate relation

$$\omega \approx \omega_0 \frac{R^2}{r^2}. \tag{2.11}$$

Equation (2.9) then implies, if the draining is so slow that the flow is quasi-steady, the existence of a radial pressure gradient

$$\frac{\partial p}{\partial r} \approx \rho\omega^2 r \approx \frac{\rho\omega_0^2 R^4}{r^3}, \tag{2.12}$$

and the argument leading to (2.10) may be used in a modified form to show that the depth of the fluid near the hole is less than the depth at large radii by

$$\zeta \approx \frac{\omega_0^2 R^4}{2r^2 g}. \tag{2.13}$$

A whirlpool or vortex is therefore to be expected, in which relatively large angular velocities are coupled with a convex profile for the surface meniscus.[2]

Such vortices are commonly observed, of course, when baths are emptied, and the theory presented above describes them with fair accuracy though they must be influenced to some extent by viscous effects. The angular momentum initially present in the water in a bath, which is responsible for the vortex and which determines the sense of the rotation associated with it, is normally the result of motions induced by the bather while scrubbing his back or while stepping out of the bath; the sense of rotation is therefore not systematically related to whether the bath is situated in the northern or southern hemisphere in the way that common superstition supposes it to be. However, experiments have been conducted on water in a circular dish with a radius of 3 feet, equipped with a central wastepipe and filled to a depth of 6 inches, and superstition has been shown to be vindicated when the water is allowed at least a day in which to become stationary before the waste valve is opened.[3] In those ideal circumstances the initial angular velocity ω_0 is the angular velocity with which the Earth is rotating about the vertical axis through the hole; at latitude λ it is $2\pi \sin\lambda$ day^{-1}, and it changes sign at the equator where $\lambda = 0$. The formation of the vortex in such circumstances may be attributed to the Coriolis force, but an explanation in terms of conservation of angular momentum is completely equivalent and rather more straightforward.

It is of some interest to examine what the vorticity of the liquid is in the two states represented by fig. 2.4(a) and fig. 2.4(b). When the drainage rate is very slow, the radial component of the velocity is negligible compared with the component ωr at right angles to the radius vector. In these circumstances, the rate of rotation of a short line embedded in the fluid is ω if the line lies at right angles to the radius vector [OP in fig. 2.5] but $\mathrm{d}(\omega r)/\mathrm{d}r$ if it lies along a radius vector [OQ in fig. 2.5]. Reference to the discussion in §1.10 and comparison of fig. 2.5 with fig. 1.8 should now convince the reader that the vorticity near O is perpendicular to the plane of that diagram and that its magnitude is

$$\Omega = \omega + \frac{\mathrm{d}(\omega r)}{\mathrm{d}r} = 2\omega + r\frac{\mathrm{d}\omega}{\mathrm{d}r}, \tag{2.14}$$

though this result can, of course, be proved in other ways, e.g. by evaluating $\nabla \wedge \boldsymbol{u}$ in cylindrical coordinates, (r, θ, z). Hence Ω is everywhere equal to $2\omega_0$ in fig. 2.4(a), while in the final state of fig. 2.4(b), if ω is then proportional to r^{-2} as (2.11) suggests, it is zero. In view of Kelvin's theorem that the vorticity of an Euler fluid

[2] It is left as an exercise for the reader to re-examine the problem discussed in §2.6 in this context, and to show that anyone adrift in a small boat which is trapped in a maelstrom had better throw the ballast overboard and stand up rather than lie down.

[3] See A. H. Shapiro, *Nature*, **197**, 1080, 1963. According to a comment on this paper in the subsequent volume of *Nature*, the first experiments of this kind were reported by an Austrian physicist called Turmlïtz in 1908.

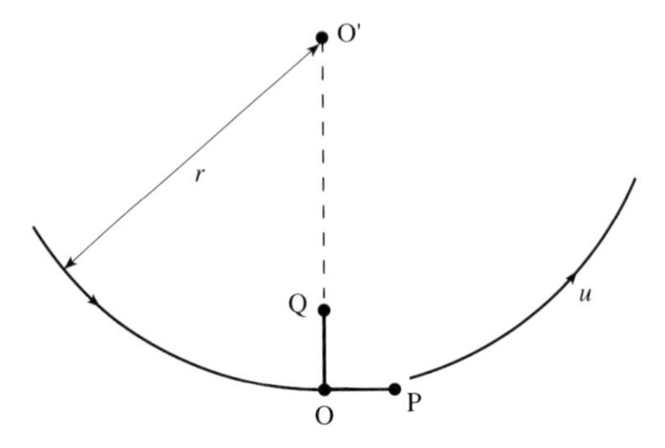

Figure 2.5 Fluid circulating about O′ with a velocity $u = \omega r$ which varies with radius. The lines OP and OQ, both embedded in the fluid, rotate at ω and $\mathrm{d}u/\mathrm{d}r$ respectively.

cannot change [§§1.10 and 4.2], these two results are inconsistent with one another. The inconsistency is a reminder that (2.11) is only approximately correct.

2.8 Bernoulli's theorem revisited

Let us now return to the longitudinal component of the pressure gradient which Euler's equation implies for steady flow. As we have seen [(1.12)], the longitudinal acceleration when $\partial \boldsymbol{u}/\partial t$ is zero is $\partial(\frac{1}{2}u^2)/\partial l$, while the longitudinal component of $\boldsymbol{f}$ is given according to (2.3) by

$$-\frac{1}{\rho}\frac{\partial p}{\partial l} - \frac{\partial(gz)}{\partial l}.$$

By equating the two and integrating along a line of flow we may show, as in §1.7, that everywhere on this line

$$\frac{p}{\rho} + gz + \frac{1}{2}u^2 = \text{constant}. \tag{2.15}$$

This statement of Bernoulli's theorem differs from (1.13) only by inclusion of a gravitational term which was not taken into account in chapter 1.

The theorem, as has been stressed already, is no more than a statement of the law of conservation of energy, i.e. of the first law of thermodynamics, as applied to an Euler fluid in which mechanical energy and thermal energy are uncoupled. It may readily be proved without integrating the equation of motion. Consider for example the column of fluid enclosed within a fixed flow tube which is sketched in fig. 2.6. At one instant it stretches from P to Q, and a moment later it stretches

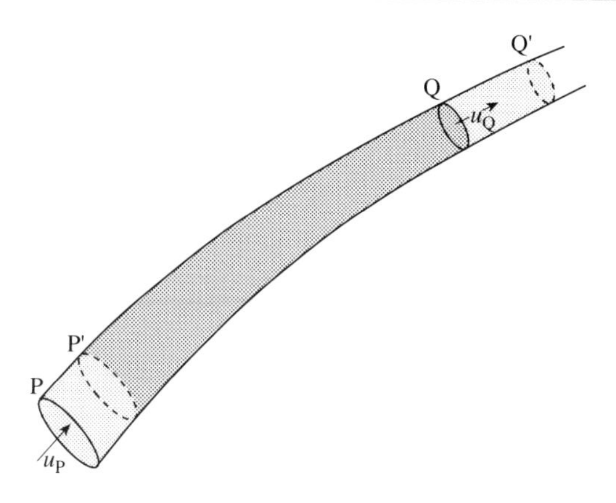

Figure 2.6 A fluid column moving through a flow tube of varying cross-section defined by lines of flow.

from P′ to Q′. The difference between these states is that a small mass of fluid Δm which was between P and P′ has been replaced by an equal mass between Q and Q′, so the increase of kinetic and gravitational potential energy is clearly

$$\Delta m \left\{ \left(\frac{1}{2} u^2 \right)_{\mathrm{Q}} - \left(\frac{1}{2} u^2 \right)_{\mathrm{P}} + (gz)_{\mathrm{Q}} - (gz)_{\mathrm{P}} \right\}. \tag{2.16}$$

But since the volume through which the ends of the column have been displaced is $\Delta m/\rho$, the work done upon it is

$$\frac{\Delta m}{\rho} (p_{\mathrm{P}} - p_{\mathrm{Q}}). \tag{2.17}$$

Provided that there is no external agency between P and Q which does work upon the fluid or extracts work from it – this is a proviso which we shall need to bear in mind in §2.13 and §2.16 below – we may arrive at Bernoulli's theorem by equating (2.16) to (2.17).

It was stated in chapter 1 that the 'constant' which features in Bernoulli's theorem is the same on every flowline when the flow is vorticity-free – as it proves to be in the great majority of situations where the theorem is useful. The proof of this statement is straightforward. If $\boldsymbol{\nabla} \wedge \boldsymbol{u}$ is everywhere zero, then

$$\frac{\partial u_2}{\partial x_1} = \frac{\partial u_1}{\partial x_2}, \quad \frac{\partial u_3}{\partial x_1} = \frac{\partial u_1}{\partial x_3},$$

which means that the component of $(\boldsymbol{u} \cdot \boldsymbol{\nabla})\boldsymbol{u}$ in an arbitrarily chosen x_1 direction can be simplified as overleaf:

$$
\begin{aligned}
(\boldsymbol{u}\cdot\boldsymbol{\nabla})u_1 &= u_1\frac{\partial u_1}{\partial x_1} + u_2\frac{\partial u_1}{\partial x_2} + u_3\frac{\partial u_1}{\partial x_3} \\
&= u_1\frac{\partial u_1}{\partial x_1} + u_2\frac{\partial u_2}{\partial x_1} + u_3\frac{\partial u_3}{\partial x_1} \\
&= \frac{\partial(\frac{1}{2}u^2)}{\partial x_1}.
\end{aligned}
$$

Similar results hold for the components in the x_2 and x_3 directions, so that

$$(\boldsymbol{u}\cdot\boldsymbol{\nabla})\boldsymbol{u} = \boldsymbol{\nabla}\left(\frac{1}{2}u^2\right).$$

Hence Euler's equation (2.3), with $\mathrm{D}\boldsymbol{u}/\mathrm{D}t$ given by (2.7), becomes

$$\frac{\partial \boldsymbol{u}}{\partial t} + \boldsymbol{\nabla}\left(\frac{p}{\rho} + gz + \frac{1}{2}u^2\right) = 0 \tag{2.18}$$

for vorticity-free flow. If the flow is also steady, then

$$\boldsymbol{\nabla}\left(\frac{p}{\rho} + gz + \frac{1}{2}u^2\right) = 0,$$

which may be integrated in any direction to show that the constant in (2.15) is the same everywhere.

The two types of circulation about a vertical axis depicted in fig. 2.4(a) and fig. 2.4(b) serve to illustrate the distinction between Bernoulli's theorem when vorticity is present and the same theorem when it is not. In the first case the liquid is rotating with uniform angular velocity, as though it were a solid body. The pressure p and the velocity u both increase with radius r for fixed height z, so that $\{(p/\rho) + gz + \frac{1}{2}u^2\}$ is certainly not the same on all the lines of flow – which are, of course, just circles around the axis of rotation in this case. That is no surprise, because the motion is not vorticity-free: $\Omega = 2\omega_0$, as we have seen. In the second case, where $\omega \propto r^{-2}$ and $u \propto r^{-1}$, the motion *is* vorticity-free. We may therefore use Bernoulli's theorem, bearing in mind that the 'constant' is now independent of r, to deduce the transverse pressure gradient whose existence we have previously deduced in a different way. We have

$$\frac{\partial}{\partial r}\left(\frac{p}{\rho} + gz + \frac{1}{2}u^2\right) = 0,$$

whence

$$\frac{1}{\rho}\frac{\partial p}{\partial \mathrm{r}} = -u\frac{\partial u}{\partial r} = +\frac{u^2}{r}$$

as in (2.9).

In the remaining sections of this chapter Bernoulli's theorem is applied to a selection of problems of interest to physicists which illustrate its power and utility. Several other illustrations will be found in later chapters.

2.9 Hydrostatic equilibrium and excess pressure

The simplest application of all is to fluid which is at rest in a state of what may be called *hydrostatic equilibrium*. Such fluid is clearly vorticity-free, so it follows from (2.15), with $u = 0$ everywhere, that

$$p(z) = p(0) - \rho g z. \tag{2.19}$$

Equation (2.19) describes the well-known way in which the pressure in stationary water, say, increases with depth. It can of course be obtained without reference to Bernoulli's theorem if preferred, by considering the balance of forces on a vertical column of the water.

The pressure exerted by the Earth's atmosphere is nearly always totally irrelevant in fluid dynamics, and the increase of pressure with depth is often irrelevant too; it is responsible for the Archimedean upthrust experienced by a solid body which is floating or fully immersed in water but has nothing whatever to do with the additional drag force which arises when the body is moved. It is often appropriate, therefore, to set these terms in the pressure on one side and to focus on the local *excess pressure* p^*. This is defined by

$$p^* = p + \rho g z - p_o, \tag{2.20}$$

where p_o is a constant, normally chosen so that p^* vanishes in a remote region where the fluid is either stationary or moving with some uniform velocity, say U. In so far as Bernoulli's theorem for vorticity-free fluid in steady flow is applicable, it tells us that the excess pressure elsewhere is related to the local velocity u by the equation

$$p^* = \frac{1}{2}\rho(U^2 - u^2). \tag{2.21}$$

This form of the theorem proves especially convenient, not so much in the present chapter as in chapters 4 and 6.

2.10 Devices for measuring rates of flow

When hydraulic engineers want to be able to monitor continuously the rate of flow in an open channel, they may build a *weir* across it and observe the rise and fall of the water level upstream from the weir, where the current is very slow. Weirs of

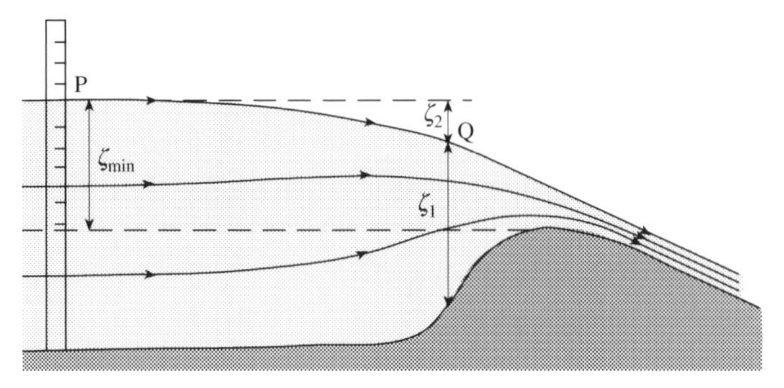

Figure 2.7 Flow over a broad-crested weir: a measurement of ζ_{min} gives the discharge rate.

the *broad-crested* type, such that the radius of curvature of the crest is much larger than the depth of water just above it, are the easiest to discuss from a theoretical point of view and they provide a nice example of Bernoulli's theorem in action.

A cross-section through such a weir, with lines of flow to indicate the way in which water moves over it, is sketched in fig. 2.7, and the quantity from which it is possible to deduce the volume flow rate per unit width of the channel, say q, is the vertical distance which is labelled on the figure as ζ_{min}. It is labelled thus because it is clearly the minimum value of a level distance $\zeta = \zeta_1 + \zeta_2$ which varies continuously with x in the neighbourhood of the crest. Here ζ_1 is the depth of water below some arbitrary point Q on a surface flowline, while ζ_2, as the figure suggests, is the amount by which this point is below a point P which lies on the same flowline in the relatively stagnant region upstream. The pressure is atmospheric at both P and Q, and it follows at once from Bernoulli's theorem [(2.15)] that if $u_P \ll u_Q$ then

$$\frac{1}{2} u_Q^2 \approx g\zeta_2.$$

But if the radius of curvature of the flowlines, like that of the top surface of the weir, is very large, then the transverse gradient of p^* as described by (2.9)[4] should be negligible, in which case we may infer from Bernoulli's theorem in the form of (2.21) that $\partial u/\partial z$ is also negligible in the neighbourhood of Q. Hence

$$q \approx u_Q\zeta_1 \approx \zeta_1\sqrt{2g\zeta_2},$$

and we may therefore write

$$\zeta = \zeta_1 + \zeta_2 \approx \zeta_1 + \frac{q^2}{2g\zeta_1^2}.$$

[4] With p replaced by p^*. See footnote on p. 41.

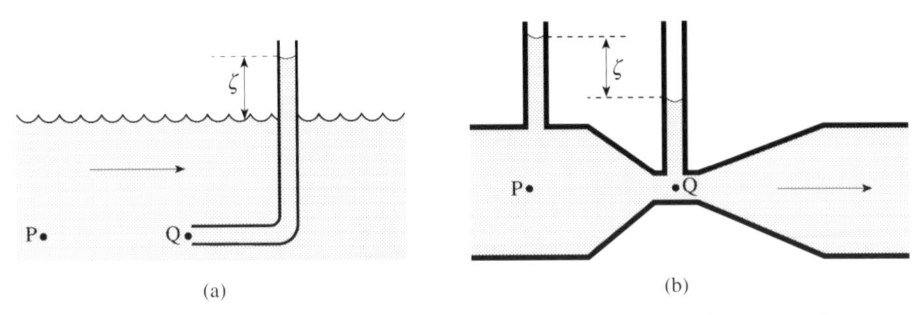

Figure 2.8 Schematic diagrams, not to the same scale, of (a) a Pitot tube, (b) a Venturi meter.

The expression on the right-hand side of this equation passes through a minimum where $\zeta_1^3 = q^2/g$, and its minimum value must describe ζ_{min}. Consequently, the flow rate per unit width is related to ζ_{min} by the equation

$$\zeta_{\mathrm{min}} \approx \frac{3}{2}\left(\frac{q^2}{g}\right)^{1/3}. \tag{2.22}$$

A more portable flowmeter is the *Pitot tube*, originally devised to monitor the flow of the River Seine. In its primitive form it consists of a tube open at both ends, which is bent into the shape of the letter L and inserted into a river with its long arm vertical and its short arm facing upstream, as in fig. 2.8(a). The tube fills with water, and because this water is not moving the point labelled Q in the figure is a *stagnation point*, i.e. a point where the fluid velocity u is zero. Bernoulli's theorem in the form of (2.15) may be used to compare the pressure at Q with the pressure at a reference point P, which is on the same level as Q and in principle on the same line of flow but some way upstream where the velocity is U. It tells us that

$$p_{\mathrm{Q}} = p_{\mathrm{P}} + \frac{1}{2}\rho U^2,$$

and hence that the height ζ above the surface of the river to which the water rises in the long arm of the tube is related to the flow velocity by the equation

$$U^2 = 2g\zeta. \tag{2.23}$$

A device which is almost as straightforward to construct and use as the Pitot tube but harder to discuss is the *Venturi meter*. This consists of a length of pipe with a tapered constriction in it, as in fig. 2.8(b), designed to be inserted in a longer pipe of matching cross-section to enable the discharge rate Q to be determined. A manometer to enable the pressure drop in the constriction to be measured forms part of the device, as the figure suggests. Let us initially suppose the fluid – an

Euler fluid with zero viscosity – to be vorticity-free, in which case u is uniform across the whole cross-sectional area A of the pipe, both at Q where the constriction is and upstream at P. In that case we may deduce from Bernoulli's theorem that

$$\rho g \zeta = p_P - p_Q = \frac{1}{2}\rho(u_Q^2 - u_P^2)$$

$$= \frac{\rho Q^2}{2A_Q^2}(1 - C^2), \tag{2.24}$$

where C is a *contraction coefficient* defined as A_Q/A_P. Alternatively, and perhaps more realistically, we may suppose that the fluid is *not* vorticity-free, and that its velocity profile near P has the parabolic form to be expected for a viscous fluid which is undergoing steady laminar flow along a pipe, with

$$u_P = \frac{2Q}{A_P}\left(1 - \frac{r^2}{a_P^2}\right)$$

[(1.32) and (1.33)]. It is left as a (non-trivial) exercise for the reader to show, using the fact that the pressure drop must be the same on every flowline connecting points near P to points near Q, that in this case the velocity profile near Q is a flatter parabola, which does not go to zero at $r = a_Q$, described by the equation

$$u_Q = \frac{2Q}{A_Q}\left(\frac{1 + C^2}{2} - C^2\frac{r^2}{a_Q^2}\right), \tag{2.25}$$

and that

$$\rho g \zeta = \frac{\rho Q^2}{2A_Q^2}(1 - C^2)^2. \tag{2.26}$$

Equation (2.26) may be more realistic than (2.24) but it cannot tell the whole story for a real, as opposed to an Euler, fluid. Real fluids must obey the no-slip boundary condition at Q as well as P, and moreover they dissipate energy as heat, at a rate per unit volume which turns out to be equal in this context to $\eta(\partial u/\partial r)^2$ [(6.26)]. To provide the energy which is dissipated as well as the extra kinetic energy which the fluid possesses in the constriction, the pressure drop must be somewhat larger than the drop which (2.26) describes. The total rate of dissipation within the length L over which the radius tapers from a_P to a_Q may be estimated on the crude assumption that (2.25), with C replaced by $a\{x\}^2/a_P^2$, provides an adequate description of the velocity profile for all values of x within this length, and the result, for a typical Venturi meter in which $a_Q \approx 0{\cdot}5a_P$, i.e. $C \approx 0{\cdot}25$, is about $10\eta L Q^2/A_P^2$. If the reader cares to refer back to the simple energy balance argument based on fig. 2.6, he will find that this implies an additional pressure drop of about

$$10\,\frac{\eta LQ}{A_P^2} = \frac{5L}{2\pi Re a_P}\,\frac{\rho Q^2}{2A_Q^2},$$

where Re is the upstream Reynolds Number, i.e. $2\rho Q/\pi a_P \eta$. The angle of taper on the upstream side of the constriction in a Venturi meter normally exceeds 20°, so $5L/2\pi a_P$ is normally less than 2. In that case the additional pressure drop should for most purposes be negligible compared with the drop described by (2.26) when Re is of the order of 10^2 or more.

In fact Re normally exceeds 10^4 in the circumstances where Venturi meters are used, which means that the flow is turbulent rather than laminar [§1.15]. Now the energy balance argument based on fig. 2.6 shows that Bernoulli's theorem can be applied to turbulent flow, with time-averaged velocity $\overline{u}$ in place of u, *provided* that the intrinsic energy density of the fluid, i.e. the energy density measured in the co-moving reference frame, is virtually the same at the two points where pressure is to be compared. It seems that this condition is adequately satisfied in Venturi meters which are being used to monitor turbulent flow, for it is found that the simple equation (2.24) is obeyed to within a few parts per cent. The profile of $\overline{u}\{r\}$ in turbulent flow is much flatter than the profile of $u\{r\}$ in laminar flow at the same value of Q [§9.8 and fig. 9.10], which explains why (2.24) works rather better than (2.26) in this context.[5]

Incidentally, the angle of taper on the downstream side of the constriction in a Venturi meter is normally only 6°. The reason for making it small is that where a fluid stream emerges suddenly from a small pipe into a larger one it emerges as what is called a *submerged jet* [§2.13]. This jet in due course shares its momentum with the surrounding fluid, but in doing so it creates extra turbulence [§7.7]. As a result, the intrinsic energy density of the fluid downstream from the enlargement is greater than it is upstream – the extra energy is stored initially as kinetic energy of fluctuating eddies, though further downstream still it may be dissipated as heat – and because of this increase the pressure drop over the length of pipe in which it occurs is greater than it would be otherwise. From a hydraulic engineer's point of view, losses of pressure head are always undesirable and sudden enlargements in a pipe are therefore to be avoided.

The question of *why* fluid tends to emerge from a constriction in the form of a submerged jet is illuminated by some remarks about fans in §2.13 below and further discussed in §7.7.

[5] Equation (2.24) is essentially the same as one of the ingredients, namely (1.35), of our answer to the syringe problem posed in chapter 1. In that context u_P^2 was so much smaller than u_O^2 as to be negligible, i.e. C^2 was negligible in the factor $(1 - C^2)$ which appears in (2.24). That being so, our answer would have been no different had we chosen to rely on (2.26) instead.

2.11 Cavitation

By forcing water through a constriction which is very narrow, the pressure p_Q within it may quite easily be reduced to zero; according to (2.24) this requires only that

$$\frac{1}{2}\rho(u_Q^2 - u_P^2) \approx \frac{1}{2}\rho u_Q^2 \approx p_P,$$

and if p_P is about 1 atmosphere this corresponds to a quite modest value for u_Q, of order 10 m s^{-1}. If u_Q is made greater still, then p_Q may actually become negative. The concept of a negative pressure may be absurd for gases, but it is certainly not absurd for solids, which can withstand large tensile stresses without cracking, and it is not absurd for liquids either. Some readers may imagine that any specimen of liquid is bound to transform into its vapour phase when the pressure within it falls below its saturated vapour pressure, but unless the specimen has a free surface across which evaporation can occur without impediment this is not the case; in a specimen which is completely enclosed by solid walls the phase transformation does not occur unless the liquid *cavitates*, either where it meets the walls or somewhere in its interior. The cavitation process is akin to cracking in a solid; it involves the formation and subsequent growth of a small vapour-filled bubble, and it is impeded by an energy barrier for which surface tension is responsible. Experiments have shown that extremely pure water at room temperatures can withstand pressures as low as -10^7 N m^{-2} (i.e. -100 atmospheres) without cavitating.

The ability of water to withstand reductions of pressure is less marked than that in normal circumstances. Irregularities on the surface of the solids with which it is in contact, or small particles of impurity which are present in suspension, often serve as nuclei from which cavities may grow, and the energy barrier is reduced by the presence of dissolved air, which tends to come out of solution at low pressures and to collect in any cavity that forms. Thus cavitation does quite often occur when water flows through pipes. For example, it occurs when water squeezes through a tap which is not quite closed. The tiny bubbles which are created by cavitation in the constriction represented by the tap collapse when the water emerges and the pressure rises again. In collapsing they emit sound, and the familiar hissing noise produced by slightly open taps is to be explained in this way.

The collapse of a vapour-filled cavity can be very rapid and can generate very large velocities in the liquid surrounding the cavity. Consider the response of a spherical bubble of radius a_o, which has formed at a pressure which is so low as to be negligible, when the ambient pressure rises suddenly to some value p_o of the order of one atmosphere. The liquid moves radially inwards with a velocity u which, if the liquid is effectively incompressible, must vary like R^{-2}, where R is distance from the bubble's centre, in order to satisfy the continuity condition

Figure 2.9 Two helical chains of bubbles in a moving column of water, which have formed by cavitation at the tips of a rotating propeller blade. [Photograph by G. L. Taylor, British Maritime Technology, Feltham, from D. H. Trevena, *Cavitation and Tension in Liquids*, Adam Hilger, 1987.]

$\nabla \cdot \boldsymbol{u} = 0$. If $u\{a\} = -\mathrm{d}a/\mathrm{d}t$ is the velocity at the surface of the bubble, where $R = a$, then

$$u\{R\} = u\{a\}\left(\frac{a}{R}\right)^2.$$

Hence the kinetic energy stored in the liquid during collapse is

$$\frac{1}{2}\rho\int_a^\infty u^2\{R\}4\pi R^2\mathrm{d}R = 2\pi\rho a^4u^2\{a\}\int_a^\infty R^{-2}\mathrm{d}R = 2\pi\rho a^3u^2\{a\}.$$

We may equate this to the work done by the ambient pressure during the collapse, which is $\frac{4}{3}\pi p_o(a_o^3 - a^3)$, to arrive at the approximate result

$$u\{a\} \approx \left(\frac{2p_o}{3\rho}\right)^{1/2}\left(\frac{a_o}{a}\right)^{3/2} \approx 10\left(\frac{a_o}{a}\right)^{3/2} \text{ m s}^{-1}.$$

It is no more than an approximation: in deriving it we have ignored a^3 in comparison with a_o^3, ignored the contents of the bubble, and ignored viscous dissipation. It suffices to indicate, however, that $u\{a\}$ may become very considerable before the bubble is finally eliminated. Thus when a hemispherical bubble is formed by cavitation at a liquid–solid interface, its subsequent collapse may

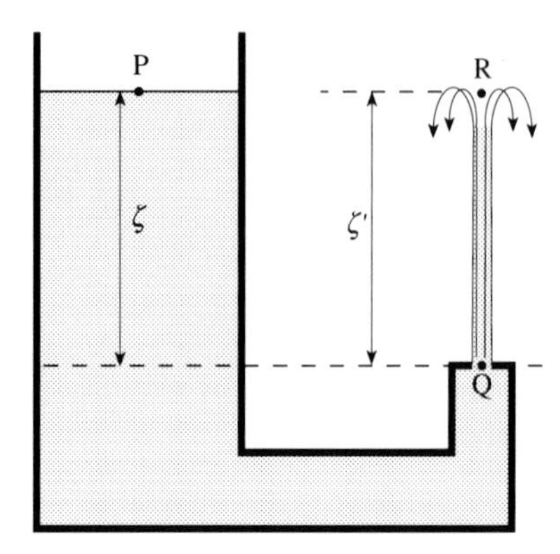

Figure 2.10 For an Euler fluid, the height to which the fountain rises (ζ') is equal to the head of water in the reservoir (ζ).

deliver a localised impulse to the solid which is sufficient to cause pitting; it may well create thereby a site at which corrosion may begin.

Cavitation is liable to occur not only in pipe flow but also – as is shown by fig. 2.9 – in the vicinity of rapidly rotating turbine blades and ships' propellers and in other situations where water acquires a velocity which locally exceeds 10 m s^{-1} or thereabouts. Engineers take steps to prevent it where they can, because of the damage which it does.

2.12 Water jets, sheets and bells

Figure 2.10 is a schematic illustration of a gravity-fed fountain. Its reservoir is filled to a height ζ above the orifice through which the water emerges. The emergent jet, with velocity U, rises to a height ζ'. At each of the points P, Q and R, which in principle lie on a single line of flow (though we need not insist that they do so since the flow should be vorticity-free), the pressure is atmospheric. It then follows at once from Bernoulli's theorem that, if viscous dissipation is negligible,

$$2g\zeta = U^2 = 2g\zeta'. \tag{2.27}$$

These results are credited to Torricelli.

Figure 2.11(a) shows an enlarged view of a jet near an orifice – in this case the jet is a horizontal one – and reveals a complication which was concealed from the reader during the discussion of the syringe problem in chapter 1: over a short distance beyond the orifice which is comparable with its diameter the jet narrows, to form what has been known for centuries as the *vena contracta* – the contracted vein. It is bound to narrow in this way because the lines of flow where they enter the orifice are curved. This implies the existence of a transverse pressure gradient; the pressure at S, on the axis of the jet, is higher than the atmospheric pressure

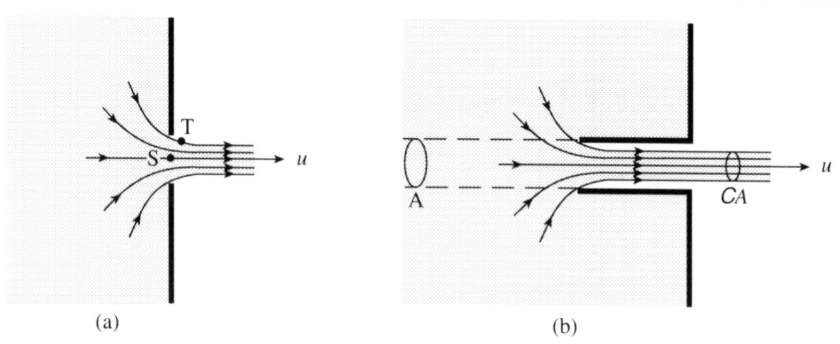

Figure 2.11 (a) Jet with *vena contracta*; (b) Borda's mouthpiece.

which acts at T. The pressure excess at S means that the water on the axis is still accelerating longitudinally as it leaves the orifice. Only in the *vena contracta* does the pressure become atmospheric all the way through the jet; only there is the velocity uniform and related to the pressure head ζ by (2.27). If the cross-sectional area of the orifice is A, the discharge rate is therefore

$$Q = CA(2g\zeta)^{1/2}, \tag{2.28}$$

where the *contraction coefficient* C is here the ratio of the area of the *vena contracta* to the area of the orifice. This contraction coefficient depends upon the design of the orifice, and because of viscous effects which are not taken into account in the Euler model it depends upon the Reynolds Number too, though this dependence is slight when Re is large. If the orifice is a circular hole punched in a thin plate, then C is normally about 0·62.

Figure 2.11(b) shows an orifice equipped with what is called *Borda's mouthpiece*, a tube with a length significantly greater than its diameter which juts backwards into the reservoir from which the water is emerging. For this re-entrant mouthpiece the contraction coefficient for large Re is exactly $\frac{1}{2}$, and the argument which shows it to be so provides a nice example of how answers to what appear to be difficult questions can often be obtained in fluid mechanics, as in other branches of physics, by simple means. The mouthpiece ensures that the region where the lines of flow are converging on the orifice is far from the walls of the reservoir. Hence where water is in contact with these walls its velocity is negligibly small and, at the level of the orifice, its pressure exceeds that of the atmosphere by $\rho g\zeta$, where ζ is the head of water in the reservoir as before. Now closed reservoirs experience no net force whatever the pressure of their contents, but our reservoir is different because it is has a hole of area A in its right-hand wall. The pressure head exerts an outwards force of magnitude $\rho g\zeta A$ on an area A on the left-hand wall of the reservoir, immediately opposite the hole, and there is no force on the right-hand wall to compensate for this. We may deduce that the water inside the reservoir experiences a reaction to the right of magnitude $\rho g\zeta A$, and it must be this

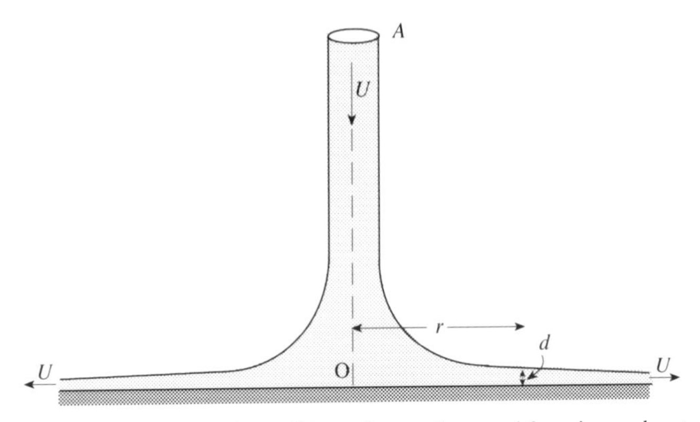

Figure 2.12 Water jet striking a large plate and forming a sheet.

force that imparts momentum to the jet. Since the momentum given to the jet per unit time is ρQU, it follows from (2.27) and (2.28) that

$$\rho g\zeta A = \rho QU = \rho CAU^2 = 2\rho g\zeta CA,$$

and hence that

$$2C = 1.$$

What happens if a jet of Euler liquid, the cross-section of which is a circle of area A, impinges with velocity U on a flat plate at normal incidence? The answer to this question may in practice depend to some extent on the direction in which the gravitational acceleration acts, but let us assume for simplicity that gravity is negligible. In that case the answer, if the plate is a large one, is that the liquid expands over the surface in a flat sheet. Within this sheet the lines of flow are virtually straight and the pressure must therefore be atmospheric, as it is in the jet before impact. We know from Bernoulli's theorem, therefore, that the velocity within the sheet, which is directed along radii diverging from the centre of impact O [fig. 2.12], is everywhere equal to the velocity within the jet, namely U. This radial motion is driven by the excess pressure which exists immediately around O; O itself, of course, is a stagnation point where $u = 0$, and Bernoulli's theorem tells us that at this point there is an excess pressure of $\frac{1}{2}\rho U^2$. Since the discharge rate Q must be the same at all values of radius r (this is the continuity condition for this problem), we know that

$$Q = AU = 2\pi r d U,$$

where d is the thickness of the sheet, so

$$d = \frac{A}{2\pi r}.$$

Figure 2.13 A vertical jet with diameter 3 mm and velocity about 3 m s^{-1} forms a horizontal sheet which breaks up into drops at its rim.
[From G. I. Taylor, *Proc. Roy. Soc. A*, **253**, 313, 1959.]

At large radii the sheet becomes so thin that its surface tension becomes important, and it then breaks up into drops.

If the plate is not a large one but has an area which is equal to that of the jet, the force which the excess pressure exerts upon it cannot exceed $\frac{1}{2}\rho U^2$A. The corresponding reaction on the liquid is then not sufficient to destroy all the forward momentum of the jet; for that to happen a reaction of $\rho U^2 A$ would be needed. The residual momentum carries the sheet forward as a cone, which is then brought inwards towards the axis again by surface tension. *Water bells* are formed in this way.

These effects were first investigated by Savart as long ago as 1833. He produced flat sheets by aiming two equal jets of water at one another, and thereby avoided the viscous retarding effects at liquid–solid interfaces which otherwise limit the validity of predictions based upon the Euler model. Savart's results were not fully understood, however, until G. I. Taylor published four papers on this topic in 1959 and 1960. Figures 2.13 and 2.14, which show a flat sheet breaking up into drops at its rim and a water bell respectively, are reproduced from those papers.

Taylor's simple analysis of the drop-formation process deserves a few words of explanation here, though it does not involve Bernoulli's theorem and is therefore unrelated to the main theme of this chapter. Consider fig. 2.15, which shows in

Figure 2.14 A water bell formed by impact of a jet on a rod of similar diameter; the tip of the rod in this case is conical in shape.
[From G. I. Taylor, *Proc. Roy. Soc. A*, **253**, 289, 1959.]

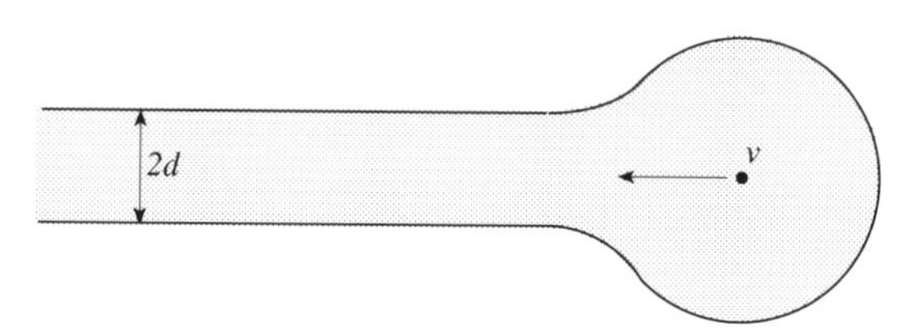

Figure 2.15 Cross-section through cylindrical globule forming the edge of a plane sheet of stationary water.

cross-section a sheet of uniform thickness $2d$ (produced, perhaps, by the collision of two jets) which is terminated at its right-hand edge by a globule of water; the edge of the sheet may be thought of initially as straight, and the globule may be thought of as a cylinder having mass m per unit length perpendicular to the diagram. The frame of reference of the diagram is such that the water in the sheet is stationary, while the globule is moving to the left with velocity v and increasing in mass as it does so. Conservation of mass per unit length requires

$$\frac{\mathrm{d}m}{\mathrm{d}t} = 2\rho d v,$$

while conservation of momentum per unit length requires

$$\frac{\mathrm{d}(mv)}{\mathrm{d}t} = v\,\frac{\mathrm{d}m}{\mathrm{d}t} + m\,\frac{\mathrm{d}v}{\mathrm{d}t} = 2\sigma,$$

where σ is the surface tension of the liquid surface. It follows that

$$m\frac{\mathrm{d}v}{\mathrm{d}t} = 2\sigma - 2\rho d v^2,$$

and hence that a globule which is initially stationary in this frame of reference accelerates until it reaches a limiting velocity given by $\sqrt{\sigma/\rho d}$. The globule is then stationary in a frame of reference in which the water in the sheet is moving to the right with that velocity. Hence if the sheet is a disc formed by the collision of two jets, in which the radial velocity is U and $d = A/2\pi r$ as above, we may expect it to be terminated by a stationary globule where

$$d = \frac{\sigma}{\rho U^2},$$

i.e. where

$$r = \frac{A\rho U^2}{2\pi\sigma}.$$

That argument implies a globule in the form of a torus which steadily increases in mass as more water flows into it. By condensing into spherical drops, however, the torus can reduce its surface area and therefore its surface free energy. The so-called *varicose instability* which leads it to condense is discussed in §8.3.

2.13 Fans and windmills

The jets of water discussed in the previous section are known as *free* jets because their surfaces are free. Jets which are surrounded by more or less stationary fluid are called *submerged*. Submerged jets occur not only when water emerges from a narrow pipe into a larger one [§2.10] but also behind ships' propellers and domestic fans, and underneath hovering helicopters. A discussion of an idealised helicopter will serve to show how Bernoulli's theorem may in principle be applied to such devices, though the results are of limited practical significance.

Let us suppose that when the helicopter blades rotate they draw air in from the atmosphere above and discharge it as a submerged jet below, along lines of flow which have the form sketched in fig. 2.16. Provided that the flow velocity is everywhere much less than the velocity of sound we may treat the air as incompressible, and the continuity condition then requires that immediately above and immediately below the blades, at P and Q, u is the same. However, work is done upon the airstream when it passes from P to Q, and the 'constant' in Bernoulli's theorem increases between these points in consequence. The pressure is therefore greater at Q than at P. Evidently, air will not be drawn in unless p_{P} is less than the atmospheric pressure p_{A} well away from the blades, and if p_{Q} is

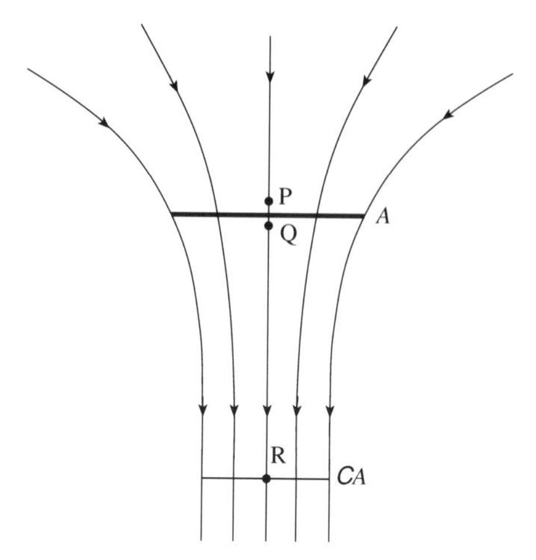

Figure 2.16 Lines of flow above and below an idealised helicopter.

greater than atmospheric pressure, as may well be the case, then the jet will accelerate below the blades and narrow to a *vena contracta* as the figure suggests. Because the density of air is small, we may ignore the variation of p_{A} with height, and we may ignore the $\rho g z$ term in Bernoulli's theorem for the same reason. Thus application of the theorem along a central line of flow above the blades tells us that

$$p_{\mathrm{P}} + \frac{1}{2}\rho u_p^2 = p_{\mathrm{A}},$$

while application of the theorem along the line of flow which links Q to a point R in the *vena contracta* tells us that

$$p_{\mathrm{Q}} + \frac{1}{2}\rho u_{\mathrm{Q}}^2 = p_{\mathrm{Q}} + \frac{1}{2}\rho u_{\mathrm{P}}^2 = p_{\mathrm{A}} + \frac{1}{2}\rho u_{\mathrm{R}}^2.$$

By subtraction we then have

$$p_{\mathrm{Q}} - p_{\mathrm{P}} = \frac{1}{2}\rho u_{\mathrm{R}}^2. \qquad (2.29)$$

Now let us suppose (a) that the pressure difference over the whole area A which the blades sweep out is the pressure difference which (2.29) describes, and (b) that the velocity at R and at adjacent points throughout the *vena contracta* is vertically downwards; the latter assumption is obviously suspect because the blades surely endow the fluid in the jet with some rotation about its axis. Then we may deduce, by equating the upthrust on the helicopter to the downwards momentum given to the jet per unit time, that

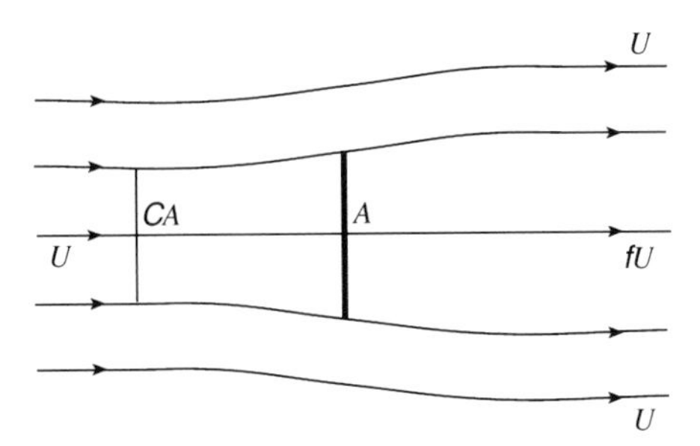

Figure 2.17 Lines of flow passing through or round an idealised windmill.

$$(p_Q - p_P)A = \frac{1}{2}u_R^2 A = \rho u_R^2 CA, \qquad (2.30)$$

where CA is the area of the *vena contracta*. Hence in this situation also [cf. the jet emerging from Borda's mouthpiece, §2.12] the contraction coefficient should be $\frac{1}{2}$.

The action of a windmill may be idealised in a similar fashion. Figure 2.17 shows the blades of the windmill as slowing down a column of air which has an incident velocity U and a cross-sectional area CA, somewhat less than the area A which the blades or sails sweep out, so that behind the sails it forms a column with velocity fU ($f < 1$) and area CA/f. A modified version of the argument above shows the thrust on the windmill to be

$$\frac{1}{2}\rho U^2 A(1 - f^2),$$

which may be equated to the reduction per unit time in the momentum of the airstream, namely

$$\rho U^2 CA(1 - f).$$

Hence

$$C = \frac{1}{2}(1 + f),$$

which means that the kinetic energy removed from the air per unit time is

$$\frac{1}{2}\rho U^3 CA(1 - f^2) = \frac{1}{4}\rho U^3 A(1 - f^2)(1 + f).$$

Presumably f is a quantity which may be varied, e.g. by varying the setting of the sails or the speed at which they are allowed to rotate. The value of f which maximises $(1 - f^2)(1 + f)$ is $\frac{1}{3}$, and the corresponding rate at which kinetic energy is removed from the air is

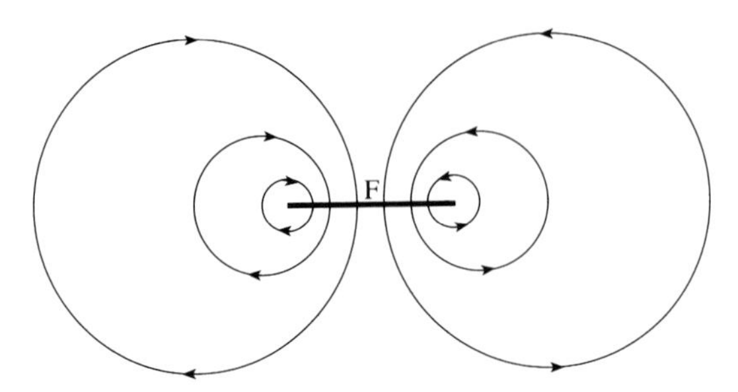

Figure 2.18 Hypothetical flow pattern generated by fan F, which cannot occur in practice.

$$\frac{8\rho A U^3}{27}. \tag{2.31}$$

The power generated by a windmill when the wind velocity is U must always be less than this.

Before leaving this topic we should consider the puzzling question of why it is that fans generate jets at all, rather than a flow pattern of the sort suggested by fig. 2.18. The answer is a subtle one, which has to do with the viscosity which distinguishes real fluids from idealised Euler fluids. Flow patterns are always reversible in Euler fluids, and the symmetrical 'dipolar' pattern sketched in fig. 2.18, in which the outflow corresponds exactly to the inflow in reverse, represents a legitimate solution to Euler's equation of motion. Since, however, the flowlines which leave the bottom of the fan in this figure return again to the top, we can deduce from Bernoulli's theorem that the pressures above and below the fan are here equal to one another. The fan is doing no work on the Euler fluid and imparting no momentum to it; it is, indeed, completely redundant. But if the fluid has viscosity, however slight, the same flow pattern dissipates energy as heat, and the fan is needed to supply this. In principle it is able to supply the energy required, because although the mean pressure p to which Bernoulli's theorem refers is necessarily the same above and below the fan the longitudinal component of pressure is not; it includes a term proportional to η which is negative where lines of flow converge and positive where they diverge [§1.9]. Hence in the situation suggested by fig. 2.18 the fan could, in the case of a real fluid, experience a finite upthrust and exert a corresponding downwards reaction; in transferring fluid from the region of convergence above the fan to the region of divergence below, it would then do work. However, it would necessarily transfer momentum to the fluid in a continuous fashion as well as energy. We can see at once that the localised circulation pattern of fig. 2.18, for which the net fluid momentum is zero, would not survive.

A fuller understanding of the fundamental asymmetry between inflow and outflow for real fluids – of why it is that one can extinguish the candles on a

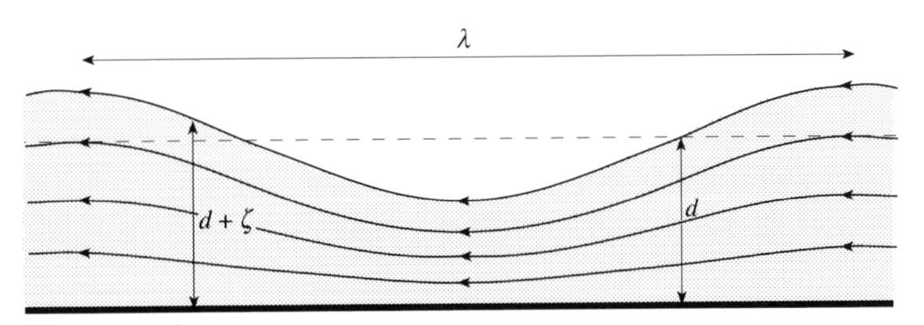

Figure 2.19 Lines of flow in the co-moving frame for a gravity wave in shallow water of mean depth d.

birthday cake by blowing but not by sucking – requires discussion of the phenomenon of *boundary layer separation*, and we return to the question in §7.7. There it is remarked that submerged jets normally become turbulent downstream from their point of origin, but this does not necessarily invalidate the results obtained above.

2.14 Gravity waves on shallow water

When waves propagate over the surface of a lake or ocean the motion of the water is vorticity-free except in boundary layers adjacent to the top and bottom surfaces which are often thin enough to be neglected, as we have seen in §1.12. The flow pattern is not normally a steady one, but it can be made at least quasi-steady by choosing a frame of reference which moves with the waves so that the waves themselves appear to be stationary, and in that frame one can apply Bernoulli's theorem with useful results. Here we consider only situations where the wavelength λ is several centimetres or more, in which case the effects of surface tension are irrelevant [§2.15]. We also restrict ourselves by supposing (a) that the water is shallow, in the sense that its depth d is much less than λ (though much larger than the boundary layer thickness), and (b) that the wave amplitude is very small, in the sense that the vertical displacement of the surface for which the wave is responsible, to be denoted by ζ, is everywhere much less than d. Effects which appear when the amplitude is *not* very small are examined in §2.16.

In a frame of reference in which the waves are stationary the flowlines have the general appearance shown in fig. 2.19. Granted the restrictions laid down above, we are justified[6] in ignoring the vertical component of fluid velocity u_z and in treating the horizontal component u_x ($\approx u$) as independent of z. However, u must vary slightly with x so as to satisfy the continuity condition

$$u(d + \zeta) = (\langle u \rangle + \Delta u)(d + \zeta) = \text{constant}.$$

[6] See the remarks in §2.10 concerning the variation of u with z in the neighbourhood of a broad-crested weir.

Hence to first order in the small quantities ζ and Δu we have

$$d\Delta u + \langle u \rangle \zeta \approx 0. \tag{2.32}$$

However, we also know from Bernoulli's theorem applied to points on the surface, where the pressure is everywhere atmospheric, that

$$\frac{1}{2}u^2 + g(d + \zeta) = \text{constant},$$

from which it follows that

$$\langle u \rangle \Delta u + g\zeta = 0 \tag{2.33}$$

to first order. Equations (2.32) and (2.33) may be combined to yield the result

$$\langle u \rangle = \sqrt{gd}.$$

In the frame of reference to which fig. 2.19 applies, this decribes the mean rate at which the water is drifting to the left. A simple transformation back into the frame in which the water is *not* drifting, tells us at once that in this frame

$$c = \sqrt{gd} \tag{2.34}$$

is the velocity with which the waves propagate to the right.

When ocean waves approach a shelving beach at an angle to it, the wave crests travel with a velocity which increases with depth, according to (2.34), and which therefore increases with distance from the shore. As a result the wave crests are *refracted*; they swing round as they travel forwards, so that by the time they break they are usually almost parallel to the shore.

2.15 When is surface tension negligible?

It is a well known fact that inside a spherical drop of liquid the pressure is greater than it is outside. One need only consider the forces that act across any plane which divides the drop into two equal halves to see that, if its radius is R and if the surface tension of the liquid is σ, the pressure jump across the interface, Δp, must be such that

$$\pi R^2 \Delta p = 2\pi R\sigma,$$

or

$$\Delta p = \frac{2\sigma}{R}. \tag{2.35}$$

Similarly, the pressure jump across a cylindrical interface of radius r is

$$\Delta p = \frac{\sigma}{r}. \tag{2.36}$$

Now when the surface of water is disturbed by a wave of small amplitude, such that (with z chosen as the vertical axis) the height of the surface varies sinusoidally in say the x direction with wavevector k $(= 2\pi/\lambda)$ but does not vary in the y direction, it acquires cylindrical curvature, and its local radius of curvature (to be counted as negative where the surface is concave rather than convex) is such that

$$\frac{\sigma}{r} = -\sigma\frac{\partial^2\zeta}{\partial x^2} = \sigma k^2\zeta. \tag{2.37}$$

The pressure immediately underneath the surface is greater than atmospheric by this amount, and (2.33) needs to be modified in consequence. Bernoulli's theorem applied to points just below the surface now tells us that

$$\langle u\rangle\Delta u + \left(g + \frac{\sigma k^2}{\rho}\right)\zeta = 0. \tag{2.38}$$

The effects of surface tension on waves on shallow water are therefore negligible provided that the wavevector k is small compared with $\sqrt{\rho g/\sigma}$, i.e. provided that

$$\lambda = \frac{2\pi}{k} \gg 2\pi\sqrt{\frac{\sigma}{\rho g}} \approx 1{\cdot}7 \text{ cm}$$

for water at normal temperatures. This condition is adequately satisfied in the context of the phenomena discussed in what remains of this chapter. We return to surface waves of wavelength less than 1·7 cm, i.e. to *ripples*, in §5.4.

2.16 Bores and hydraulic jumps

Equation (2.34) tells us that the velocity of gravity waves on the surface of shallow water is independent of their wavelength. Such waves are therefore expected to be *non-dispersive*; any disturbance of the surface which can be generated by Fourier synthesis of sinusoidal waves, all travelling in the same direction and all having wavelengths much greater than the depth, should travel with them at velocity $\sqrt{gd}$ without changing shape as it does so. That prediction proves to be correct, however, only in the limit of infinitesimal amplitudes. The general theory of finite-amplitude wave propagation on shallow water is far from easy because of non-linear terms in the relevant differential equations, but some interesting results can be uncovered by very simple arguments.

Consider the particular two-dimensional disturbance represented in cross-section by fig. 2.20(a), a disturbance which is intended to be infinitesimal but which has been exaggerated in the figure for the sake of clarity. The surface of a

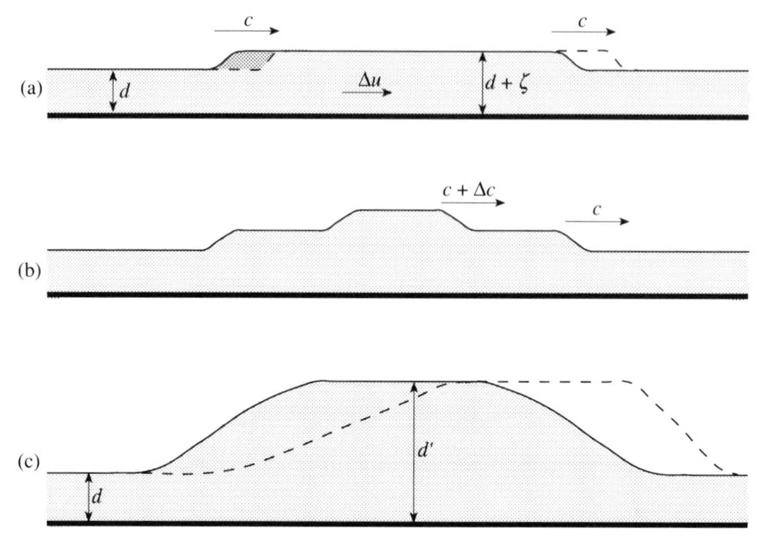

Figure 2.20 Localised disturbances travelling from left to right through shallow water which is stationary on either side of the disturbance; in all of these diagrams, the step heights have been exaggerated for clarity.

layer of stationary water of depth d has two steps in it – in view of the restriction placed upon the wavelengths of the Fourier components which make up the disturbance they must be much more gradual than they appear to be in the figure – and between these steps the depth of water is $d + \zeta$ with $\zeta \ll d$. The steps are travelling in the x direction, from left to right in the diagram, with the wave velocity $c = \sqrt{gd}$, and broken lines in the figure show where they are likely to be a moment later. Evidently their motion must be accompanied by a transfer of water, away from the region indicated by darker shading on the left of the diagram and towards the equivalent unshaded region on the right. Between the steps, therefore, the water must be flowing from left to right with a small velocity Δu, such that

$$(d + \zeta)\Delta u = \zeta\sqrt{gd}; \tag{2.39}$$

this is, in effect, a continuity condition.

Now consider the more complicated disturbance suggested by fig. 2.20(b), where there are four infinitesimal steps rather than two, all of them travelling to the right. The velocity with which the two middle ones travel is

$$c + \Delta c = \sqrt{g(d + \zeta)} + \Delta u. \tag{2.40}$$

It is greater than c on two counts, (a) because the depth of the water over which these two steps are moving is greater than d, and (b) because they are moving over a moving base. Consequently, the second step catches up the first, while the fourth step lags behind the third.

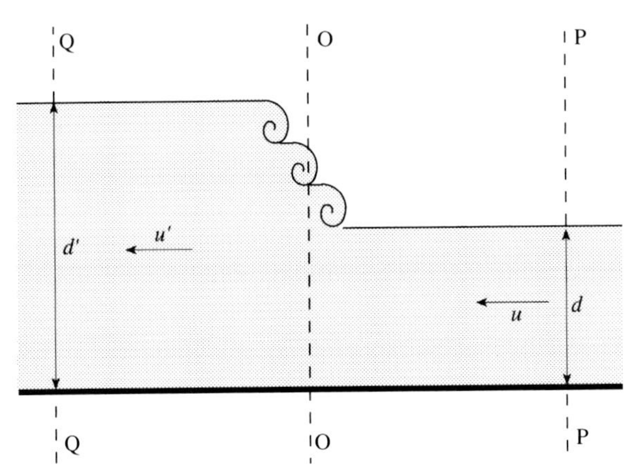

Figure 2.21 Formalised representation of a stationary turbulent foaming front at which depth increases from d to d' and velocity decreases from u to u'.

In the light of that deduction, consider a finite two-dimensional disturbance which initially has the form represented by the full line in fig. 2.20(c); here there are two steps travelling to the right whose height, $(d' - d)$, is not necessarily small compared with d. It is presumably legitimate to regard this finite elevation of the surface as built up from infinitesimal elevations of the type suggested by fig. 2.20(b). If so, the step on the right must surely get steeper as time passes and the step on the left must get less steep, in the way that broken lines suggest.

As the leading edge of the disturbance sketched in fig. 2.20(c) gets steeper it inevitably acquires Fourier components for which λ is no longer much greater than d, and the velocity of propagation of these turns out to be somewhat less than $\sqrt{gd}$ [§5.6]. They show up behind the edge itself, where undulations tend to develop in the surface. What happens next is revealed only by a much more sophisticated analysis than can be attempted here. If the ratio d'/d is less than about 1·3 the edge, with undulations attached, acquires a smooth profile which is stable against further change; if the ratio is greater than about 1·3, however, the edge continues to get steeper, until it breaks into what is often called a *foaming front* within which there is violent turbulence. In both cases propagation continues with some new velocity C, but as there is continuous dissipation of energy, either in the front itself if it is turbulent or behind the front due to the small viscous attenuation to which any undulation is subject, we cannot use Bernoulli's theorem to find what this velocity is. The law of conservation of momentum, however, provides the answer.

Figure 2.21 shows a foaming front in a frame of reference such that the front itself is stationary. The water ahead of the front moves towards it in this frame with a velocity u, while the water behind the front moves away from it with velocity u'. The continuity condition is

$$ud = u'd'. \tag{2.41}$$

Now consider the leftward momentum of all the water between the vertical planes PP and QQ in the diagram. Per unit width in the y direction this is

$$\rho(\text{QO})u'd' + \rho(\text{OP})ud,$$

so that, if we choose to regard PP and QQ as embedded in the water and moving with it, the leftward momentum stored between them per unit width is decreasing with time at a rate

$$-\rho \frac{\mathrm{d(QO)}}{\mathrm{d}t} u'd' - \rho \frac{\mathrm{d(OP)}}{\mathrm{d}t} ud = \rho(-u'^2d' + u^2d).$$

But since the pressure at a depth z below the surface is greater than the atmospheric pressure p_A by ρgz, the horizontal force to the right to which the water is subject per unit width is

$$\int_0^{d'} (p_\text{A} + \rho gz)\mathrm{d}z - \int_0^{d} (p_\text{A} + \rho gz)\mathrm{d}z - p_\text{A}(d' - d) = \frac{1}{2}\rho g(d'^2 - d^2). \tag{2.42}$$

On equating force per unit width to rate of change of momentum per unit width one obtains

$$g(d'^2 - d^2) = 2(-u'^2d' + u^2d),$$

which reduces with the aid of (2.41) to

$$\frac{g}{d}(d'^2 - d^2) = 2\frac{u^2}{d'}(d' - d).$$

Since the speed u at which water flows towards a stationary foaming front is also, in the frame of reference in which the water is stationary, the speed C at which the front propagates, we have

$$C = \sqrt{\frac{gd'(d' + d)}{2d}}. \tag{2.43}$$

This result is also valid when the front is followed by undulations and does not foam. The expression on the right-hand side of it evidently reduces to the small-amplitude wave velocity $c = \sqrt{gd}$ when the difference between d and d' tends to zero. Fronts of finite height, however, travel with a velocity which is not only greater than c but also greater than $c' = \sqrt{gd'}$, the small-amplitude velocity behind the front. The differences of velocity may be considerable; if $d'/d = 4$, for example, we have $c:c'::C::1:2:\sqrt{10}$.

A brief digression may be in order at this point. The planes PP and QQ in fig. 2.21 constitute what is sometimes called, particularly in engineering texts on fluid dynamics, a *control surface*, the space which they enclose being a *control volume*. Another example of a control volume enclosed by a control surface is the column of fluid in a flow tube sketched in fig. 2.6 above. It is frequently necessary to apply Newton's second law (force equals rate of change of momentum) to a control volume, and there are two ways of going about this. One way is to suppose that the control surface is embedded in the fluid and moves with it, in which case the control volume is changing with time but its content is always the same. The other way is to treat the control volume as fixed, and to say that the momentum stored within it is changing not only because of an applied force $\boldsymbol{F}$ (say) but also because there is a flux of matter, and therefore of momentum, across the control surface. The two approaches are, of course, completely equivalent. Both of them lead to an equation which *in cases of steady flow* may in general be written as

$$\boldsymbol{F} = \int \rho\boldsymbol{u}(\boldsymbol{u}\cdot\mathrm{d}\boldsymbol{\Sigma}), \tag{2.44}$$

where the integral is over the control surface in its instantaneous configuration, and (2.42) is merely a special case of this. In later applications of Newton's second law we adopt whichever approach seems the more transparent in the context of the problem in hand.

The most dramatic examples of the phenomena treated in this section are the bores which may sometimes be observed in the tidal reaches of rivers which have long estuaries containing relatively shallow water, like the Bristol Channel in England. The increase of water level between low tide and high tide at the mouth of the estuary causes a disturbance to travel up it, which is a gradual one initially but which steepens as it goes. In the case of the Bristol Channel, a bore forms at the top of the estuary near Sharpness, about 100 km from its mouth, whenever the tidal range is large; from there it travels up the River Severn as far as Gloucester, taking about $2\frac{1}{2}$ hours to do so. Initially, where the river is still wide and shallow, d'/d normally exceeds 1·3 and the bore has a turbulent foaming front [fig. 2.22], but as it nears Gloucester, where the river is narrow and relatively deep, it becomes undular [fig. 2.23]. Anyone wishing to observe it should plan to do so when the tide is particularly high, i.e. during the vernal or autumnal springs.

What is effectively a bore, though normally on a smaller scale and produced in a different way, may be seen any day on a sandy beach, where a spent wave moves forward as a foaming front over a receding layer of water left by the previous wave. Often one may see two such foaming fronts moving forward simultaneously, one behind the other, and the second front almost always catches up the first. The reason why it does so is very much the same as the reason why the second small elevation catches up the first small elevation in fig. 2.20(b).

Stationary bores, usually referred to by hydraulic engineers as *hydraulic jumps*, may often be seen where river water spills over a weir to form a smooth and rapidly flowing sheet on a pavement below, or where it emerges as a smooth sheet

Figure 2.22 The Severn bore: turbulent foaming front where the river is broad and $d'/d > 1{\cdot}3$.
[From a transparency by D. H. Peregrine.]

from underneath a sluice gate. In such situations there is frequently a hydraulic jump at which the sheet increases in thickness and slows down, in just the way that fig. 2.21 suggests. A similar effect occurs in a kitchen sink when the tap is left running; the jet of water from the tap spreads out into a sheet where it hits the base of the sink, and around the point of impact there is a ring-shaped hydraulic jump with a radius of a few centimetres. In all such cases the jump is caused by some impediment to flow downstream from the jump itself. In a kitchen sink, for example, the sides of the sink impede indefinite expansion of the sheet (before, incidentally, it has a chance to reach the thickness at which it is liable to break up into drops under the influence of surface tension, as described in §2.12). The impediment creates a miniature bore which then propagates upstream, until it reaches a point where C is exactly equal to the velocity with which the water in the sheet is moving towards it.

One final comment on turbulent and undular fronts may be helpful, if only to prepare the reader for the discussion of shock waves in chapter 3. Because they dissipate kinetic energy as heat, in a way that is characteristic of real fluids rather than of ideal Euler fluids, the fluid motion associated with them is never reversible. Relative to a front which is stationary, the water on the shallow side has to move towards the front; reversal of its motion would cause the front to broaden and disperse.

Figure 2.23 The Severn bore higher up the river, where $d'/d < 1{\cdot}3$. [From a transparency by D. H. Peregrine.]

2.17 The Coanda effect

Hold a finger horizontally so that it intercepts, not quite centrally, a thin stream of falling grains of sand. If the grains of sand strike the finger on its right side, they bounce off it with a deflection to the right. Now repeat this experiment using a thin jet of water falling from a tap. The water curls round the finger and is deflected to the left instead. This curious phenomenon is an example of the *Coanda effect*, named after a Romanian aeronautical engineer. The effect frequently manifests itself in everyday life, in the tendency of liquids to curl round the rims of vessels from which they are being poured and to drip where they are not wanted.

One may approach an explanation based upon the Euler model by considering a two-dimensional analogue. Figure 2.24 shows in cross-section a plane sheet of liquid of thickness d, of indefinite extent in a direction perpendicular to the diagram, striking a stationary cylinder of radius a ($\gg d$) with velocity U. Although the liquid is shown as coming from above, gravity plays no part in this particular explanation and may be ignored for the time being. Where it strikes the cylinder the liquid spreads sideways into two sheets, having thickness d' and d'' as marked. We know from Bernoulli's theorem that at any rate on the outer surface of each sheet, where $p = p_A$, the velocity is U, so to an approximation which is good if $a \gg d$ the continuity condition is

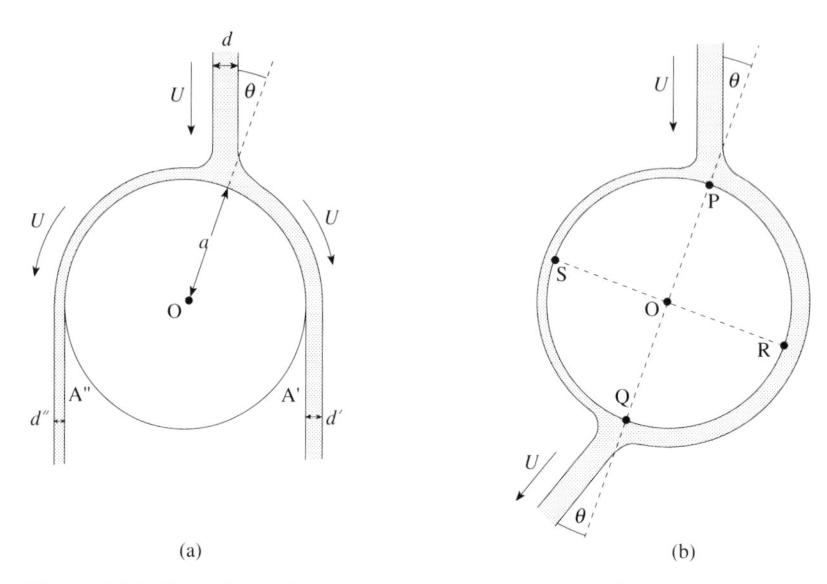

Figure 2.24 Two-dimensional flow of a jet in the form of a plane sheet of water, seen here in cross-section, striking a solid cylinder; (b) represents a possible outcome but (a) does not.

$$dU = d'U + d''U.$$

It may readily be shown, however, that momentum is not conserved on impact in a direction tangential to the cylinder unless, to the same approximation,

$$\rho U^2 d \sin\theta = \rho U^2 d' - \rho U^2 d''.$$

Hence the two thicknesses should be given by

$$d' = \frac{1}{2} d(1 + \sin\theta), \quad d'' = \frac{1}{2} d(1 - \sin\theta). \tag{2.45}$$

Now for the same reason that a water bell forms when a jet impinges on a small plate [§2.12] – namely, in order to conserve momentum in the radial direction – both sheets have an initial tendency to follow the contours of the cylinder's surface. They cannot adopt the cylinder's curvature, however, unless there is a transverse pressure gradient to deflect the lines of flow [§2.5]. Moreover, when the flowlines become circles centred on the axis of the cylinder through O the velocity must vary slightly across the thickness of each sheet, from U at its outer surface where $r = a + d'$ or $r = a + d''$ to $U\{1 + (d'/a)\}$ or $U\{1 + (d''/a)\}$ at $r = a$, as the case may be; we know this because the flow is vorticity-free, and circulatory flow is not vorticity-free unless $u \propto r^{-1}$ [§2.7]. That being so, can either sheet leave the cylinder and carry straight on, as suggested by fig. 2.24(a)? No, because once the fluid velocity has become greater than U at $r = a$ a longitudinal pressure gradient is needed to remove this extra velocity if the flowlines are to straighten out again. There seems no way in which either sheet can leave the cylinder, while remaining vorticity-free throughout, unless the two sheets meet up again as shown in fig. 2.24(b). If they do that, the conditions at impact are exactly reproduced and the liquid can leave as a single sheet with the initial thickness d, at the same inclination to the cylinder's surface as the inclination θ at which it arrived. We obtain, as might have been expected, a flow pattern which is completely reversible.

In fig. 2.24(b), the two sheets are shown as meeting one another at a point Q which is diametrically opposite to the point of impact P, in which case the total deflection to the left is 2θ. That seems the most likely possibility, because when the sheets first form, immediately after the 'tap' is turned on and the first particles of liquid arrive, they will expand around the cylinder at the same rate; Q is therefore where they first encounter one another. It is not, however, the only possibility. By temporarily diverting one sheet, e.g. with the aid of blotting paper applied to one side of the cylinder for a short period only, the meeting point could presumably be shifted from Q in either direction, and a quite different total deflection would result.

Whatever its total deflection, the change of momentum of the liquid stream can be accounted for without difficulty in terms of the transverse pressure gradients mentioned above. In the particular case which fig. 2.24(b) represents, the pressure which the liquid exerts on the cylinder is

$$p_{\mathrm{R}} = p_{\mathrm{A}} - d' \frac{\mathrm{d}p}{\mathrm{d}r} \approx p_{\mathrm{A}} - \frac{\rho U^2}{2a} d(1 + \sin\theta)$$

on one side – the side which includes R – and

$$p_{\mathrm{S}} = p_{\mathrm{A}} - d'' \frac{\mathrm{d}p}{\mathrm{d}r} \approx p_{\mathrm{A}} - \frac{\rho U^2}{2a} d(1 - \sin\theta)$$

on the other. The difference implies a force per unit length on the cylinder in the direction SR of magnitude

$$2a(p_{\mathrm{S}} - p_{\mathrm{R}}) = 2\rho U^2 d \sin\theta.$$

The equal and opposite reaction on the liquid corresponds exactly (in the limit $a \gg d$) to its rate of change of momentum.

Anyone who studies the deflection of a falling water stream by a finger, or by a cylinder of larger radius such as a bottle, will soon decide that the above discussion has rather limited relevance to what he observes. He will see the incident jet spreading out initially into a sheet in which the water travels away from the point of impact in all directions, some of it uphill. He will also see that at some

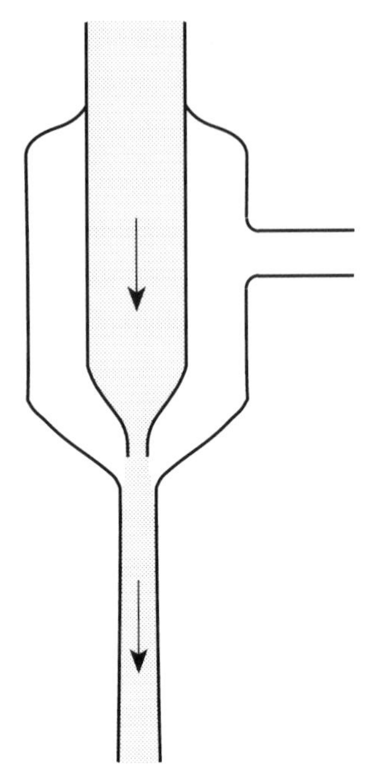

Figure 2.25 Schematic diagram of a water jet aspirator.

millimetres from the point of impact there is a hydraulic jump [§2.16], for which gravity must be responsible. Liquid flows back downhill in the relatively thick band of fluid which forms outside this hydraulic jump, and it seems more than likely that viscosity as well as gravity plays a part in determining the details of the motion. Nevertheless, there is a tendency for the water to leave the cylinder at a point diametrically opposite its point of arrival.

A jet of water from a tap is normally surrounded by – *submerged* in – air, which is obliged to move with the water, not just where the two phases are directly in contact but throughout a boundary layer of finite thickness [§1.12]. This phenomenon of what is called *entrainment* of the air by the jet is exploited in the *water jet aspirator* or *filter pump*, which is a common piece of equipment in chemical laboratories. A sketch of this device is shown in fig. 2.25. As the water jet disappears into the throat below, it carries with it entrained air and more air is sucked in to replace this through the side port. The suction can be considerable, as anyone who has used the device will know. It is mentioned here because entrainment may in some circumstances be involved in the Coanda effect. Were two sheets of water travelling round a cylinder to leave it in the manner suggested by fig. 2.24(a), the pressure of the air in the regions labelled A′ and A″ would be reduced by this effect, and the resulting transverse pressure gradient in the water would encourage the lines of flow there to curve in again.

The density difference between air and water is so great that entrainment is probably irrelevant to the Coanda effect as so far described. The effect also occurs, however, with submerged jets of air in air. It accounts for the astonishing stability of the small pithballs which may be seen, supported on rising jets of air, in fairground shooting galleries; if one of these balls strays off centre in one direction the air jet is deflected in that direction, and the deflection reveals the existence of a reaction on the ball which discourages it from straying further. In such circumstances entrainment may well be significant.

Further reading

For more information concerning the practical devices discussed in this chapter – flowmeters, windmills and such like – consult one of the many excellent textbooks intended for use by engineers: *Mechanics of Fluids*, Van Nostrand Reinhold (UK), Wokingham, 4th edition, by B. S. Massey, is especially recommended.

3

Gas dynamics

3.1 Introduction to compressible flow

When the amount Δp by which pressure varies from place to place in a moving fluid is no longer small compared with p, the compressibility of the fluid cannot be ignored. Bernoulli's theorem does not have to be abandoned in these circumstances but it does need to be reformulated, and reformulation requires knowledge of the equation which relates pressure to density. The necessary equation is well known for ideal gases, and it is on gases – and on air in particular, which at normal temperatures and pressures conforms closely to the ideal model – that we focus attention in this chapter. As we shall see, the reformulated version of Bernoulli's theorem can be applied in an elementary way to a number of interesting phenomena which have to do with compressible flow of air. Among these are the shock fronts which develop following explosions and which accompany supersonic projectiles, and it is chiefly because these shock fronts are in many respects analogous to the tidal bores and hydraulic jumps discussed in §2.16 that this chapter stands where it does.

For the time being we shall continue to ignore viscosity. Neglect of viscosity is usually justified when the Reynolds Number is large compared with unity, for reasons which were outlined in §1.9, and most of the phenomena to be discussed below occur at flow rates where Re is 10^5 or more. The point was made in §§1.9 and 1.10 that in a fluid which is vorticity-free Bernoulli's theorem describes the mean pressure p exactly, irrespective of the magnitude of Re. Close examination of (1.17) reveals that it is only when the fluid is effectively incompressible that this is the case, so the reformulated version of Bernoulli's theorem which applies to the steady compressible flow of an ideal gas is always an approximation. It is, however, a very good approximation when Re is large.

Euler's equation [(2.3)] holds whether or not the fluid is compressible as long as viscous shear stresses may be neglected, and in situations where the gravitational term is insignificant and where the flow is steady, so that $\partial u/\partial t$ is everywhere zero, the longitudinal component of Euler's equation is

$$-\frac{1}{\rho}\frac{\partial p}{\partial l} = u\frac{\partial u}{\partial l}. \tag{3.1}$$

If the fluid is compressible, local variations of density along each line of flow may be described in terms of a local compressibility

$$\beta = -\frac{1}{V}\frac{\partial V}{\partial p} = \frac{1}{\rho}\frac{\partial \rho}{\partial p},$$

(β may well be a function of pressure and therefore of l) by the equation

$$\frac{\partial \rho}{\partial l} = \beta\rho\frac{\partial p}{\partial l}. \tag{3.2}$$

It follows at once from (3.1) and (3.2) that

$$\frac{\partial(\rho u)}{\partial l} = \rho\frac{\partial u}{\partial l} + u\frac{\partial \rho}{\partial l} = \frac{1}{u}\frac{\partial p}{\partial l}(\beta\rho u^2 - 1),$$

and hence that

$$(1 - \beta\rho u^2)\frac{\partial p}{\partial l} = -u\frac{\partial(\rho u)}{\partial l} \tag{3.3}$$

in steady flow where gravity is negligible. Note how the term which involves β becomes irrelevant when u is small compared with $\sqrt{1/\beta\rho}$, a result anticipated in §1.8. Without this term, and with ρ treated as independent of l, (3.3) integrates, of course, to give the form of Bernoulli's theorem we have used in chapters 1 and 2.

A first justification for the formula for the velocity of sound c_s which was quoted in 1.4, namely

$$c_s = \sqrt{\frac{1}{\beta\rho}}, \tag{3.4}$$

may be based upon (3.3). The argument is similar to the argument used in §2.14 to find the velocity of gravity waves in shallow water. We consider a plane sound wave which is propagating with velocity c_s in say the x_1 direction through fluid whose mean velocity is zero, and reduce it to rest by transforming into a reference frame in which the wave is stationary and the fluid moves backwards with mean velocity $u_1 = -c_s$[fig. 3.1]. In the latter frame the flow is at least quasi-steady. Since sound waves in fluids are always longitudinal rather than transverse, the flow is also *unidirectional*, u_2 and u_3 being zero everywhere. The continuity

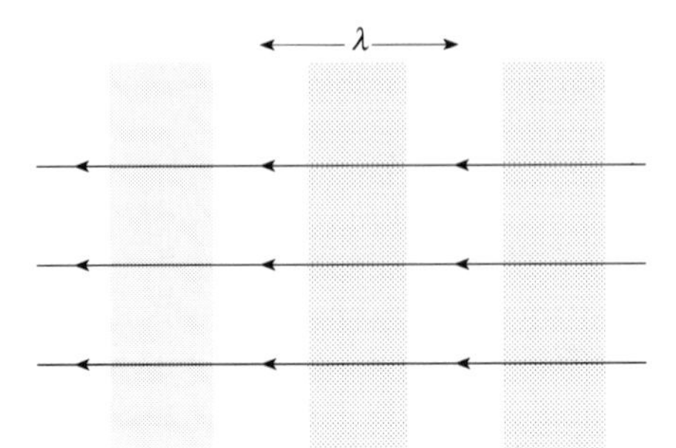

Figure 3.1 A sound wave in a fluid, as seen in a reference frame which moves with it. The grey bands represent regions in which the fluid is compressed by the wave, there being regions of rarefaction in between. The fluid velocity u varies slightly about a mean value which is equal to c_s.

condition tells us in these circumstances that ρu_1 is independent of x_1 for steady flow and hence that

$$\frac{\partial(\rho u)}{\partial l} = 0,$$

so (3.3) reduces to

$$(1 - \beta \rho u_1^2) \frac{\partial p}{\partial x_1} = 0.$$

One way to satisfy this is to make $\partial p/\partial x_1$ zero, but that is a trivial possibility; it is obvious enough that in the absence of pressure or density gradients a fluid can move steadily in any direction at any speed. Otherwise, we must choose

$$u_1 = \pm \sqrt{\frac{1}{\beta \rho}},$$

which provides the justification required.

The argument is a slapdash one, because it pays no attention to the fact that β, ρ and u_1 are not independent of x_1 but oscillate slightly about their mean values. Correspondingly, the right-hand side of (3.3) oscillates about its mean value of zero, by an amount which turns out to be proportional to the square of the amplitude of the sound wave when this amplitude is small. Terms which are of second order in amplitude may safely be ignored in the limit where the amplitude becomes infinitesimal. Their existence tells us, however, that at finite amplitudes the flow cannot be strictly steady in the frame of reference in which the sound wave is stationary; $\partial \rho/\partial t$ and $\partial u/\partial t$ are not exactly equal to zero in that frame. Just as the profile of elevation ζ for a finite-amplitude wave on shallow water changes with time, so too does the density profile of a finite-amplitude sound wave change.

A fuller discussion of sound propagation in fluids, which follows a more conventional route by seeking to identify differential equations which describe the variations of p, ρ or u in a plane sound wave, will be found in the appendix at the end of the book. It is relegated to an appendix partly on account of the amount of mathematics involved, but also because the principal results, especially those that relate to sound attenuation, lie somewhat outside the domain of traditional fluid dynamics; they are of considerable interest to physicists but are not quoted elsewhere in this book. The appendix includes a brief account of the classical method for treating one-dimensional sound waves of arbitrary amplitude, due originally to Riemann, which deserves inclusion for its elegance alone. Riemann's results can be anticipated, however, on physical grounds [§3.6].

Returning to (3.3), let us briefly consider what it tells us about steady flow which is *not* unidirectional. When lines of flow are converging or diverging the continuity condition is

$$\rho u \Delta A = \text{constant},$$

where ΔA is the cross-sectional area of a flow tube, which means that

$$\frac{\partial(\rho u)}{\partial l} = -\frac{\rho u}{\Delta A}\frac{\partial(\Delta A)}{\partial l}$$

and therefore, in view of (3.3) and (3.4), that

$$\left(1 - \frac{u^2}{c_s^2}\right)\frac{\partial p}{\partial l} = \frac{\rho u^2}{\Delta A}\frac{\partial(\Delta A)}{\partial l}. \tag{3.5}$$

Hence whereas $\partial p/\partial l$ and $\partial(\Delta A)/\partial l$ have the same sign when u is less than the velocity of sound they have opposite signs when the flow is supersonic, a result which, taken in conjunction with (3.1), carries the curious implication that supersonic streams slow down as they enter constrictions. In that case, why isn't the rate of transmission of fluid mass smaller at the constriction than elsewhere? Because an increase of density at the constriction, due to the increase of pressure there, compensates for the decrease in both A and u. The moral of these remarks at this stage is merely that the differences between compressible and incompressible flow are more than skin deep. We take a more careful look at compressible flow through constrictions in §3.5 below.

3.2 Isothermal *versus* adiabatic flow

Newton derived (3.4) for the velocity of sound, but in using it to predict the velocity of sound in air he made a mistake over the compressibility and therefore obtained an answer which was too small by about 20%. His mistake lay in using the *isothermal* compressibility β_T, on the assumption that the temperature of a fluid remains constant as a sound wave passes by. In so far as air conforms to the ideal gas model and obeys the well-known ideal equation of state, namely

$$p = \frac{\rho k_B T}{m_{mol}} \tag{3.6}$$

where k_B is the Boltzmann constant, T is absolute temperature and m_{mol} is mean molecular mass, one has

$$\beta_T = \frac{1}{\rho}\left(\frac{\partial \rho}{\partial p}\right)_T = \frac{1}{p}. \tag{3.7}$$

It seems to have been Laplace who first appreciated that sound waves in gases are in fact adiabatic rather than isothermal; the quantity which stays constant in each element of gas during the compressions and rarefactions associated with sound is heat content rather than temperature. That means, since the compressions and rarefactions are thermodynamically reversible, that the entropy of each element remains constant, and hence that the relevant compressibility is the *isentropic* one, β_S. For an ideal gas, the proportionality law which links pressure and density in an isentropic process is

$$p \propto \rho^\gamma, \tag{3.8}$$

where γ is the specific heat ratio c_p/c_V, say 1·4 for air. It follows from (3.8), or else from a familiar thermodynamic identity, that the adiabatic compressibility for an ideal gas is

$$\beta_S = \frac{1}{\rho}\left(\frac{\partial \rho}{\partial p}\right)_S = \frac{\beta_T}{\gamma} = \frac{1}{\gamma p}, \tag{3.9}$$

and hence from (3.4) and (3.6) that

$$c_s = \sqrt{\frac{\gamma p}{\rho}} = \sqrt{\frac{\gamma k_B T}{m_{mol}}}. \tag{3.10}$$

The sound velocity is evidently very similar to the mean velocity of the molecules in the gas, which is

$$\bar{c}_{mol} = \sqrt{\frac{8 k_B T}{\pi m_{mol}}}. \tag{3.11}$$

It is not difficult to confirm that Laplace was right and Newton wrong. Consider a plane sound wave propagating in the x_1 (or x) direction[1] through fluid whose local velocity is u_1 (or u), in a frame of reference such that $\langle u \rangle = 0$. Suppose that

[1] In situations such as this where the flow is one-dimensional, the x_2 and x_3 coordinates being quite irrelevant, numerical suffices are not needed and may be dropped.

associated with the sound wave there is some periodic variation, say ΔT, in temperature. The rate of change of temperature of an element of the gas – the rate of change following the motion of this element – is

$$\frac{\mathrm{D}(\Delta T)}{\mathrm{D}t} = \frac{\partial(\Delta T)}{\partial t} + u\,\frac{\partial(\Delta T)}{\partial x}, \tag{3.12}$$

but the second term on the right-hand side is of second order in the amplitude of the wave and may be ignored when the amplitude is infinitesimal. Then, if we take ΔT to vary like $\exp\{\mathrm{i}(kx - \omega t)\}$, we have

$$\frac{\mathrm{D}(\Delta T)}{\mathrm{D}t} \approx \frac{\partial(\Delta T)}{\partial t} = -\mathrm{i}\omega\Delta T. \tag{3.13}$$

This rate of change is due partly to work done upon (or by) the element as its volume changes, and partly to conduction of heat into (or out of) it. Now the longitudinal heat flux per unit area in the x direction is

$$H = -\,\kappa\frac{\partial(\Delta T)}{\partial x},$$

where κ is thermal conductivity, so the rate at which energy accumulates due to thermal conduction in an element of length Δx is given, per unit area perpendicular to the x direction, by

$$\begin{aligned} -\frac{\partial H}{\partial x}\Delta x &= \frac{\partial}{\partial x}\left\{\kappa\,\frac{\partial(\Delta T)}{\partial x}\right\}\Delta x \\ &\approx -\,\kappa k^2\Delta T\Delta x. \end{aligned}$$

The second line of this equation is approximate in so far as κ depends upon the temperature of the gas, but the approximation is legitimate when the amplitude is infinitesimal; the periodic variation of κ contributes a second-order correction. It follows that the rate of change of temperature due to thermal conduction alone is

$$\left\{\frac{\mathrm{D}(\Delta T)}{\mathrm{D}t}\right\}_{\mathrm{cond}} \approx -\,\frac{\kappa k^2\Delta T\Delta x}{\rho c_V\Delta x}; \tag{3.14}$$

the factor ρ on the right-hand side is needed because c_V, and likewise c_p, represent heat capacities per unit mass of the fluid. Thermal conduction is evidently negligible, and the flow associated with the sound wave may be treated as adiabatic, provided that

$$\left|\frac{\mathrm{D}(\Delta T)}{\mathrm{D}t}\right|_{\mathrm{cond}} \ll \left|\frac{\mathrm{D}(\Delta T)}{\mathrm{D}t}\right|$$

for all values of x. It is apparent from (3.14) and (3.13) that this condition is satisfied if

$$\frac{\kappa k^2}{\rho c_V} \ll \omega. \tag{3.15}$$

At the other extreme, when

$$\frac{\kappa k^2}{\rho c_V} \gg \omega, \tag{3.16}$$

$\{D(\Delta T)/Dt\}_{cond}$ and $\{D(\Delta T)/Dt\}$ necessarily become almost equal to one another for all x, which is not possible (because (3.13) contains a factor i which is absent from (3.14)) unless both of them are almost zero. This other extreme corresponds, therefore, to isothermal propagation.

These results apply to liquids and solids as well as to gases, and examination of tables of thermal conductivity, specific heat and velocity of sound will show that, whatever the medium, the inequality labelled (3.15) is normally satisfied more than adequately at the sort of frequencies and wavelengths which are accessible by ultrasonic techniques. If we restrict our attention to gases, however, we have no need to consult tables, since it is a standard result of elementary kinetic theory of gases that

$$\frac{\kappa}{\rho c_V} \approx \overline{c}_{mol}\ell, \tag{3.17}$$

where ℓ is the mean free path.[2] Using this result, and using (3.10) and (3.11) to relate ω/k, which is the same thing as c_s, to $\overline{c}_{mol}$, we may show that the condition for sound propagation to be adiabatic in a gas for which γ is about 1·4 may be re-expressed in terms of wavelength λ as

$$\lambda \gg 10\ell. \tag{3.18}$$

That this condition is well satisfied in practice is immediately apparent; the wavelength of sound in air at a frequency of 1 kHz is about 0·3 m, whereas ℓ is about 10^{-7} m. The mean free path may of course be increased by reducing the pressure of the gas, but if it is increased to such an extent that it becomes comparable with λ one enters the Knudsen regime, where the gas no longer behaves like a continuous medium and sound propagation is impossible.

It is perhaps worth stressing, since a few texts are misleading in this respect, that Laplace's ideal of completely adiabatic propagation is approached in the limit where λ tends to infinity, and hence where frequency tends to zero, rather than the reverse. The k^2 factor in $\{D(\Delta T)/Dt\}_{cond}$ and the ω factor in $\{D(\Delta T)/Dt\}$ both

[2] More advanced theories suggest that the right-hand side of (3.17) should include the numerical factor $25\pi/64$, but this is close to unity and of no significance here.

tend to zero as this limit is approached, but the former does so more rapidly. At high frequencies – where $\{\mathrm{D}(\Delta T)/\mathrm{D}t\}_{\text{cond}}$, though small compared with $\{\mathrm{D}(\Delta T)/\mathrm{D}t\}$, is no longer completely negligible – thermal conduction contributes to the attenuation of sound, but attenuation need not concern us here; it is discussed in the appendix.

As a rule, compressible flow is effectively adiabatic whether it is oscillatory, as in a sound wave, or steady. There are exceptions to this rule – for example, cases of thermal convection [§8.5], in which temperature gradients are imposed on a fluid by hot or cold bodies with which it is in contact, and in which the thermal conductivity of the fluid is certainly not negligible – and no general proof of it can therefore be given. But consider, as an illustrative example, the case of steady flow into a narrow orifice, as discussed in the context of incompressible flow in §1.9. Here lines of flow converge on a common centre, and the radial velocity of the fluid u depends only on radius measured from that centre, R [fig. 1.7]. If u becomes comparable with the velocity of sound then temperature gradients may be generated in the fluid by compression, but they will be longitudinal and not transverse. Now the longitudinal heat flux in a flow tube of cross-sectional area ΔA is $-\kappa\Delta A(\partial T/\partial l)$, and when ΔA varies the rate of change of temperature of an element of fluid due to thermal conduction alone, when transverse temperature gradients are absent, is

$$\left(\frac{\mathrm{D}T}{\mathrm{D}t}\right)_{\text{cond}} = \frac{1}{\Delta A\rho c_V}\frac{\partial}{\partial l}\left(\kappa\Delta A\frac{\partial T}{\partial l}\right). \tag{3.19}$$

If the flow is steady, however, $\partial T/\partial t$ is zero and the total rate of change of temperature of the element of fluid is

$$\frac{\mathrm{D}T}{\mathrm{D}t} = u\frac{\partial T}{\partial l}. \tag{3.20}$$

Thermal conduction is therefore negligible, and the flow is effectively adiabatic, when

$$\left|\frac{1}{\kappa\Delta A(\partial T/\partial l)}\frac{\partial}{\partial l}\left(\kappa\Delta A\frac{\partial T}{\partial l}\right)\right| \ll \frac{\rho c_V u}{\kappa}. \tag{3.21}$$

The left-hand side of this inequality turns out to be of the same order as

$$\left|\frac{1}{\Delta A}\frac{\partial(\Delta A)}{\partial l}\right| = \frac{2}{R}.$$

It then follows from (3.17) that if the fluid is a gas the inequality labelled (3.21) is satisfied provided that

$$R \gg \frac{c_{\mathrm{s}}\ell}{u},$$

or, since no temperature gradients arise unless the factor c_s/u is of the order of, or less than, unity, provided that

$$R \gg \ell.$$

As in the case of sound propagation, therefore, it is the mean free path of the gas which matters; we can safely assume the flow to be adiabatic except in the Knudsen regime.

3.3 Bernoulli's theorem for compressible gas flow

Granted that the steady flow of gases is adiabatic, and granted that it is also reversible in the thermodynamic sense, the integration along a line of flow which yields Bernoulli's theorem is straightforward. Note first that when (3.8) is satisfied

$$\begin{aligned}\frac{\partial(p/\rho)}{\partial l} &= \frac{1}{\rho}\frac{\partial p}{\partial l} - \frac{p}{\rho^2}\frac{\partial \rho}{\partial l}\\ &= \frac{1}{\rho}\frac{\partial p}{\partial l}\left(1 - \frac{1}{\gamma}\right).\end{aligned} \tag{3.22}$$

This result means that Euler's equation for steady flow, including the gravitational term omitted from (3.1), can be re-written as

$$\frac{\gamma}{(\gamma - 1)}\frac{\partial(p/\rho)}{\partial l} + \frac{\partial(gz)}{\partial l} + \frac{\partial(\frac{1}{2}u^2)}{\partial l} = 0. \tag{3.23}$$

Along any line of flow, therefore, we have

$$\frac{\gamma}{(\gamma - 1)}\frac{p}{\rho} + gz + \frac{1}{2}u^2 = \text{constant}. \tag{3.24}$$

Here is the form of Bernoulli's theorem which describes the steady compressible flow of an effectively ideal gas. It looks rather different from (2.15), which states Bernoulli's theorem for incompressible flow, but it is readily shown that in circumstances where u is much less than c_s it describes the same rate of variation of p with u. Since $\{\gamma/(\gamma - 1)\}(p/\rho)$ and (p/ρ) are different quantities, however, the numerical values of the 'constants' in (3.24) and (2.15) cannot be the same.

There is an elementary thermodynamic identity which tells us, in view of (3.6), that for an ideal gas

$$c_p - c_V = -\frac{T}{\rho^2}\left(\frac{\partial p}{\partial T}\right)_\rho \left(\frac{\partial \rho}{\partial T}\right)_p = \frac{k_B}{m_{mol}}, \tag{3.25}$$

and hence that

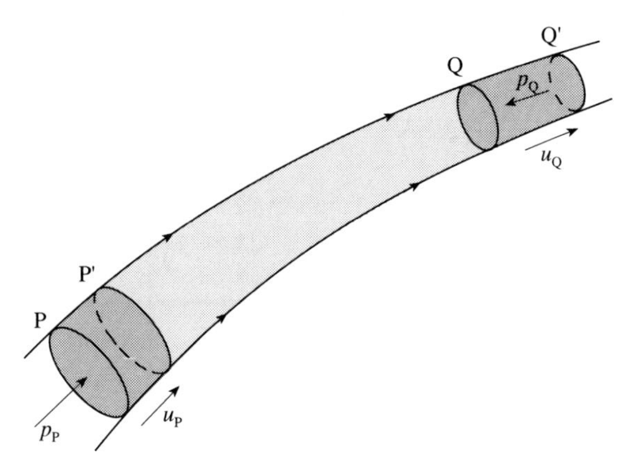

Figure 3.2 A fluid column moving through a flow tube.

$$\frac{\gamma}{(\gamma - 1)} = \frac{c_p}{(c_p - c_V)} = \frac{m_{\text{mol}} c_p}{k_B} = \frac{\rho c_p T}{p}. \tag{3.26}$$

Thus Bernoulli's theorem for gases can also be expressed by the equation

$$c_p T + gz + \frac{1}{2} u^2 = \text{constant}, \tag{3.27}$$

which for an ideal gas is completely equivalent to

$$h + gz + \frac{1}{2} u^2 = \text{constant}, \tag{3.28}$$

where h is the *intrinsic enthalpy* of the gas per unit mass, i.e. the enthalpy per unit mass as measured in a frame of reference such that the local values of z and u are zero.

Bernoulli's theorem for incompressible flow can readily be derived from the law of conservation of energy rather than by integration of the equation of motion [§2.8] and the same is true of (3.28). Figure 3.2 represents, like fig. 2.6, a column of fluid enclosed by a fixed flow tube, which stretches from P to Q at one instant and from P′ to Q′ an instant later. The energy of the second state is higher than that of the first by

$$\Delta m \left\{ \left(e + gz + \frac{1}{2} u^2 \right)_Q - \left(e + gz + \frac{1}{2} u^2 \right)_P \right\},$$

where Δm represents the mass of fluid which lies between P and P′ in the first state and between Q and Q′ in the second, and where e is the *intrinsic internal energy* of the fluid per unit mass. We have only to equate this increase to the work done on the column by the pressures acting on its two ends, which is

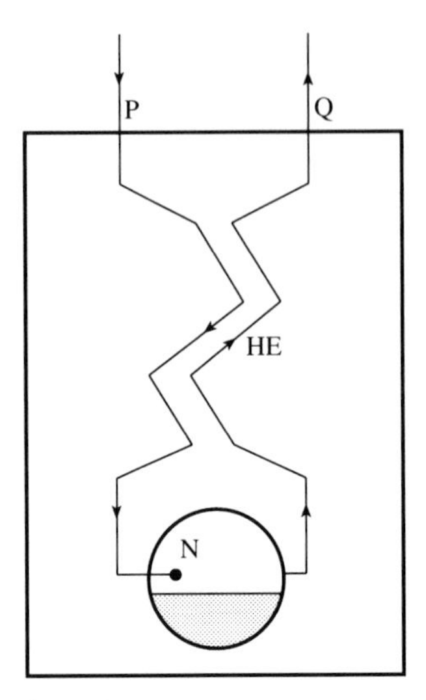

Figure 3.3 Schematic view of a Joule–Kelvin liquefier: HE, heat exchanger; N, expansion nozzle.

$$\Delta m\left\{\left(\frac{p}{\rho}\right)_{\mathrm{P}} - \left(\frac{p}{\rho}\right)_{\mathrm{Q}}\right\},$$

and to remember that

$$h = e + \frac{p}{\rho}, \tag{3.29}$$

to arrive at (3.28). This alternative derivation shows that (3.28) is a general result for steady flow, not confined to ideal gases. More significantly, it shows that the result holds even when the fluid undergoes an irreversible process between P and Q, provided always the column does not gain or lose energy in ways that have not been allowed for above.

Consider, for example, what goes on in the Joule–Kelvin gas liquefier which is shown schematically in fig. 3.3. A stream of non-ideal gas at a relatively high pressure and at a temperature which is below its inversion temperature enters at P, then flows through a heat exchanger, then flows through an expansion nozzle N across which there is a sudden drop of pressure and also a drop in temperature, and flows back through the heat exchanger to the exit at Q. We may suppose the stream to have been flowing for long enough for the system to have reached a steady state where the temperature at Q is no longer falling, in which case only a fraction $(1 - f)$ of the gas which enters at P returns to Q; the remainder accumulates below the nozzle as stationary liquid. The heat exchanger depends entirely for its effect on transverse thermal conduction in the incoming and outgoing streams, so the flow cannot be adiabatic in the normal sense. Nor is it reversible: thermal conduction is an essentially irreversible process and so is sudden expansion through a nozzle. Nevertheless, enthalpy is conserved in the sense that

$$\left(h + gz + \frac{1}{2}u^2\right)_{\text{gas,P}} = (1 - f)\left(h + gz + \frac{1}{2}u^2\right)_{\text{gas,Q}} + f(h + gz)_{\text{liquid}},$$

provided only that heat fed into the system from the outside world, for example by thermal conduction along the metal tubes which contain the gas, may be neglected. In practice one may use the simpler expression

$$h_{\text{gas,P}} = (1 - f)h_{\text{gas,Q}} + fh_{\text{liquid}},$$

since in Joule–Kelvin liquefiers the other terms are normally small by comparison.

Later in this chapter we consider the steady flow of an ideal gas through a stationary shock front, across which there is a finite change of pressure and temperature. The temperature gradient within the front itself may well be too large for longitudinal thermal conduction to be negligible there, so flow through a shock front, like flow through a Joule–Kelvin liquefier, is unlikely to be adiabatic. Even if it were effectively adiabatic it would still be irreversible, as we shall see below, so we cannot safely assume that p is everywhere proportional to ρ^γ and we do not, therefore, know how to integrate the equation of motion. Nevertheless $(h + gz + \frac{1}{2}u^2)$ must be the same on both sides of the shock, and the quantities represented by the left-hand sides of (3.24) and (3.27) must be the same on both sides also.

3.4 The atmospheric lapse rate

This section concerns a simple example of compressible flow where the compressibility happens to be relevant because the gravitational potential varies greatly along the lines of flow, rather than because u is comparable with the velocity of sound. Indeed, u is small enough compared with c_s for the $\frac{1}{2}u^2$ terms which feature in all three of the versions of Bernoulli's theorem derived in the previous section to be ignored.

Within the layer of air adjacent to the Earth's surface which is known as the *troposphere*, a layer whose thickness is normally about 10 km, our atmosphere is almost continuously stirred, either by winds or by local convection currents stimulated by the effect of sunlight in warming the ground below. Although thermal conduction undoubtedly plays a role in transferring heat from ground to air its effects are negligible in other respects, and we may reasonably apply Bernoulli's theorem to the rising and falling air currents.

Equation (3.27) is perhaps the most illuminating form of the theorem to use in this context. Given that the $\frac{1}{2}u^2$ term may be ignored, it tells us immediately that $(c_pT + gz)$ is independent of height, and hence that

$$\frac{\mathrm{d}T}{\mathrm{d}z} = -\frac{g}{c_p} = -\frac{(\gamma - 1)}{\gamma}\frac{m_{\mathrm{mol}}g}{k_{\mathrm{B}}}. \tag{3.30}$$

Equation (3.30) describes the so-called *adiabatic lapse rate* of temperature in the troposphere. For completely dry air, in which the mean molecular mass is about 30 atomic mass units while γ is about 1·4, this lapse rate would be about 10 K km^{-1}. In practice, observations of temperature over different parts of the Earth's surface and at different heights suggest a mean lapse rate of about 6·5 K km^{-1} instead. The discrepancy is attributable to the presence of water vapour in the atmosphere. This tends to condense as the air rises and cools; it releases latent heat as it does so, thereby reducing the rate of cooling.

Condensed water vapour is liable to precipitate, of course, in the form of rain or snow. Suppose that precipitation occurs at the top of a mountain range across which an initially moisture-laden stream of air is flowing. As the stream ascends on one side of the range it cools at say 6·5 K km^{-1}. As it falls on the other side it warms up again at a higher rate, perhaps as high as 10 K km^{-1} if it has been deprived of almost all its water content. The result is a noticeable increase of temperature. In the European Alps, the word *föhn* is used to describe weather conditions which arise in this way.

3.5 Choked flow through a constriction

In §2.10 we encountered the *Venturi meter* and found a formula [(2.24)] for the drop in pressure of an incompressible fluid as it passes through a constriction in a pipe by combining Bernoulli's theorem with the distinctly dubious assumption that the velocity of the fluid is uniform over the whole cross-section of the pipe at every point along it. We are now to examine, using the same assumption, the equivalent but significantly more complicated problem in compressible flow. A stream of gas flowing steadily along a duct of non-uniform cross-sectional area A encounters a constriction or *throat* of area A_{T}, in the neighbourhood of which its velocity u becomes comparable with the local velocity of sound c_{s}. In this neighbourhood the gas is significantly less dense than it is elsewhere, and it is also colder. We are therefore not entitled to assume, as we did in §2.10, that u is inversely proportional to A and we cannot treat c_{s} as uniform. How do these quantities vary along the duct, and how too does p vary?

Euler's equation in the forms of (3.1) and (3.5) tells us that as long as u is everywhere less than c_{s} the compressibility of the gas makes no qualitative difference. It remains true, as for an incompressible fluid, that the gas accelerates into the throat and decelerates on its way out, that the pressure falls to a minimum value p_{T} at the narrowest part of the throat and then increases, and that if the velocity well downstream from the throat returns to the value it had upstream then the pressure returns to its original value too (always provided that the neglect of viscous effects is justified). We may, of course, lower the downstream pressure by connecting the duct to a pump of some sort and thereby increase the rate of flow, but as long as it remains subsonic at all points we will not generate in this way any difference of pressure between two points upstream and downstream where A and u are the same; it is where the gas first flows into the duct that the pump creates a pressure gradient, rather than in the duct itself.

Suppose, however, that the flow rate becomes so high that the velocity u_T at the narrowest part of the throat becomes just equal to the local sound velocity there, $c_{s,T}$. At that stage there seem, on the basis of (3.5), to be two possibilities: (a) the flow reverts to being subsonic immediately the throat is passed and the pressure rises again as described above, or (b) the pressure continues to fall in a monotonic fashion beyond the throat and the gas continues to accelerate. It is because the coefficient which relates $\partial p/\partial l$ to $\partial(\Delta A)/\partial l$ changes sign once u exceeds c_s [(3.5)] that (b) is a possibility. However, at points downstream from the throat there is one (and only one) value of p consistent with (a) and a different value consistent with (b), as suggested by fig. 3.4. What happens if the pump is not adjusted to maintain precisely either of these? What happens in particular if, starting with conditions which correspond to (a), the rate of pumping is suddenly increased? A full answer to these questions involves shock waves and expansion fans and must be deferred to §3.13, but a partial answer to the second is as follows. News of the increase in pumping rate propagates upstream as a reduction of pressure which is a sound wave of a sort. Its leading edge propagates at the local velocity of sound relative to the gas and therefore with velocity $(c_s - u)$ relative to the throat. Since $(c_s - u)$ is zero at the throat itself the news can never get through it. Consequently, the news never reaches the gas upstream from the throat and the rate of flow through the throat remains the same, however hard the pump tries to increase it. The flow is said to be *choked*.[3]

Let us now apply Bernoulli's theorem to the problem. In the form of (3.24) it tells us, since the gravitational term is irrelevant in this context, that

$$\frac{\gamma}{(\gamma - 1)}\frac{p}{\rho} + \frac{1}{2}u^2 = \frac{\gamma}{(\gamma - 1)}\frac{p_o}{\rho_o}, \tag{3.31}$$

or else, in view of (3.10), that

$$c_s^2 + \frac{1}{2}(\gamma - 1)u^2 = c_{s,o}^2, \tag{3.32}$$

where p_o, ρ_o and $c_{s,o}$ are respectively the pressure, density and velocity of sound in some hypothetical reservoir which supplies the gas, within which u is zero. Since

[3] There is a close analogy between flow of gas through a constriction and flow of water over a weir [§2.10]. In the latter case the flow is choked at the crest of the weir where, at any rate in the situation depicted by fig. 2.7, the velocity u of the water is equal to the velocity $\sqrt{g\xi_1}$ of waves on its surface.

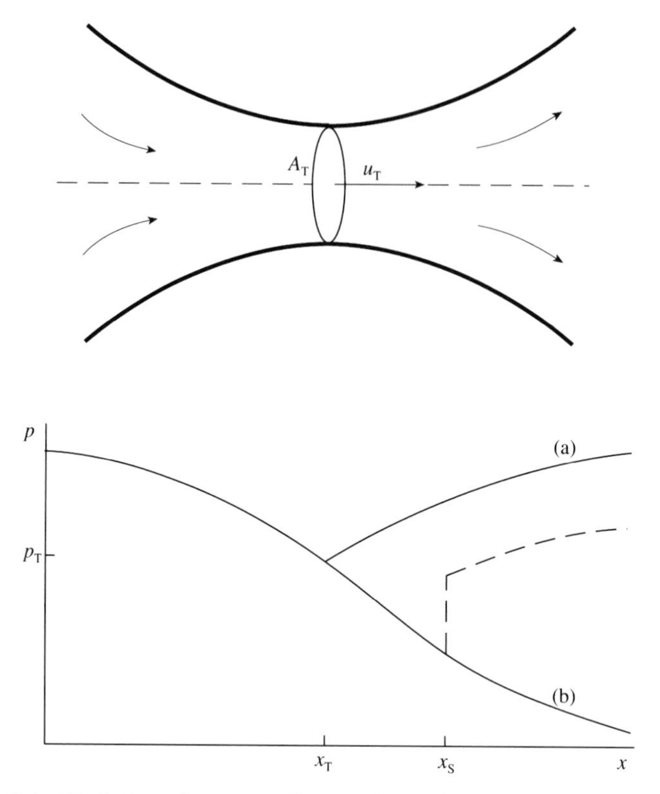

Figure 3.4 Variation of pressure (lower diagram) along the axis of a duct with a constriction in it (upper diagram). Gas accelerates into the constriction from left to right, until at $x = x_T$ its velocity u_T is just equal to the local velocity of sound. If the downstream exit pressure lies on curve (a) u is subsonic for all $x > x_T$, while if it lies on curve (b) u is supersonic for all $x > x_T$. The broken curve represents an intermediate possibility, where the flow changes from supersonic to subsonic in crossing a normal shock front situated at $x = x_S$ [§3.13].

the flow is isentropic as well as isenthalpic (upstream from the throat at any rate) it is subject to the additional constraint that ρ is proportional to $p^{1/\gamma}$. This means that

$$\left(\frac{c_S}{c_{S,o}}\right)^2 = \frac{(p/\rho)}{(p_o/\rho_o)} = \left(\frac{p}{p_o}\right)^{(\gamma-1)/\gamma}, \tag{3.33}$$

and we may eliminate $c_{S,o}$ from (3.32) and (3.33) to arrive at

$$\frac{1}{2}(\gamma - 1)u^2 = c_S^2\left\{\left(\frac{p_o}{p}\right)^{(\gamma-1)/\gamma} - 1\right\}. \tag{3.34}$$

From (3.34) it can be seen that u_T becomes equal to $c_{S,T}$ when

$$p_{\mathrm{T}} = p_{\mathrm{o}} \left(\frac{2}{\gamma + 1} \right)^{\gamma/(\gamma-1)}, \tag{3.35}$$

i.e. when p_{T} is about $0{\cdot}53 p_{\mathrm{o}}$ if γ is $1{\cdot}4$. The mass discharge rate is then

$$\begin{aligned} A_{\mathrm{T}} \rho_{\mathrm{T}} u_{\mathrm{T}} &= A_{\mathrm{T}} \rho_{\mathrm{T}} c_{\mathrm{s,T}} = A_{\mathrm{T}} \rho_{\mathrm{o}} c_{\mathrm{s,o}} \left(\frac{\rho_{\mathrm{T}} p_{\mathrm{T}}}{\rho_{\mathrm{o}} p_{\mathrm{o}}} \right)^{1/2} \\ &= A_{\mathrm{T}} \rho_{\mathrm{o}} c_{\mathrm{s,o}} \left(\frac{2}{\gamma + 1} \right)^{(\gamma+1)/2(\gamma-1)}. \end{aligned} \tag{3.36}$$

This, being the rate associated with choked flow, is the maximum rate at which gas can be extracted from the reservoir through a throat of area A_{T}. Even if the pump is capable of reducing the downstream pressure effectively to zero, it cannot, according to the simplified one-dimensional theory presented above, extract more.

It is of interest to compare that result with the *rate of effusion* of gas through a small hole of area A_{T} in the wall of a reservoir, as predicted by kinetic theory. When the pressure and density are p_{o} and ρ_{o} inside the reservoir and zero outside, and when the size of the hole is small compared with the mean free path (i.e. in the limiting case of Knudsen flow), the mass of gas which escapes per unit time is just

$$\frac{1}{4} A_{\mathrm{T}} \rho_{\mathrm{o}} \bar{c}_{\mathrm{mol,o}} = \sqrt{\frac{1}{2\pi\gamma}}\, A_{\mathrm{T}} \rho_{\mathrm{o}} c_{\mathrm{s,o}}. \tag{3.37}$$

The coefficient on the right-hand side of (3.37) is about $0{\cdot}34$ when $\gamma = 1{\cdot}4$, whereas the coefficient on the right-hand side of (3.36) is about $0{\cdot}58$. The discrepancy is not surprising, since it can readily be shown that when the size of the hole is *not* small compared with the mean free path the argument on which (3.37) is based must lead to an underestimate of the true answer.

Of course, the assumption that u is uniform over the whole cross-section is not remotely realistic if the 'duct' takes the form of a hole punched in a thin sheet, and it is scarcely realistic in the case discussed briefly in §3.13, of flow through the expansion nozzle of a rocket. Nevertheless, measured values of the pressure in the neighbourhood of the throat of a rocket nozzle agree quite closely with values predicted using the one-dimensional theory outlined above. Perhaps, as may be the case with Venturi meters [§2.10], the fact that the flow is turbulent improves the agreement. Turbulence is to be expected in rocket nozzles, not only because the Reynolds Number is so high – when $u \approx c_{\mathrm{s}}$ the Reynolds Number in a duct of diameter d is about d/ℓ, where ℓ is the mean free path [(3.10), (3.11) and (A.35)] – but also because turbulence is surely present in the combustion chamber.

3.6 The development and decay of a shock front

Our formula for the velocity of sound [(3.4)] does not include the frequency or wavelength of the sound wave in question, so sound waves, like gravity waves on

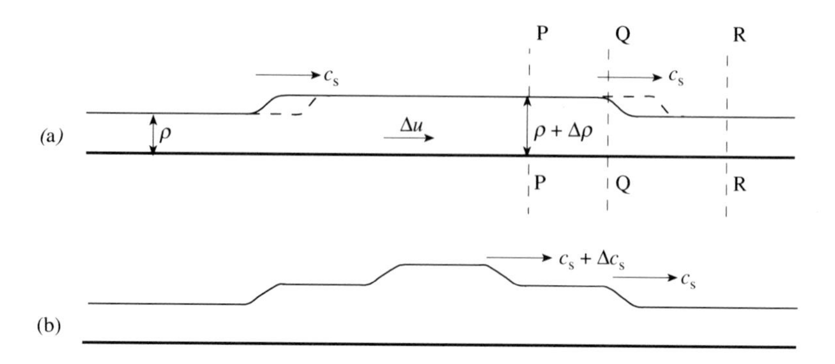

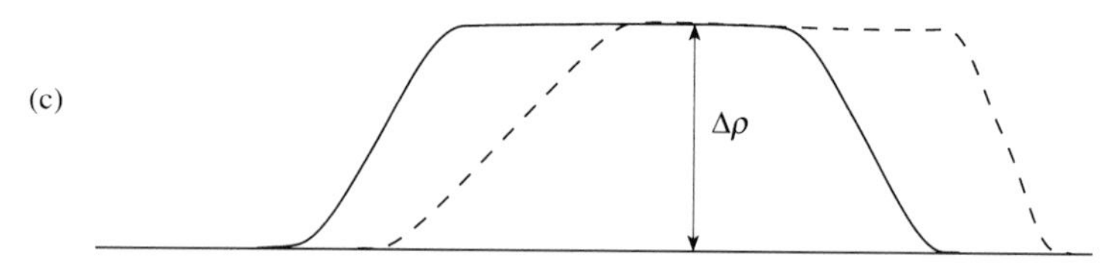

Figure 3.5 Density profiles for layers of enhanced density propagating from left to right through stationary gas.

shallow water, should be non-dispersive when their amplitude is small. Hence when parallel plane sound waves, all travelling in the same direction, are superimposed to create a localised density disturbance, this disturbance might be expected to propagate through the fluid at velocity c_s without changing its form.

Consider, however, the hypothetical one-dimensional disturbance suggested by the density profile sketched in fig. 3.5(a) – a figure which is deliberately drawn to resemble fig. 2.20 in order to emphasise that the argument here is very closely related to the argument applied to gravity waves in shallow water in §2.16. The profile has two steps in it which are propagating from left to right through stationary fluid of uniform density ρ – the broken line represents their positions at a slightly later time than the full line – and between these steps the density is $\rho + \Delta\rho$. Although the figure makes the excess density $\Delta\rho$ look quite sizable compared with ρ, it is intended at this stage to be an infinitesimal quantity. Per unit area of the band the mass of fluid lying between the planes PP and RR, which constitute a *fixed* control surface [§2.16], is

$$(\rho + \Delta\rho)\mathrm{PQ} + \rho\mathrm{QR},$$

and this mass is increasing with time at a rate

$$(\rho + \Delta\rho)\frac{\mathrm{d(PQ)}}{\mathrm{d}t} + \rho\frac{\mathrm{d(QR)}}{\mathrm{d}t} = \Delta\rho c_s.$$

The continuity condition therefore dictates that behind the first step the fluid is not completely stationary but is drifting from left to right with an infinitesimal velocity Δu such that

$$\rho \Delta u \approx \Delta \rho c_{\mathrm{s}}. \tag{3.38}$$

This result provides a clue as to how the disturbance might be generated, say in a long cylinder of uniform cross-section equipped with a movable piston and filled with gas of uniform density: it could be generated by displacing the piston inwards for a limited time at the constant speed Δu which (3.38) describes. The disturbances represented by figs 3.5(b), 3.5(c) and 3.6 could likewise be generated by piston displacement, and it is worth bearing in mind that in each case the area under the profile of excess density, i.e. $\int \Delta \rho \, \mathrm{d}x$, is the mass of gas displaced by the piston per unit area and carried away by the disturbance. This area is a conserved quantity; the profile may change its shape as time passes but the area underneath it must stay the same.

Figure 3.5(b) represents a disturbance such that between the two density steps of fig. 3.5(a) there are two more steps, likewise propagating from left to right, between which the density is augmented by a further infinitesimal amount. The inner steps propagate with velocity $c_{\mathrm{s}}\{\rho + \Delta\rho\} + \Delta u$, which is greater than the velocity $c_{\mathrm{s}}\{\rho\}$ of the outer steps on two counts, because c_{s} increases with ρ and because Δu is positive. Hence the second step is bound to catch up the first one, while the fourth lags behind the third. Then, since any one-dimensional disturbance of finite amplitude can presumably be treated as a sequence of infinitesimal steps in density, we may infer that the leading edge of the disturbance suggested by fig. 3.5(c) (where $\Delta\rho$, no longer infinitesimal, is plotted rather than ρ) gets steeper as it propagates, while the trailing edge gets less steep. A broken line in that figure suggests what the excess density profile may look like a moment later.

Let us look for equations to describe in mathematical terms the evolution of a travelling density disturbance of arbitrary form and amplitude. The task is straightforward enough, provided that only one dimension is relevant, i.e. provided that the density is a function of say x_1 (or x) but is not a function of x_2 or x_3. Equation (3.38), transformed into a frame of reference in which the fluid in front of the density step is not stationary but has some arbitrary velocity u in the x direction, tells us the increase in u associated with an infinitesimal increase in ρ; it tells us, in fact, that at any instant of time

$$\frac{\partial u}{\partial x} = \frac{c_{\mathrm{s}}}{\rho} \frac{\partial \rho}{\partial x}. \tag{3.39}$$

To integrate (3.39) we need to know how c_{s} depends upon density. For a gas in which p is proportional to ρ^{γ} we may relate c_{s} to its value $c_{\mathrm{s,o}}$ in some reference region where the density of the gas is ρ_{o} and its velocity is u_{o} by the equation

$$c_s = c_{s,o}\left(\frac{\rho}{\rho_o}\right)^{(\gamma-1)/2} \tag{3.40}$$

[(3.33)], in which case the integral of (3.39) is

$$u - u_o = \frac{2c_{s,o}}{(\gamma - 1)}\left\{\left(\frac{\rho}{\rho_o}\right)^{(\gamma-1)/2} - 1\right\}. \tag{3.41}$$

Thus the velocity at which each infinitesimal step in the density profile propagates along x is

$$u + c_s = c_{s,o}\left\{\frac{(\gamma + 1)}{(\gamma - 1)}\left(\frac{\rho}{\rho_o}\right)^{(\gamma-1)/2} - \frac{2}{(\gamma - 1)}\right\} + u_o. \tag{3.42}$$

These equations look more transparent if they are expanded in powers of the dimensionless quantity

$$s = \frac{(\rho - \rho_o)}{\rho_o}, \tag{3.43}$$

which is conventionally referred to as the *condensation* of the gas. To first order in this quantity they become

$$c_s \approx c_{s,o}\left\{1 + \frac{1}{2}(\gamma - 1)s\right\}, \tag{3.44}$$

$$u \approx u_o + c_{s,o}s, \tag{3.45}$$

and

$$u + c_s \approx u_o + c_{s,o}\left\{1 + \frac{1}{2}(\gamma + 1)s\right\}. \tag{3.46}$$

Applied to the disturbance represented by fig. 3.5(c), with ρ_o chosen to be the density of the gas on either side of the disturbance and with u_o chosen to be zero, (3.46) tells us two things. Firstly, that the leading and trailing edges of the disturbance, where s is zero, move at the same speed $c_{s,o}$, so that its total length is constant. Secondly, that between these end points the profile of excess density – or the profile of condensation s, which amounts to much the same thing – is subject to continuous shear, at a rate which to first order in s is uniform. The broken curve in fig. 3.5(c) is a sheared version of the full one.

If the shear were to continue indefinitely, a section of the leading edge of the profile near the point of inflection would first become vertical and then start to lean over. This process is illustrated by curves (a) and (b) in fig. 3.6, which charts the development of a pulse of condensation with a profile which is initially symmetrical but lacking the flat top of fig. 3.5(c). Now a vertical profile implies

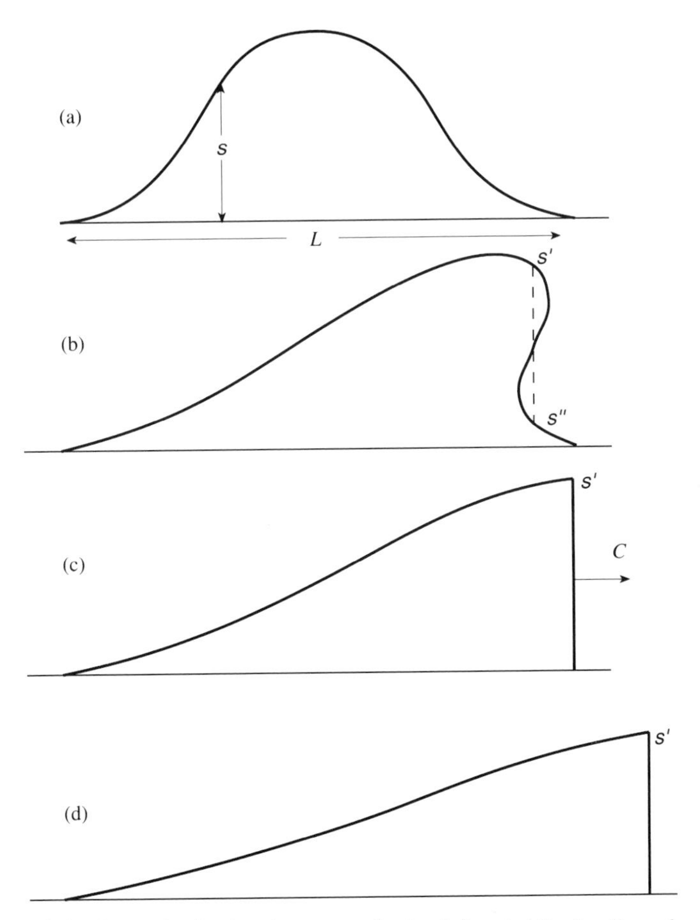

Figure 3.6 Stages in the development of a shock front at the leading edge of a pulse of condensation s. After the shock has reached the leading edge (stage (c)) the pulse length L starts to increase and the strength of the shock, s', starts to diminish. Note that all these profiles have the same area underneath them.

that at some value of x the density of the gas takes a range of values, and this is clearly impossible; it is equally impossible for the density to be triple-valued as curve (b) suggests. Something must go wrong with the theory developed above, therefore, when the leading edge of the profile becomes very steep, and we do not have far to look to see what the trouble is: the temperature gradient in the gas becomes so large that thermal conduction ceases to be negligible, and it turns out that viscous dissipation ceases to be negligible too. What happens, in fact, is that a *shock front* develops, across which the condensation jumps almost discontinuously, from s' to s'' in the case of fig. 3.6(b). Shock fronts are the analogues in terms of sound propagation of tidal bores and hydraulic jumps.

The dissipative processes within a shock front are irreversible, so that the entropy per unit mass of the gas is in principle greater behind the front than ahead of it: the pressure p is still proportional to ρ^γ on either side of the front, but the constant of proportionality is different. It is shown in §3.8, however, that to first – and indeed to second – order in s the increase of entropy is negligible. For weak disturbances incorporating shocks which are also weak, therefore, (3.46) remains a credible result; it accurately describes the propagation velocities $(u' + c_s')$ and $(u'' + c_s'')$ of the profile just behind the shock and just ahead of it, where the condensations are s' and s'' respectively. The velocity with which the front itself propagates is presumably intermediate between these. If so, the front will overtake those parts of the profile which lie ahead of it at the stage represented by curve (b) and will be overtaken by those parts which lie behind it. In the process the shock front will move to the head of the disturbance and the jump between s' and s'' will increase. The profile of curve (b) will transform, in fact, into that of (c).

When it heads the disturbance, the shock front propagates into the stationary gas with a velocity C which is greater than $c_{s,o}$, so that at this stage the length of the pulse is no longer constant; it is increasing, at a rate

$$\frac{dL}{dt} \approx (C - c_{s,o}). \tag{3.47}$$

In so far as the area under the excess density profile is proportional to Ls', we may infer from the constancy of area that s' is decreasing, at a rate

$$\frac{ds'}{dt} = -\frac{s'}{L}(C - c_{s,o}) \propto -s'^2(C - c_{s,o}). \tag{3.48}$$

The decrease occurs because the profile behind the front, which is still subject to shear, continues to overtake it, with a relative velocity which to first order is [(3.46)]

$$c_{s,o}\left\{1 + \frac{1}{2}(\gamma + 1)s'\right\} - C.$$

We can therefore express the rate of decrease in a different way, as

$$\frac{ds'}{dt} \approx -\left[c_{s,o}\left\{1 + \frac{1}{2}(\gamma + 1)s'\right\} - C\right]\left(\frac{\partial s}{\partial x}\right)', \tag{3.49}$$

where $(\partial s/\partial x)'$ is the instantaneous slope of the condensation profile immediately behind the front. To the extent that curves (c) and (d) may be treated as triangles of base L, we have $(\partial s/\partial x)' \approx s'/L$. For (3.48) and (3.49) to be consistent, therefore, we require

$$c_{s,o}\left\{1 + \frac{1}{2}(\gamma + 1)s'\right\} - C \approx C - c_{s,o},$$

or

$$C \approx c_{s,o}\left\{1 + \frac{1}{4}(\gamma + 1)s'\right\} \tag{3.50}$$

to first order. Read in conjunction with (3.48), this tells us that ds'/dt is proportional to $-s'^3$, and hence that the strength of the shock decays like $t^{-1/2}$ at large times.

An alternative method for deriving most of the above results, due to Riemann, is outlined in the appendix. It sheds light, in particular, on the distinction between complex disturbances which contain components travelling in opposite directions and so-called *simple waves*, which travel, as we have assumed the profiles in figs. 3.5 and 3.6 to travel, in one direction only. Shock fronts which are not necessarily weak enough to justify the approximations upon which we have relied so far are discussed in §3.8, and the validity of (3.50) for C is there confirmed. Shock fronts which expand in three dimensions rather than one are discussed in §3.9. Shock fronts which lie at an angle to the fluid velocity are discussed in §3.10. Shock fronts turn out, incidentally, to have a finite thickness, which in gases may be as much as 1 mm if the shock is very weak, though the limiting thickness for a very strong shock is comparable with the molecular mean free path. But their thickness has no bearing on the way in which they propagate, and there is no need for us to consider it.

Shock fronts are not visible in the world about us in the way that tidal bores and hydraulic jumps are, but they are certainly audible. Anyone who has experienced a thunderclap has heard one. Aircraft in supersonic flight trail weak shock fronts behind them, as described in §§3.11 and 3.12 below, and these are audible as *sonic booms*. If the aircraft's flight path is curved, there may be places at ground level at which shock fronts originating from different points along the path arrive simultaneously, and the resultant shock may then be strong enough to break window panes. Strong shocks are also generated in air when bombs explode, and the boundary of the notorious mushroom cloud which appeared over New Mexico on a fateful day in 1945 coincided with an expanding shock front. Shock fronts can be generated explosively in liquids and solids too, and experimental studies of the manner in which they propagate supply information concerning the properties of liquids and solids at very high densities which cannot be obtained in any other way.

3.7 Momentum transfer by sound waves

This section is something of a digression, but the problem which it addresses is of too much interest to physicists to be buried in the appendix. Every physicist knows that light waves – or photons, if the language of quantum mechanics is to be preferred – carry momentum as well as energy, and that in free space the ratio between the two is just $1/c$, where c is the velocity of light. Consequently, light exerts pressure on anything that intercepts it. Do sound waves – or *phonons* – likewise carry momentum as well as energy? If so, is the ratio between them $1/c_s$?

One school of thought, represented by more than one reputable textbook, maintains that both questions are to be answered in the affirmative, but the elementary arguments advanced to support this point of view, associated with the name of Larmor, do not stand up to scrutiny. Another school of thought has it that the answer to the first question is no, so that the second question does not arise. This school relies on the fact that a sound wave as normally defined, unlike the disturbances discussed in §3.6, carries no excess mass along with it. Suppose, for example, that at one end of a long cylinder full of gas of uniform density ρ_o there is a quartz transducer rather than a movable piston. Stimulation of the transducer for a short interval of time generates a pulse of sound which, if friction at the walls may be ignored, travels down the tube in say the x direction as a train of one-dimensional sound waves. But the transducer reverts to its original position once stimulation ceases, and each 'particle' of the gas likewise reverts to its original position once the wave train has passed by it. This implies that the mean velocity of each particle, averaged over the time for which the wave train affects it, must be zero. It appears to follow that an average of velocities over all the particles which lie within the wave train at any particular instant of time must be zero too. In that case, is not the net momentum of the gas within the wave train zero, just as it is zero for the stationary gas in front of the wave train and behind it?

That argument does not stand up to scrutiny either, because in circumstances where the wave changes its character as it propagates an average over t for fixed x does not yield the same result as an average over x for fixed t. The momentum associated with the wave train is in fact best calculated from a space average, to be indicated in what follows by angle brackets: if the two ends of the train are instantaneously at $x = X$ and $x = X + L$ then the mean momentum per unit volume is

$$\frac{1}{L}\int_X^{X+L} \rho u \,\mathrm{d}x = \rho_o\langle(1 + s)u\rangle. \tag{3.51}$$

To obtain a useful answer we must first improve upon (3.45), by expanding (3.41) to second order in s rather than to first order only. With $u_o = 0$ the necessary improvement is

$$u \approx c_{s,o}\left\{s + \frac{1}{4}(\gamma - 3)s^2\right\}. \tag{3.52}$$

To second order, therefore, the mean momentum per unit volume is

$$\rho_o c_{s,o}\left\{\langle s\rangle + \frac{1}{4}(\gamma + 1)\langle s^2\rangle\right\}.$$

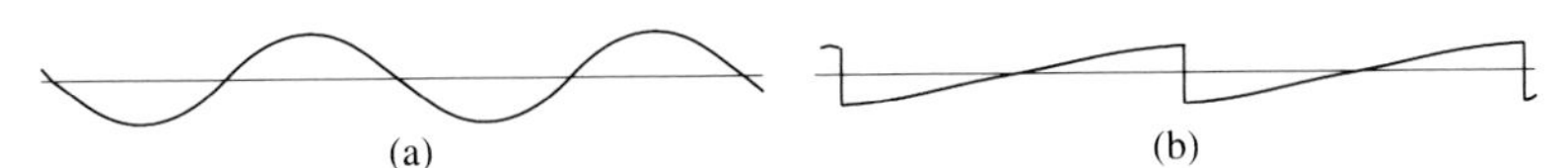

Figure 3.7 Variation of condensation with x in a plane sound wave: (a) initial sinusoidal variation, (b) after development of shock fronts.

or

$$\frac{1}{4}(\gamma + 1)\rho_o c_{s,o}\langle s^2\rangle \tag{3.53}$$

if the sound wave indeed carries no mass with it, i.e. if $\langle s\rangle = 0$. The mean kinetic energy density in the wave is then $\frac{1}{2}\langle\rho u^2\rangle$, which to second order in s is the same thing as $\frac{1}{2}\rho_o c_{s,o}^2\langle s^2\rangle$. Since the total energy is twice the kinetic energy, as is always the case in wave propagation problems, we arrive at the answer that for sound waves of modest amplitude in ideal gases

$$\frac{\text{momentum}}{\text{total energy}} = \frac{(\gamma + 1)}{4c_{s,o}}. \tag{3.54}$$

It is possible to continue this line of analysis to find to second order in s the *time* average of velocity which determines the net displacement of a particle, and hence to show that the ratio of net displacement to the length of the wave train L is just $\langle s\rangle$; the terms involving $\langle s^2\rangle$ cancel in this instance. Since this answer is necessarily correct to all orders in s – it follows from elementary considerations based on the law of conservation of mass – the details will not be given here.

The arguments developed in §3.6 show that given time a travelling one-dimensional sound wave in which the initial variation of s is sinusoidal should distort to form a sequence of shock fronts, as suggested by fig. 3.7. However, a plane wave of normal amplitude, readily audible but not uncomfortably loud, would need to travel through air for several kilometres in order to reach this stage.

3.8 Normal shock fronts in gases

An analysis of shock propagation in gases which is valid to all orders in s requires algebraic manipulations which are entirely straightforward but nevertheless cumbersome, so let us concentrate here on results rather than on derivations. Let us imagine the shock front to be a plane across which pressure increases suddenly from p to p', and density from ρ to ρ'. If we choose our frame of reference so that this plane is stationary, then the gas is to be pictured as flowing towards it on the low pressure side and away from it on the high pressure side, with velocities $\boldsymbol{u}$ and $\boldsymbol{u}'$ respectively. Initially we may suppose the shock front to be normal to $\boldsymbol{u}$, and in that case it has to be normal to $\boldsymbol{u}'$ also, as in fig. 3.8; the pressure gradient within

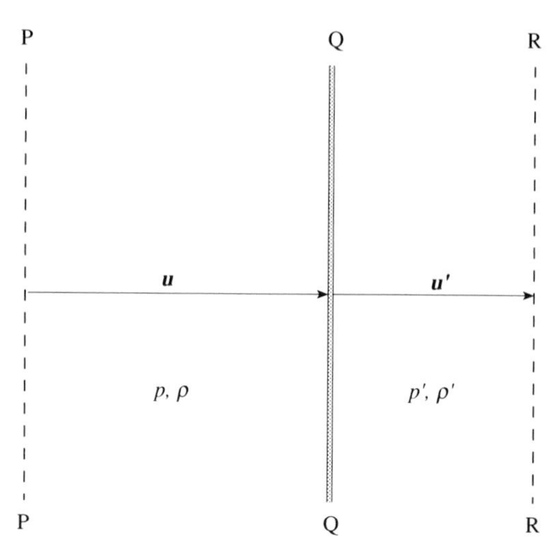

Figure 3.8 Fluid velocities ($u' < u$) on either side of a normal shock front QQ, in a frame of reference in which the front is stationary. (Shock fronts are represented, here and in later figures in this chapter, by thin bands of shading between parallel lines.)

the front has no component perpendicular to $\boldsymbol{u}$ to accelerate the gas in that direction. Generalisation to situations where the shock front is *oblique* and the two velocities are not parallel to one another proves to be trivial but is deferred to §3.10. Transformation of the results to other frames of reference is also trivial, of course; in the frame in which the gas in front of the shock is stationary the front evidently moves with a velocity $\boldsymbol{C}$ which must equal $-\boldsymbol{u}$. In fig. 3.8 the gas moves from left to right, this being the mode of representation conventionally applied to flow past obstacles etc., so the shock would move from right to left if the gas in front of it were stationary, i.e. in the opposite direction to the shock fronts sketched in figs. 3.6 and 3.7.

The six variables indicated in fig. 3.8 are linked by three equations which collectively constitute the *Rankine–Hugoniot* relations for shock propagation in an ideal gas. The first of them expresses the fact that when gas flows through a shock front its change of intrinsic enthalpy is given by Bernoulli's theorem, i.e. by (3.24); the reasons why Bernoulli's theorem is applicable despite the irreversible processes which occur within the front itself are explained in §3.3 above. It is marginally easier to handle if written in terms of the Mach Numbers $M = u/c_s$ and $M' = u'/c_s'$, where c_s and c_s' are the velocities of sound on either side of the shock front, rather than in terms of velocities. It then becomes [(3.32)]

$$\frac{p}{\rho}\left\{1 + \frac{1}{2}(\gamma - 1)M^2\right\} = \frac{p'}{\rho'}\left\{1 + \frac{1}{2}(\gamma - 1)M'^2\right\}. \tag{3.55}$$

The second equation represents the elementary continuity condition that $\rho u = \rho' u'$. Written in terms of Mach Numbers this becomes, using (3.10) for the velocity of sound and after squaring both sides,

$$p\rho M^2 = p'\rho' M'^2. \tag{3.56}$$

In §3.5 above we based a third condition upon the rule that p is normally proportional to ρ^γ during an adiabatic compression, but since this applies only to *reversible* adiabatic compressions we cannot rely upon it here. Momentum considerations, however, provide us with a substitute. Per unit area of the shock front the rate of change of momentum of the gas contained between the two planes labelled PP and RR in fig. 3.8, which in this instance are both to be regarded as embedded in the gas and moving with it, is

$$\rho u \frac{\mathrm{d(PQ)}}{\mathrm{d}t} + \rho' u' \frac{\mathrm{d(QR)}}{\mathrm{d}t} = -\rho u^2 + \rho' u'^2.$$

This may be equated to $(p - p')$, which is the force per unit area acting on the gas in question, to yield a third equation involving Mach Numbers:

$$p(1 + \gamma M^2) = p'(1 + \gamma M'^2). \tag{3.57}$$

Manipulation of the Rankine–Hugoniot relations shows that the two Mach Numbers are uniquely determined by a single dimensionless parameter which represents the strength of the shock. A more convenient choice for this parameter than the condensation s defined by (3.43) is

$$\sigma = \frac{(p' - p)}{p}, \tag{3.58}$$

and in terms of this we obtain

$$M^2 = 1 + \frac{(\gamma + 1)}{2\gamma}\sigma, \tag{3.59}$$

$$M'^2 = 1 - \frac{(\gamma + 1)}{2\gamma}\frac{\sigma}{(1 + \sigma)}. \tag{3.60}$$

It follows at once from (3.59) and (3.60), since σ is a positive quantity when $p' > p$ as we have supposed, that on the low pressure side of a stationary normal shock the flow is always *supersonic*, whereas on the high pressure side it is always *subsonic*. Both values of the Mach Number tend to unity, of course, in the limit where the shock strength σ becomes infinitesimal. If the strength is finite but small enough to justify neglect of all but first-order terms in σ, then

$$u = Mc_s \approx c_s\left\{1 + \frac{(\gamma + 1)}{4\gamma}\sigma\right\},$$

[(3.59)] which of course implies that the speed with which a normal shock propagates into gas which is stationary ahead of it is

$$C \approx c_s \left\{1 + \frac{(\gamma + 1)}{4\gamma}\sigma\right\}, \tag{3.61}$$

a result equivalent to (3.50). For very strong shocks, however,

$$C \approx c_s \sqrt{\frac{(\gamma + 1)p'}{2\gamma p}} = \sqrt{\frac{(\gamma + 1)p'}{2\rho}}. \tag{3.62}$$

The jump in density across a normal shock front is uniquely determined by M, or else by the jump in pressure, according to the equations

$$\frac{\rho'}{\rho} = \frac{u}{u'} = \frac{(\gamma + 1)M^2}{(\gamma - 1)M^2 + 2}, \tag{3.63}$$

$$= 1 + \frac{\sigma}{\gamma + \frac{1}{2}(\gamma - 1)\sigma}. \tag{3.64}$$

There is also a jump in temperature, of course, which can likewise be expressed in terms of σ since

$$\frac{T'}{T} = (1 + \sigma)\frac{\rho}{\rho'}. \tag{3.65}$$

Note that there is no upper limit to σ and no upper limit to T'. No matter how strong the shock, however, ρ' cannot exceed $\rho(\gamma + 1)/(\gamma - 1)$ and M'^2 – as is apparent from (3.60) – cannot fall below $(\gamma - 1)/2\gamma$.

The amount by which the entropy of the gas per unit mass increases as it crosses the shock front is

$$s' - s = \frac{k_B}{(\gamma - 1)m_{mol}} \ln\left(\frac{p'\rho^{\gamma}}{p\rho'^{\gamma}}\right). \tag{3.66}$$

Equation (3.66) serves to confirm that s' is always greater than s, whatever the strength of the shock provided that $\sigma > 1$, and thereby to confirm our assumption that when gas flows through a stationary front it flows *from* the low pressure side *to* the high pressure side. Gas could not flow in the reverse direction without contravening the second law of thermodynamics. However, when the logarithmic expression in (3.66) is expanded in powers of σ using (3.64) the terms in σ and σ^2 are found to cancel. It follows that if gas flows through an infinite succession of shocks of infinitesimal strength, such that the resultant change of density and pressure is finite, the resultant change in its entropy per unit mass is infinitesimal. This confirms that gas flow through a region in which pressure and density are changing in a continuous fashion *is* reversible, both in the thermodynamic sense and in the sense that the flow may be in either direction.

3.9 Shock fronts generated by explosions

Detonation in the open air of a conventional high-explosive bomb is followed by the rapid release of a quantity of hot gas at high pressure, and the sudden expansion of this gas launches a spherical shock front into the surrounding atmosphere. An atomic bomb releases little gas, of course, but it heats air in the vicinity, and expansion of this air generates a shock front which does not differ greatly in kind from the shock front generated by a conventional bomb though it certainly differs in scale.

Granted a few assumptions, not all of them wholly plausible, we may deduce how the radius R_F of the shock front varies with time t using dimensional analysis alone. The assumptions needed are (i) that the shock is very strong, (ii) that the available energy in the bomb, E, is released virtually instantaneously at $t = 0$, and (iii) that air continues to behave like an ideal gas with $\gamma \approx 1{\cdot}4$, even under the extreme conditions that obtain near the centre of a strong explosion. When assumption (i) is justified the density ρ' just inside the front takes the limiting value of $\rho\{(\gamma + 1)/(\gamma - 1)\} \approx 6\rho$, where ρ is the density of the undisturbed atmosphere, and the front's velocity C is given by (3.62). The pressure and temperature outside the front are then irrelevant, not only to ρ' and C but to everything that happens inside the front. This means that R_F must be completely determined by t, E, ρ and γ. The combination $Et^2/\rho R_F^5$ is dimensionless and so is γ, but no other independent dimensionless combinations of the relevant variables exist. Hence we may assert that

$$R_F = \left\{\frac{Et^2}{\rho}\right\}^{1/5} f\{\gamma\}, \tag{3.67}$$

and a full analysis indicates that $f\{1{\cdot}4\}$ is about $1{\cdot}03$. G. I. Taylor reached this conclusion in 1941, and when photographs showing the expansion of the first mushroom cloud became available to him a few years later he was able not only to verify his theory [fig. 3.9] but also to estimate E with considerable accuracy. His success is said to have disconcerted the security experts of the day.

Naturally the shock strength diminishes as R_F increases. From (3.67) we know that dR_F/dt is proportional to $t^{-3/5}$, i.e. to $R_F^{-3/2}$, and since dR_F/dt is the same as C we may deduce from (3.62) that the pressure p' immediately behind the front diminishes like R_F^{-3}. At some radius which depends upon the energy released in the explosion, therefore, assumption (i) must fail. What happens to the propagating disturbance thereafter can be computed, though full analytical solutions of the relevant equations of motion are not available, and it is known that a *suction phase*

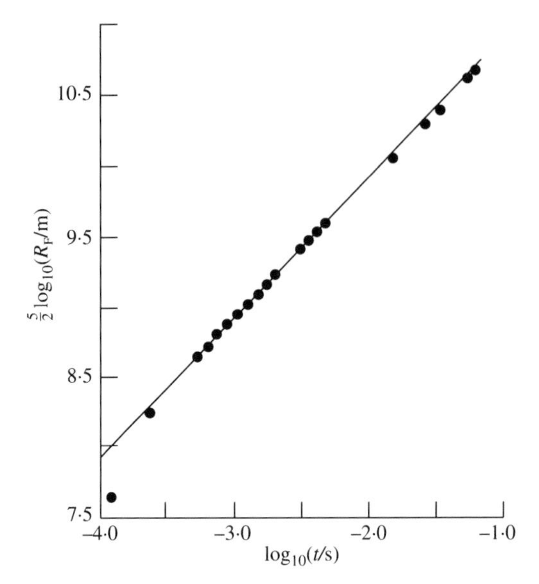

Figure 3.9 Variation with time of the radius R_F of a spherical shock front. [Based on a figure in G. I. Taylor, *Proc. Roy. Soc. A*, **201**, 175, 1950.]

develops behind the advancing shock front, i.e. a region in which the pressure is *less* than that of the undisturbed atmosphere. By the time that the shock has become weak enough for the associated increase of entropy to be ignored, the excess density $\Delta\rho$ in the neighbourhood of the front varies with radius R in the manner suggested by fig. 3.10.

The area underneath the excess density profile of fig. 3.10 represents, like the area under the profiles in fig. 3.5, the mass of gas being displaced outwards per unit area of the shock front. If the total mass displaced by the explosion is M, then the area underneath the profile in fig. 3.10 is

$$\int_0^\infty \Delta\rho \, \mathrm{d}R \approx \frac{M}{4\pi R_F^2}.$$

It is because this tends to zero for large R_F, whereas the corresponding area for the disturbances illustrated by fig. 3.5(c) and fig. 3.6 is independent of x, that a suction phase develops when the shock front is expanding in three dimensions but not when it is planar and propagates in one dimension only.

The existence of a suction phase is confirmed by experimental observations, among them the observation that when windows have been broken by an explosion the fragments of glass sometimes lie *outside* the building rather than inside. A disturbance of the form represented by fig. 3.10 delivers two impulses to

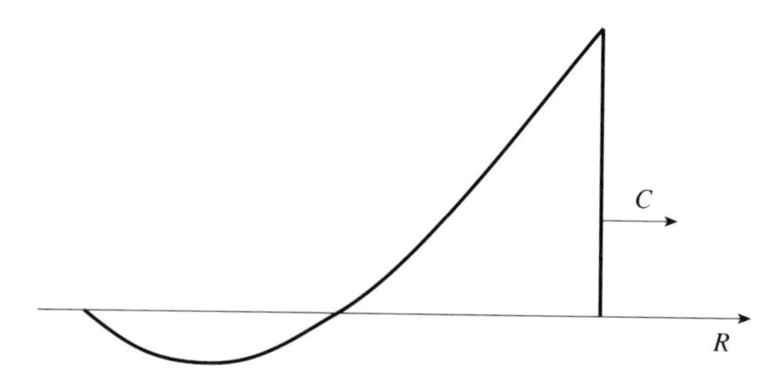

Figure 3.10 Profile of excess density behind an expanding spherical shock front.

a window pane, an inwards one due to the shock front and to the high pressure region immediately behind it and an outwards one due to the suction phase. The two impulses may be nearly equal in magnitude. When the first one arrives, however, the pane is still intact and much of the impulse is transferred through the glass to the window frame. By the time that the suction phase arrives the glass is in pieces.

3.10 Oblique shock fronts

A plane shock front which is stationary in one frame of reference is also stationary in a frame which moves with respect to the first at some velocity $\boldsymbol{v}$, provided that $\boldsymbol{v}$ lies in the plane. Now if we add such a velocity $\boldsymbol{v}$ to the velocities $\boldsymbol{u}$ and $\boldsymbol{u}'$ which are indicated in fig. 3.8 we arrive at velocities $\boldsymbol{w}$ and $\boldsymbol{w}'$ which are no longer normal to the shock front and no longer parallel to one another, as in fig. 3.11. We arrive, in fact, at a diagram which represents an *oblique* shock front. In terms of angles β and β' which are defined by the figure, it must be described by the equation

$$w \cos \beta = w' \cos \beta' = v \tag{3.68}$$

and also, since we may substitute $w \sin \beta$ for u and $w' \sin \beta'$ for u' in (3.59) and (3.63), by

$$M^2 \sin^2 \beta = 1 + \frac{(\gamma + 1)}{2\gamma} \sigma \tag{3.69}$$

and

$$\frac{w \sin \beta}{w' \sin \beta'} = \frac{(\gamma + 1)M^2 \sin^2 \beta}{(\gamma - 1)M^2 \sin^2 \beta + 2}, \tag{3.70}$$

in which M represents w/c_s rather than u/c_s as heretofore. We have only to combine (3.68) and (3.70) to see that

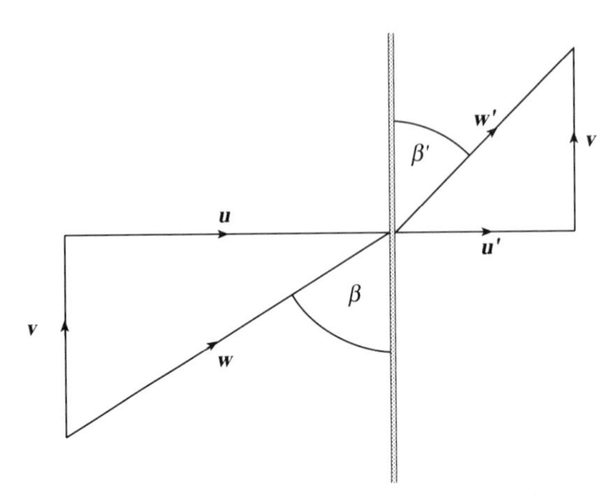

Figure 3.11 The shock front of fig. 3.8 in a different frame of reference. The fluid velocities ($\boldsymbol{w}' < \boldsymbol{w}$) on either side of the front are oblique to it, and they are not parallel to one another.

$$\frac{\tan \beta'}{\tan \beta} = \frac{(\gamma - 1)M^2 \sin^2 \beta + 2}{(\gamma + 1)M^2 \sin^2 \beta}. \tag{3.71}$$

Thus if M and the angle β are given, the strength of the shock is fully determined [(3.69)], and the angle through which $\boldsymbol{w}'$ is deflected from $\boldsymbol{w}$, namely $\delta = \beta - \beta'$, is also fully determined [(3.71)]. The way in which δ varies with β for various values of M is shown by the curves in fig. 3.12. It vanishes, as does σ, when β equals the so-called *Mach angle* μ defined by

$$M \sin \mu = 1, \tag{3.72}$$

and it also vanishes when the shock front becomes normal, i.e. when $\beta = \pi/2$. In between these zeros it passes through a maximum, the height of which increases with M.

In §3.12 we consider oblique shock fronts for which β is only slightly greater than the Mach angle, and for which both δ and σ are small. It may readily be shown from (3.71) and (3.69) that

$$\delta \approx \sin \beta \cos \beta \frac{2\sigma}{2\gamma + (\gamma + 1)\sigma},$$

when $\delta \ll 1$, so that since $\sigma \ll 1$ as well we have

$$\delta \approx \sin \mu \cos \mu \frac{\sigma}{\gamma} = \frac{\sqrt{M^2 - 1}}{M^2} \frac{(p' - p)}{\gamma p}. \tag{3.73}$$

This relation between the jump in pressure across an oblique shock front and the angle through which a supersonic air stream is deflected in passing through it

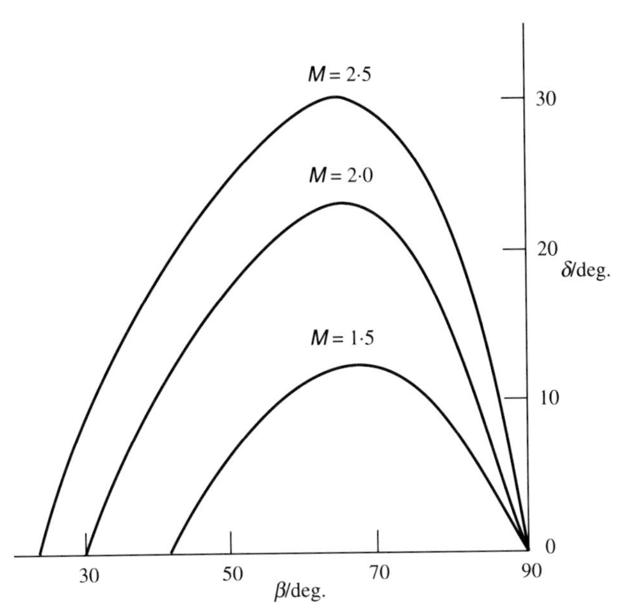

Figure 3.12 Variation of the angle $\delta = \beta - \beta'$ through which a fluid stream is deflected in passing through an oblique shock front, as a function of β for different values of M.

enables us to discuss in a quantitative fashion the forces which act on the wings of aircraft in supersonic flight.

3.11 Mach's construction

By way of introduction to the topic of oblique shock fronts attached to supersonic aircraft two well known diagrams, first drawn by Mach, are reproduced in figs. 3.13(a) and (b). The point labelled O in each of these figures represents the position at time $t = 0$ of the leading edge of a rectangular plate which is advancing through otherwise stationary fluid from right to left with uniform velocity U. It could equally well represent the tip of a rod-shaped projectile – an arrow, in fact – but the case of a plate is slightly simpler to discuss because the flow pattern in the surrounding fluid does not depend upon the third dimension. The points O′, O″ etc. represent the position of the leading edge at earlier times $t = -\tau$, -2τ etc., and the circles centred on these points represent disturbances originating from the edge at those times which have spread out through the fluid at the velocity of sound c_s. The velocity U is less than c_s in fig. 3.13(a) and greater than c_s in fig. 3.13(b).

If the edge were an *active* source of sound, emitting waves at a steady frequency τ^{-1}, say, and remaining audible even when stationary, then the circles in the figure

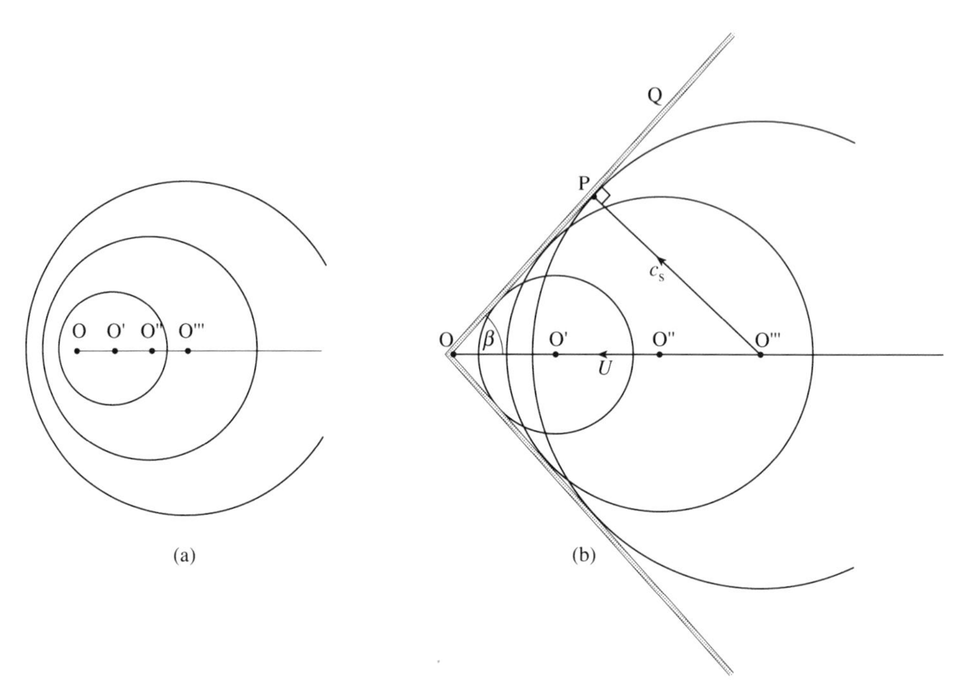

Figure 3.13 Mach's construction: (a) $U < c_S$; (b) $U > c_S$.

might represent cross-sections through cylindrical wave crests. In that case fig. 3.13(a) would serve to explain the well-known *Doppler effect*, whereby the wavelength of sound directly in front of a moving emitter is reduced by the motion from λ to $\lambda\{1 - (U/c_s)\}$. In fact, however, it is a *passive* source, which disturbs the fluid only by displacing it. Each circle therefore represents a thin cylindrical shell of fluid whose density has been minutely increased, and to which a minute radial velocity has been imparted, by a small forward movement of the plate while its edge was at O′, O″ etc. We need to imagine many such shells, and the diagram should really include a host of other circles besides those drawn, with centres intermediate between the points O′, O″ etc.

When $U < c_s$, the flow pattern which results from superposition of the motion associated with individual shells does not in any way correspond to the emission of sound by the plate. It corresponds instead to a pattern of continuous backflow of displaced fluid, which is of no special interest here. When $U > c_s$, however, there are two planes passing through O to which all the shells are tangential, and on these planes the motions associated with different shells must reinforce one another. The amplitude of the net disturbance may then be large enough for non-linear effects to be significant, in which case the idea that we can reconstruct the disturbance by straightforward superposition may let us down. In view of the discussion in §3.6, however, it seems likely that a shock front exists on these planes, i.e. that pressure increases abruptly across them from left to right by a factor $(1 + \sigma)$, with a corresponding increase in density. If so, the excess density profile behind the shock front presumably has the form of curve (d) in fig. 3.6. As the front propagates perpendicular to itself, the point labelled P in fig. 3.13 – a point to be regarded as attached to the front rather than to the gas through which the front is propagating – moves outwards, and some time later it will be in the position relative to the projectile that Q occupies now. Since a plane shock behind which the density varies in the manner shown in fig. 3.6(d) gets weaker as it propagates [§3.6], we may infer that the shock is weaker at Q than at P.

Mach's construction suggests that the angle β at which the shock front is inclined to the trajectory of the projectile is $\sin^{-1}(c_s/U)$. We have only to view the front in the frame of reference in which the projectile is at rest and the gas is moving from left to right with velocity U to see that c_s/U is equivalent in the notation of §3.10 to M^{-1}, and that $\sin^{-1}(c_s/U)$ is the Mach angle μ already defined by (3.72). However, (3.69) tells us that the inclination of the shock front must in fact depend upon its strength, i.e. that

$$\sin\beta = \left\{1 + \frac{(\gamma + 1)}{2\gamma}\sigma\right\}^{1/2} \sin\mu. \tag{3.74}$$

The two predictions are inconsistent except at large distances from the projectile where σ tends to zero. Mach's construction does not reveal the whole truth, therefore, and close to the projectile it is particularly inadequate.

3.12 Supersonic flow past thin plates

Figure 3.14 shows an enlarged view of the front of the plate to which fig. 3.13 refers, which is here revealed as bevelled to form a wedge with a sharp leading edge through O. The diagram is drawn for the frame of reference in which the plate is stationary, and lines of flow show how in this frame the air stream splits as it encounters the plate. It splits abruptly, each line of flow remaining undeflected until it crosses one of two oblique shock fronts which originate at the leading edge, where the angle of deflection, up or down as the case may be, is necessarily the semi-angle δ of the wedge.[4] If δ is small enough to justify the application of (3.73), we may infer the strength of these shock fronts to be

$$\sigma \approx \frac{\gamma\delta M^2}{\sqrt{M^2 - 1}}. \tag{3.75}$$

[4] When U is subsonic, the lines of flow in a situation otherwise similar to that depicted in fig. 3.14 start to separate well before they reach the obstacle itself. They can only do this, however, because pressure disturbances which signal the presence of the obstacle are able to propagate upstream against the current, and when U exceeds c_s such upstream propagation is impossible. Compare the remarks about choked flow on p. 91.

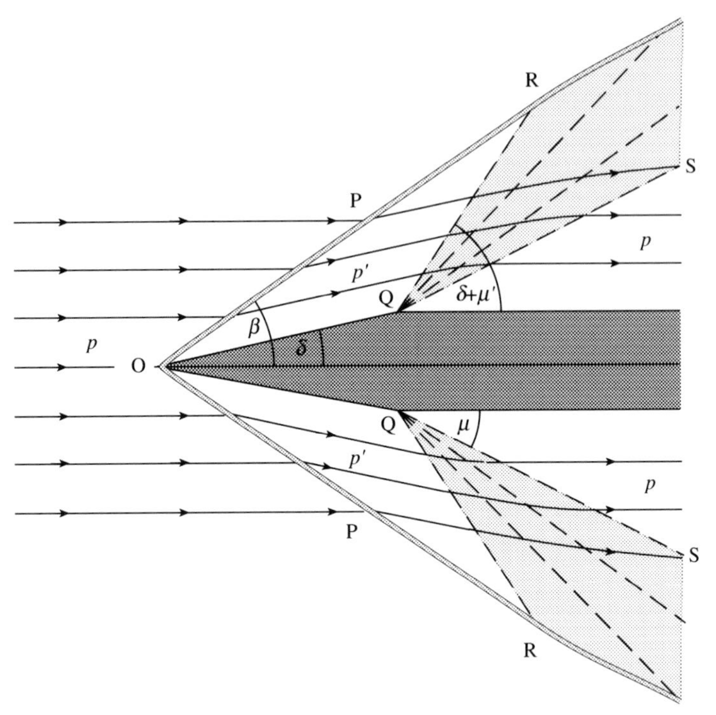

Figure 3.14 Two-dimensional supersonic flow from left to right past a stationary plate with a bevelled leading edge: OPR, shock fronts; RQS, expansion fans.

Their angle of inclination β, given in terms of σ by (3.74), is slightly greater than the Mach angle μ. Behind them there is a uniform excess pressure, $p' - p = \sigma p$.

When the air stream passes the corners labelled Q in fig. 3.14 it is deflected for the second time, so as to run parallel to its initial direction of motion and parallel to the plate. This inwards deflection is in the wrong sense to be associated with a shock front of positive σ, across which pressure abruptly increases. Instead, it is associated with a fall in pressure which is gradual rather than abrupt. The broken lines which radiate from Q in fig. 3.14, within lightly shaded regions which correspond to what are called *Prandtl–Meyer expansion fans*, represent isobars which chart the fall in pressure, from p' on the lines QR back to p on the lines QS. They may also be regarded, however, as members of a sequence of shock fronts, each of infinitesimal strength, for which σ is negative. As such, each of these lines must be inclined with respect to the lines of flow which cross it at the local Mach angle. This is $\sin^{-1}(1/M_{\text{local}})$; its value μ' in the regions where the pressure is p', i.e. between OPR and OQR, is greater than μ because M', though not necessarily less than unity as it would be for a normal shock front, is certainly less than M. With respect to the plate the lines QR are inclined at angles $\pm(\delta + \mu')$, while the lines QS are inclined at angles $\pm\mu$.

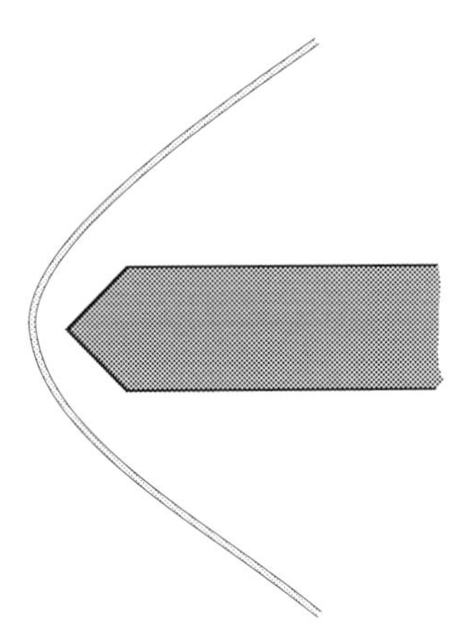

Figure 3.15 Detached shock front formed in supersonic flow from left to right past a plate with a blunt leading edge.

It may be shown that $(\delta + \mu')$ exceeds the inclination of the shock fronts OPR, i.e. the angle β defined by (3.74) and (3.75). Thus the expansion fans overlap the shock fronts at some distance from the projectile, as the figure suggests. At larger distances the strength of the shock fronts σ diminishes and so does their inclination, at such a rate that (3.74) is always satisfied though (3.75) is not; the shock fronts curve, in fact, so that at large distances they approach asymptotically the inclination μ suggested by Mach's construction.

Because the leading edge of the plate experiences a pressure p' which is greater than the ambient pressure p it experiences what is called a *drag force* – a force directly opposing its motion – which according to (3.75) is given by

$$(p' - p)A \approx \frac{\gamma p A \delta M^2}{\sqrt{M^2 - 1}} = \frac{\delta \rho U^2 A}{\sqrt{M^2 - 1}}, \tag{3.76}$$

where A is the plate's cross-sectional area in a plane perpendicular to U. This force can evidently be reduced by reducing the angle δ, while for given δ it is at a minimum when $M = \sqrt{2}$. The implication of (3.76) that it becomes infinite where $U = c_s$ and $M = 1$ is not to be taken seriously. As M tends to unity from above, the maximum angle through which a gas stream can be deflected by a planar oblique shock tends to zero [fig. 3.12], and fig. 3.14 becomes misleading once that maximum angle is less than the semi-angle of the edge. At that stage, in fact, the two planar shock fronts become detached from the edge and move slightly upstream to form a single curved front, as suggested by fig. 3.15, and the calculation of the drag force then becomes more complicated; (3.76) no longer applies. Nevertheless, it is true that the drag force experienced by a projectile is particularly large for values of U which are just above the velocity of sound c_s, and

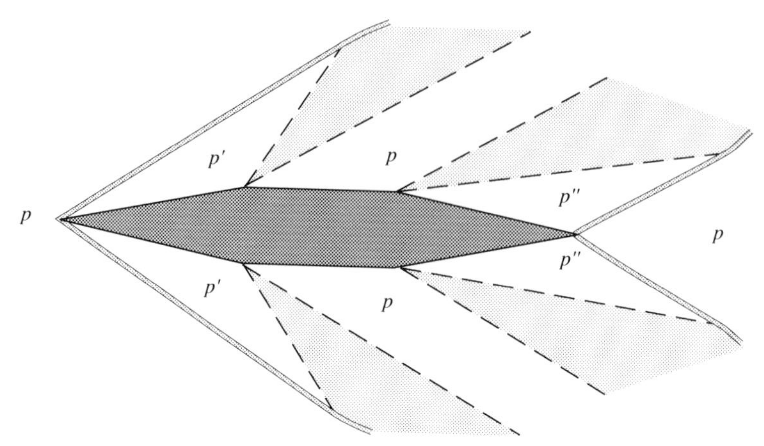

Figure 3.16 Shock fronts and expansion fans originating at the vertices of a stationary plate whose trailing edge is bevelled to match the leading edge. The flow is supersonic and from left to right.

this accounts for the so-called *sound barrier* which appears to inhibit the acceleration of aircraft from subsonic to supersonic speeds. The extra work which the engines of an aircraft must do to surmount the sound barrier supplies, of course, the energy which at supersonic speeds is stored in the accompanying shocks.

So far we have not considered the rear, or *trailing*, edge of the projectile. If this is also bevelled, as in fig. 3.16, its vertices will be the source of further shock fronts and expansion fans, between which the pressure p'' must be *less* than p. Figure 3.16 illustrates a symmetrical case, such that the shock fronts originating at the leading and trailing edges deflect the air stream through the same angle δ. In that case $(p - p'')$ is the same as $(p' - p)$ to first order in δ, and the total drag force experienced by the projectile is therefore about double that exerted on the leading edge alone.

Finally, suppose that the plate itself – now of such small thickness that we have no need to discuss whether its edges are bevelled or not – is inclined to the air stream at a small angle α, as shown in fig. 3.17. Attached to its front edge there is then a shock front below and an expansion fan above, while at the rear edge the situation is reversed. Hence the pressure p' which acts on the sheet from below is greater than p, while the pressure p'' which acts on it from above is less than p, the difference being given in each case, to first order in α, by $\alpha\rho U^2/\sqrt{M^2 - 1}$ [(3.76)]. Hence an almost horizontal plate which is moving horizontally through stationary gas with supersonic velocity U may be expected to experience a vertical *lift force*, and the magnitude of this lift force per unit area of the plate should be approximately

$$p' - p'' \approx \frac{2\alpha\rho U^2}{\sqrt{M^2 - 1}}. \tag{3.77}$$

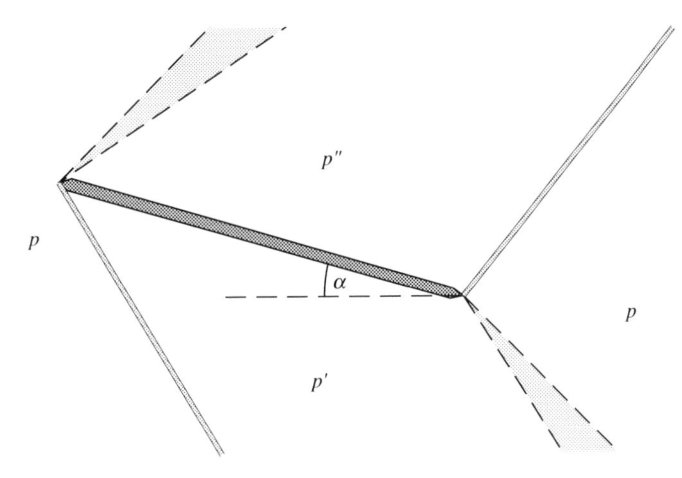

Figure 3.17 Shock fronts and expansion fans originating at the edges of an inclined stationary plate. The flow is supersonic and from left to right.

The aerofoils which constitute the wings of supersonic aircraft are not thin rectangular plates, of course; if they were, they would not support these aircraft while they are accelerating towards the sound barrier at values of U which are subsonic, as we shall discover in §7.11. However, the modifications which are needed to the first-order theory outlined in this section, before the results can be applied with confidence to thick plates of arbitrary cross-section, involve no new physical principles.

The existence of the shock fronts and expansion fans discussed in this section can be verified by a variety of photographic techniques. A striking example of the many photographs recorded in the literature is reproduced in fig. 3.18.

3.13 Rockets

The engines that are used to propel rockets and modern aeroplanes develop thrust, of course, by generating gas at high pressure in a combustion chamber and ejecting it at the rear. Rocket engines differ from so-called *jet engines* in that all the matter ejected is stored within the rocket before the flight starts, whereas in a jet engine stored fuel is allowed to ignite with air which has been drawn in from the surroundings, and which is normally compressed before ignition. In both cases the ejected gas emerges through an *expansion nozzle*, the standard form of which is sketched in fig. 3.19; such nozzles are referred to as *de Laval nozzles*, after the engineer who first devised them to supply jets of steam to turbines. The gas accelerates into the nozzle's throat and is choked there [§3.5], and the designer's object is to ensure that beyond the throat it continues to expand in an isentropic fashion and continues to accelerate, before emerging as a non-divergent jet of

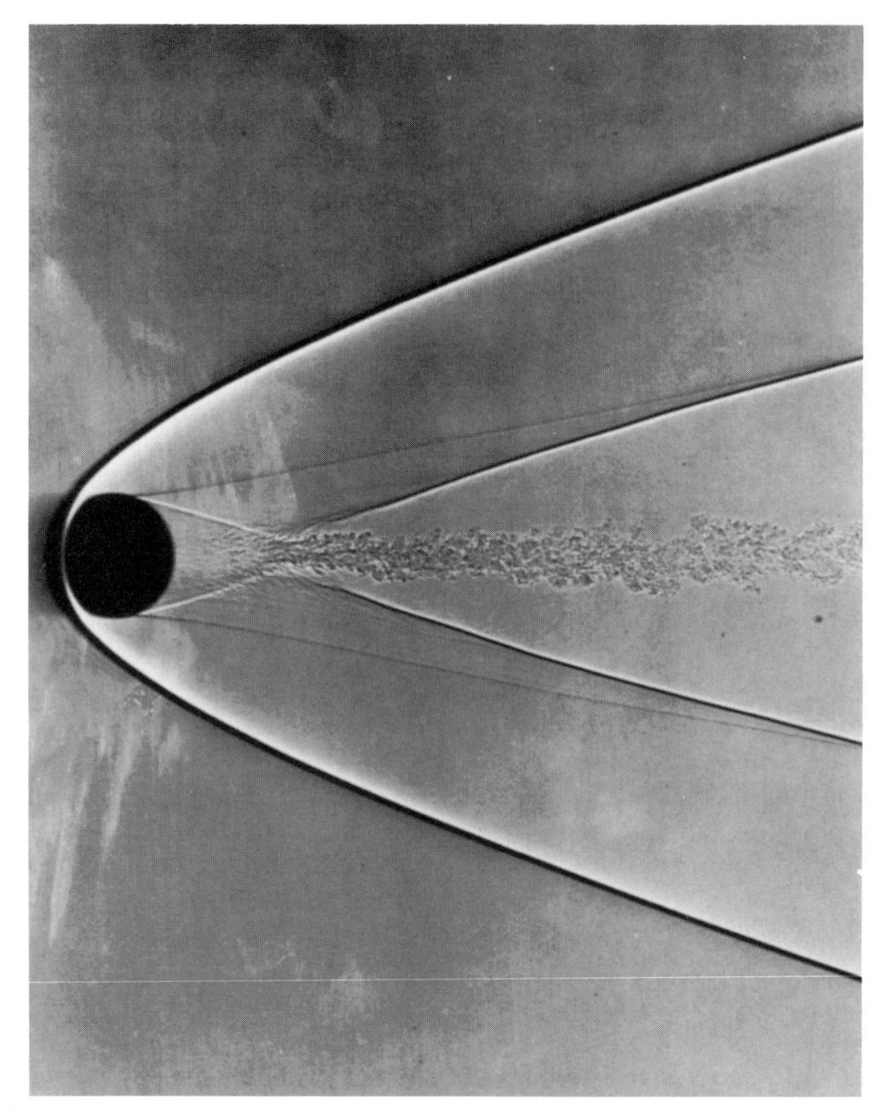

Figure 3.18 Spark photograph of a sphere (diameter 0·5 inch) travelling from right to left through air at $M = 4{\cdot}01$. Two relatively strong shock fronts are visible, one in front of the sphere and the other behind it; they resemble the shock fronts sketched in fig. 3.16, except in so far as both of them are detached from the projectile itself. The front one is detached because the projectile's nose is in this case blunt [cf. fig. 3.15]. The rear one is detached because the back half of the sphere is in effect extended by a separated boundary layer [§7.3], which shows up in the photograph as two white lines originating from points just behind the sphere's equator. A third much weaker shock front originates from these points of separation.
[Photograph by A. C. Charters, Ballistic Research Laboratory (now Army Research Laboratory), Maryland.]

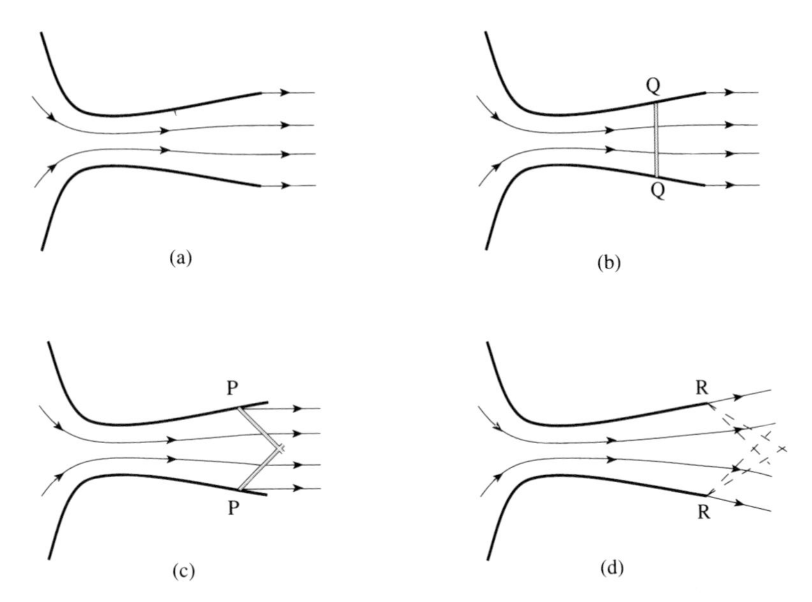

Figure 3.19 Lengthwise cross-section through a de Laval nozzle. (a) Ideal operation, where the gas accelerates continuously along the nozzle and exits at atmospheric pressure. (b) The exit pressure is here too high to match the rate of flow through the nozzle and a normal shock front has formed (QQ); to the right of this the flow is subsonic. (c) Another possible outcome when the exit pressure is too high: the flowlines separate prematurely from the walls of the nozzle at PP, with formation of a shock front which is oblique to the flow. (d) The exit pressure is here too low, and the lines of flow diverge through an expansion fan originating from the rim of the nozzle RR.

area A_E and velocity u_E relative to the nozzle, at an *exit pressure* p_E which is equal to the pressure of the surrounding atmosphere. Provided that that object is achieved, the thrust ($\rho_E A_E u_E^2$) is uniquely determined – at any rate in the case of a rocket – by the rate of combustion, since that fixes not only the rate at which mass is ejected per unit time ($\rho_E A_E u_E$) but also the rate of generation of kinetic energy ($\frac{1}{2}\rho_E A_E u_E^3$).

Figure 3.19(a) suggests what the lines of flow look like when the exit pressure lies, as the designer intended, on the curve labelled (b) in fig. 3.4. When this state is disturbed by a sudden increase in exit pressure (or by a sudden reduction in the rate of combustion which has the same effect) news of the increase travels upstream into the nozzle in the form of a shock wave. If the increase is sufficient to bring the exit pressure up to the curve labelled (b) in fig. 3.4, then the shock wave will reach the nozzle and the flow will become subsonic throughout. If the increase is not as large as that, however, then the shock wave may stabilise as shown in fig. 3.19(b), at a point where its upstream velocity relative to the gas, C, is just equal to the downstream velocity of the gas relative to the nozzle, u; this stabilised normal shock front marks an abrupt change from supersonic to subsonic flow [§3.8], and on the subsonic side the pressure increases with distance along the nozzle in the manner suggested by the broken curve in fig. 3.4. There are, however, other possible outcomes of an increase in exit pressure, and one of them is indicated in fig. 3.19(c).[5] Whatever the precise outcome a shock front is involved, and as the gas flows through this its entropy increases; available energy is lost unnecessarily as heat, and in consequence the thrust is less than it might be.

What if the atmospheric pressure is too low to match the design value, or if the rate of combustion is too high? In that case the supersonic gas stream passes through an expansion fan as it emerges, which allows the pressure to drop but at the cost of some divergence [fig. 3.19(d)]; again the thrust is less than it might be, but on account of the divergence rather than because of an increase in entropy.

Further reading

The book suggested as a reference for chapter 2, *Mechanics of Fluids*, by B. S. Massey, contains useful sections on compressible flow. A fuller text on the subject, also with an engineering slant, is *Compressible-Fluid Dynamics*, McGraw-Hill, New York, 1972, by P. A. Thompson. Thompson's book includes chapters on the theory of sound propagation, but the brief section devoted to the momentum carried by sound waves is fallacious. The same could be said of another excellent book, addressed to theoretical physicists rather than engineers: *Fluid Dynamics*, Pergamon Press, Oxford, 2nd edition 1987, by L. D. Landau and E. F. Lifshitz. These two books may be taken as representatives of the two schools of thought regarding sound momentum which are referred to in §3.7.

[5] Figure 3.19(c) has been deliberately simplified by premature termination of the oblique shock front. In fact, this feature continues within the jet for some distance to the right of the figure, undergoing periodic reflection due to a mismatch of acoustic impedance at the jet's boundaries, until the last traces of it are finally extinguished by dissipative effects. Figure 3.19(d) has been simplified in a similar fashion.

4

Potential flow

4.1 The use of potentials to describe flow

Suppose that, given a flow pattern described by a velocity field $\boldsymbol{u}\{\boldsymbol{x}, t\}$, we set out to define a scalar quantity $\phi\{\boldsymbol{x}, t\}$ in such a way that for infinitesimal excursions in space at any given instant in time we have

$$\mathrm{d}\phi = u_1 \mathrm{d}x_1 + u_2 \mathrm{d}x_2 + u_3 \mathrm{d}x_3. \tag{4.1}$$

In what circumstances will the finite difference in ϕ which may be derived by integrating this expression along a path connecting two points a finite distance apart be always the same, whichever of the infinite number of possible paths is chosen? In what circumstances, in fact, will ϕ be *single-valued*? The answer is well known to be that ϕ must be a *perfect differential*, i.e. that the equation

$$\frac{\partial^2 \phi}{\partial x_1 \partial x_2} = \frac{\partial^2 \phi}{\partial x_2 \partial x_1}, \tag{4.2}$$

together with two similar equations obtained by exchanging x_3 with x_1 or x_2, are everywhere satisfied. This condition may be seen to correspond to

$$\frac{\partial u_2}{\partial x_1} - \frac{\partial u_1}{\partial x_2} = (\boldsymbol{\nabla} \wedge \boldsymbol{u})_3 = \Omega_3 = 0,$$

and the two related conditions dictate that Ω_1 and Ω_2 must likewise be zero. The quantity ϕ is single-valued, therefore, if and only if the flow pattern is everywhere vorticity-free. In these circumstances the introduction of ϕ as a *flow potential*, related to the velocity field through the equation

$$\boldsymbol{u}\{\boldsymbol{x}, t\} = \boldsymbol{\nabla}\phi\{\boldsymbol{x}, t\}, \tag{4.3}$$

evidently makes sense. It turns out, as we shall see later in this chapter, to make sense also for certain flow patterns which are vorticity-free almost but not quite everywhere; the potential may then be multi-valued, but the values which it takes at any one point are discrete and separated from one another by a well defined 'quantum'. When vorticity permeates the whole region of interest, however, any attempt to generate a potential from (4.1) results in a function which is totally ambiguous and of no use whatever.

Where flow is vorticity-free and (4.3) applies, we may use the continuity condition in the form of (2.6) to deduce that

$$\rho\boldsymbol{\nabla}\cdot\boldsymbol{u} = \rho\nabla^2\phi = -\frac{\mathrm{D}\rho}{\mathrm{D}t}. \tag{4.4}$$

However, the excursion into the realm of compressible flow which occupied us in chapter 3 is now over, and we are once more concentrating on flow at speeds much less than the velocity of sound, such that the fluid may be treated as incompressible. If ρ is effectively constant, $\mathrm{D}\rho/\mathrm{D}t$ may be set equal to zero in (4.4), which then becomes

$$\nabla^2\phi = 0. \tag{4.5}$$

Thus the potential obeys *Laplace's equation*.

In the analysis of potential flow which will occupy us for the remainder of this chapter we appear to ignore viscosity as well as compressibility, by choosing Euler's equation of motion as our starting point. The results obtained turn out to be valid for real fluids as well as for idealised Euler fluids, however, in circumstances where they are genuinely free of vorticity. The reason, as explained briefly in §1.9, is that the viscous terms in the more realistic equation of motion due to Navier and Stokes disappear when Ω is everywhere equal to zero. The reader must take this essential point on trust until the viscous terms are formulated, in chapter 6.

4.2 Kelvin's circulation theorem

Potential flow would be of relatively little interest were it not for the theorem that fluid which is vorticity-free at one instant of time remains vorticity-free thereafter. The formal proof of this theorem, advanced as a conjecture on plausibility grounds in §1.10, requires discussion of what Kelvin christened the *circulation* round a closed loop embedded in moving fluid. This quantity, denoted by K, is a line integral round the loop of the instantaneous fluid velocity, i.e.

$$K\{t\} = \oint \boldsymbol{u}\{\boldsymbol{x}, t\}\cdot\mathrm{d}\boldsymbol{l}, \tag{4.6}$$

where $\mathrm{d}\boldsymbol{l}$ is an element of arc length round the loop. It is related to vorticity Ω through Stokes's theorem, which tells us that

$$\oint \boldsymbol{u}\cdot\mathrm{d}\boldsymbol{l} = \int \boldsymbol{\Omega}\cdot\mathrm{d}\boldsymbol{\Sigma}. \tag{4.7}$$

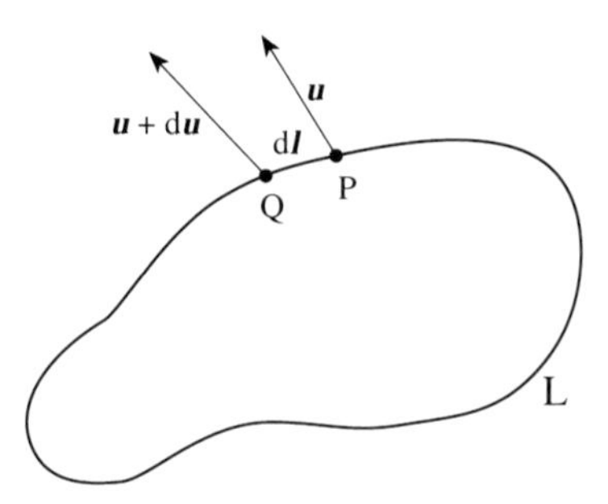

Figure 4.1 The loop L moves with the fluid in which it is embedded. The relative velocity of Q with respect to P is $\mathrm{d}\boldsymbol{u} = \mathrm{D}(\mathrm{d}\boldsymbol{l})/\mathrm{D}t$.

Evidently K must vanish for all loops if $\boldsymbol{\Omega}$ is zero everywhere. We set out to prove that K remains zero for all loops as time passes; the result that $\boldsymbol{\Omega}$ remains zero is a straightforward corollary.

Because every loop is embedded in fluid and moves with it, changing in shape as it does so no doubt, the rate of change of circulation we require is $\mathrm{D}K/\mathrm{D}t$, where $\mathrm{D}/\mathrm{D}t$ is the differential operator discussed in §2.4. Now

$$\frac{\mathrm{D}K}{\mathrm{D}t} = \oint \left\{ \frac{\mathrm{D}\boldsymbol{u}}{\mathrm{D}t} \cdot \mathrm{d}\boldsymbol{l} + \boldsymbol{u} \cdot \frac{\mathrm{D}(\mathrm{d}\boldsymbol{l})}{\mathrm{D}t} \right\}. \tag{4.8}$$

The quantity $\mathrm{D}(\mathrm{d}\boldsymbol{l})/\mathrm{D}t$, being the rate of change as the fluid moves of the vector which joins two nearby points on the loop (say P and Q in fig. 4.1), is just the instantaneous relative velocity of these two points; it can also be written as $(\partial \boldsymbol{u}/\partial l)\mathrm{d}l$, so

$$\boldsymbol{u} \cdot \frac{\mathrm{D}(\mathrm{d}\boldsymbol{l})}{\mathrm{D}t} = \boldsymbol{u} \cdot \frac{\partial \boldsymbol{u}}{\partial l}\,\mathrm{d}l = \frac{\partial(\frac{1}{2}\boldsymbol{u}\cdot\boldsymbol{u})}{\partial l}\,\mathrm{d}l = \mathrm{d}\left(\frac{1}{2}u^2\right).$$

Since one complete passage round the loop leaves u^2 unchanged, the second term in the integrand on the right-hand side of (4.8) necessarily integrates to zero. As regards the first term, we know from Euler's equation in the form of (2.3) that when viscous stresses are negligible

$$\frac{\mathrm{D}\boldsymbol{u}}{\mathrm{D}t} = -\left\{\frac{1}{\rho}\boldsymbol{\nabla}p + \boldsymbol{\nabla}(gz)\right\}. \tag{4.9}$$

If the fluid is effectively incompressible, i.e. if ρ is effectively uniform, this can be written as

$$\frac{\mathrm{D}\boldsymbol{u}}{\mathrm{D}t} = -\boldsymbol{\nabla}\left(\frac{p}{\rho} + gz\right), \tag{4.10}$$

in which case

$$\frac{\mathrm{D}\boldsymbol{u}}{\mathrm{D}t}\cdot\mathrm{d}\boldsymbol{l} = -\,\mathrm{d}\left(\frac{p}{\rho} + gz\right). \tag{4.11}$$

Evidently this also integrates to zero round a closed loop. *Kelvin's theorem*, which states that

$$\frac{\mathrm{D}K}{\mathrm{D}t} = 0 \tag{4.12}$$

for all loops provided that the flow is vorticity-free, is therefore proved for incompressible flow, together with the corollary that no vorticity arises at later times.

The applications of Kelvin's theorem which are of interest later in this book all concern flow which is slow enough to be incompressible, but in fact the theorem is equally valid for compressible flow. Since ρ is a single-valued function of pressure it must always be possible to find another single-valued function $f\{p\}$ such that $(1/\rho)\boldsymbol{\nabla}p$ can be expressed as $\boldsymbol{\nabla}f$, and that is sufficient for the proof to hold. If the fluid is an ideal gas, the appropriate function whatever the flow velocity is $\{\gamma/(\gamma-1)\}(p/\rho)$ [(3.22)].

The implication of Kelvin's theorem that fluid which is initially stationary, and therefore vorticity-free, can never be set into a state of rotation with uniform angular velocity ω and vorticity $\Omega = 2\omega$ is clearly absurd. It is common experience that the tea in a teacup can be made to rotate by stirring it with a spoon, or else by rotating the cup. Kelvin's theorem is misleading in these circumstances, as we have seen in §1.12, because it pays inadequate attention to what happens at the fluid's boundaries. If one rotates the teacup while the tea is stationary, the frictional shear stresses which arise at the interface exert a torque on the layer of tea immediately adjacent to the cup which is bound to accelerate it, though this acceleration is not described by (4.10). Then this layer starts to exert a torque on the fluid layer inside it, and so on. In this way vorticity is able, despite Kelvin's theorem, to diffuse radially inwards. To begin with it is confined to a boundary layer the thickness δ of which is about $\sqrt{2\eta t/\rho}$, where t is the time which has elapsed since the diffusion began [§1.12], but in due course all the tea is affected.

If we are to apply Kelvin's theorem, and the theory of potential flow which rests upon it, to real fluids with non-zero viscosities, we must be sure that such boundary layers as are present are too thin to matter.

4.3 Bernoulli's theorem for unsteady potential flow

We have already seen [(2.18)] that when flow is everywhere vorticity-free the fluid acceleration may be written as

$$\frac{\mathrm{D}\boldsymbol{u}}{\mathrm{D}t} = \frac{\partial \boldsymbol{u}}{\partial t} + \boldsymbol{\nabla}\left(\frac{1}{2}u^2\right).$$

Since such flow may be described, not just at one instant of time but at all later times according to Kelvin's theorem, by a potential which satisfies (4.3), the term $\partial \boldsymbol{u}/\partial t$ may clearly be expressed in gradient form, as $\boldsymbol{\nabla}(\partial\phi/\partial t)$, and Euler's equation for an incompressible fluid [(4.10)] therefore becomes

$$\boldsymbol{\nabla}\left(\frac{p}{\rho} + gz + \frac{1}{2}u^2 + \frac{\partial\phi}{\partial t}\right) = 0.$$

It may at once be integrated over space to show that

$$\frac{p}{\rho} + gz + \frac{1}{2}u^2 + \frac{\partial\phi}{\partial t} = \text{constant}, \tag{4.13}$$

or in terms of the excess pressure defined by (2.20)

$$\frac{p^*}{\rho} + \frac{1}{2}u^2 + \frac{\partial\phi}{\partial t} = \text{constant}, \tag{4.14}$$

the 'constant' being the same on every streamline. Here is a useful generalisation of Bernoulli's theorem for incompressible fluids. It is not restricted, as is (2.15), to steady flow, though its validity is of course limited by the requirement that the flow be vorticity-free.

In the fields of electrostatics and magnetostatics, where potentials which satisfy Laplace's theorem are also very frequently used, there exists a well-known *uniqueness theorem*. We turn to this for assurance that when we have found one solution of Laplace's equation which satisfies the boundary conditions of a problem we have found the only possible solution – apart from an arbitrary term ϕ' which is independent of position and which is of no physical significance. A similar uniqueness theorem holds for flow potentials also, though no proof of it will be given here. The term ϕ' may depend upon t, and we may choose it so that, in cases of unsteady flow, the 'constant' on the right-hand side of (4.14) is genuinely constant, i.e. independent of t as well as $\boldsymbol{x}$.

4.4 Sources and sinks

The potentials of electrostatics and magnetostatics are so well known to be proportional to $1/R$ round isolated point charges or point poles, situated at the origin O from which radius R is measured, that most readers will be prepared to accept without formal proof that a flow potential of the form

$$\phi = -\frac{Q}{4\pi R} \tag{4.15}$$

satisfies Laplace's equation. It describes an isotropic flow velocity of magnitude $Q/4\pi R^2$ which points away from O if Q is positive and towards O if Q is negative. It

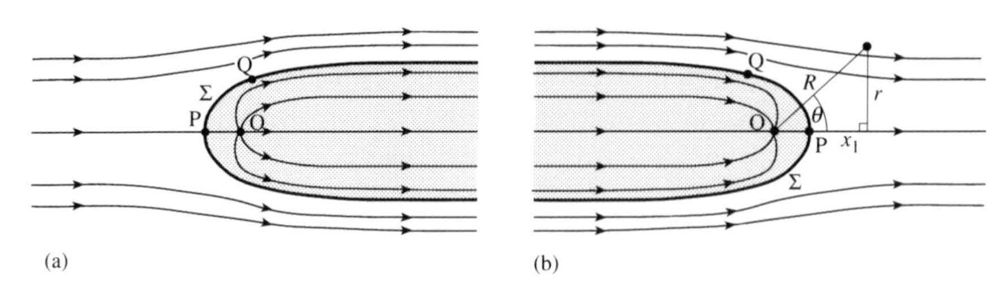

Figure 4.2 Lines of flow past (a) a point source, (b) a point sink. The surface of revolution Σ encloses all the fluid coming from, or destined for, the source or sink respectively.

implies the existence at the origin of in one case a *source* of fluid and in the other a *sink*. Evidently Q is the discharge rate, i.e. the volume of fluid which emerges from the origin per unit time.[1]

Another flow potential which is clearly a solution of Laplace's equation is

$$\phi = Ux_1. \tag{4.16}$$

This describes uniform motion in the x_1 direction with speed U. Now Laplace's equation is linear, which means that any superposition of possible solutions is itself a solution. We may therefore add the uniform flow which (4.16) describes to the radial flow described by (4.15) to arrive at an allowed pattern of steady vorticity-free flow such that in a mixed coordinate system

$$u_1 = U + \frac{Q}{4\pi R^2}\cos\theta, \quad (u_2^2 + u_3^2)^{1/2} = \frac{Q}{4\pi R^2}\sin\theta,$$

or, using spherical polar coordinates only,

$$u_R = U\cos\theta + \frac{Q}{4\pi R^2}, \quad u_\theta = -\,U\sin\theta. \tag{4.17}$$

The angle θ is defined in fig. 4.2, where corresponding flowlines are plotted for flow past (a) a source ($Q > 0$) and (b) a sink ($Q < 0$) at the origin O. In both cases there is a *stagnation point* P where u vanishes, which lies on a surface (labelled Σ in the figures) separating fluid which comes from the source in one case, or which disappears into the sink in the other, from fluid which comes from a large distance on the left and continues on to a large distance on the right. In three dimensions, of course, both flow patterns have rotational symmetry about the x_1 axis OP.

[1] A good approximation to the ideal of a point sink is provided in practice by the open end of a thin tube through which fluid is continuously sucked. Point sources are harder to realise because, for reasons that were touched on in §2.13, when fluid is pushed out of a tube it nearly always emerges as a jet rather than isotropically. However, the tendency for a jet to form may in principle be frustrated by enclosing the end of the tube with a spherical cap, made of some solid material which is uniformly permeable by the fluid.

If the excess pressure p^* is zero at infinite values of R where $u = U$, then according to Bernoulli's theorem in the form of (2.21) its value elsewhere is

$$p^* = \frac{1}{2}\rho(U^2 - u_R^2 - u_\theta^2) = -\frac{\rho UQ\cos\theta}{4\pi R^2} - \frac{\rho Q^2}{32\pi^2 R^4}.$$

Let us evaluate the total force in the x_1 direction exerted by this excess pressure on the fluid inside a spherical control surface centred on O, of arbitrary radius R. The required integral is

$$-\int_0^\pi p^* 2\pi R^2 \sin\theta \, \mathrm{d}\theta \cos\theta,$$

where $2\pi R^2 \sin\theta \, \mathrm{d}\theta$ represents the area of a ring-shaped element on the surface of the sphere and the $\cos\theta$ factor is needed because we are resolving along x_1. The total force is therefore

$$\frac{1}{2}\rho UQ \int_0^\pi \left(\cos^2\theta \sin\theta + \frac{Q\cos\theta\sin\theta}{8\pi R^2 U}\right)\mathrm{d}\theta = \frac{1}{3}\rho UQ. \tag{4.18}$$

As the rate of change of momentum in the x_1 direction of the fluid within the sphere is [(2.44)]

$$\int_0^\pi \rho u_1 u_R 2\pi R^2 \sin\theta \, \mathrm{d}\theta$$

$$= \int_0^\pi \left\{U^2\cos\theta + \frac{UQ(1+\cos^2\theta)}{4\pi R^2} + \frac{Q^2\cos\theta}{16\pi^2 R^4}\right\} 2\pi R^2 \sin\theta \, \mathrm{d}\theta$$

$$= \frac{4}{3}\rho UQ, \tag{4.19}$$

we may deduce by subtraction that the fluid inside the sphere experiences an additional force in the x_1 direction of magnitude ρUQ. Since this can only be exerted by the source (or sink, as the case may be) the latter must experience a corresponding reaction; evidently the reaction is described in vector form by the equation

$$\boldsymbol{F} = -\rho \boldsymbol{U} Q. \tag{4.20}$$

It is an instructive exercise to integrate $p^* \cos\theta$ over the surface Σ rather than over a large sphere, in order to find what net force in the x_1 direction is exerted across this surface, *by* the fluid outside *on* the fluid inside. It looks initially as though the answer could have either sign; the excess pressure is clearly positive $(+\frac{1}{2}\rho U^2)$ at the stagnation point P, but it is negative at other points on Σ such as Q, where the fluid velocity exceeds U. It turns out, however, to be zero. The integration is not performed here, because we can use (4.20) to prove in one line

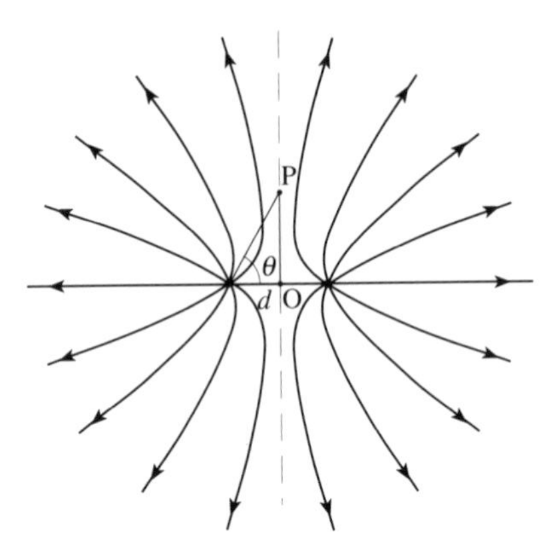

Figure 4.3 Representation in two dimensions of the three-dimensional flow pattern associated with two equal point sources.

that the net force across Σ is bound to vanish. In the circumstances depicted in fig. 4.2(a), the force exerted on the fluid inside Σ by the source alone corresponds exactly to the momentum which this fluid carries away downstream per unit time; there is no room for any additional force.

It follows that if one end of an infinitely long solid cylinder is rounded and tapered so that it corresponds exactly in shape to the surface Σ, and if this streamlined cylinder is then inserted head on into a uniform fluid stream, it should not – according to potential theory – experience any drag force. Streamlined shields known as *fairings* are sometimes used to reduce drag forces on the front of solid obstacles, and so-called *Rankine fairings* correspond in shape to Σ, to the extent that this is possible in a fairing of finite length. We shall discover in §§7.8 and 7.9, however, that in practice drag forces are influenced much more by what happens behind the obstacle than by what happens in front of it.

Equation (4.20) has been proved for the case in which $\boldsymbol{U}$ describes a flow field which is uniform. Let us now consider the case of two equal sources in otherwise stationary fluid, a distance $2d$ apart, each producing at the other a diverging flow field of magnitude $U = Q/4\pi(2d)^2$. The lines of flow to be expected in this situation are sketched in fig. 4.3. On the plane which bisects the line joining the two sources the normal component of $\boldsymbol{u}$, being the superposition of equal but opposite components due to the two sources, vanishes. The radial component, however, i.e. the component in the direction of OP, is

$$\frac{2Q \sin \theta}{4\pi(d \sec \theta)^2}.$$

Hence if the excess pressure is defined to be zero at large distances its value at P is given in terms of the angle θ defined in fig. 4.3 by

$$p^*\{\theta\} = -\frac{\rho Q^2 \sin^2\theta \cos^4\theta}{8\pi^2 d^4}.$$

The fluid to the left of the bisecting plane therefore experiences a force to the right due to excess pressure, of magnitude

$$-\int_0^\infty p^*\{\theta\} 2\pi d \tan\theta \, \mathrm{d}(d\tan\theta) = \frac{\rho Q^2}{4\pi d^2}\int_0^{\pi/2} \sin^3\theta \cos\theta \, \mathrm{d}\theta = \frac{\rho Q^2}{16\pi d^2}.$$

Since the effect of the flow is to transfer fluid to infinity from whatever reservoirs supply the sources, and since the fluid at infinity, like that in the reservoirs, has zero momentum, the total momentum of the fluid on either side of the bisecting plane does not change with time. It follows that the whole of the force calculated above must be transferred by the fluid to the source enclosed within it. Hence the source on the left is drawn to the right (and *vice versa*), and the strength of the attraction is

$$\frac{\rho Q^2}{16\pi d^2} = \rho U Q.$$

This result is completely consistent with (4.20). It illustrates the fact that that equation is valid whether $\boldsymbol{U}$ is uniform or not.

The inverse-square-law force of attraction which acts between two sources, and equally between two sinks, is relevant to the behaviour of bubbles in a liquid which is being de-gassed by subjection to ultrasonic vibration. It plays no part in the process whereby the gas emerges from solution in the first place, but it accounts for the way in which bubbles draw together and coalesce, after which they rise to the surface more quickly than they would do otherwise. The pressure in the fluid oscillates when the ultrasonic transducer is switched on, and any bubbles which are present expand and contract in sympathy. While they are expanding they act as sources, and while they are contracting they act as sinks. Providing that the distance separating two bubbles is small compared with the ultrasonic wavelength they oscillate in phase with one another, and the force between them is then attractive at all times.

4.5 Magnetostatic analogies

The parallels between the results obtained for fluid sources in the previous section and results which are very well known for point electric charges and point magnetic poles will not have escaped the reader. The parallels between vorticity-free flow and magnetostatics (as opposed to electrostatics) turn out to be particularly close, and it is worth trying to construct a systematic scheme for the translation of results from one area to the other. Given such a scheme, a physicist

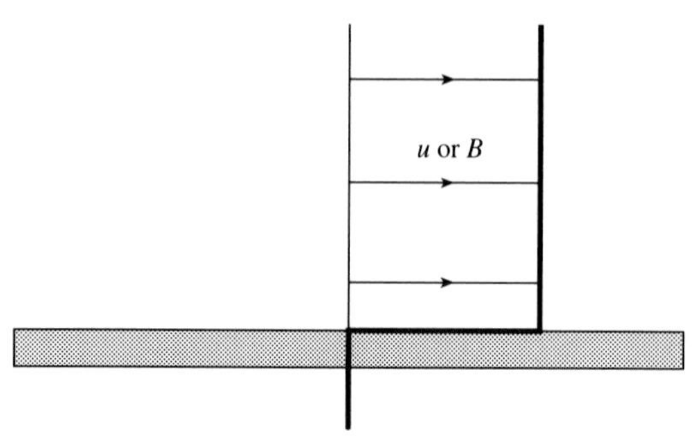

Figure 4.4 Profile of *either* fluid velocity *or* magnetic field in the neighbourhood of a confining wall. In the first case the wall is composed of any solid material and the force exerted on it is upwards; in the second case it is superconducting and the force is downwards.

who already possesses a working knowledge of magnetostatics can obtain answers with rather little effort to flow problems which appear intractable at first sight.

More than one consistent scheme can be constructed. It seems appropriate to start with the requirement that flow potential should translate directly into magnetostatic potential in free space, since it is, after all, the fact that both these potentials obey Laplace's equation which provides the parallels we are seeking to exploit. In that case, however, fluid velocity $\boldsymbol{u}$ may be translated into either $\boldsymbol{B}$ or $\boldsymbol{H}$; in free space, after all, these fields differ only by the constant factor μ_0, and if one can be described as the gradient of a potential which obeys Laplace's equation then so can the other. The advantage of translating $\boldsymbol{u}$ into $\boldsymbol{B}$ is that this allows the velocity, say $\boldsymbol{U}'$, of an impermeable solid body which is immersed in a fluid to be translated in the same way, into the field, say $\boldsymbol{B}'$, inside a (stationary) body of the same shape. The boundary condition for $\boldsymbol{B}$ at the surface of a magnetised body is, of course, that its normal component is the same as the normal component of $\boldsymbol{B}'$, and the normal components of $\boldsymbol{u}$ and $\boldsymbol{U}'$ must likewise be the same if the fluid surface and the solid surface keep in touch with one another and no cavitation occurs. A solid inclusion which is *not* moving ($\boldsymbol{U}' = 0$) may be translated into a body within which $\boldsymbol{B}' = 0$, i.e. into a superconductor.

Fluid sources and sinks translate into the effective magnetic poles which appear at the ends of long thin current-carrying solenoids. Since a point magnetic pole of strength m (positive for a north pole and negative for a south pole) generates at radius R a diverging magnetic field such that

$$B = \frac{\mu_0 m}{4\pi R^2},$$

it follows that if u and B are equivalent quantities then so are Q and $\mu_0 m$. Now consider the situation represented by fig. 4.4, which shows *either* the velocity

Table 4.1

Magnetostatic quantity		Flow equivalent	
Field in free space,	$\boldsymbol{B}$	Velocity of fluid,	$\boldsymbol{u}$
Field inside body,	$\boldsymbol{B}'$	Velocity of solid inclusion,	$\boldsymbol{U}'$
Free-space permittivity,	μ_o	Inverse density*,	$-1/\rho$
Pole strength,	m	Mass discharge rate*,	$-\rho Q$
Current,	i	Density × vortex strength*,	$-\rho \boldsymbol{K}$

* with sign reversed

profile of fluid moving at uniform speed past a stationary flat plate *or* the profile of B for a magnetic field running parallel to a superconductor. In the first case the excess pressure acting on the solid is *less* than the excess pressure where u is zero by an amount $\frac{1}{2}\rho u^2$, whereas the pressure acting on the superconductor in the second case is *more* than it would be otherwise by an amount $B^2/2\mu_o$. To make these results equivalent we need to translate μ_o by $-\rho^{-1}$. Hence m $(= \mu_o m/\mu_o)$ translates into $-\rho Q$, and (4.20) therefore translates into

$$\boldsymbol{F} = +\, m\boldsymbol{B}, \tag{4.21}$$

which is immediately recognisable in the magnetostatic context. The difference in sign between (4.20) and (4.21) is significant. It accounts for the fact that like sources attract whereas like poles repel, and for some related distinctions between potential flow and magnetostatics which will come to light later in this chapter. It has nothing whatever to do with the fact that we have chosen to define flow potentials in such a way that $\boldsymbol{u} = +\boldsymbol{\nabla}\phi$, whereas in magnetostatics it is the normal practice to write $\boldsymbol{B}$ (or $\boldsymbol{H}$) $= -\boldsymbol{\nabla}\phi$. That sign difference is purely a matter of convention, of no significance whatever.

Thus one internally consistent translation scheme is the one expressed by table 4.1. The final line of this table may be ignored for the time being, but it contains a strong hint as to why it is that comparisons with magnetostatics are in general more apposite than comparisons with electrostatics.

4.6 Some analytical solutions of Laplace's equation

We are shortly to embark on problems to do with potential flow round spheres and cylinders set transverse to the direction of flow, which are best handled using spherical polar, or two-dimensional circular polar, coordinates. Some relevant solutions of Laplace's equation are collected here for ease of reference, with a few reminders as to how these and other solutions may be obtained. Many readers will be already familiar with the material in this section, in which case they may pass it by.

(i) *Two-dimensional circular polar coordinates (r, θ)*

In this system Laplace's equation becomes

$$r\frac{\partial}{\partial r}\left\{r\frac{\partial \phi}{\partial r}\right\} + \frac{\partial^2 \phi}{\partial \theta^2} = 0.$$

Single-valued solutions in which the variables are separated can readily be found. They are:

$$\phi = \text{constant},$$

$$\phi \propto \phi_0 = \ln r, \tag{4.22}$$

$$\phi \propto \phi_n = r^n \cos(n\theta), \quad \textit{or} \quad \phi \propto \psi_n = r^n \sin(n\theta) \tag{4.23}$$

$$[n = \pm 1, \pm 2, \pm 3 \text{ etc.}].$$

Since Laplace's equation is linear in ϕ, solutions may be *superposed*, i.e. any linear combination of the separated solutions having the form

$$\phi = \text{constant} + A_0\phi_0 + \Sigma(A_n\phi_n + B_n\psi_n) \tag{4.24}$$

is also a solution. The separated solutions are orthogonal to one another and form a complete set, which means that all possible single-valued solutions can be expressed in this way.

(ii) *Complex potentials in two dimensions*

A powerful general method for handling Laplace's equation in two dimensions rests on some elementary results in the theory of functions of complex variables. Let

$$\phi + \mathrm{i}\psi = f\{x_1 + \mathrm{i}x_2\}, \tag{4.25}$$

where ϕ and ψ are both real and where $f\{Z\}$ is any sensible, differentiable, function of its argument Z. Then

$$\frac{\partial^2 \phi}{\partial x_1^2} = f'', \quad \frac{\partial^2 \phi}{\partial x_2^2} = \mathrm{i}^2 f'' = -f'',$$

so

$$\frac{\partial^2 \phi}{\partial x_1^2} + \frac{\partial^2 \phi}{\partial x_2^2} = 0,$$

i.e. ϕ is a solution of Laplace's equation in the two-dimensional space covered by the cartesian coordinates (x_1, x_2), and the same is true of ψ. Now

$$\begin{aligned}(\nabla\phi)\cdot(\nabla\psi) &= \frac{\partial\phi}{\partial x_1}\frac{\partial\psi}{\partial x_1} + \frac{\partial\phi}{\partial x_2}\frac{\partial\psi}{\partial x_2} \\ &= \mathrm{i}^{-1}(f')^2 + \mathrm{i}(f')^2 \\ &= 0.\end{aligned}$$

Hence the two-dimensional gradient vectors $\nabla\phi$ and $\nabla\psi$ are everywhere orthogonal to one another, and, since they are orthogonal to the contours of constant ϕ and ψ respectively, these contours must be orthogonal to one another also. If we choose to regard the quantity ϕ defined by (4.25) as a two-dimensional flow potential such that $\nabla\phi = \boldsymbol{u}$, then $\boldsymbol{u}$ is tangential everywhere to the contours of constant ψ, and these contours therefore serve to describe the streamlines associated with ϕ or, in cases of steady flow, the lines of flow. We are equally entitled, however, to regard ψ as the flow potential, in which case the contours of constant ϕ are streamlines.

The separated solutions expressed by (4.22) and (4.23) provide elementary examples of the application of (4.25). With

$$x_1 + \mathrm{i}x_2 = r \exp(\mathrm{i}\theta)$$

they correspond to

$$\begin{aligned}\phi_0 &= \mathrm{Re}\{\ln(x_1 + \mathrm{i}x_2)\}, \\ \phi_n &= \mathrm{Re}(x_1 + \mathrm{i}x_2)^n, \\ \psi_n &= \mathrm{Im}(x_1 + \mathrm{i}x_2)^n.\end{aligned}$$

We may note that

$$\psi_0 = \mathrm{Im}\{\ln(x_1 + \mathrm{i}x_2)\} = \theta \tag{4.26}$$

is also a valid separated solution of Laplace's equation. It was not included in subsection (i) because it is not single-valued, but we shall need to refer to it later in this chapter.

(iii) *Conformal mapping*

The complex potential approach enables Laplace's equation in two dimensions to be solved analytically in a great variety of situations where the boundary conditions are such that expansions in the form of (4.24) are not helpful. Given a set of equipotentials and streamlines described by the complex potential $f_1\{r\exp(\mathrm{i}\theta)\}$ say, one aims to develop a set more appropriate to the problem in hand by transforming to a new complex potential of the form $f_2\{f_1\{r\exp(\mathrm{i}\theta)\}\}$. The *method of conformal transformation*, as it is often called, or else of *conformal mapping*, is not used in this book, but readers should be aware of its existence.

(iv) *Three-dimensional spherical polar coordinates* (R, θ, ϕ)

Laplace's equation in spherical polars has separated solutions which form a complete set, like the two-dimensional solutions described by (4.22) and (4.23). We need not list them fully here, because we shall be concerned only with problems in which the flow is axially symmetric, i.e. in which the flow potential does not vary with the azimuthal angle ϕ.[2] In these circumstances Laplace's equation simplifies to

$$\frac{\partial}{\partial R}\left(R^2 \frac{\partial \phi}{\partial R}\right) + \frac{1}{\sin\theta}\frac{\partial}{\partial \theta}\left(\sin\theta \frac{\partial \phi}{\partial \theta}\right) = 0,$$

and its separated solutions may be written as

$$\phi \propto \phi_n^+ = R^n \, \mathrm{P}_n\{\cos\theta\}, \tag{4.27}$$

$$\phi \propto \phi_n^- = R^{-(n+1)} \, \mathrm{P}_n\{\cos\theta\}, \tag{4.28}$$

$$[n = 0, +1, +2, +3 \text{ etc.}].$$

The Legendre functions $\mathrm{P}_n\{\cos\theta\}$ may be expanded as polynomials in their argument, and we shall need the following expressions in particular:

$$\mathrm{P}_0\{\cos\theta\} = 1, \tag{4.29}$$

$$\mathrm{P}_1\{\cos\theta\} = \cos\theta, \tag{4.30}$$

$$\mathrm{P}_2\{\cos\theta\} = \frac{1}{2}(3\cos^2\theta - 1). \tag{4.31}$$

The full functions ϕ_n^+ and ϕ_n^- are properly called *zonal solid harmonics*. They are orthogonal to one another, and all other solutions of Laplace's equation in three dimensions which share their symmetry (or asymmetry) may be expressed as linear combinations of them [cf. (4.24)].

Some of the solutions described by (4.27) and (4.28) are of course trivial. Thus $\phi_0^+ = 1$ for all values of R and θ. As for

$$\phi_1^+ = R\cos\theta = x_1$$

and

$$\phi_0^- = R^{-1},$$

we have already met these, or potentials proportional to them, in §4.4. In magnetostatics the potential ϕ_0^-, when multiplied by $\mu_0 m/4\pi$, describes the field round an isolated pole of strength m, situated at the origin, and it may be described as *monopolar* for that reason. Now were that isolated pole to be moved

[2] The fact that the same symbol is used for azimuthal angle and potential is unfortunate, but since the azimuthal angle is for our purposes irrelevant the clash will not recur.

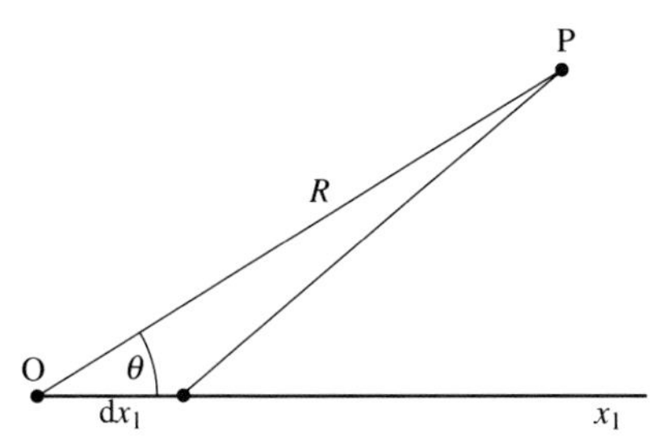

Figure 4.5 Displacement of O through Δx_1 changes the length of OP by $\Delta R = -\cos\theta\,\Delta x_1$ and changes its inclination by $\Delta\theta = +\sin\theta\,\Delta x_1/R$.

from the origin to $(\frac{1}{2}\Delta x_1, 0, 0)$, and were a pole of the same strength but opposite sign to be added at the point $(-\frac{1}{2}\Delta x_1, 0, 0)$, we would have a magnetic dipole at the origin, of strength $M = m\Delta x_1$. Evidently the potential at the point $\boldsymbol{x}$ would then be

$$\frac{\mu_o m}{4\pi}\left\{-\frac{\partial\phi_0^-\{\boldsymbol{x}\}}{\partial x_1}\left(\frac{1}{2}\Delta x_1\right)+\frac{\partial\phi_0^-\{\boldsymbol{x}\}}{\partial x_1}\left(-\frac{1}{2}\Delta x_1\right)\right\} = -\frac{\mu_o M}{4\pi}\frac{\partial\phi_0^-\{\boldsymbol{x}\}}{\partial x_1},$$

and the right-hand side of this expression, transformed into spherical polar coordinates [fig. 4.5 may be helpful in this connection], amounts to

$$-\frac{\mu_o M}{4\pi}\left(\cos\theta\,\frac{\partial\phi_0^-}{\partial R}-\frac{1}{R}\sin\theta\,\frac{\partial\phi_0^-}{\partial\theta}\right) = \frac{\mu_o m\Delta x_1}{4\pi}\frac{1}{R^2}\cos\theta = \frac{\mu_o M}{4\pi}\phi_1^-.$$

Thus in so far as ϕ_0^- may be described as *monopolar* ϕ_1^- is *dipolar*, and since ϕ_2^- may be related in a similar fashion to $\partial\phi_1^-/\partial x_1$ – there exists a useful recurrence relation

$$\phi_{n+1}^- = -\frac{1}{n+1}\frac{\partial\phi_n^-}{\partial x_1} = -\frac{1}{n+1}\left(\cos\theta\,\frac{\partial\phi_n^-}{\partial R}-\frac{1}{R}\sin\theta\,\frac{\partial\phi_n^-}{\partial\theta}\right) \qquad (4.32)$$

which may readily be shown to hold for all values of n – ϕ_2^- is *quadrupolar*. This terminology derives from magnetostatics and electrostatics, but it is used in fluid dynamics too.

4.7 Potential flow round a sphere

Armed with the results quoted in §4.6(iv), let us consider the case of a solid sphere of radius a which is moving with uniform speed U' through fluid which at large distances from the sphere is moving in the opposite direction with uniform speed $U'' = U - U'$. We shall be principally concerned with the frames of reference in which either $U' = 0$ and $U'' = U$ or *vice versa*, but when we come to consider the possibility that the sphere is accelerating it will prove advantageous to be able to refer to a more general solution. So as not to be distracted prematurely by worries

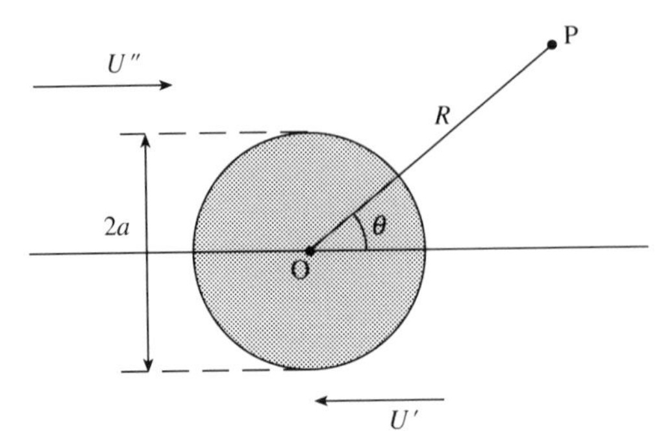

Figure 4.6 Coordinate system for discusssion of flow past a sphere. The sphere moves to the left with velocity U' and the fluid at large distances from the sphere moves to the right with uniform velocity U''. Their relative velocity is $U' + U'' = U$.

about boundary layers, we shall assume the fluid to be an ideal Euler fluid with zero compressibility and zero viscosity. In that case, we can be sure from Kelvin's theorem that because it is vorticity-free well ahead of the sphere it remains vorticity-free throughout. Potential theory undoubtedly applies.

To find a potential which describes flow round a moving sphere, we first expand in zonal solid harmonic functions, thus:

$$\phi = \Sigma(A_n^+ \phi_n^+ + A_n^- \phi_n^-).$$

To satisfy the boundary condition that at large values of R the fluid's velocity in what we may take to be the $\theta = 0$ direction [fig. 4.6] is uniformly equal to U'' we must clearly choose $A_1^+ = U''$, and the same boundary condition dictates that A_n^+ must be zero for all n greater than 1. At the surface of the sphere, where $R = a$, the radial component of the fluid velocity must equal the velocity with which the surface of the sphere is moving in a radial direction, to ensure that fluid and solid are always in contact, so

$$\left(\frac{\partial \phi}{\partial R}\right)_{R=a} = - U' \cos \theta. \tag{4.33}$$

We have no hope of satisfying (4.33) for all values of θ if we include a monopolar or quadrupolar term in the potential, and higher-order multipole terms can be ruled out for the same reason. The required potential must therefore be of the form

$$\phi = A_0^+ + U''R \cos \theta + A_2^- R^{-2} \cos \theta,$$

in which (4.33) serves to fix A_2^-. To the extent that A_0^+ can be any function of time the answer is not unique. However, we may set $A_0^+ = 0$ without loss of generality, in which case

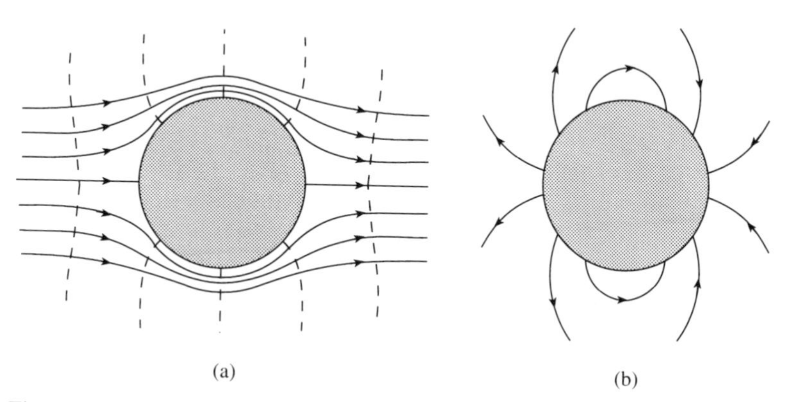

Figure 4.7 (a) Lines of flow and equipotentials round a stationary sphere ($U' = 0$). (b) The same flow pattern in a frame of reference such that $U'' = 0$ and the sphere is moving.

$$\phi = U''R\cos\theta + \frac{1}{2}a^3(U' + U'')\frac{\cos\theta}{R^2}. \tag{4.34}$$

The corresponding velocity components of the fluid are

$$\begin{aligned} u_R &= \frac{\partial\phi}{\partial R} = U''\cos\theta - U\left(\frac{a}{R}\right)^3\cos\theta \\ u_\theta &= \frac{1}{R}\frac{\partial\phi}{\partial\theta} = -\,U''\sin\theta - \frac{1}{2}U\left(\frac{a}{R}\right)^3\sin\theta, \end{aligned} \tag{4.35}$$

where $U = U' + U''$ is the velocity of the sphere relative to the fluid stream, as specified above.

In the frame of reference in which the sphere is stationary ($U' = 0$, $U'' = U$) the flow pattern is a steady one, with lines of flow and equipotentials as shown in fig. 4.7(a) – a figure which could be used equally well to depict lines of force and equipotentials round a superconducting sphere subjected to a uniform external magnetic field. In this frame, we may deduce how the excess pressure varies using Bernoulli's theorem in the form of (2.21). With p^* defined to be zero at large distances we have

$$p^* = \frac{1}{2}\rho(U^2 - u_R^2 - u_\theta^2),$$

so that in contact with the sphere

$$p^*_{R=a} = \frac{1}{2}\rho U^2\left(1 - \frac{9}{4}\sin^2\theta\right). \tag{4.36}$$

The excess pressure is positive at the sphere's two poles, where $\theta = 0$ or π, but it is negative near the equator, where $\theta = \pi/2$.

In the frame of reference in which the sphere is moving through fluid which is stationary at large distances ($U'' = 0$, $U' = U$) the pattern of dipolar backflow described by (4.35) is not a steady one, so the lines which represent this pattern in fig. 4.7(b) are instantaneous streamlines rather than lines of flow [§1.6]. The magnetostatic analogue for which this figure could serve as an illustration concerns the field distribution round a spherical magnet in the absence of any external field. Since the flow is unsteady, we now have to use the generalised version of Bernoulli's theorem, (4.14), to find the pressure distribution. The calculation is slightly laborious, because the fact that O is moving in this frame of reference means that the coordinates (R, θ) of a point which is fixed in space are changing continuously with time; the $\partial\phi/\partial t$ term which distinguishes (4.14) from (2.21) is non-zero on this account and needs to be taken into consideration. The calculation will not be performed here, because it leads to the result for $p^*_{R=a}$ which we have derived already using a different frame, (4.36). It is bound to do so, because pressure is always invariant under a Galilean transformation from one inertial frame to another.

Because the excess pressure at $R = a$ is completely symmetrical about the equatorial plane, a sphere which is in uniform motion relative to fluid experiences no force, apart from its own weight and the hydrostatic upthrust which we have suppressed. This is an example of *d'Alembert's paradox* [§7.8], though there is nothing really paradoxical about it. If a drag force did exist, then the fluid would experience a corresponding reaction, which would feed momentum continuously into the fluid and which, in the frame of reference to which fig. 4.7(b) refers, would feed in kinetic energy at the same time. The momentum associated with the flow pattern in fig. 4.7(b) is ambiguous, as we shall see in the following section, but the kinetic energy is not, and since it certainly remains constant as time passes a non-zero drag force is out of the question.

Throughout the above discussion we have paid no attention to the no-slip boundary condition [§1.11]. This would require u_θ to vanish in the frame for which $U' = 0$, and it is clearly not satisfied by (4.35). In the ideal world where fluids have no viscosity it may make no difference whether there is slip or not, because solids in that world may well be perfectly smooth, but what about the real world, in which fluids are always viscous and solids are always rough? As we have already noted, the existence of viscosity does not prevent real fluids from obeying Euler's equation, provided that they are vorticity-free, and equations (4.35) therefore describe patterns of flow which are, in principle, possible for real fluids. However, the existence of viscosity means, as we have seen in §1.9, that where flow is convergent or divergent the pressure is not isotropic, and it also means that where fluid undergoes shear there is continuous dissipation of energy as heat. It turns out [(6.11)] that if the fluid in fig. 4.7(a) and fig. 4.7(b) is a real one then the transverse pressure which acts upon the sphere is greater than the mean pressure which (4.36) describes by a term

$$-2\eta\left(\frac{\partial u_R}{\partial R}\right)_{R=a} = -\frac{6\eta U\cos\theta}{a},$$

which implies a drag force on the sphere of magnitude $8\pi\eta aU$. However, the solution of Euler's equation recorded in (4.35) is valid for a real fluid when, and only when, energy is being fed continuously into the fluid by the sphere to make good the viscous dissipation. It presupposes, in fact, that the sphere has a rough surface which somehow contrives to move at the same speed u_θ as the fluid in contact with it. Supposing that to be possible, the sphere's surface would exert a shear stress on the fluid in the direction of u_θ of magnitude $\eta R\{\partial(u_\theta/R)/\partial R\}_{R=a}$, and the corresponding reaction on the sphere implies an additional drag force of $-8\pi\eta aU$. Just as in the case of ideal fluids, therefore, the net drag force associated with steady potential flow past a sphere is zero – a result which here again is obvious from the constancy of the fluid's kinetic energy in fig. 4.7(b).

Real spheres are not equipped with moving surfaces, however. If the flow pattern of fig. 4.7(a) could somehow be established instantaneously in a real fluid moving past a real sphere, it would immediately start to deteriorate by the diffusion of vorticity into the fluid from the sphere's surface and subsequent convection of this vorticity downstream in a wake of some sort. One might hope that at large values of the Reynolds Number the boundary layer attached to the sphere, and the continuation of this boundary layer in the downstream wake, would be too thin to matter [§1.12], but in practice the boundary layer *separates* from the sphere [§7.3]; the flow pattern then loses its symmetry and the momentum of the fluid in the downstream direction decreases continuously with time. The results we have derived with such labour have very little relevance, therefore, to real fluids and real spheres so long as the relative velocity U is constant.

4.8 The virtual mass of an accelerating solid body

Suppose that the sphere in fig. 4.7(a) is instantaneously accelerating from rest, while the counter-velocity U'' of the fluid stream remains constant. In that case the flow pattern is not perfectly steady and it follows from (4.34) (the calculation is elementary provided $U' = 0$) that

$$\frac{\partial\phi}{\partial t} = \frac{a^3\cos\theta}{2R^2}\frac{\mathrm{d}U'}{\mathrm{d}t}.$$

We may immediately deduce from Bernoulli's theorem in its generalised form that

$$p^*_{R=a} = \frac{1}{2}\rho U''^2\left(1 - \frac{9}{4}\sin^2\theta\right) - \frac{1}{2}\rho a\cos\theta\,\frac{dU'}{dt}. \qquad (4.37)$$

Hence while it is accelerating the sphere experiences a pressure which is greater on its upstream side than on its downstream side. It therefore experiences a drag force of magnitude

$$F = -\int_0^\pi 2\pi a^2 p^* \cos\theta\sin\theta\,d\theta$$

$$= \pi\rho a^3\frac{dU'}{dt}\int_0^\pi \cos^2\theta\sin\theta\,d\theta$$

$$= \frac{1}{2}m\frac{dU'}{dt},$$

where

$$m = \frac{4}{3}\pi\rho a^3, \qquad (4.38)$$

and the external force required to cause the acceleration is larger on that account. The sphere behaves, in fact, as though in addition to its own inertial mass it carried a *virtual mass* equal to half the mass m of fluid displaced. The virtual mass is evidently independent of U''.

The work done on the fluid while the sphere is accelerating through it from rest to some final velocity U is

$$\int FU'dt = \frac{1}{4}mU^2.$$

For the case in which the fluid is otherwise stationary (i.e. $U'' = 0$), it is easy to verify that this is the amount of energy stored in the dipolar backflow which equations (4.35) and the streamlines in fig. 4.7(b) describe. All that is necessary is to integrate the energy density

$$\frac{1}{2}\rho u^2 = \frac{1}{2}\rho(u_R^2 + u_\theta^2) = \frac{1}{2}\rho U^2\left(\frac{a}{R}\right)^6\left(\cos^2\theta + \frac{1}{4}\sin^2\theta\right)$$

over the volume outside the sphere.

During the acceleration the force $\boldsymbol{F}$ must transfer momentum to the fluid as well as energy, of course, and the total momentum transfer is clearly

$$\int \boldsymbol{F}\,dt = \frac{1}{2}m\boldsymbol{U}.$$

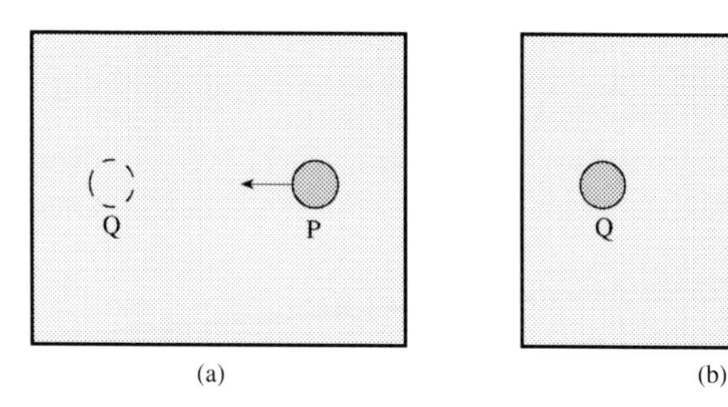

Figure 4.8 The sphere situated at P in (a) has moved to Q in (b). An equal volume of fluid has moved from Q to P.

We cannot verify this in a similar fashion, by integrating $\rho(u_R \cos\theta - u_\theta \sin\theta)$, because the integral over θ generates a factor of zero, while the integral over R diverges logarithmically at infinity. The divergence is a sign that the momentum stored in the fluid depends on the precise nature of the boundary conditions at infinity, in a way that the kinetic energy of the fluid does not. If the fluid is enclosed by a very large box with rigid walls, any displacement of the sphere in one direction must be accompanied by the displacement of an equal volume of fluid in the other direction [fig. 4.8]; the momentum of the fluid in these circumstances must be $-m\boldsymbol{U}$ rather than $+\frac{1}{2}m\boldsymbol{U}$. That only means, however, that in addition to the impulse of $+\frac{1}{2}m\boldsymbol{U}$ which the sphere administers to the fluid as it accelerates an impulse of $-\frac{3}{2}m\boldsymbol{U}$ must be administered by the walls of the box. The negative impulse delivered by the walls is absorbed by virtually the whole body of fluid; it results in an infinitesimal backflow, superposed on the finite but localised backflow which the equations (4.35) describe. Because the mass of fluid involved is so large, the kinetic energy associated with the backflow (being $P^2/2M$ for a mass M carrying momentum P) is negligible.

It proves fruitful to refer at this stage to the magnetic analogue for the flow situation depicted by fig. 4.7(b), i.e. to the case of a uniformly magnetised sphere in the absence of any external field. If the strength of the field inside the sphere is B' [the analogue of U' – see §4.5], then

$$B' = \mu_o(H' + I) = \mu_o(1 - n)I, \qquad (4.39)$$

where I is the magnetisation of the sphere, i.e. its magnetic moment per unit volume, and $H' = -nI$ is what is known as the *depolarising field*; n is the so-called *depolarisation factor*, which is further discussed below. Now the work which would have to be done to produce on a spherical surface of radius *a in free space* the distribution of magnetic poles which would produce outside that surface the same field distribution as our uniformly magnetised sphere is given, in view of (4.39), by

$$\frac{4}{3}\pi a^3\left(-\frac{1}{2}\mu_o H'I\right) = \frac{4}{3}\pi a^3\left(\frac{1}{2}\mu_o n I^2\right) = \frac{4}{3}\pi a^3 \frac{n}{(1-n)^2}\frac{B'^2}{2\mu_o}. \tag{4.40}$$

This energy could be regarded as stored in the magnetic field with density $B^2/2\mu_o$, and since B would be given inside the spherical surface by $\mu_o H'$ the energy stored inside that surface would be

$$\frac{4}{3}\pi a^3\left(\frac{1}{2}\mu_o H'^2\right) = \frac{4}{3}\pi a^3 \frac{n^2}{(1-n)^2}\frac{B'^2}{2\mu_o}. \tag{4.41}$$

We may deduce by subtraction of (4.41) from (4.40) that if we integrate $B^2/2\mu_o$ over the volume outside the sphere the result will be

$$\frac{4}{3}\pi a^3 \frac{n}{(1-n)}\frac{B'^2}{2\mu_o}.$$

That tells us that if we similarly integrate $\rho u^2/2$ in the equivalent flow situation to find the kinetic energy stored in the fluid the answer will be

$$\frac{4}{3}\pi a^3 \frac{n}{(1-n)}\frac{\rho U'^2}{2},$$

and hence that the virtual mass is a fraction f of the mass m of fluid displaced, where

$$f = \frac{n}{(1-n)}. \tag{4.42}$$

The point of the above argument, which is bound to mystify readers who have not previously met the concepts of depolarising fields and depolarisation factors, is that it can be applied without modification to bodies which are ellipsoids rather than spheres. Ellipsoids have three depolarisation factors, associated with polarisation along their three principal axes, and formulae which allow these to be determined once the axial ratios are known have been readily available since the time of Maxwell. The three always add up to unity, and symmetry therefore ensures that for a sphere each of them is equal to $\frac{1}{3}$. In that case, (4.42) tells us that f is $\frac{1}{2}$, the result we have previously established in a more direct way. For a prolate spheroid which is so long that it looks like a cylinder, i.e. with axial ratio $c/a \gg 1$, n is almost zero when the polarisation lies along the c axis (i.e. the axis of rotational symmetry), and it follows that when the polarisation is transverse to the c axis $n \approx \frac{1}{2}$. Equation (4.42) therefore tells us that for a cylinder which is accelerating through fluid in a transverse direction $f = 1$. For a thin oblate spheroid with $c/a \ll 1$ the depolarisation factor along the c axis is approximately

$$1 - \frac{\pi c}{2a}, \tag{4.43}$$

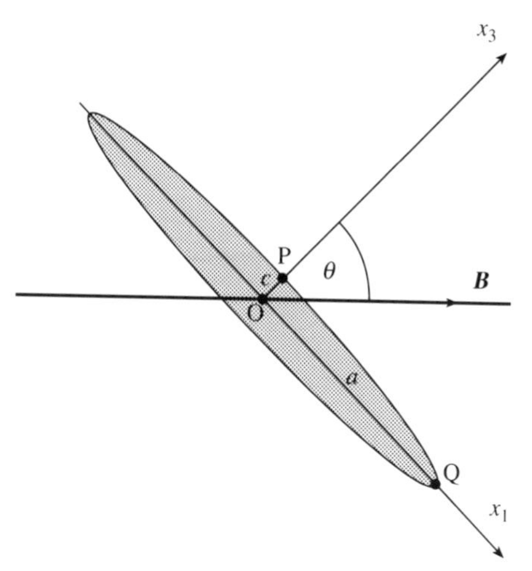

Figure 4.9 Cross-section through a thin oblate superconducting spheroid (effectively a disc) in a uniform magnetic field $\boldsymbol{B}$.

which corresponds to $f \approx 2a/\pi c$, i.e. to a virtual mass of $8\rho a^3/3$. Since the thickness of the oblate spheroid does not feature in that answer it can presumably be applied to discs of radius a, whether they are strictly elliptical in cross-section or not. The implication that the force required to accelerate a disc through fluid is significantly larger when the direction of acceleration is normal to the plane of the disc than when it lies within this plane is unsurprising.

If the fluid is a real one with viscosity, and if the acceleration of the sphere or ellipsoid is constant, the range of practical validity of the above results is rather small. They apply only to the early stages of acceleration, before boundary layers have had time to grow and separate. If the velocity U' is *oscillatory*, however, the boundary layer thickness does not necessarily grow with time; when the amplitude of the oscillation is reasonably small this thickness should be of order $\sqrt{\eta/\omega\rho}$ [§1.12]. In such circumstances potential theory should be applicable provided that $\sqrt{\eta/\omega\rho}$ is small compared with a, i.e. provided that the frequency $\omega/2\pi$ is large compared with say $0{\cdot}1(\eta/\rho a^2)$. For a sphere of radius 1 cm in water, $0{\cdot}1(\eta/\rho a^2)$ is only 1 mHz.

4.9 The Rayleigh disc

A flow problem which can be solved in a few lines by reference to the magnetostatic analogue concerns the torque experienced by an oblate spheroid set at an angle to a fluid stream. The analogue is illustrated by fig. 4.9. Here we have a superconducting spheroid whose thickness along the axis of revolution OP is $2c$ and whose radius in the direction OQ is a; OP lies at an angle θ to a uniform external field $\boldsymbol{B}$. What is the torque about O?

That question may be answered by decomposing both $\boldsymbol{B}$ and the magnetic moment of the spheroid $\boldsymbol{M}$ into components along the principal axes OQ and OP, i.e. into (B_1, B_3) and (M_1, M_3) respectively. The (clockwise) torque about the third principal axis, perpendicular to the plane of the diagram, is

$$G_2 = B_1 M_3 - B_3 M_1. \tag{4.44}$$

Now inside a magnetisable ellipsoid which is subjected to a uniform external field B along one of its principal axes the field is

$$B' = B + \mu_o I(1 - n)$$

[(4.39)], which means, since $B' = 0$ inside a superconductor, that

$$\mu_o I = -\frac{B}{(1 - n)}.$$

Hence, since $M = \frac{4}{3}\pi a^2 c I$, (4.44) may be recast as

$$G_2 = \frac{4\pi a^2 c}{3\mu_o} B_1 B_3 \left\{ \frac{1}{(1 - n_1)} - \frac{1}{(1 - n_3)} \right\}. \tag{4.45}$$

The appropriate formula for n_3 has been quoted in the last section [(4.43)], and it implies

$$n_1 = n_2 \approx \frac{\pi c}{4a}.$$

On combining these results with (4.45) we find, with $B_3 = B\cos\theta$ and $B_1 = B\sin\theta$,

$$G_2 \approx -\frac{4a^3}{3\mu_o} B^2 \sin(2\theta). \tag{4.46}$$

This torque is zero for $\theta = 0$ and $\theta = \pi/2$. Since it is negative between these settings, a thin superconducting spheroid is in stable equilibrium at $\theta = \pi/2$ and in unstable equilibrium at $\theta = 0$.

Translated into terms that apply to the analogous flow problem, (4.46) tells us that a stationary thin oblate spheroid – in effect a disc of radius a – in a fluid stream which has uniform velocity U at large distances experiences a torque

$$G_2 \approx \frac{4}{3}\rho a^3 U^2 \sin(2\theta), \tag{4.47}$$

at any rate when boundary layer effects may be ignored. The positive sign means that the disc is in stable equilibrium at $\theta = 0$ rather than at $\theta = \pi/2$; it tends to set so that its plane is normal to the flow.

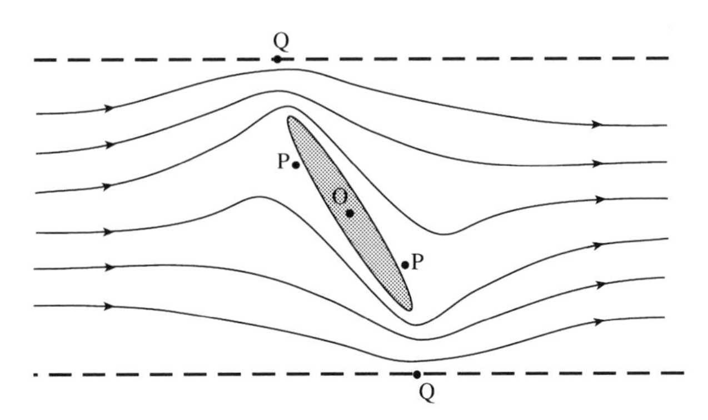

Figure 4.10 Two-dimensional potential flow past an inclined strip. The pressure acting on the strip is relatively high near the stagnation points P. If the fluid is constrained by walls (broken lines) the pressure acting on these is relatively low near Q.

For reasons given in the previous section, boundary layer effects tend to be negligible when U is oscillatory. In such circumstances (4.47) is sufficiently trustworthy to provide a basis for the absolute determination of mean square velocity $\overline{U^2}$ from measurements of mean torque as a function of θ. A device known as the *Rayleigh disc*, which is used to determine the absolute intensity of sound waves in fluids, operates on this principle.

Figure 4.10 is a sketch of the streamline pattern associated with two-dimensional oscillatory potential flow past an inclined rectangular strip which is long in the direction perpendicular to the diagram, which may be calculated using the method of conformal mapping. Inevitably, it is symmetrical to the extent that the diagram would look the same if rotated through an angle π in its own plane about an axis through O. There are stagnation points at P, and one may view the clockwise torque to which the plate is subject as the consequence of relatively high pressure at these points. The fluid must be subject to an equal and opposite torque, of course, which tends to increase its anticlockwise angular momentum about O. But if the fluid is constrained above and below the plate by plane solid surfaces – suggested by broken lines in the figure – it is impossible for the angular momentum of the fluid to change as it flows past the plate, because each streamline is obliged to return to exactly the same level on the extreme right of the diagram as it starts from on the extreme left. In these circumstances, therefore, the fluid must be subject to an additional torque about O exerted by the constraining surfaces, which exactly cancels the torque exerted by the inclined plate. The origin of this additional torque, and the fact that it is clockwise, may be understood in terms of the relatively low pressure at Q, where the fluid squeezes past the sides of the plate with relatively high velocity. If the constraining surfaces are absent, so that the fluid effectively extends to infinity above and below the

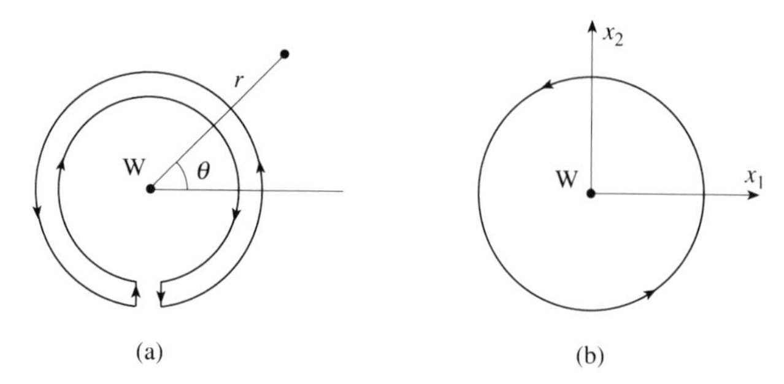

Figure 4.11 Two possible loops in the neighbourhood of a current-carrying wire which passes through W in a direction normal to the plane of the diagram.

plate, the streamlines *are* displaced in passing the plate, though in so far as the displacement affects an infinite body of fluid it is infinitesimal.

4.10 Multi-valued potentials in electromagnetism

Around a long straight wire which carries a steady current i along the x_3 axis there is, of course, a circulating magnetic field with components given in terms of cylindrical coordinates (r, θ, x_3) by

$$B_\theta = \frac{\mu_0 i}{2\pi r}, \quad B_r = B_3 = 0. \tag{4.48}$$

This field is such that the integral of $\boldsymbol{B}\cdot \mathrm{d}\boldsymbol{l}$ round the closed loop of two linked circles which is indicated in fig. 4.11(a) is zero, whatever the radius of these circles. It therefore follows from Stokes's theorem [(4.7)] that everywhere outside the wire $\boldsymbol{\nabla} \wedge \boldsymbol{B}$ is zero – a result which may also be deduced from the fact that when $B_r = 0$

$$(\boldsymbol{\nabla} \wedge \boldsymbol{B})_3 = \frac{B_\theta}{r} + \frac{\mathrm{d}B_\theta}{\mathrm{d}r} \tag{4.49}$$

[(2.14)]. Hence the conditions required for $\boldsymbol{B}$ to be describable by a potential are satisfied, and granted the sign convention that $\boldsymbol{B} = -\boldsymbol{\nabla}\phi$ an appropriate potential is clearly

$$\phi = -\frac{\mu_0 i\theta}{2\pi} = -\frac{\mu_0 i\psi_0}{2\pi} \tag{4.50}$$

[(4.26)]. However, the angle θ is multi-valued and so is ϕ. Round any closed loop like the single circle in fig. 4.11(b), which encircles the current in a way that the loop sketched in fig. 4.11(a) does not, we have

$$\Delta\phi = -\oint \boldsymbol{B}\cdot \mathrm{d}\boldsymbol{l} = -\mu_0 i, \tag{4.51}$$

while for a loop which encircles the current twice we have $\Delta\phi = -2\mu_0 i$, and so on. Ampère's law asserts that (4.51) is true whether the current-carrying wire is straight or not. To find the magnetic field associated with an arbitrary distribution of current we may apply the equivalent law of Biot and Savart.

The above results are cornerstones of electromagnetism with which every student of physics is familar, so the analogous situation in which a multi-valued flow potential exists should need little explanation.

4.11 Vortex lines

The direct analogue of a straight current-carrying wire is a long straight cylindrical spindle, say of radius a, which is set rotating about its own axis with steady angular velocity ω_0 in a large body of initially stationary fluid. The no-slip boundary condition at solid–fluid interfaces ensures that circulation is imparted to the fluid immediately adjacent to the cylinder; this fluid picks up a velocity $u_\theta = \omega_0 a$. Vorticity then diffuses out into the rest of the fluid. Initially only a boundary layer of thickness $\sqrt{2\eta t/\rho}$ is affected, and the strength of the vorticity within this layer is of order $\omega_0 a\sqrt{\rho/2\eta t}$. As time passes, however, the volume of circulating fluid becomes ever greater and the strength of the vorticity within it becomes correspondingly smaller. The end result is a situation in which, except at such large values of r that the fluid is still not affected, the velocity of circulation is

$$u_\theta = \omega r = \omega_0 \frac{a^2}{r}. \tag{4.52}$$

In this situation the spindle exerts a torque on the fluid, which turns out [(6.54)] to be given per unit length along the x_3 axis by

$$g_3 = -2\pi\eta r^3 \frac{\mathrm{d}\omega}{\mathrm{d}r} = 4\pi\eta a^2\omega_0. \tag{4.53}$$

All this torque is transferred, however, to the fluid at very large distances. Closer in, each cylindrical element coaxial with the spindle experiences a torque on its inner surface which tends to accelerate it, described by (4.53), and an equal and opposite decelerating torque on its outer surface [(6.54)]. The state of motion described by (4.52) is therefore a steady one, not susceptible to further change so long as the spindle is kept rotating. It is, however, a state which involves continuous dissipation of energy as heat throughout the circulating fluid. This energy is provided, of course, by whatever agency is turning the spindle against the reactive torque of $-4\pi\eta a^2\omega_0$.

The steady state of circulation around a rotating spindle therefore involves a velocity which varies with r in exactly the same way as the magnetic field round a long straight current-carrying wire – compare (4.52) with (4.48). The velocity field is such that Ω is zero everywhere, i.e. it is vorticity-free, and indeed the fact that viscous stresses no longer accelerate the fluid once this state has been reached is another example of the general principle, still to be proved, that the viscous terms in the equation of motion for real fluids disappear when the vorticity vanishes everywhere. However, although the circulation K is therefore zero round any loop embedded in the fluid which does not encircle the spindle, it is *not* zero for all loops. Round any loop which encircles the spindle once and once only we have

$$\Delta\phi = +\oint \boldsymbol{u}\cdot\mathrm{d}\boldsymbol{l} = 2\pi\omega_0 a^2.$$

A rotating spindle round which fluid is circulating in the manner described above is said to carry a *bound vortex line*. Where the same pattern of circulation exists without a spindle the vortex line is said to be *free*.[3] In either case the strength of the vortex line is defined as the circulation

$$K = \oint \boldsymbol{u}\cdot\mathrm{d}\boldsymbol{l} \tag{4.54}$$

round any closed loop which encircles it once, and a comparison of (4.54) with (4.51) will show why, with $\boldsymbol{u}$ equivalent to $\boldsymbol{B}$ and ρ equivalent to $-\mu_0^{-1}$, $-\rho\boldsymbol{K}$ was given in table 4.1 [§4.5] as the translation for current $\boldsymbol{i}$. Current is a vector in electromagnetism because it has direction as well as magnitude, and the strength of a vortex line can be treated as a vector in the same way. Naturally, an analogue of the Biot–Savart law exists, which can be used to determine the velocity field associated with vortex lines which are not straight.

The strength of an ideal vortex line, whether it is bound or free, is a *conserved* quantity in two respects. Firstly, its magnitude K cannot change with time, even when the line is being stretched or deformed in some way. This is a straightforward consequence of Kelvin's theorem that $\mathrm{D}K/\mathrm{D}t$ is zero round any closed loop which is embedded in vorticity-free fluid. The validity of Kelvin's theorem depends only on an assumption that the fluid is uncontaminated by vorticity in the immediate neighbourhood of the loop itself; granted that assumption, it applies whether K is zero or not zero. Secondly, K cannot vary along the length of the line unless the line branches. This can be proved by reference to fig. 4.12. The continuous loop L goes round the vortex line V at two points A and B where the line strength is K_A and K_B respectively. Since it goes round in a clockwise sense at one point and in an anticlockwise sense at the other, the total circulation round this loop is $K_L = \pm(K_A - K_B)$; contributions to K_L from the two almost coincident lengths of the loop which run between A and B cancel one another. However, the vortex line does *not* pass through the loop. This means that K_L is zero, and hence that $K_A = K_B$. A similar argument applies to currents, of course, which are continuous in the same way.

[3] Vortex lines are referred to in some texts as *line vortices*. This point of nomenclature is discussed in §7.1.

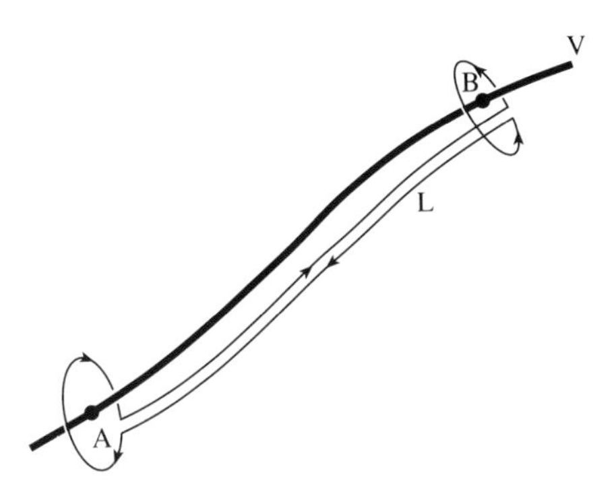

Figure 4.12 Perspective sketch of a loop L which almost encircles a vortex line V at two different points along its length.

4.12 The Magnus effect

Before looking more closely at vortex lines which are free, let us consider the case of a vortex line of uniform strength K which is bound to a long straight spindle surrounded by fluid which is moving relative to it with a velocity $\boldsymbol{u}$ which, at large distances, is uniform and equal to $\boldsymbol{U}$. The magnetic analogue is a long straight current-carrying wire in a uniform external magnetic field $\boldsymbol{B}$, and it is well known that such a wire experiences a force per unit length $\boldsymbol{f} = \boldsymbol{i} \wedge \boldsymbol{B}$. According to the dictionary in table 4.1 [§4.5], the vortex-carrying spindle should experience a force per unit length given by

$$\boldsymbol{f} = -\rho \boldsymbol{K} \wedge \boldsymbol{U} = \rho \boldsymbol{U} \wedge \boldsymbol{K}. \tag{4.55}$$

Can we verify that prediction, by evaluating the excess pressure which acts upon the spindle?

Let us take the spindle to be a cylinder of radius a whose axis lies in the x_3 direction and, for simplicity, take the fluid velocity at large distances to be *transverse*, e.g. to be in the x_1 direction and uniformly equal to U_1. In this geometry, and in the case when $K = 0$ and there is no circulation around the cylinder, the potential which describes vorticity-free flow past it may be found in the same way that we found a solution of Laplace's equation to describe flow past a sphere in §4.7. Apart from an irrelevant constant, it must be expressible [(4.24)] in the coordinate system defined by fig. 4.13 as

$$\phi = U_1 r \cos\theta + A_{-1} r^{-1} \cos\theta,$$

and with A_{-1} fixed by the boundary condition that $\partial\phi/\partial r = 0$ at $r = a$ it becomes

$$\phi = U_1 \cos\theta \left(r + \frac{a^2}{r} \right). \tag{4.56}$$

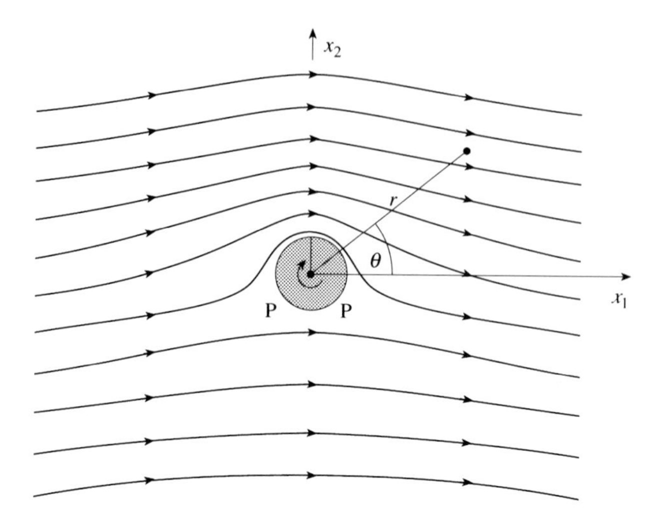

Figure 4.13 Potential flow past a transverse cylinder of radius a when the circulation round the cylinder is $K_3 = -2\pi Ua$. These lines of flow are contours of constant ψ; and the value of ψ increases from one line to the next by 1.

If we now add to that the potential, also a solution of Laplace's equation, which describes the circulation associated with a bound vortex line of strength K_3, we arrive at

$$\phi = U_1 \cos\theta\left(r + \frac{a^2}{r}\right) + \frac{K_3\theta}{2\pi},$$

and hence at

$$u_r = U_1 \cos\theta\left(1 - \frac{a^2}{r^2}\right), \quad u_\theta = -U_1 \sin\theta\left(1 + \frac{a^2}{r^2}\right) + \frac{K_3}{2\pi r}. \tag{4.57}$$

The steady flow pattern to which equations (4.57) correspond depends upon the magnitude of K_3; the lines of flow which are included in fig. 4.13, which are contours of constant ψ computed from the equation

$$\psi = U_1 \sin\theta\left(r - \frac{a^2}{r}\right) + \frac{K_3}{2\pi}\ln\left(\frac{r}{a}\right)$$

[§4.6(ii)], correspond to a value of -2π for the dimensionless ratio K_3/Ua; the negative sign implies circulation in a clockwise sense in fig. 4.13.

With the aid of Bernoulli's theorem, we may deduce from (4.57) that in contact with the cylinder the excess pressure is

$$p^*_{r=a} = \frac{1}{2}\rho(U_1^2 - u_\theta^2) = \frac{1}{2}\rho\left\{U_1^2 - \left(-2U_1\sin\theta + \frac{K_3}{2\pi a}\right)^2\right\}$$

$$= \frac{1}{2}\rho\left(\frac{K_3}{2\pi a}\right)^2 + \frac{1}{2}\rho U_1^2(1 - 4\sin^2\theta) + \frac{\rho U_1 K_3}{\pi a}\sin\theta.$$

The symmetry of the first two terms on the right-hand side is such that they contribute nothing to the net force on the cylinder. The final term, however, makes the excess pressure higher (since K_3 is negative) over the bottom half of the cylinder's surface than over the top half (it reaches a maximum value at the two stagnation points labelled P in the diagram). The final term therefore contributes a force in the x_2 direction – a *lift* force as opposed to a *drag* force – whose magnitude per unit length of the cylinder is

$$-\frac{\rho U_1 K_3}{\pi a}\int_0^{2\pi} a\sin^2\theta\,\mathrm{d}\theta = -\rho U_1 K_3.$$

This result is fully consistent, in its direction as well as its magnitude, with expectations based upon (4.55).

It is instructive to see how the existence of the transverse force may also be established by evaluating excess pressure and momentum flux over a control surface which is far removed from the spindle. Consider, for example, a control surface in the form of a cylinder, coaxial with the spindle but of greater radius b. On this surface the excess pressure is

$$p^*_{r=b} = \frac{1}{2}\rho(U_1^2 - u_r^2 - u_\theta^2),$$

in which the relevant term is

$$\frac{\rho U_1 K_3}{2\pi b}\left(1 + \frac{b^2}{r^2}\right)\sin\theta.$$

The excess pressure therefore exerts a lift force on the fluid inside the control surface of magnitude

$$-\frac{1}{2}\rho U_1 K_3\left(1 + \frac{a^2}{b^2}\right) \tag{4.58}$$

per unit length (which reduces to $-\rho U_1 K_3$, of course, when $b = a$). However, the rate of change of momentum of the contained fluid, in the direction of the lift force and per unit length, is the value at $r = b$ of

$$\int_0^{2\pi} \rho u_r(u_r\cos\theta + u_\theta\sin\theta)r\,\mathrm{d}\theta$$

[(2.44)], and when this integral is evaluated using (4.57) it is found to amount to

$$\frac{1}{2}\rho U_1 K_3\left(1-\frac{a^2}{b^2}\right). \tag{4.59}$$

The difference between (4.58) and (4.59) reveals that the fluid must be subject to an additional force of magnitude $\rho U_1 K_3$ per unit length, which only the spindle can exert upon it. Hence the spindle must experience the reaction which (4.55) describes.

The force which acts on a current-carrying wire depends *only* on $\boldsymbol{i}$ and $\boldsymbol{B}$, the area and shape of the wire's cross-section being quite irrelevant, and the lift force which acts on a bound vortex line is likewise independent of the cross-section of the solid object to which it is bound. To attempt to prove this by repeating the above analysis for non-circular cross-sections of increasing complexity would be absurd. If one were to make such an attempt, however, one would find oneself having to include in the potential a number of additional terms proportional to ϕ_n with $n < -1$, and therefore proportional [(4.23)] to r^{-2}, r^{-3} etc. Because the lift force may be inferred from the velocity distribution over a control surface which is far removed from the solid object, and because the additional terms in ϕ become less and less significant by comparison with the ϕ_{-1} term as r increases, these additional terms are bound to make no difference to the final answer.

It is readily confirmed by experiment that when fluid moves with uniform velocity $\boldsymbol{U}$ past a cylinder which is spinning about its axis with steady angular velocity $\boldsymbol{\omega}_0$ the cylinder does experience a force in the direction of $\boldsymbol{U} \wedge \boldsymbol{\omega}_0$. This constitutes the *Magnus effect*, and the force is sometimes referred to as a Magnus force. A particularly spectacular demonstration involves projecting a spinning cardboard tube through air with a catapult of some sort – a long stretched rubber band, one end of which is fastened to a length of tape wound round the tube, is effective. It should not be assumed, however, that the magnitude of the force is given by (4.55) with $\boldsymbol{K} = 2\pi\boldsymbol{\omega}_0 a^2$, as the argument in §4.11 suggests. Equations (4.57) may describe with reasonable accuracy the flow pattern which results when a cylinder is endowed with translation as well as rotation, at any rate if the Reynolds Number $(2a\rho U/\eta)$ is large enough to ensure that the boundary layer is relatively thin, but the strength of the bound vortex is affected by U as well as by ω_0 [§7.12].

4.13 Free vortex lines with cores

When there is no spindle along the axis of a vortex line and the whole space in the neighbourhood of the line, right in to $r = 0$, is occupied by fluid, it is not possible for all this fluid to be vorticity-free. The rule that u_θ is proportional to r^{-1} must break down for small values of r, to avoid unphysical singularities on the axis itself; the fluid velocity can surely never be infinite in the real world, and in so far as it reverses in direction between two points which lie close to the axis but on

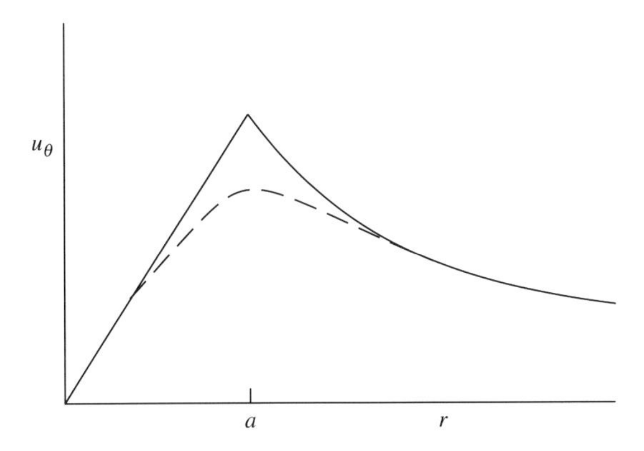

Figure 4.14 Variation of fluid velocity u_θ with radius r according to the idealised Rankine model of a free vortex line. This velocity distribution cannot persist if the fluid has viscosity, and the broken curve suggests what it might look like at a later moment.

opposite sides of it the change cannot be a discontinuous one. Thus every free vortex line must have a *core* of some sort.

The simplest possibility, illustrated by fig. 4.14, is that the velocity associated with a free vortex line which lies along the x_3 axis, of strength K_3, is described by the equations

$$u_\theta = \frac{K_3}{2\pi r} \quad [r > a]; \qquad u_\theta = \frac{K_3 r}{2\pi a^2} \quad [r < a]. \tag{4.60}$$

According to this model – known as the *Rankine model* – the core has a well-defined radius a within which the fluid rotates as a solid body would, with uniform angular velocity $\omega_o = K_3/2\pi a^2$. All the fluid outside the core is vorticity-free, but inside the core the vorticity is

$$\Omega_3 = 2\omega_o = \frac{K_3}{\pi a^2} \tag{4.61}$$

[(2.14)]. The Rankine model has its limitations, and we shall encounter alternatives in §6.9(iii). It provides a sufficient basis, however, for the ensuing discussion.

The kinetic energy per unit length associated with a straight Rankine vortex of strength K_3 is

$$\frac{1}{2}\rho\int_0^{r_{\max}} u^2 2\pi r \mathrm{d}r = \frac{\rho K_3^2}{4\pi}\left\{\frac{1}{4} + \ln\left(\frac{r_{\max}}{a}\right)\right\}. \tag{4.62}$$

It diverges, of course, if we let the upper limit to the integral tend to infinity, just as the energy stored in the magnetic field round a long straight current-carrying wire diverges, or the strain energy round a long straight dislocation in a crystal. As in those analogous situations, the energy is prevented from diverging in practice by the finite size of the fluid sample, or else by the fact that it contains other vortex lines, and we cut off the integral in recognition of that fact; the actual magnitude of r_{max} is of rather little interest. Now the rate at which energy is dissipated as heat outside the core is the rate at which, if the vortex were bound to a spindle of radius a rotating with angular velocity ω_o, the spindle would supply energy, namely

$$g_3\omega_o = 4\pi a^2\eta\omega_o^2 = 4\pi a^2\eta\left(\frac{K_3}{2\pi a^2}\right)^2 = \frac{\eta K_3^2}{\pi a^2}$$

per unit length, and in the absence of any spindle the kinetic energy must be decreasing at this rate. Since the strength of the vortex cannot change [§4.11], the decrease of kinetic energy presumably results from an increase in the core radius a. On differentiating (4.62) with respect to time and equating the result to $-\eta K_3^2/\pi a^2$ one obtains

$$a\frac{\mathrm{d}a}{\mathrm{d}t} = \frac{4\eta}{\rho},$$

which suggests that if a is very small at time $t = 0$ it is given at later times by

$$a \approx 2\sqrt{\frac{2\eta t}{\rho}}. \tag{4.63}$$

That argument is not entirely sound, because it involves an assumption that if the equations labelled (4.60) hold at $t = 0$ they hold at later times also, though with larger values of a. In fact that is not the case; the velocity profile becomes rounded, as suggested by a broken curve in fig. 4.14. A fuller analysis confirms, however, that in so far as a core radius can be defined (the boundary of the core is never really sharp in practice) it grows like $\sqrt{2\eta t/\rho}$ [§6.9(iii), and (6.58) in particular]. In water, for which the kinematic viscosity η/ρ is about 10^{-6} m^2 s^{-1} at normal temperatures, it takes about 1 minute for this radius to become 1 cm. In air at normal temperatures the kinematic viscosity is about 15 times greater, and $\sqrt{2\eta t/\rho}$ is about 4 mm after 1 second or 1 m after 10 hours. These figures provide some indication of the lifetimes to be expected for the vortices which occur in nature.

The growth of the core is a slow, diffusive, process, which is irrelevant in circumstances where a vortex line is undergoing relatively rapid extension. In such circumstances the vorticity-free fluid remains vorticity-free, which means that the total volume occupied by the core does not change. Hence if the length L of the line increases its radius a must decrease, at such a rate that La^2 is constant. Since the strength of the line cannot change, it follows from (4.61) that $\omega_o a^2$ is also constant, and hence that ω_o increases like L. This increase is merely the increase of angular velocity to be expected in any rotating body when the material composing it is brought nearer to the axis of rotation, thereby decreasing its moment of inertia, while its angular momentum remains constant. During the extension work must be done to supply the fluid with more kinetic energy, so a vortex line exerts a tension on whatever is stretching it. The magnitude of the tension is

$$\frac{\mathrm{d}}{\mathrm{d}L}\left[\frac{\rho K^2 L}{4\pi}\left\{\frac{1}{4}+\ln\left(\frac{r_{\max}}{a}\right)\right\}\right]=\frac{\rho K^2}{4\pi}\left\{\frac{3}{4}+\ln\left(\frac{r_{\max}}{a}\right)\right\} \tag{4.64}$$

[(4.62) – remember that, since a^2 is proportional to L^{-1}, $2\mathrm{d}a/\mathrm{d}L = -a/L$]. If the vortex line terminates at a fluid–solid interface, tension is exerted on the solid as a result of the excess pressure near the line being negative. It may be evaluated, with the same result, as

$$\int_0^{r_{\max}} p^* 2\pi r \mathrm{d}r,$$

using [(2.9)]

$$p^* = -\frac{1}{2}\rho\left(\frac{K}{2\pi r}\right)^2 \quad [r > a],$$

$$\frac{\mathrm{d}p^*}{\mathrm{d}r} = \rho\omega_0^2 r = \rho\left(\frac{K}{2\pi a^2}\right)^2 r \quad [r < a].$$

How may a free vortex line be generated? Here are a few possible procedures, all of them depending in some degree upon the presence of viscosity and all of them ineffective, therefore, in an ideal Euler fluid.

(i) Use a rotating spindle as described in §4.11, and then withdraw it along its axis.

(ii) Communicate angular momentum to a body of liquid by rotating a container outside it and then allow this liquid to drain from a central plug-hole, forming a whirlpool as it does so. The velocity of circulation may be almost proportional to $1/r$ by the time that most of the liquid has gone, for reasons outlined in §2.7. In the Earth's atmosphere, vortex lines known as tornadoes arise in very much this way [fig. 4.15], though in this case the fluid is removed along the axis by convection upwards rather than by draining downwards.

(iii) Push a solid plate through fluid in a direction normal to its plane, or blow fluid past the edge of plate which is stationary; as a result of boundary layer separation, eddies form behind the edge and are shed as free vortices [§7.3]. Vortices may readily be produced by this method in a cup of tea, using a teaspoon, or in a bath, using one's hand.

There is no good reason to suppose that the velocity distribution around free vortex lines always conforms to the ideal model we have discussed above, in which

Figure 4.15 A tornado.
[Photograph by courtesy of the National Oceanic and Atmospheric Administration Photo Library, Rockville, Maryland, USA.]

the fluid has no vorticity whatever outside a well-defined core of comparatively small radius. Nevertheless, the ideal model seems to describe rather successfully the way in which real vortex lines behave. It is to the ideal model, therefore, that the remaining sections of this chapter refer.

Figure 4.16 A loop L_1 embedded in fluid which is moving from left to right becomes L_2 at a later time. In (a) the vortex line V_1 is free and moves with it to V_2. In (b) the vortex line is pinned and the loop L_2 is snagged at the stagnation point P.

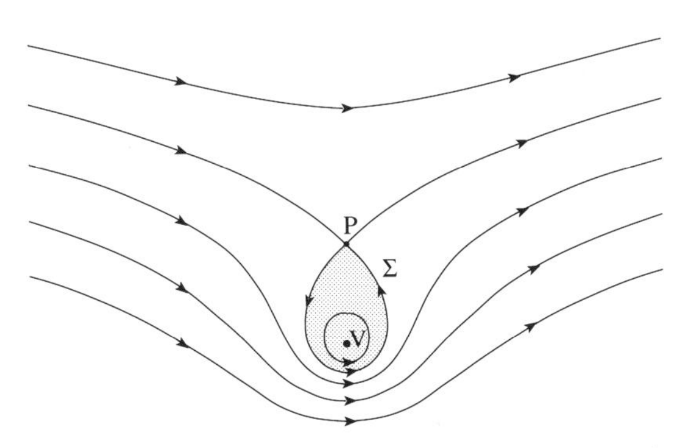

Figure 4.17 Surface Σ round a pinned vortex line V, separating fluid which stays with the vortex line from fluid which does not.

4.14 Behaviour of free vortex lines

Figure 4.16(a) represents a situation where all the fluid within the field of view is moving in an irregular fashion from left to right, and the two closed curves in this figure are snapshots, separated in time, of a closed loop which is embedded in the fluid and moving with it. In the first of these snapshots the loop encloses a vortex line, whose core is represented by a small black circle, as a result of which the circulation round the loop is non-zero. Kelvin's theorem tells us that the circulation round the loop in the second snapshot is just the same, and hence that the vortex line is within that one also. In other words, the vortex line is moving with the loop; *it must be embedded in the fluid which surrounds it*, in the same way that the loop itself is embedded.

In fig. 4.16(b) the fluid as a whole is again moving from left to right but the vortex line is pinned in some way, e.g. by a solid spindle to which it is bound. Kelvin's theorem still tells us that any loop which encloses the vortex at one time must continue to enclose it, and it follows that a part of the loop must at some stage get snagged, e.g. at the point labelled P in the figure. This must be a stagnation point, at which the velocity of the fluid is zero, and it lies on a surface Σ which

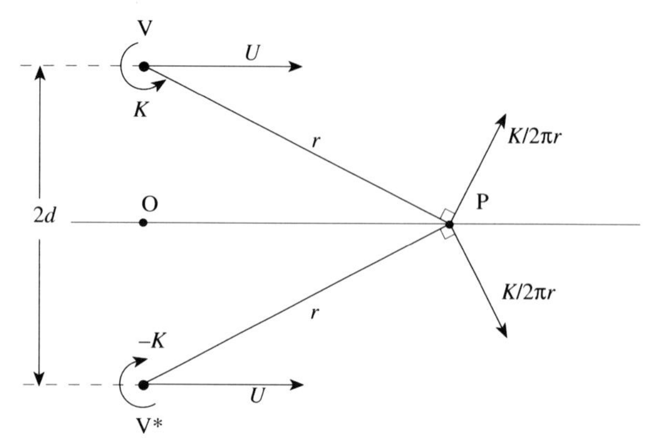

Figure 4.18 Two vortex lines of equal but opposite strength, or a single vortex line V and its image V*, propel each other forward with velocity $U = K/4\pi d$. The lines V and V* contribute velocities $K/2\pi r$, as indicated by two arrows in the diagram, to the fluid near P, and the resultant velocity of the fluid in this region lies in the plane OP.

separates fluid which is free to pass the vortex line from fluid which is trapped and remains near it. In the particular case where the vortex line is straight and where the fluid velocity at large distances is uniform, the shape of this separating surface can be calculated without difficulty; it is sketched in fig. 4.17.

Figure 4.18 illustrates a situation where two vortex lines of equal but opposite strength, $\pm K$, lie parallel to one another a distance $2d$ apart. In the magnetostatic analogue for this situation we have two unlike currents which repel one another with a force of magnitude $\mu_0 i^2/4\pi d$ per unit length, so the translation scheme expressed by table 4.1 implies a force of attraction between the vortices, of magnitude $\rho K^2/4\pi d$ per unit length. They react to this attraction, however, by drifting to the right in fig. 4.18 at such a speed relative to the fluid at large distances that it is precisely cancelled by the Magnus force which each of them then experiences. The necessary speed U is given by

$$\rho UK = \frac{\rho K^2}{4\pi d},$$

or

$$U = \frac{K}{4\pi d}. \tag{4.65}$$

This result is obvious on other grounds: each vortex line imposes a velocity of $K/4\pi d$ on the fluid surrounding the other line, and since both of them are embedded in the surrounding fluid they propel each other along at that rate.

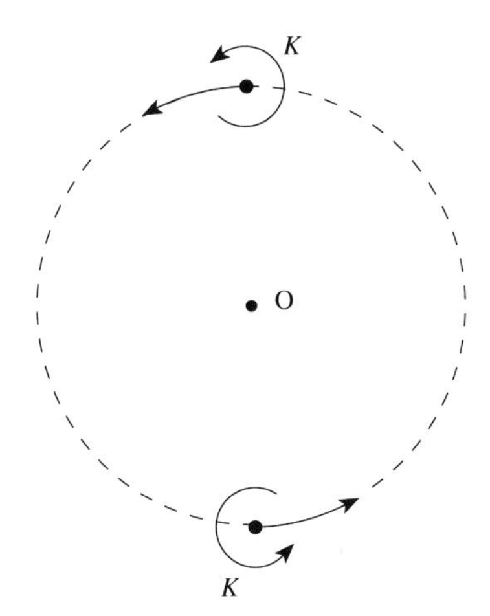

Figure 4.19 Two identical vortex lines propel each other round a circle.

At any point P on the plane which bisects at right angles the line joining the two vortices in fig. 4.18 the contribution made by one vortex to the fluid velocity normal to this plane cancels the contribution made by the other; the velocity on the bisecting plane, therefore, is along OP. This means that one could insert a solid sheet on this plane without affecting the velocity field, provided that the no-slip boundary condition is negligible, i.e. provided that the boundary layer is thin compared with d. Hence a single vortex line which is parallel to a plane solid surface at a distance from it of d behaves as though, instead of this surface, there were an *image* vortex line on the other side of the plane with equal but opposite circulation. It moves steadily over the plane at the speed given by (4.65), in just the opposite direction to the direction in which a solid cylinder would roll, if set rotating in the same sense.

The two parallel vortex lines in fig. 4.19 have strengths which are equal in direction as well as magnitude. Here too each vortex propels the other one along with a velocity of $K/4\pi d$, and the result is that they rotate steadily about a line half way between them, with angular velocity $K/4\pi d^2$. If the two vortices happen to be pinned in one place but not everywhere, the rotation will twist them up into a double helix.

4.15 Smoke rings

Because K cannot change along the length of a vortex line, vortex lines cannot suddenly terminate except at the fluid's boundaries. However, a vortex line is not obliged to extend to these boundaries; its ends may join up to form a localised vortex ring. Smoke rings are vortex rings, made visible by smoke particles which are embedded in the fluid just as the vortices are and which therefore move with them. In a frame of reference in which the air at large distances is stationary smoke rings propel themselves forward, in very much the same way as the two unlike vortices in fig. 4.18 propel each other forward, at a drift velocity $\boldsymbol{U}$ which is directed along the axis of the ring.

The magnetostatic analogue for an ideal vortex ring is a circular loop of radius b, say, made out of wire of radius a which carries a uniformly-distributed current i. According to the Biot–Savart law, the field strength $B\{r\}$ in the plane of a loop for which $b \gg a$ is $\mu_0 i/2b$ at $r = 0$, the loop's centre, but increases with radius until r reaches $b - a$, at the inner edge of the wire. In the interior of the wire itself the field decreases rapidly with r in a linear fashion, passing though a value which is given to a good approximation by

$$B\{b\} \approx \frac{\mu_0 i}{4\pi b}\left\{\ln\left(\frac{8b}{a}\right) - \frac{1}{4}\right\} \tag{4.66}$$

at the wire's centre. At the outer edge it is reversed in sign, and at large radii it falls off like r^{-3}. Equivalent results hold for the fluid velocity $u\{r\}$ in the plane of an ideal vortex ring with a core of radius a, as measured in a frame of reference such that the fluid at large distances is stationary, and it is the analogue of (4.66), namely

$$U = u\{b\} \approx \frac{K}{4\pi b}\left\{\ln\left(\frac{8b}{a}\right) - \frac{1}{4}\right\}. \tag{4.67}$$

which describes the ring's drift velocity. When $b/a \approx 300$, U exceeds $u\{0\}$ by about 20%, and fig. 4.20 applies to this case. Note that $u\{r\}$ varies so rapidly with r in the interior of the vortex core that the lines in fig. 4.20(a) which represent its variation are virtually horizontal in this region.

The appearance of smoke rings may be partly understood in terms of fig. 4.20(b), which refers to the same case as fig. 4.20(a) but to a frame of reference in which the ring is stationary and the flow pattern a steady one (at any rate for an ideal fluid with no viscosity); the lines of flow in fig. 4.20(b) are only roughly sketched, but for the purposes of this discussion they should suffice. At the points labelled P the fluid velocity in the co-moving frame is zero, and these points lie on a toroidal surface Σ which separates fluid which passes through or around the ring from fluid which is trapped and circulates endlessly about the vortex. Smoke rings normally form in regions where all the air is smoke-laden, but it is only the smoke that lies within Σ initially which gets carried forward with the ring and which makes it visible. Note that when $b/a \approx 300$ the region occupied by smoke is a great deal larger than the region occupied by the vortex core.

In a real fluid with viscosity the core radius a increases as time passes, in the manner discussed in §4.13. That means, according to (4.67), that U decreases – in the frame of reference of fig. 4.20(a) the vortex ring slows down – and hence that

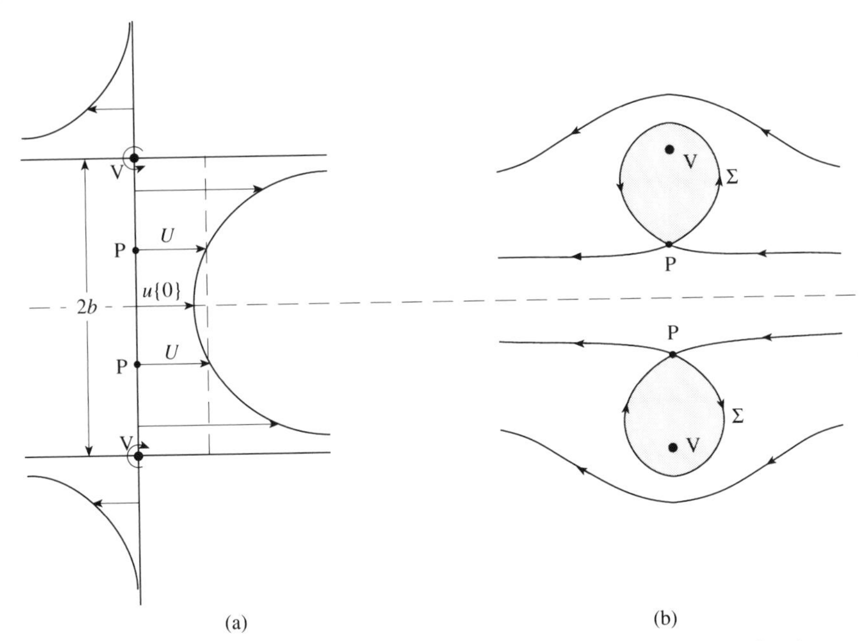

Figure 4.20 Cross-section through a circular vortex ring V of radius b, when the core radius a is such that $b/a \approx 300$. (a) Profile of fluid velocity in the plane of the ring, in a frame of reference such that at large distances the fluid is stationary; the velocity of the ring itself is U. (b) Lines of flow (sketch only) in a frame of reference in which the ring is stationary; fluid in the shaded area is trapped by the vortex and circles round it.

the points labelled P in fig. 4.20 move inwards to smaller values of r. The surface Σ enlarges, therefore, and an increasing volume of fluid is trapped by the ring and obliged to move with it. We have noted in §4.13 that the core radius of a vortex in air is likely to be at least 4 mm after only 1 second has elapsed. At that stage, therefore, the ratio b/a for the sort of smoke ring that an adept pipe smoker produces is perhaps only 10 rather than 300. If so, then U is certainly less than $u\{0\}$, which means that the two stagnation points in the flow pattern equivalent to the one sketched in fig. 4.20(b) lie on the axis of the ring and at some distance from its plane, and that the volume of trapped air is several times larger than it was when the ring first formed. The smoke in a smoke ring diffuses too slowly through air to render this enlargement visible, but it does reveal that the ring is slowing down.

When a smoke ring of strength K is blown against a wall it sees an image of itself, i.e. a vortex ring of strength $-K$, and it interacts with its image in very much the same way that the vortex line in fig. 4.18 does: each segment of the ring moves away from the axis, so that b increases while the plane of the ring stays parallel to the wall. This effect is easily demonstrated by anyone who can blow smoke rings.

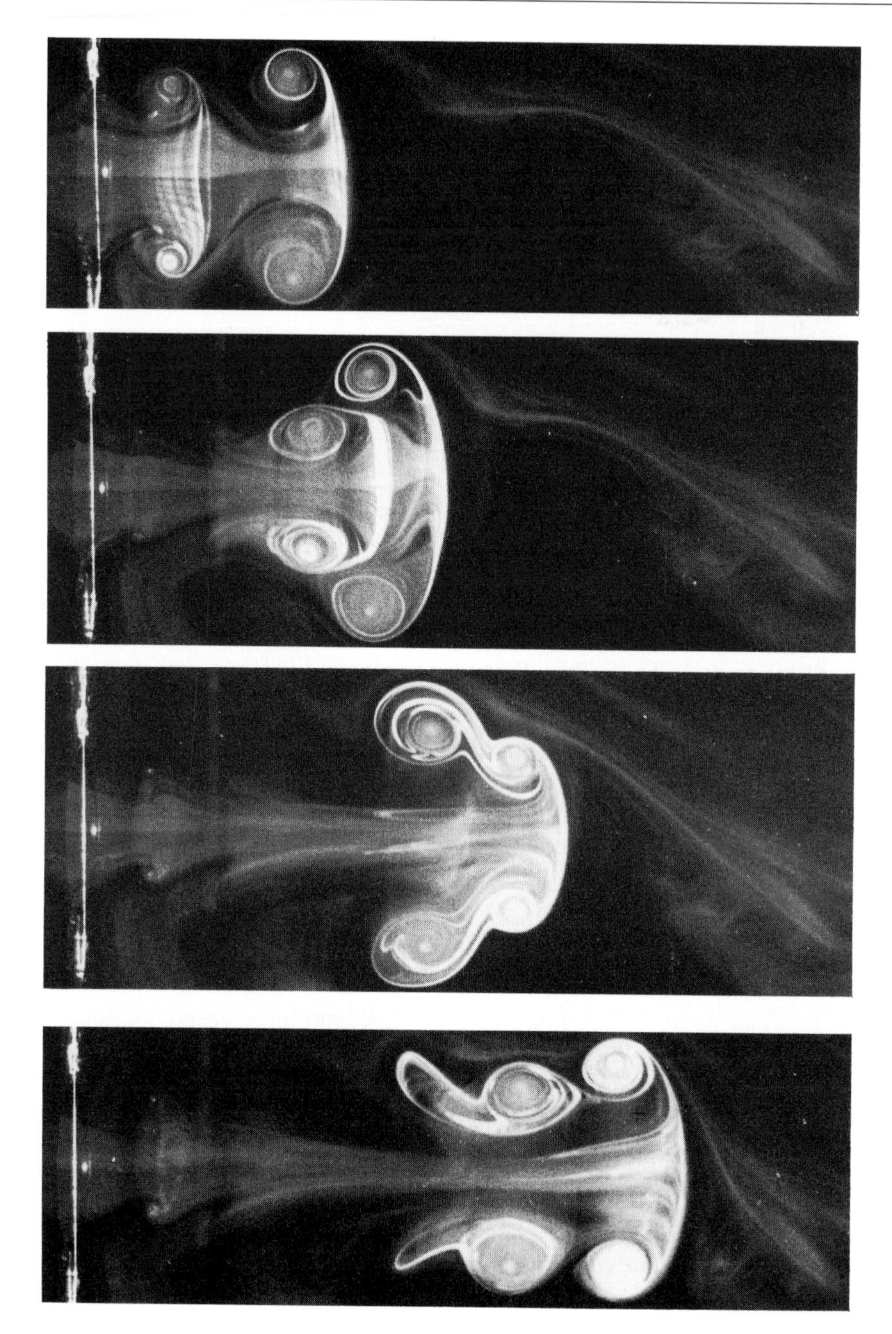

Figure 4.21 Two similar smoke rings moving from left to right through air; the one which is second at the start (top photograph) moves through the other and becomes the first ring (bottom photograph), whereupon the process is repeated.
[Photographs by H. Yamada & T. Matsui. [*Phys. Fluids*, **21**, 292, 1978.]]

As b increases a decreases, according to the argument in §4.13, but the observable consequences of that are slight.

One can generate smoke rings more reproducibly in the laboratory than one can by mouth, and it is possible with care to produce two rings of the same size and strength which closely follow one another. The velocity field associated with the second of these rings causes the one in front to increase in radius b, whereupon, as is apparent from (4.67), its drift velocity U decreases. As the first ring gets larger and slower, however, the second ring gets smaller and faster. The second ring passes through the middle of the first one, in fact, and becomes first in its turn, whereupon the whole process is repeated. One way to describe this remarkable leap-frogging process, which is illustrated by fig. 4.21, is to say that adjacent segments of the two rings circle round one another, like the two straight vortex lines in fig. 4.19.

Vortex rings can, of course, be generated in liquids as well as gases, and the reader can study some of their properties with no more equipment than a cup of tea and a teaspoon; when a spoon with a rounded bowl, only partly submerged, is drawn gently through tea and then removed it leaves half a vortex ring behind it, i.e. a vortex line in the form of a semicircle, the two ends of which are clearly visible as dimples in the surface. Visible vortex rings – whole ones – may also be generated in cold water, by allowing a drop of milk or thin cream to fall into it from a small height. Rings generated in the latter fashion turn out to be unstable, but the instabilities to which they are prone are too complex for discussion here.

Further reading

The classic text on potential flow is *Hydrodynamics*, by H. Lamb, which first appeared (under a different title) in 1879. Reviewing the fourth edition in 1916 Lord Rayleigh remarked: 'During the last few years much work has been done in connexion with artificial flight. We may hope that before long this may be coordinated and brought into closer relation with theoretical hydrodynamics. In the meantime one can hardly deny that much of the latter science is out of touch with reality.' Even the sixth and last edition (Cambridge University Press, 1932) is out of touch with reality in some respects, but it nevertheless remains an invaluable source. A reader wishing to check, for example, the result concerning the virtual mass of an ellipsoid which is quoted above as (4.42) should turn first to Lamb.

A modern classic, to be recommended without any reservation, is *An Introduction to Fluid Dynamics*, by G. K. Batchelor, Cambridge University Press, 1967. *A First Course in Fluid Mechanics*, by A. R. Paterson, Cambridge University Press, 1983 and *An Informal Introduction to Theoretical Fluid Mechanics*, by J. Lighthill, Cambridge University Press, 1978, contain accessible accounts – with illustrative applications – of the theory of complex potentials and conformal mapping.

5

Surface waves

5.1 The propagation of wave groups in one dimension

This chapter is devoted to waves on fluid interfaces, and more particularly to waves on the surface of canals or lakes or oceans where the water is not necessarily shallow enough for the theory developed in §§2.14–2.16 to apply. Such waves turn out to be *dispersive*: that is to say, the velocity at which their crests and troughs propagate is not a fixed quantity but depends upon wavelength. One cannot understand their behaviour fully without first understanding the concepts of *group velocity* and *spreading*. Some readers may welcome an introduction to these concepts; others may prefer to move straight to §5.2.

A one-dimensional 'monochromatic' wave, i.e. a plane sinusoidal wave characterised by a single wavelength λ and a single frequency $1/\tau$, propagating in say the x direction, may conveniently be represented by a complex expression, of which only the real part has physical significance:

$$\zeta = \zeta_k \exp\left\{2\pi \mathrm{i}\left(\frac{x}{\lambda} - \frac{t}{\tau}\right)\right\} = |\zeta_k|\, \mathrm{e}^{\mathrm{i}\chi_k}\, \mathrm{e}^{\mathrm{i}(kx-\omega t)}. \tag{5.1}$$

Here ζ denotes the quantity which varies sinusoidally in space and time – in the context of surface waves it is a vertical displacement of a liquid surface – with wavevector k $(= 2\pi/\lambda)$ and angular frequency ω $(= 2\pi/\tau)$. The amplitude of the wave, $\zeta_k = |\zeta_k| \exp(\mathrm{i}\chi_k)$, may be complex, but it is normally possible to adjust the origins of x or t so as to make it real, i.e. so as to make the phase shift χ_k vanish. The velocity of this wave on its own is

$$c = \frac{\lambda}{\tau} = \frac{\omega}{k}. \tag{5.2}$$

Suppose, however, that two waves of the same amplitude ($\zeta_{k_1} = \zeta_{k_2} = \zeta_o$) but slightly different wavevectors k_1 and k_2, both travelling in the x direction, are superposed. In that case

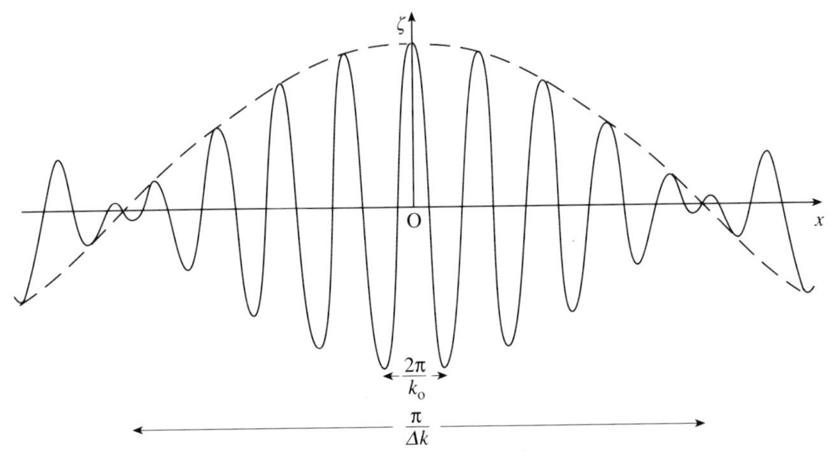

Figure 5.1 The beat pattern for two waves of equal amplitude which are in phase at O but which have different k values, $k_o \pm \Delta k$. In the case illustrated, $\Delta k = k_o/18$.

$$\begin{aligned}\zeta &= \zeta_o\{e^{i(k_1 x - \omega_1 t)} + e^{i(k_2 x - \omega_2 t)}\} \\ &= 2\zeta_o \cos(\Delta k\, x - \Delta\omega\, t)\, e^{i(k_o x - \omega_o t)},\end{aligned} \tag{5.3}$$

where

$$k_{2,1} = k_o \pm \Delta k, \quad \omega_{2,1} = \omega_o \pm \Delta\omega.$$

The two waves *beat* with one another to produce a single travelling wave [$\exp\{i(k_o x - \omega_o t)\}$], often called a *carrier wave* in contexts of this sort, whose wavelength and frequency are intermediate between those of the two components and whose amplitude [$2\zeta_o \cos(\Delta k\, x - \Delta\omega\, t)$] is *modulated* by a sinusoidal factor of relatively long wavelength [fig. 5.1]. The velocity with which the crests and troughs of the modulation travel is

$$\frac{\Delta\omega}{\Delta k} \approx \left(\frac{d\omega}{dk}\right)_{k=k_o} = c_g. \tag{5.4}$$

When ω is not simply proportional to k and the waves are dispersive, the velocities c and c_g associated with $k = k_o$, as defined by (5.2) and (5.4) respectively, may be totally different. For gravity waves on deep water, for example, ω turns out to be proportional to $\sqrt{k}$ [(5.25)]. In that case

$$\frac{d\omega}{\omega} = \frac{dk}{2k},$$

so that

$$c_g = \frac{d\omega}{dk} = \frac{\omega}{2k} = \frac{1}{2}c. \tag{5.5}$$

Thus when two gravity waves of similar wavelength, both travelling in the same direction, are superposed, the crests of the resultant carrier wave [fig. 5.1, full curve] travel twice as fast as the crests of the modulating profile [broken curve]; individual crests of the carrier wave wax and wane in amplitude as they travel through the profile with relative velocity $(c - c_g)$.

In general, wave-like disturbances contain many more than just two 'monochromatic' components. Sometimes the components are excited in such a way that their phases are completely uncorrelated, and the result is the sort of random disturbance often described as *noise*. In other cases, however, the excitation process is such that, at some moment in time which may be chosen as $t = 0$, a *group* of components which are all travelling in the same direction and have wavevectors distributed over a range of k between say k_1 and k_2 $(> k_1)$ are precisely in phase with one another – with phase angles of zero – at some point in space which may be chosen as $x = 0$. The group can be described as an amplitude-modulated carrier wave, but at $t = 0$ the amplitude is normally appreciable only within some localised region of space centred on $x = 0$ within which constructive interference occurs.[1] The range, say $L\{0\}$, over which the amplitude is appreciable at $t = 0$ is related to the width of the group by the approximate equation

$$L\{0\} \approx \frac{1}{k_2 - k_1}, \tag{5.6}$$

so that large values of $L\{0\}$ imply small bandwidths in terms of k, and *vice versa*. As time passes the localised disturbance associated with a wave group moves, and to a first approximation it does so with the velocity c_g which (5.4) describes. This velocity is therefore called the *group velocity*, whereas c is called the *phase velocity*. While the disturbance moves, however, it normally changes in appearance: in particular, it tends to *spread* over a range $L\{t\}$ which gets steadily bigger than the initial range $L\{0\}$.

Those statements are readily confirmed. Let us represent the group by a sum over k,

$$\zeta = \sum_k \zeta_k \, e^{i(kx-\omega t)} = \sum_k \zeta_k \, e^{i\chi_k} \, e^{i(k_o x - \omega_o t)},$$

where k_o lies in the range between k_1 and k_2, where ω_o is the angular frequency which corresponds to k_o, and where the initial conditions ensure that within this

[1] The two components of the disturbance in fig. 5.1 are in phase not only near $x = 0$ but also near $x = \pm\pi/\Delta k, \pm 2\pi/\Delta k, \ldots$ Once there are many more than two components however, the chance of these reinforcing one another in more than one locality is remote.

range the amplitude ζ_k of each component is real. The phase angle χ_k is not a constant shift here; it varies with x, t and $\Delta k = (k - k_o)$ or $\Delta\omega = (\omega - \omega_o)$ thus:

$$\chi_k\{x, t\} = \Delta k\, x - \Delta\omega\, t,$$

or, if we expand $\Delta\omega$ as a Taylor series in powers of Δk and use a suffix zero to denote derivatives taken at $k = k_o$,

$$\begin{aligned}\chi_k\{x, t\} &= \left\{x - \left(\frac{d\omega}{dk}\right)_o t\right\} \Delta k - \frac{1}{2}\left(\frac{d^2\omega}{dk^2}\right)_o t(\Delta k)^2 - \ldots \\ &= (x - c_{g,o}t)\Delta k - \frac{1}{2}\left(\frac{dc_g}{dk}\right)_o t(\Delta k)^2 - \ldots . \end{aligned} \qquad (5.7)$$

Now for constructive interference to occur the components of the group must be as far as possible in phase with one another, i.e. have the same value of χ_k. Constructive interference is therefore to be expected where χ_k is *stationary with respect to variations in* k, i.e. where $\partial\chi_k/\partial(\Delta k)$ is zero. Equation (5.7) tells us at once to expect constructive interference where

$$x = c_{g,o}t. \qquad (5.8)$$

It also tells us, however, that constructive interference occurs only between those components whose wavevectors are sufficiently close to k_o for $\frac{1}{2}(dc_g/dk)_o t(\Delta k)^2$ to be small compared with 2π, i.e. only over a band of wavevectors whose width is of order

$$\left\{\left(\frac{dc_g}{dk}\right)_o t\right\}^{-1/2}. \qquad (5.9)$$

Once this has become smaller than the total width of the group $(k_2 - k_1)$, i.e. once

$$t > \frac{1}{(k_2 - k_1)^2|(dc_g/dk)_o|} \approx \frac{L\{0\}^2}{|(dc_g/dk)_o|},$$

the amplitude of the propagating disturbance must be less than it was initially, when all the components of the group were in phase, and it follows from (5.9) that it falls like $t^{-1/2}$ thereafter. Since the energy stored in the disturbance, which is proportional to the square of its amplitude and to its length, is normally conserved, we may infer that at large times the length of the group $L\{t\}$ increases like t.

The fact that $L\{t\}$ increases like t is obvious on other grounds. Equation (5.8) describes the point at which the amplitude of a carrier wave of wavevector k_o is large, but we are at liberty to choose any value for k_o as long as it lies between k_1 and k_2. At one extreme, with $k_o = k_1$, the point of maximum amplitude is seen to move with velocity $c_{g,1}$. At the other extreme, with $k_o = k_2$, it is seen to move with

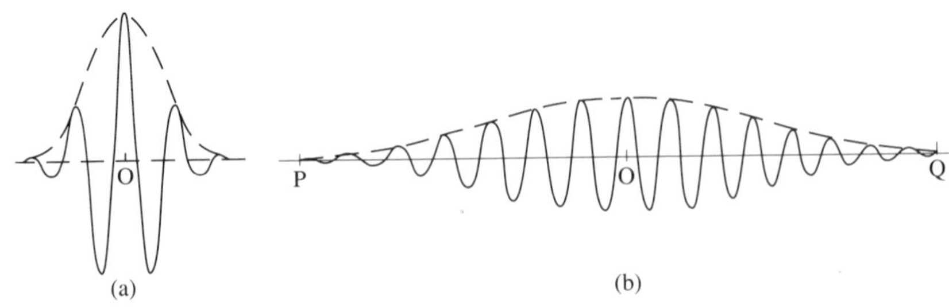

Figure 5.2 After travelling for some distance through a dispersive medium in which c_g increases with k, i.e. in which $d^2\omega/dk^2 > 0$, the initially Gaussian wave group shown in (a) has spread to four times its original length and halved in height as in (b). Moreover, it is no longer symmetrical: the Fourier components of larger k have worked their way to the front of the group, and the wavelength is shorter there than at the back. Thus there is room for 7 maxima between O and Q but for only 6 between O and P.

velocity $c_{g,2}$. These two velocities normally correspond to the velocities of the leading and trailing edges of the disturbance, and the rate of spreading is the difference between them, i.e.

$$\frac{dL\{t\}}{dt} = |c_{g,2} - c_{g,1}| \approx (k_2 - k_1)\left|\frac{dc_g}{dk}\right| \approx \frac{1}{L\{0\}}\left|\frac{dc_g}{dk}\right|. \qquad (5.10)$$

If dc_g/dk is negative, as is the case for gravity waves on deep water, then the wavelength of the carrier wave which shows up at the front of the disturbance when t is large is the largest wavelength in the group, namely $2\pi/k_1$, while the wavelength which shows up at the back is $2\pi/k_2$; in between, the wavelength varies continuously. If dc_g/dk is positive, however, it is the short wavelengths present in the group which get ahead [fig. 5.2].

The behaviour of groups made up of components which are propagating in more than one direction is more complicated, as we shall see in §5.11 below.

5.2 Boundary conditions

Imagine a layer of water of uniform depth – as it were, the water in an idealised lake. In the absence of any disturbance it is at rest, in contact with a solid on the horizontal plane $z = -d$, say,[2] and in contact with the atmosphere at its 'free' surface on the horizontal plane $z = 0$. Since the water is vorticity-free in this condition it should, according to Kelvin's theorem, remain vorticity-free when its surface is ruffled by a breeze or by the passage of a boat. That being so, the motion

[2] Throughout this chapter we use (x, y, z) rather than (x_1, x_2, x_3) to label our cartesian axes, because gravity plays such an essential role in the phenomena to be discussed.

associated with waves on its surface can be described by a potential which obeys Laplace's equation at all times.

To simplify the discussion, let us suppose the waves to be propagating in the x direction and to have amplitudes which do not vary with y. In that case, the potential must satisfy Laplace's equation in its two-dimensional form,

$$\frac{\partial^2 \phi}{\partial x^2} + \frac{\partial^2 \phi}{\partial z^2} = 0. \tag{5.11}$$

Solutions of this equation in circular polar coordinates were discussed in §4.6(i). In cartesian coordinates, the separated solutions which replace the ones labelled (4.23) are

$$\phi \propto \phi_k = e^{ik_x x}\, e^{ik_z z},$$

where k_x and k_z are in general complex quantities satisfying the condition

$$k_x^2 + k_z^2 = 0.$$

In the present context, however, k_x has to be real, since otherwise the wave amplitude diverges to infinity at large values of x or $-x$. Hence k_z has to be imaginary. With $k_x = k$ and $k_z = \pm ik$, we have

$$\phi \propto \phi_k^{\pm} = e^{\pm kz}\, e^{ikx}. \tag{5.12}$$

All other solutions associated with propagation in the x direction are expressible in Fourier series form, as

$$\phi = \sum_k (a_k^+ \phi_k^+ + a_k^- \phi_k^-). \tag{5.13}$$

The coefficients a_k^+ and a_k^- are liable, of course, to vary with time t. The way in which they do so is determined by the following boundary conditions.

(i) The water must remain in contact with the solid at $z = -d$, so the vertical component of its velocity must vanish there, i.e.

$$\left(\frac{\partial \phi}{\partial z}\right)_{z=-d} = 0. \tag{5.14}$$

(ii) In mathematical terms, the free surface is defined by an equation of the form

$$z = \zeta\{x, t\}, \tag{5.15}$$

which describes its vertical displacement ζ from the plane $z = 0$ in which it lies when the water is undisturbed, and at a fixed value of x the surface moves up or down with velocity $\partial\zeta/\partial t$. However, the vertical velocity of the surface is also that of the layer of water which lies immediately beneath it, and this may be expressed in terms of the potential ϕ. The situation is slightly complicated by the fact that particles of water within the surface layer are liable to move horizontally with velocity $u_x = (\partial\phi\partial x)_{z=\zeta}$ as well as vertically with velocity $u_z = (\partial\phi/\partial z)_{z=\zeta}$. From (2.5) we know that *following the motion* of any one such particle the rate of change of ζ is

$$\frac{\mathrm{D}\zeta}{\mathrm{D}t} = \frac{\partial\zeta}{\partial t} + u_x\frac{\partial\zeta}{\partial x} + u_y\frac{\partial\zeta}{\partial y} + u_z\frac{\partial\zeta}{\partial z},$$

which simplifies, because ζ does not vary with y or z, to

$$\frac{\mathrm{D}\zeta}{\mathrm{D}t} = \frac{\partial\zeta}{\partial t} + u_x\frac{\partial\zeta}{\partial x}.$$

The second boundary condition is obtained by equating u_z to this rate of change of vertical displacement following the particle's motion; it is

$$\left(\frac{\partial\phi}{\partial z}\right)_{z=\zeta} = \frac{\partial\zeta}{\partial t} + \left(\frac{\partial\phi}{\partial x}\right)_{z=\zeta}\frac{\partial\zeta}{\partial x}. \tag{5.16}$$

(iii) The pressure difference which exists across the layer of infinitesimal thickness which constitutes the free surface must be only what is needed to balance the effect of surface tension as described by (2.37). Now the generalised version of Bernoulli's theorem which applies to vorticity-free flow [(4.13)] tells us that at all points in the water

$$p + \rho\left(gz + \frac{1}{2}u^2 + \frac{\partial\phi}{\partial t}\right) = p_\mathrm{o},$$

where the constant p_o is the pressure at some point where z, u and $\partial\phi/\partial t$ are all zero. Since these quantities are all zero at the free surface in regions of the lake which are still undisturbed, p_o is clearly the pressure which the atmosphere exerts. To balance surface tension we need

$$p_{z=\zeta} - p_\mathrm{o} = -\sigma\frac{\partial^2\zeta}{\partial x^2},$$

[(2.37)] so the third boundary condition is

$$g\zeta + \left(\frac{1}{2}u^2 + \frac{\partial\phi}{\partial t}\right)_{z=\zeta} = \frac{\sigma}{\rho}\frac{\partial^2\zeta}{\partial x^2}. \tag{5.17}$$

These three conditions will do to be going on with. In principle we should recognise the finite viscosity of water by taking into account (iv) the no-slip boundary condition at $z = -d$, where u_x must vanish as well as u_z, and (v) the condition that, because the viscosity of air is so much smaller than that of water, the shear stress exerted on the water over the surface $z = \zeta$ is effectively zero. We cannot do so, however, within the framework of potential theory; there must be boundary layers at the bottom and top of the lake within which the water is contaminated by vorticity and no potential can be defined. Fortunately, these boundary layers are normally very thin [§1.12] and we shall neglect them in what follows, though by doing so we lose sight of terms in the dispersion relation for surface waves which describe attenuation. There must be some attenuation, of course, because the flow associated with surface waves involves shear and is therefore dissipative, not just in the boundary layers but also in the vorticity-free interior of the water. It turns out that nearly all the dissipation occurs in the interior, and that the overall rate of it – and therefore the rate of attenuation – can be estimated quite precisely without taking the boundary layers into account at all. The calculation [based on (6.26) and on approximate equations for ϕ to be derived below] is not difficult, but because the attenuation is in practice insignificant, except at wavelengths which are too short to be of interest here, we shall not attempt it.

Yet more complications arise if we pause to consider whether, by including an atmospheric pressure p_o which subsequently cancels out, we have paid the atmosphere enough attention. Should we not treat it as a second fluid of smaller density ρ', say, within which, when a surface wave is excited, there are motions described by a potential ϕ', expressible in the form of (5.13) with a new set of coefficients $a_k^{+\prime}$ and $a_k^{-\prime}$? The answer, in principle, is yes. We shall discover in §5.5, however, that the only result is an additional factor $(\rho - \rho')/(\rho + \rho')$ in the dispersion relation and, since ρ' for air at normal pressures and temperatures is smaller than ρ for water by a factor of about 10^{-3}, this is not a material change.

Two of the three important boundary conditions, (5.16) and (5.17), contain terms which are of second (or higher) order in the wave amplitude, but as long as we are content to consider only waves whose amplitude is infinitesimal we may discard these. Equation (5.16) may then be simplified by omission of the term $(\partial\phi/\partial x)_{z=\zeta}(\partial\zeta/\partial x)$, and also by neglect of the distinction between $\partial\phi/\partial\zeta$ evaluated at $z = \zeta$ and the same quantity evaluated at $z = 0$; thus linearised it becomes

$$\left(\frac{\partial\phi}{\partial z}\right)_{z=0} \approx \frac{\partial\zeta}{\partial t}. \tag{5.18}$$

The linearised version of (5.17) is evidently

$$g\zeta + \left(\frac{\partial\phi}{\partial t}\right)_{z=0} \approx \frac{\sigma}{\rho}\frac{\partial^2\zeta}{\partial x^2}. \tag{5.19}$$

If the wavelength is large compared with

$$\lambda^* = 2\pi\sqrt{\frac{\sigma}{\rho g}}, \tag{5.20}$$

then $\sigma k^2/\rho g \ll 1$, in which case the term on the right-hand side of (5.19) is much smaller than either of the terms on the left-hand side. Surface tension is then negligible [§2.15], and (5.19) may be replaced by the even simpler condition

$$\left(\frac{\partial \phi}{\partial t}\right)_{z=0} \approx -g\zeta. \tag{5.21}$$

5.3 Gravity waves, 1 ($|\zeta| \ll \lambda \ll d, \lambda \gg \lambda^*$)

For water, the critical wavelength λ^* defined by (5.20) is about 1·7 cm. The restoring forces which drive surface waves whose wavelength is much longer than that are due to the weight of the fluid displaced, and such waves are therefore known as *gravity waves*. Here we discuss gravity waves whose amplitude is so small compared with their wavelength ($|k\zeta| \ll 1$) that use of the linearised boundary conditions represented by (5.18) and (5.21) is justified, and whose wavelength is so small compared with the depth d that for positive values of k – and we shall take k to be positive throughout the analysis that follows – the coefficients a_k^- in (5.13) are necessarily all zero; if a single a_k^- were to be non-zero, the potential would include a term proportional to $\exp(-kz)$ which would become so large at $z = -d$ that (5.14) could not possibly be satisfied. Gravity waves in other situations are discussed in §§5.7, 5.6 and 5.8 respectively.

When all the coefficients a_k^- are zero we have

$$\phi = \sum_k a_k^+ \, \mathrm{e}^{kz} \, \mathrm{e}^{\mathrm{i}kx} \quad [k > 0], \tag{5.22}$$

and (5.18) and (5.21) become respectively

$$\sum_k k a_k^+ \, \mathrm{e}^{\mathrm{i}kx} = \frac{\partial \zeta}{\partial t}, \tag{5.23}$$

and

$$\frac{1}{g}\sum_k \frac{\partial a_k^+}{\partial t} \, \mathrm{e}^{\mathrm{i}kx} = -\zeta. \tag{5.24}$$

On differentiating the second of these equations with respect to t and adding it to the first one obtains

$$\sum_k \left(gka_k^+ + \frac{\partial^2 a_k^+}{\partial t^2}\right) \mathrm{e}^{\mathrm{i}kx} = 0.$$

This must be satisfied for all x and t, which is possible only if

$$\frac{\partial^2 a_k^+}{\partial t^2} = -gka_k^+$$

for all k, i.e. only if

$$a_k^+ \propto e^{-i\omega t} \textit{ or } e^{+i\omega t}$$

(of which two equivalent possibilities we adopt the former), with

$$\omega^2 = gk. \tag{5.25}$$

This is the *dispersion relation* for small-amplitude gravity waves on deep water. It implies a phase velocity

$$c = \frac{\omega}{k} = \sqrt{\frac{g}{k}} = \sqrt{\frac{g\lambda}{2\pi}}, \tag{5.26}$$

and a group velocity which is half as big [(5.5)].

To the extent that linearised boundary conditions suffice, components of different k are not in any way coupled together, and good solutions of Laplace's equation exist for which only one coefficient a_k^+ is non-zero. The surface elevation described by such a solution may be found from (5.23) or (5.24). With the one non-zero coefficient written as $a\exp(-i\omega t)$, where a is *not* time-dependent, it is

$$\zeta = -\frac{ka}{i\omega}\, e^{i(kx-\omega t)} = \frac{i\omega a}{g}\, e^{i(kx-\omega t)},$$

and if a is real the physically significant part of this is

$$\mathrm{Re}\zeta = -\frac{ka}{\omega}\sin(kx-\omega t) = -\frac{\omega a}{g}\sin(kx-\omega t).$$

The profile of the wave is therefore sinusoidal. The velocity of the water has a horizontal component given by

$$u_x = \frac{\partial\phi}{\partial x} = ika\, e^{kz}\, e^{i(kx-\omega t)},$$

of which, if a is real, the physically significant part is

$$\mathrm{Re}u_x = -kae^{kz}\sin(kx-\omega t) = \omega e^{kz}\,\mathrm{Re}\zeta. \tag{5.27}$$

The vertical component of velocity, however, is

$$u_z = \frac{\partial\phi}{\partial\zeta} = kae^{kz}\, e^{i(kx-\omega t)},$$

such that

$$\mathrm{Re}u_z = kae^{kz}\cos(kx-\omega t).$$

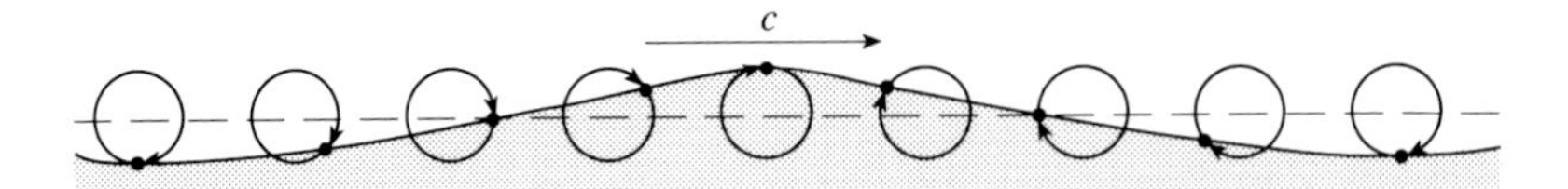

Figure 5.3 In a wave of small amplitude which is travelling from left to right with velocity c, individual particles on the surface of the water move clockwise with uniform angular velocity ω round trajectories which to first order are circles; here arrowheads are used to indicate the instantaneous direction of motion of the particles represented by small black dots. Particles below the surface also move round circles, the radii of which decrease exponentially with depth.

The two components vary sinusoidally with equal amplitude (for given ζ), but they are in quadrature. To first order in a it is true to say that each particle of water moves with angular velocity ω round a circle of radius $ka\mathrm{e}^{kz}$ as the wave passes by, where z denotes the particle's mean value of z, i.e. its level in the absence of any disturbance of the surface. Pathlines for particles of water on the surface are shown in fig. 5.3.

Two readily observed features of the behaviour of gravity waves on the ocean are accounted for by the above analysis. Firstly, there is the tendency of the large rollers which result from a storm at sea to have a shorter wavelength on the second day than on the first. The storm creates, far from the shore, a complex disturbance which propagates outwards as a group. As it does so it *spreads* [§5.1]: the many Fourier components present in the group separate from one another – hence the regularity of the rollers – and because $\mathrm{d}c_g/\mathrm{d}k$ is negative [(5.5)] the components which reach the beach first are those for which k is small and λ large. Secondly, there is the forward motion of the water near the crest of a roller and the rearward motion near its troughs, illustrated by fig. 5.3, which every swimmer has experienced.

Gravity waves carry energy, of course, and also momentum in the direction of propagation, as is dramatically apparent when large ocean waves encounter a breakwater. Both quantities depend upon the *square* of a (with additional terms proportional to higher powers), but they can be evaluated to second order using the first-order solution derived above. Thus to second order the mean kinetic energy per unit area of the surface is, if a is real,

$$\left\langle \int_{-\infty}^{\mathrm{Re}\zeta} \frac{1}{2}\,\rho\{(\mathrm{Re}u_x)^2 + (\mathrm{Re}u_z)^2\}\mathrm{d}z \right\rangle \approx \left\langle \int_{-\infty}^{0} \frac{1}{2}\,\rho\{(\mathrm{Re}u_x)^2 + (\mathrm{Re}u_z)^2\}\mathrm{d}z \right\rangle$$

$$\approx \frac{1}{2}\,\rho(ka)^2 \int_{-\infty}^{0} \mathrm{e}^{2kz}\,\mathrm{d}z$$

$$\approx \frac{1}{4}\,\rho k a^2, \tag{5.28}$$

where the angle brackets indicate an average over x. To the same order, the mean gravitational potential energy stored in the wave per unit area is

$$\left\langle \int_0^{\mathrm{Re}\zeta} \rho g z \mathrm{d}z \right\rangle = \frac{1}{2}\rho g \langle (\mathrm{Re}\zeta)^2 \rangle \approx \frac{1}{4}\rho g \left(\frac{ka}{\omega}\right)\left(\frac{\omega a}{g}\right). \tag{5.29}$$

As was to be expected, the mean kinetic and potential energies are equal. The mean horizontal momentum per unit area is

$$\begin{aligned}\left\langle \int_{-\infty}^{\mathrm{Re}\zeta} \rho \mathrm{Re} u_x \mathrm{d}z \right\rangle &\approx \rho\omega \left\langle \mathrm{Re}\zeta \int_{-\infty}^{\mathrm{Re}\zeta} \mathrm{e}^{kz}\mathrm{d}z \right\rangle \approx \frac{\rho\omega}{k}\langle \mathrm{Re}\zeta\ \mathrm{e}^{k\mathrm{Re}\zeta}\rangle \\ &\approx \rho c \langle \mathrm{Re}\zeta\,(1 + k\mathrm{Re}\zeta)\rangle \\ &\approx \frac{\rho k a^2}{2c},\end{aligned} \tag{5.30}$$

so to second order we have

$$\frac{\text{horizontal momentum}}{\text{total energy}} \approx \frac{1}{c}. \tag{5.31}$$

The equation which describes the momentum carried by sound waves in gases is not quite so simple [(3.54)].

5.4 Ripples ($|\zeta| \ll \lambda \ll d, \lambda \ll \lambda^*$)

If the wavelength of a small-amplitude wave on the surface of a liquid is much less than the wavelength λ^* defined by (5.20), then the restoring force due to surface tension outweighs that due to gravity and the wave is called a *capillary wave*, or else a *ripple*. In these circumstances (5.19) can be simplified by neglect of the gravitational term to

$$\left(\frac{\partial \phi}{\partial t}\right)_{z=0} \approx \frac{\sigma}{\rho}\frac{\partial^2 \zeta}{\partial x^2},$$

or for a disturbance which consists of a single 'monochromatic' wave of wave-vector k to

$$\left(\frac{\partial \phi}{\partial t}\right)_{z=0} \approx -\frac{\sigma k^2 \zeta}{\rho}.$$

Formally, this is precisely equivalent to (5.21) with g replaced by an effective gravitational field

$$g_{\mathrm{eff}} = \frac{\sigma k^2}{\rho}.$$

No other modification to the analysis in §5.3 is required, and we may at once conclude that the dispersion relation for ripples is

$$\omega^2 = g_{\text{eff}} k = \frac{\sigma k^3}{\rho}. \tag{5.32}$$

The group velocity for ripples is therefore

$$c_g = \frac{d\omega}{dk} = \frac{3\omega}{2k} = \frac{3}{2} c. \tag{5.33}$$

Surface waves which have wavelengths comparable with λ^*, intermediate between pure gravity waves and pure ripples, are also possible of course. To describe these, as is apparent from (5.19), we may use

$$g_{\text{eff}} = g + \frac{\sigma k^2}{\rho}$$

and obtain

$$\omega^2 = gk + \frac{\sigma k^3}{\rho}, \tag{5.34}$$

of which both (5.25) and (5.32) are limiting cases. In general, therefore, the phase velocity of small-amplitude surface waves on deep liquids is

$$c = \sqrt{\frac{g}{k} + \frac{\sigma k}{\rho}}. \tag{5.35}$$

Where

$$\frac{g}{k} = \frac{\sigma k}{\rho},$$

i.e. where $\lambda = \lambda^*$, it has a minimum value of

$$c^* = \sqrt[4]{\frac{4g\sigma}{\rho}}. \tag{5.36}$$

For water at a normal temperature, this minimum phase velocity is about 23 cm s^{-1}.

The dispersion relation represented by (5.34) is plotted in fig. 5.4. The minimum phase velocity corresponds to the slope of the line from the origin to P in that figure. Note that at P the phase and group velocities are equal.

5.5 Waves on a liquid–fluid interface

Suppose that a deep layer of stationary liquid of density ρ, with its upper surface in the horizontal plane $z = 0$, has a deep layer of stationary fluid on top of it, of

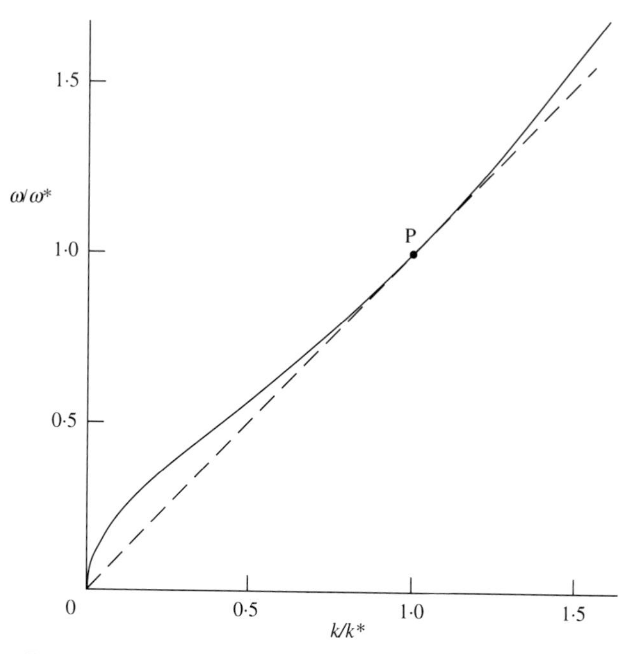

Figure 5.4 The dispersion curve for waves of small amplitude on the surface of deep water. The phase velocity is a minimum at P, where $k = k^* = 2\pi/\lambda^*$ and $\omega = \omega^* = (4\rho g^3/\sigma)^{1/4}$.

density ρ' ($<\rho$). When the interface is disturbed by a 'monochromatic' wave of vertical displacement, both fluids move. If the flow potential in the lower one is

$$\phi = a\mathrm{e}^{kz}\,\mathrm{e}^{\mathrm{i}(kx-\omega t)}\ [k > 0],$$

the potential in the upper one (also a solution of Laplace's equation) must be of the form

$$\phi' = a'\mathrm{e}^{-kz}\,\mathrm{e}^{\mathrm{i}(kx-\omega t)} + \text{constant}, \tag{5.37}$$

for unless k and ω are the same in both layers it is not possible to satisfy boundary conditions at the interface for all x and t. The boundary condition expressed by (5.16), however, is now complemented by

$$\left(\frac{\partial\phi'}{\partial z}\right)_{z=\zeta} = \frac{\partial\zeta}{\partial t} + \left(\frac{\partial\phi'}{\partial x}\right)_{z=\zeta}\frac{\partial\zeta}{\partial x}, \tag{5.38}$$

of which the linearised version is

$$\left(\frac{\partial\phi'}{\partial z}\right)_{z=0} = \frac{\partial\zeta}{\partial t}.$$

For waves of small amplitude, therefore,

$$\left(\frac{\partial \phi'}{\partial z}\right)_{z=0} = \left(\frac{\partial \phi}{\partial z}\right)_{z=0},$$

which implies

$$\phi'_{z=0} = -\phi_{z=0} + \text{constant}. \tag{5.39}$$

But the pressure difference across the interface must be

$$(p - p')_{z=\zeta} = \{(p - p_o) - (p' - p_o)\}_{z=\zeta} = -\sigma \frac{\partial^2 \zeta}{\partial x^2},$$

where σ is now the interfacial surface tension and where p_o, as above, is the pressure at the interface in a region where it is not disturbed. The boundary condition which corresponds to (5.17) may readily be derived from this with the aid of Bernoulli's theorem. In its linearised form it amounts to

$$(\rho - \rho')g\zeta + \rho\left(\frac{\partial \phi}{\partial t}\right)_{z=0} - \rho'\left(\frac{\partial \phi'}{\partial t}\right)_{z=0} = \sigma k^2 \zeta,$$

which can be reduced with the aid of (5.39) to

$$(\rho - \rho')g\zeta + (\rho + \rho')\left(\frac{\partial \phi}{\partial t}\right)_{z=0} = \sigma k^2 \zeta.$$

It corresponds to (5.21) with g replaced by

$$g_{\text{eff}} = \frac{\rho - \rho'}{\rho + \rho'} g + \frac{\sigma k^2}{\rho + \rho'},$$

so the dispersion relation for small-amplitude interfacial waves is

$$\omega^2 = \frac{\rho - \rho'}{\rho + \rho'} gk + \frac{\sigma k^3}{\rho + \rho'}. \tag{5.40}$$

The implications of this result as regards the stability of liquid–fluid interfaces are explored in §8.2.

5.6 Gravity waves, 2 ($|\zeta| \ll \lambda \approx d$)

The next case of interest is that of small-amplitude gravity waves on the free surface of a layer of water, say, which is not necessarily deep. The appropriate potential to describe a 'monochromatic' wave is now

$$\phi = (a^+ \mathrm{e}^{kz} + a^- \mathrm{e}^{-kz})\, \mathrm{e}^{\mathrm{i}(kx - \omega t)}, \tag{5.41}$$

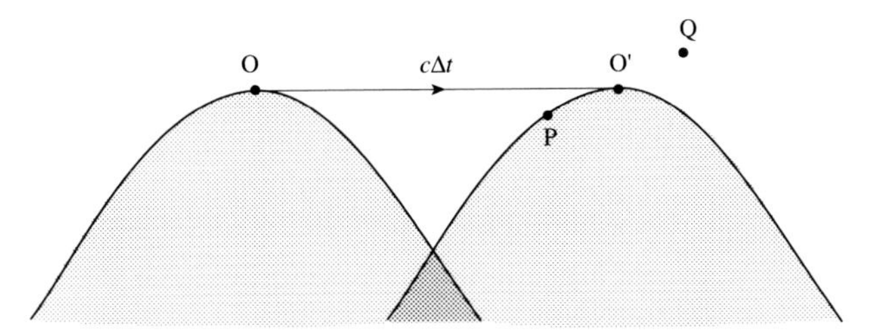

Figure 5.5 In time Δt, a wave crest moves from O to O$'$. The horizontal velocity of a particle of water initially at O must be less than the velocity of the crest, c; in that case it can drop to a point such as P and still remain on the surface. Were its horizontal velocity to exceed c the particle would rise to say Q, and that is out of the question.

and we must choose a^- so as to satisfy (5.14). This requires

$$a^- = a^+ e^{-2kd},$$

which means that (5.41) can be expressed as

$$\phi = 2a^+ e^{-kd} \cosh \{k(d + z)\} \, e^{i(kx - \omega t)}.$$

Equations (5.18) and (5.21) then become

$$2ka^+ e^{-kd} \sinh (kd) \, e^{i(kx-\omega t)} \approx \frac{\partial \zeta}{\partial t},$$

$$-2i\omega a^+ e^{-kd} \cosh (kd) \, e^{i(kx-\omega t)} \approx -g\zeta,$$

from which it follows that

$$\omega^2 = gk \tanh (kd). \tag{5.42}$$

This dispersion relation reduces to (5.25) when $kd \gg 1$, but at the other extreme, in liquid which is so shallow that $kd \ll 1$, it reduces to

$$\omega^2 = gdk^2. \tag{5.43}$$

In that limit gravity waves are non-dispersive and their velocity is $\sqrt{gd}$, a result already obtained by a more elementary argument in §2.14.

5.7 Gravity waves, 3 ($|\zeta| \approx \lambda \ll d$)

According to (5.27), the horizontal velocity at the free surface of a deep layer of water disturbed by a travelling gravity wave reaches a maximum value at the crests, where it points in the direction in which the wave is propagating and is equal to $\omega|\zeta|$. When $|\zeta|$ is small, this velocity is less than the phase velocity c with which the crest itself is moving forward. Hence a particle of water which lies on a crest at one instant of time lies a bit behind the crest a moment later, in a region where u_z is negative. Consequently, it drops from the point O in fig. 5.5 to the point P, say, still on the free surface. Suppose, however, that $\omega|\zeta| > c$. In that case the first-order solution discussed in §5.3 implies that the particle at O gets ahead of the crest into a region where u_z is positive, and that it is therefore to be found a moment later at a point such as Q. This is clearly nonsense, for Q lies above the free surface, in a region unoccupied by water. The contradiction is a sign that the first-order solution becomes inadequate for amplitudes such that

$$|\zeta| \approx \frac{c}{\omega} = \frac{\lambda}{2\pi}.$$

Waves with amplitudes of this order are not necessarily impossible, but they are governed by boundary conditions – (5.16) and (5.17) – in which the non-linear terms are certainly not negligible. These terms couple components of different k together. In consequence, a wave which has a sinusoidal profile at one instant of time does not remain sinusoidal (except in the limit of infinitesimal amplitude). Conversely, waves which propagate without change of form are not sinusoidal.

The first person to make a serious attempt to analyse the behaviour of gravity waves whose amplitude is not small was Stokes, and non-sinusoidal gravity waves which propagate without change of form are referred to as *Stokes waves* in his honour. He developed a method for expanding solutions in powers of the wave's *steepness* kZ, where Z is the difference of level between crest and trough ($= 2|\zeta|$ in the small-amplitude limit). It shows that as kZ increases the troughs become flatter and the crests become more sharply peaked.[3] It also shows that the phase velocity of a Stokes wave increases with amplitude; to second order,

$$c \approx \left(1 + \frac{1}{8}k^2Z^2\right)\sqrt{\frac{g}{k}} \tag{5.44}$$

on deep water. There is, however, a limiting value to kZ at which the peaks become cusps, and if energy is fed into a wave to increase its amplitude beyond this limit it 'breaks' to produce the foaming crests known as *white horses*. Stokes's expansion converges very slowly near this limit, but he was able to show by an independent argument that the cusp encloses an angle of 120°. The availability of very powerful computers has stimulated further work on the problem during the last twenty years, and it is now known that the limiting value of kZ for deep liquids is close to 0·887, that the limiting value of $c\sqrt{k/g}$ is close to 1·092, and that (to the surprise of experts in this field) the variation of c with kZ near the limit is oscillatory rather than monotonic.

Some computed wave profiles which include the limiting case are shown in fig. 5.6. Note that because the vertical scale is expanded relative to the horizontal scale the angle which should be 120° appears smaller.

[3] For ripples, however, the situation is reversed; as kZ increases, their crests become flatter than their troughs.

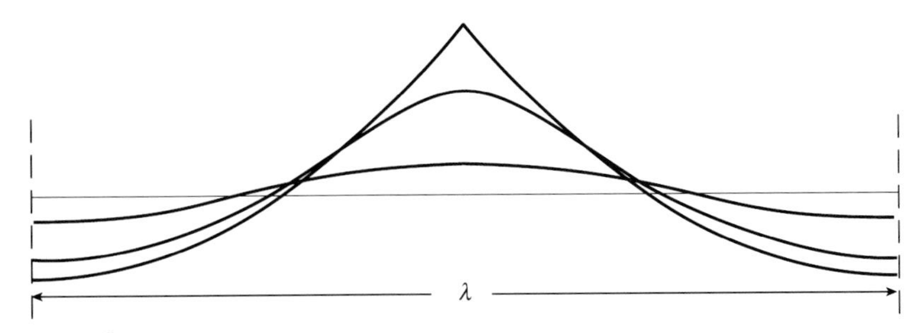

Figure 5.6 Profiles for Stokes waves of steepness $kZ = 0{\cdot}188$, $0{\cdot}628$ and $0{\cdot}887$; the vertical scale is twice as large as the horizontal scale.
[Based on a figure in L. W. Schwarz, *J. Fluid Mech.*, **62**, 553, 1974.]

5.8 Solitary waves ($|\zeta| \approx \lambda \approx d$)

Gravity waves of large amplitude on water which is not very deep compared with their wavelength are of particular practical importance in canals, and in order to understand canal waves properly we need first to understand how the sides of the canal *guide* them, in such a way that $\boldsymbol{k}$ often has no transverse component.

Most physicists are familiar with the guided electromagnetic waves which may be propagated between, say, two parallel conducting planes a distance b apart. These *do* have a transverse component of $\boldsymbol{k}$, which is restricted by the boundary conditions to certain discrete values. For TE (transverse electric) modes, for example, the condition that the electric field E_z vanishes at $y = \pm\frac{1}{2}b$ means that

$$\begin{aligned} E_z &\propto \cos\left(\frac{n\pi y}{b}\right) e^{i(k_x x - \omega t)} \; [n = 1, 3 \text{ etc.}] \\ E_z &\propto \sin\left(\frac{n\pi y}{b}\right) e^{i(k_x x - \omega t)} \; [n = 2, 4 \text{ etc.}], \end{aligned} \tag{5.45}$$

and the dispersion relation which determines ω for given k_x is affected by n; it is

$$\frac{\omega^2}{c_o^2} = k_x^2 + \frac{n^2\pi^2}{b^2}, \tag{5.46}$$

where c_o is the free-space velocity of light.

In that electromagnetic analogue $n = 0$ is forbidden, because it corresponds to E_z being zero everywhere, in which case no wave exists. The essential respect in which gravity waves differ is apparent when an ocean wave encounters a vertical wall, e.g. the sort of stone wall that is built to shelter harbours. Reflection from the wall creates a standing wave, but at the wall itself the displacement has an antinode rather than a node. That is because it is the horizontal component of $\boldsymbol{u}$ which vanishes at the wall (at any rate when the wave arrives at normal incidence), and u_x and ζ vary in quadrature as we have seen in §5.3. Consequently, for guided waves in canals with parallel, vertical sides we may expect small-amplitude waves to be possible in which the displacement varies thus:

$$\zeta \propto \cos\left(\frac{n\pi y}{b}\right) e^{i(k_x x - \omega t)} \quad [n = 0, 2 \text{ etc.}]$$
$$\zeta \propto \sin\left(\frac{n\pi y}{b}\right) e^{i(k_x x - \omega t)} \quad [n = 1, 3 \text{ etc.}]. \tag{5.47}$$

The dispersion relation is that of (5.46), though with c_0 representing $\sqrt{g \tanh(kd)/k}$ rather than the velocity of light [(5.42)]. On a plot of ω *versus* k the dispersion curve has many branches, but the lowest branch, for which n is zero, is the one of most practical importance. It is only this branch which we need to consider below.

In 1834, an engineer called Scott Russell, who was responsible for the design of barges for use on the Union Canal near Edinburgh, was observing a boat being drawn rapidly through the water by a pair of horses when the boat suddenly stopped. His graphic account of what then happened has been frequently quoted but is worth quoting again.

> The mass of water in the channel which [the boat] had put in motion ... accumulated round the prow of the vessel in a state of violent agitation, then suddenly leaving it behind, rolled forward with great velocity, assuming the form of a large solitary elevation, a rounded, smooth and well-defined heap of water, which continued its course along the channel apparently without diminution of form or change of speed. I followed it on horseback and overtook it, still rolling on at a rate of some eight or nine miles an hour, preserving its original figure some thirty feet long and a foot to a foot and a half in height.

The phenomenon which Scott Russell observed was what is now called a *solitary wave* or, less justifiably, a *soliton*. It is a most surprising phenomenon for two reasons: firstly, because (5.42) – to which (5.46) is equivalent when $n = 0$ – implies that surface waves are dispersive except in the limit $kd \gg 1$; secondly, because the argument in §2.16 suggests that such a disturbance should become steeper at its leading edge and develop a bore there. On both grounds, a hump of water having the dimensions quoted above, travelling along a canal where the water depth was presumably five or six feet, would surely have been expected *not* to preserve its original figure. Presumably the two tendencies, to dispersion and to bore formation, contrive in some circumstances to compensate for one another, though it is far from clear how they do so.

In 1895, Korteweg and de Vries introduced an approximate non-linear differential equation which they argued should describe the propagation of two-dimensional ($n = 0$) disturbances along a canal, in circumstances where $k\zeta$ and ζ/d are both significant but nevertheless a good bit less than unity. Now known as the *KdV equation*, it is

$$\frac{\partial \zeta}{\partial t} = \pm\sqrt{gd}\,\frac{\partial}{\partial x}\left(\zeta + \frac{1}{6}\,d^2\,\frac{\partial^2 \zeta}{\partial x^2} + \frac{3}{4}\,\frac{\zeta^2}{d}\right). \tag{5.48}$$

Here ζ, the elevation of the surface, is to be treated as real rather than complex, and the sign is to be chosen according to whether the disturbance is moving in the $-x$ or $+x$ directions. When $\zeta/d \ll 1$, so that the final term on the right-hand side may be ignored, the equation clearly has travelling-wave solutions of the form $\zeta \propto \sin(kx - \omega t)$ with

$$\omega = \sqrt{gdk^2}\left\{1 - \frac{1}{6}\,(kd)^2\right\}, \tag{5.49}$$

and (5.49) represents the first two non-zero terms in an expansion of (5.42) in powers of kd. Thus the middle of the three terms inside the brackets on the right-hand side of (5.48) is a plausible allowance for the dispersive tendency, and an allowance for the bore-formation tendency is similarly concealed in the final term.

The KdV equation does turn out to have solitary-wave solutions. They are expressible as

$$\zeta = \zeta_{\max}\,\mathrm{sech}^2\left(\frac{x \pm ct}{L}\right), \tag{5.50}$$

where the velocity of propagation c increases with height $\zeta_{\max}$, while the width L decreases with $\zeta_{\max}$, according to the equations

$$c = \sqrt{gd}\left(1 + \frac{\zeta_{\max}}{2d}\right), \quad L = 2\left(\frac{d^3}{3\zeta_{\max}}\right)^{1/2}. \tag{5.51}$$

The properties of these solutions have been the subject of intensive investigation during the last two decades. Perhaps the most striking of them is illustrated in fig. 5.7, a figure generated direct from the KdV equation by computer simulation. This shows a narrow solitary wave of large $\zeta_{\max}$ overtaking a lower and broader one; the two separate after the 'collision' with their profiles quite unchanged, although the narrow solitary wave has gained a bit of distance while the broad one has lost some.[4]

The great interest which solitary waves have aroused is partly due to the realisation that they occur in many other contexts. The KdV equation crops up in

[4] It is perhaps worth noting that two one-dimensional, localised, 'simple waves' of sound in a fluid which are travelling in opposite directions can also pass through one another with very little change, whatever their amplitude, as is shown in the appendix. However, such waves do not preserve their form as they propagate. They are not 'solitary waves'.

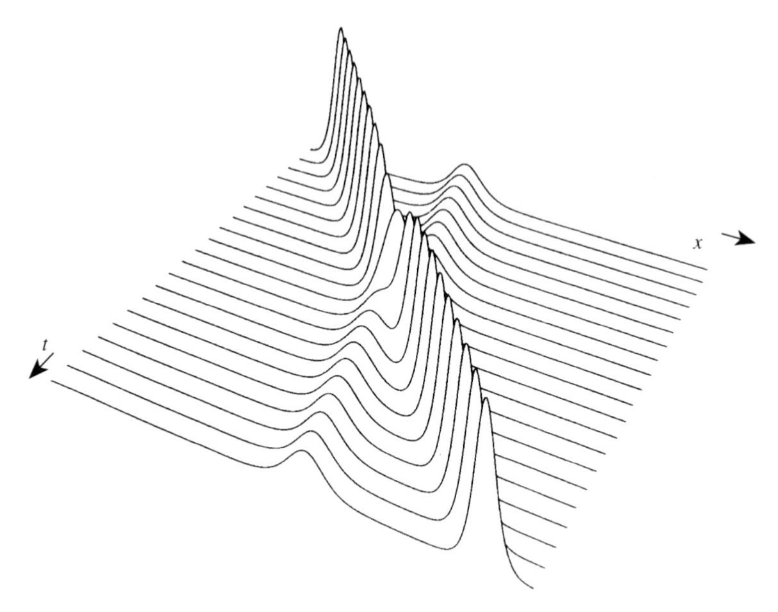

Figure 5.7 Numerical solution of the KdV equation, showing one solitary wave overtaking another.
[From R. K. Dodd, J. C. Eilbeck, J. D. Gibbon & H. C. Morris, *Solitons and Nonlinear Wave Equations*, Academic Press, New York, 1982.]

branches of physics which seem otherwise to have little connection with surface waves in canals, and moreover, it turns out to be only one of a number of non-linear differential equations which have solitary-wave solutions.

5.9 Ship waves in canals

In this section we return once more to gravity waves of small amplitude and consider how and when they are created when a solid object – perhaps a ship – moves steadily through stationary water. It is a topic of practical importance, since the creation of waves costs energy and results in an additional drag force on ships, to be discussed in the section that follows. The essential physics of the problem is most readily discerned when the waves propagate in one direction only, without spreading sideways from the object as they do if the surface of the water is unbounded. For this reason, we shall restrict our attention initially to guided waves in canals, and to the lowest ($n = 0$) branch of the dispersion curve in particular. The discussion is enlarged in §5.11.

The 'ships' used on canals are *barges*, of course, and we may picture an idealised barge as a fairly narrow object having a rectangular cross-section which does not vary along its length, and so long that initially we need to worry only about its front end, or *bow*. Its bow continuously displaces water as it advances and therefore

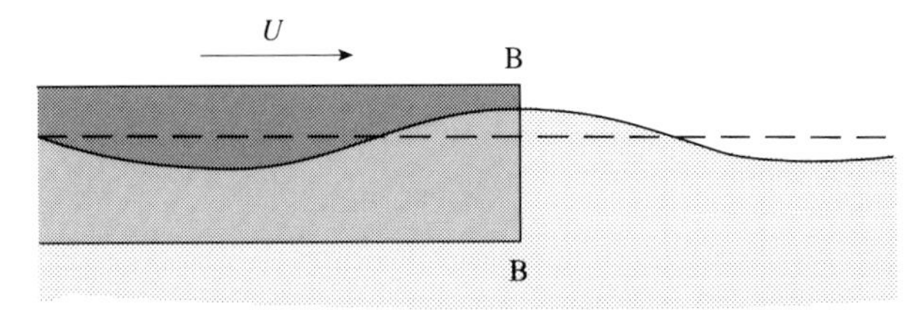

Figure 5.8 Idealised sketch of the front section of a barge moving along a canal in which the water is disturbed by a single harmonic wave. The bow BB feeds energy continuously into the wave provided that it moves with the same velocity as the crest with which it coincides.

acts as a moving source. It evidently displaces more water when it coincides with the crest of a wave, as sketched in fig. 5.8, than it does when no wave is present or when it coincides with a wave trough, so presumably it experiences a larger drag force then; this prediction is reinforced by the observation that underneath the crest of a gravity wave the pressure p for given z ($= p_0 + \rho g(\zeta - z)$) is enhanced. The extra work that has to be done to tow the barge in these circumstances is presumably fed into the wave and tends to increase its amplitude. If the wave speed is different from that of the barge, however, the bow coincides with a crest at some times but with a trough at others, and energy which is fed into the wave during one half-cycle is extracted during the other half. In that case the average rate of exchange of energy between barge and wave is zero (for small wave amplitudes at any rate), and in the absence of stimulation by some other source the wave is bound to die away because of viscous dissipation. Thus the only waves which the barge itself can stimulate are those that keep in step with it, with crest and bow coinciding at all times – or, to be more exact, coinciding over a (long) time at least equal to the time over which the wave, left to itself, would die away. Because the decay time cannot be infinite, the waves which the bow stimulates while moving from say x' to $x' + \Delta x'$ form a group with narrow bandwidth, which propagates with the group velocity. This is always less than the phase velocity for gravity waves in a canal, whatever the depth of water, and it is therefore less than the velocity of the barge. The wave group may be expected, therefore, to show up *behind the barge but not in front of it*.

Analytical verification requires contour integration but is otherwise fairly straightforward, as the following outline shows. If $\Delta x'$ is small, the immediate effect of the bow's displacement through this distance is surely a localised hump in the surface, whose height is proportional to $\Delta x'$ and whose range of localisation $L\{0\}$ is comparable, presumably, to the boat's width and depth. This hump may be treated as a superposition of Fourier components whose amplitude is more or less independent of their wavevector k, provided that k is much less than $L\{0\}^{-1}$. If we pick out the longitudinal Fourier components, i.e. those for which $n = 0$, and sum them after a time interval of t', the resultant displacement $\Delta\zeta\{x\}$ at some point x is proportional to

$$\int e^{i\{k(x-x')-\omega_k t'\}}\, e^{-\Gamma_k t'}\, dk, \tag{5.52}$$

where ω_k is the angular frequency corresponding to wavevector k and where the small positive quantity Γ_k describes attenuation. To obtain the total displacement $\zeta\{x\}$ we must now sum $\Delta\zeta\{x\}$ over all the previous displacements $\Delta x'$ which have led the barge to its 'present' position, at which the coordinates of its bow are say $x = 0$, $t = 0$. This sum involves a second integration of (5.52), over x' rather than k, from $x' = -\infty$ to zero. If the barge has been moving forward for a long time with uniform velocity U we may set $t' = -x'/U$, in which case $\zeta\{x\}$is proportional to

$$\int e^{ikx} dk \int_{-\infty}^{0} e^{\{-i(k-\omega_k/U)+(\Gamma_k/U)\}x'}\, dx' = U \int \frac{e^{ikx}\, dk}{-i(kU-\omega_k)+\Gamma_k}. \tag{5.53}$$

Now the integral over k on the right-hand side of (5.53) is dominated by contributions from Fourier components for which $(kU - \omega_k)$ is close to zero. Let us therefore write $k = k_o + \Delta k$, where k_o is the wavevector for which the phase velocity is exactly equal to the velocity of the boat, and

$$\omega_k \approx \omega_{k_o} + \left(\frac{d\omega_k}{dk}\right)_{k=k_o} \Delta k.$$

Then the right-hand side of (5.53) may to a good approximation be expressed as an integral over Δk with infinite limits as follows:

$$U e^{ik_o x} \int_{-\infty}^{\infty} \frac{e^{i\Delta kx}\, d\Delta k}{-iw_o \Delta k + \Gamma_{k_o}} \approx \frac{iU}{w_o} e^{ik_o x} \int_C \frac{e^{i\Delta kx}\, d\Delta k}{\Delta k + i\Gamma_{k_o}/w_o}, \tag{5.54}$$

where w_o – another positive quantity – is defined by

$$w_o = U - \left(\frac{d\omega_k}{dk}\right)_{k=k_o} = (c - c_g)_{k=k_o}. \tag{5.55}$$

On the right-hand side of (5.54) the integral is to be evaluated round a contour C in the complex Δk plane which runs from $-\infty$ to $+\infty$ along the real axis and is closed by a semicircle at infinity. Where $x > 0$ one must choose the semicircle in the upper half of the complex plane to ensure that the integral along this part of C is infinitesimal. In that case the integrand has no pole inside C and the integral is zero. Where $x < 0$, however, one must choose the lower semicircle. In that case C passes round the pole at $\Delta k = -i\Gamma_{k_o}/w_o$ in a clockwise sense, and it follows at once by the residue theorem that, at $t = 0$ when the bow is at $x = 0$,

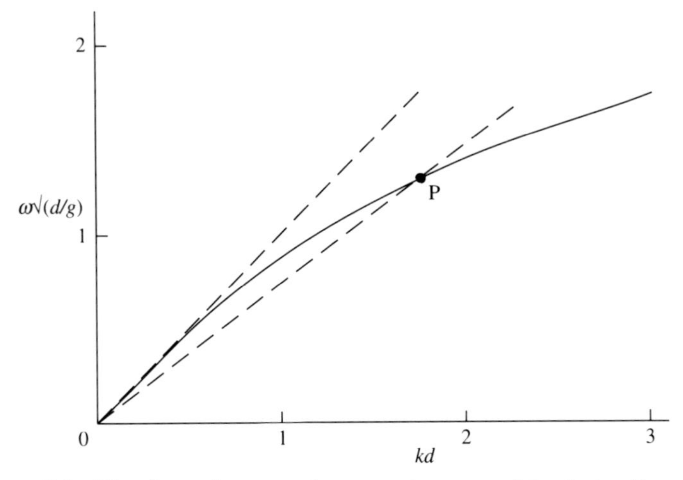

Figure 5.9 The dispersion curve for waves in water of depth d; ω/k nowhere exceeds $U_c = \sqrt{gd}$.

$$\zeta\{x\} \propto \frac{2\pi U}{w_o} \, e^{ik_o x} \, e^{\Gamma k_o x/w_o}. \tag{5.56}$$

The coefficient of proportionality depends on the width and depth of the barge, which determine the rate at which it displaces water, and is of no special interest here.

The result expressed by (5.56) is just what we anticipated. It tells us that our idealised barge is accompanied by a *bow wave* travelling in the same direction, whose wavelength is such that barge and wave remain in step with one another. A crest of the wave coincides with the bow and behind this the amplitude falls off slowly with distance. In front of the bow the amplitude drops suddenly, according to the approximate theory outlined above. A more exact treatment would show that the transition from finite to zero amplitude occurs over a length of order $L\{0\}$.

The $n = 0$ dispersion curve for gravity waves on a canal of finite depth, corresponding to (5.42), is plotted in fig. 5.9; it is straight where $kd \ll 1$ but bends over towards the deep-water dispersion curve as k increases. If the slope of the straight line from the origin to P in this figure is equal to the barge's speed, then k_o is the wavevector corresponding to the intersection at P. As U increases, k_o evidently decreases, i.e. the bow wave acquires a longer wavelength. However, there is a critical speed $U_c = \sqrt{gd}$, i.e. about 9·5 m.p.h. in water which is 6 feet deep, at which the wavelength becomes infinite. Once the barge exceeds this speed the conditions required for the existence of a purely longitudinal bow wave cannot be satisfied and no such wave is formed, as was discovered by a bargeman early in the nineteenth century in an uncontrolled experiment, when his horses

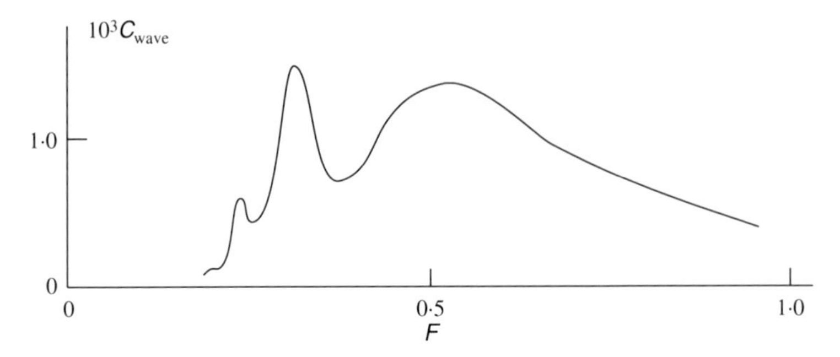

Figure 5.10 Variation of the wave resistance drag coefficient C_{wave} with Froude Number F in a typical case.

bolted. In the days of heavy canal traffic it proved enormously advantageous to maintain speeds greater than U_c, because the rate of erosion of the canals' banks was thereby much reduced.

5.10 Wave resistance

A submarine which is moving well below the surface experiences a *drag force*, and the energy which it expends against this, if not dissipated directly as heat, is taken up by eddies in the water behind it [chapter 7]. A boat which is moving on the surface also heats the water and also leaves eddies behind it, but it is liable to leave waves behind it as well. These waves take extra energy from the boat and must therefore contribute an extra term to the drag force, say ΔF_{D}, usually referred to as the *wave resistance*. Measurements on model boats in large tanks yield only the total drag force, but in some cases it is possible to estimate with fair precision the part which is *not* due to waves and to deduce the wave resistance by subtraction. Some typical results for a boat of barge-like shape, i.e. long and narrow, in deep water are sketched in fig. 5.10, where a dimensionless *drag coefficient* is plotted against the dimensionless *Froude Number*. We shall meet drag coefficients again in chapter 7, but differently defined; the quantity plotted in fig. 5.10 is

$$C_{\text{wave}} = \frac{\Delta F_{\text{D}}}{\frac{1}{2}\rho U^2 A},$$

where U is the velocity of the boat relative to the water and where A is the *wetted area* of the hull's surface, rather than an area of cross-section as in (7.21). As for the Froude Number, this is defined in terms of the boat's length L by

$$F = \frac{U}{\sqrt{Lg}}. \tag{5.57}$$

The striking thing about the curve in fig. 5.10 is that it has a number of distinct peaks in it. To explain these in detail is a complicated task, but their origin may be understood in a general way by reference to the one-dimensional case examined in the previous section, of a barge confined to a canal. In that section we dealt only with the *bow wave* attached to the front of the vessel, and we must now recognise that there is a *stern wave* attached to its rear as well. It is similar to the bow wave in all respects but one: whereas the bow acts as a *source*, in the sense that it displaces water outwards, the stern acts as a *sink*, and on this account the stern coincides with a trough of the stern wave rather than a crest. Behind the barge the bow and stern waves interfere with one another, of course, and the energy density in the combined wave system – and hence the wave resistance – is enhanced when the boat's speed is such that the interference is constructive. Because the bow and stern waves differ in phase by π at their respective origins, the condition for constructive interference is

$$L = \frac{2m+1}{2}\lambda_o = \frac{(2m+1)\pi}{k_o} \qquad [m = 0, 1, 2 \text{ etc.}],$$

which evidently corresponds to

$$F = \sqrt{\frac{1}{(2m+1)\pi}} = 0{\cdot}564,\ 0{\cdot}326,\ 0{\cdot}252 \text{ etc.} \tag{5.58}$$

if the water is deep enough for us to use (5.26) in the form

$$\frac{1}{k_o} = \frac{c_{k_o}^2}{g} = \frac{U^2}{g}.$$

The figures on the right-hand side of (5.58) correspond surprisingly closely to the values of F at which the peaks in fig. 5.10 occur.

Anyone who has sailed a small dinghy in a variable wind will have experienced the frustration of trying to get it to accelerate up to the speed which corresponds to the final peak, the one for which $m = 0$; the wave resistance, being proportional to U^2 as well as to C_{wave}, is increasing rapidly with U in this region. He will also have experienced as the wind gathers strength, however, the elation of feeling the boat begin to *plane*. When a small boat is planing it is normally well past the $m = 0$ peak and the wave disturbance behind it is relatively small. The first crest of the bow wave is apparent at the bow itself, however, and the first trough of the stern wave is apparent at the stern; the bow tends to lift in consequence and the stern tends to fall, so that the boat tilts as though it were moving uphill. In fact it is moving horizontally, of course, and because it is tilted the motion generates an increase in the pressure acting on its bottom surface. This raises the boat as a whole, bow and stern together, and thereby reduces the wetted area. The reduction may be

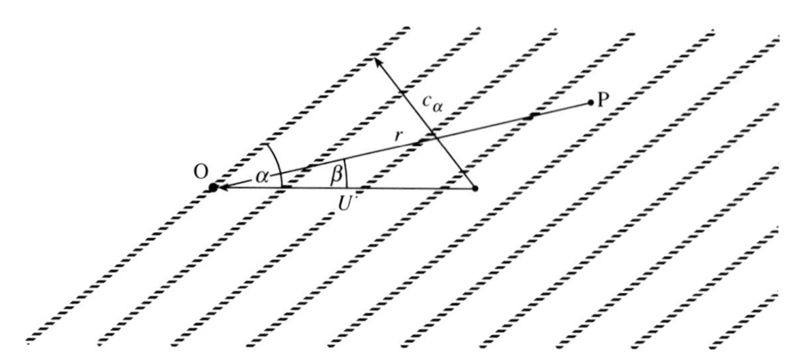

Figure 5.11 A set of wave crests accompanying a source O which is moving steadily from right to left with velocity U.

sufficient to ensure that over a limited range of speeds the total drag force actually falls as U increases.

It is not only sailing dinghies that plane, of course. Water-skiers and boats supported by hydrofoils do so too.

5.11 Ship waves on open water: the Kelvin wedge

On open water, the waves created by a moving ship cannot in any reasonable approximation be treated as exclusively longitudinal. The wavevectors $\boldsymbol{k}$ of their various Fourier components may point in any direction in the plane of the surface, and the corresponding linear wave crests may lie at any inclination α to the ship's velocity $\boldsymbol{U}$. It is still true, however, that the ship cannot feed energy continuously into a Fourier component to replace the losses due to viscous dissipation unless its bow travels with a crest (or else its stern with a trough). This 'surf-riding' condition restricts the waves which occur to those whose inclination and phase velocity c_α are such that

$$c_\alpha = U \sin \alpha \tag{5.59}$$

[fig. 5.11]. If the waves are gravity waves which satisfy (5.26), therefore, it restricts $\boldsymbol{k}$ to vectors whose inclination α and magnitude k_α are such that

$$k_\alpha = \frac{g}{U^2 \sin^2 \alpha}. \tag{5.60}$$

Figure 5.11 shows one set of linear wave crests supposed to satisfy this condition, accompanying a source O which could be the bow of a ship and which is moving steadily from right to left.[5] Note that they occupy only the half-space to

[5] Figure 5.11, like subsequent figures in this chapter but unlike fig. 5.8, conforms to the convention observed in chapters 3 and 4 that in cases of flow past obstacles it is the fluid which moves in the $+x$ direction, not the obstacle.

the right of the diagram; because group velocity is less than phase velocity for gravity waves, the amplitude of this particular component must be zero in the region ahead of the crest which passes through O itself [§5.9]. However, this is only one component out of many. We should picture the source as accompanied by many other similar sets of linear wave crests, having all possible values of α between $+\pi/2$ and $-\pi/2$ and various wavelengths to match.

In general, the phase angle χ_α associated with the set of waves of inclination α, which is here defined to be zero at O and at all other points along the crest which passes through O, varies so rapidly with α or k_α at points well behind this crest that the displacements of the surface due to different sets of waves tend to cancel. However, at points where χ_α happens to be stationary with respect to variation of k_α about some value k_{α_o} we may expect constructive interference over a band of wavevectors centred on k_{α_o}, and a periodic disturbance of wavelength $2\pi/k_{\alpha_o}$ will be apparent; the wave crests in this region will have the inclination α_o which corresponds to k_{α_o}.

Consider the arbitrary point P in fig. 5.11, which lies at a distance r from O along a line of inclination β as shown. Here

$$\chi_\alpha = \boldsymbol{k}_\alpha \cdot \boldsymbol{r} = -k_\alpha r \sin(\alpha - \beta), \tag{5.61}$$

so

$$\frac{d\chi_\alpha}{dk_\alpha} = -r \sin(\alpha - \beta) - k_\alpha r \cos(\alpha - \beta) \frac{d\alpha}{dk_\alpha}. \tag{5.62}$$

From (5.60) we have

$$\frac{dk_\alpha}{k_\alpha} = -2 \frac{\cos\alpha \, d\alpha}{\sin\alpha},$$

i.e.

$$k_\alpha \frac{d\alpha}{dk_\alpha} = -\frac{1}{2} \tan\alpha. \tag{5.63}$$

It follows from (5.62) and (5.63), when $d\chi_\alpha/dk_\alpha$ is set equal to zero, that at P the dominant inclination α_o is such that

$$\frac{\tan\alpha_o}{2 + \tan^2\alpha_o} = \tan\beta; \tag{5.64}$$

it depends only on β, not on r. The way in which the left-hand side of this equation varies with α_o in the range between $-90°$ and $+90°$ is represented graphically in fig. 5.12; the curve has extrema at

$$\alpha_o = \pm\tan^{-1}\sqrt{2} = \pm 54{\cdot}7°, \tag{5.65}$$

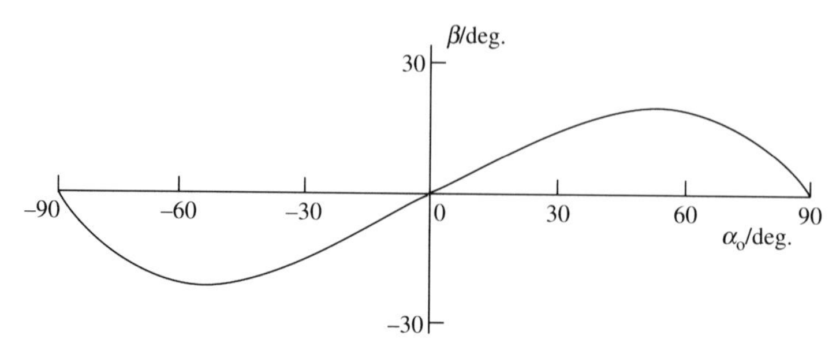

Figure 5.12 Variation of β with αo according to (5.64). The angle β is the inclination of a line through O in fig. 5.11 on which constructive interference is to be expected between sets of waves whose inclination is close to αo.

where

$$\beta = \pm \tan^{-1}\left(\frac{1}{2\sqrt{2}}\right) = \pm \sin^{-1}\left(\frac{1}{3}\right) = \pm 19{\cdot}5^\circ. \tag{5.66}$$

Evidently, constructive interference occurs only within a wedge-shaped region behind the source, with the source at its apex, for which $-19{\cdot}5^\circ \leqslant \beta \leqslant +19{\cdot}5^\circ$. For given β within this wedge there are in general two roots for α_o, so that two distinct wave systems arise. The crests for each system are naturally continuous but they curve, as shown in fig. 5.13, to ensure that α_o and the corresponding wavelength $2\pi/k_{\alpha_o}$ change with β in the manner dictated by (5.64) and (5.60). At the boundaries of the wedge the two systems come together. The amplitude of the disturbance is particularly large in this region because the bandwidth over which χ_α is virtually constant is particularly large: where $\beta = \pm\sin^{-1}(\frac{1}{3})$, $\mathrm{d}^2\chi_\alpha/\mathrm{d}k_\alpha^2$ vanishes as well as $\mathrm{d}\chi_\alpha/\mathrm{d}k_\alpha$. It should be emphasised that the angles which characterise the wave pattern in fig. 5.13, and in particular the angles described by (5.65) and (5.66), are completely independent of the boat's speed. The wavelengths, however, scale like U^2 [(5.60)].

The so-called *Kelvin wedge*, named after Lord Kelvin who was the first to analyse this problem, is a familiar phenomenon. It is illustrated not only by fig. 5.13 but also by the photograph in fig. 5.14; two overlapping sets of curved wave crests are clearly visible in this photograph. Behind any boat or ship such as the one shown in fig. 5.14 there must be bow waves and stern waves superposed on one another, of course, and not infrequently other wave systems may be discerned which appear to originate from parts of the hull between the bow and stern. The effects of such superposition, not to mention non-linear effects which can become significant when the amplitude is large, may complicate the wave pattern behind a moving ship to such an extent that its resemblance to the idealised Kelvin pattern of fig. 5.13 seems rather slight. Nevertheless the waves are normally confined, or virtually confined, to the predicted wedge.

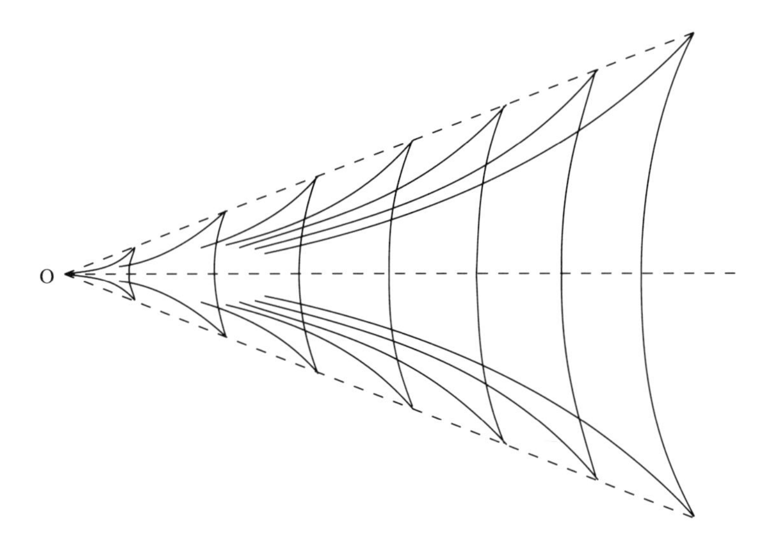

Figure 5.13 Wave crests in the Kelvin wedge.

On a smaller scale, something like a Kelvin wedge can be created in a bath, by dipping a finger into the surface of the water and moving it steadily sideways at a reasonably brisk speed. Alternatively, the finger may be dipped into a river and held stationary while the water moves past it, which amounts to the same thing though the frame of reference is different. Anyone who tries one of these experiments will immediately observe, however, that there are waves accompanying the finger in front of it as well as behind. Their existence can be understood by reference to fig. 5.15, which reproduces the deep-water dispersion curve of fig. 5.4. The line OPQ, whose slope is equal to the speed U, intersects the dispersion curve in two places, and (5.59) can be satisfied for any wavevector in the range between them, i.e. for any wavelength λ in the range between $2\pi/k_{\mathrm{P}}$ and $2\pi/k_{\mathrm{Q}}$, a range which embraces the critical wavelength defined by (5.20), $\lambda^* \approx 1{\cdot}7$ cm for water. Where $\lambda > 1{\cdot}7$ cm, the group velocity c_{g} is less than the phase velocity c, and the wave amplitude is then zero except behind the crest which passes through the source, as we have seen in §5.9. Where $\lambda < 1{\cdot}7$ cm, however, $c_{\mathrm{g}} > c$ and the reverse is true.

But if U is less than the minimum phase velocity, namely $c^* \approx 23\ \mathrm{cm\ s^{-1}}$ for water [(5.36)], then there are no values of k and α which satisfy the surf-riding condition of (5.59). In that case, as is readily confirmed by observation, no waves appear.[6]

[6] The creation of waves on the surface of a lake by a current of wind over it is a more complicated phenomenon than the creation of waves by a moving stick. There is a critical wind speed below which no waves appear but it is substantially larger than c^*. We return to this problem in 8.12.

Figure 5.14 Waves generated by a moving boat.
[Photograph from W. H. Munk *et al.*, *Proc. Roy. Soc. A*, **412**, 231, 1987, and previously from Naval Oceans System Center Technical Report no 978, 1985.]

5.12 Mach's construction revisited

The fact that the angle enclosed by the Kelvin wedge is independent of the speed of the source comes as a surprise to many physicists when it is first drawn to their attention, probably because they are familiar with Mach's construction [§3.11] and with the fact that the shock front cone which it predicts encloses a semi-angle $\alpha = \sin^{-1}(c/U)$ which is distinctly U-dependent. Mach introduced his construction, however, to describe the emission of sound waves by a moving projectile and it is based on an assumption that the waves are non-dispersive. If they are dispersive instead, the disturbances associated with infinitesimal displacements of the source near the points O′, O″ etc. in fig. 3.13 are not confined to the circles drawn in that diagram. The construction is then no longer useful.

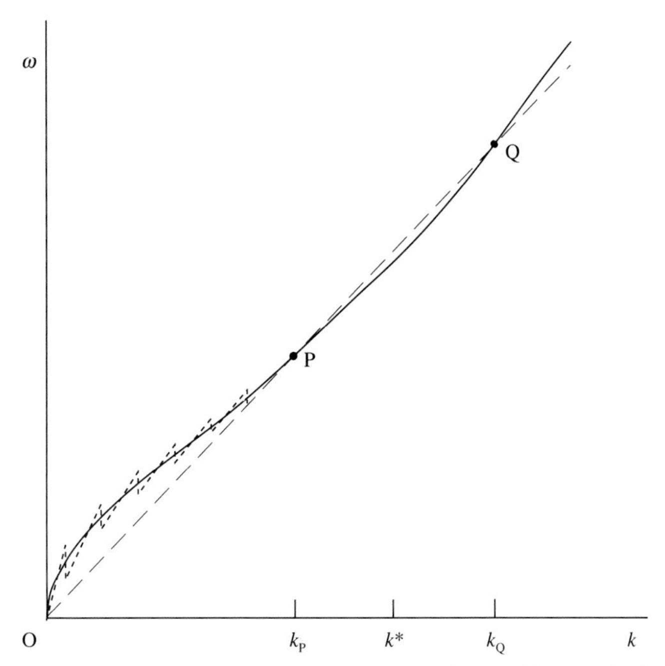

Figure 5.15 A source moving through water with a uniform velocity U, which corresponds to the slope of the line OPQ, creates waves which travel in the same direction with the same velocity, having wavevectors k_P and k_Q. The zigzag line represents an attempt to construct part of the dispersion curve out of non-dispersive segments; the construction in fig. 5.16 is based upon this.

It is possible to modify the construction so that it correctly predicts the angle of the Kelvin wedge, though the modified version rests on rather shaky foundations. Imagine a universe where the dispersion curve for gravity waves consists of a number of non-dispersive straight line segments, as suggested by a zigzag line in fig. 5.15, rather than a continuous curve. Wave emission within a bandwidth covered by a single segment can in that universe be treated by the normal Mach construction. Alternatively, it can be treated by the method outlined for the case where $\boldsymbol{k}$ is confined to one dimension in §5.9 above and extended to two dimensions in §5.11; the difference (w_o) between c and c_g is zero when the dispersion curve is a straight line through the origin, and (5.56) then describes just what the Mach construction describes – a disturbance limited to two lines passing through the source with inclinations $\pm\sin^{-1}(c/U)$. Two such lines, OP and OQ, are drawn in the bottom half of fig. 5.16, with different inclinations corresponding to the values of c associated with two different segments of the hypothetical dispersion curve. On these two lines, the parts of the disturbance attributable to emission by the source when it was at O‴ lie near the normals through O‴, i.e. near the points P and Q. From the geometry of the figure it is clear that these points,

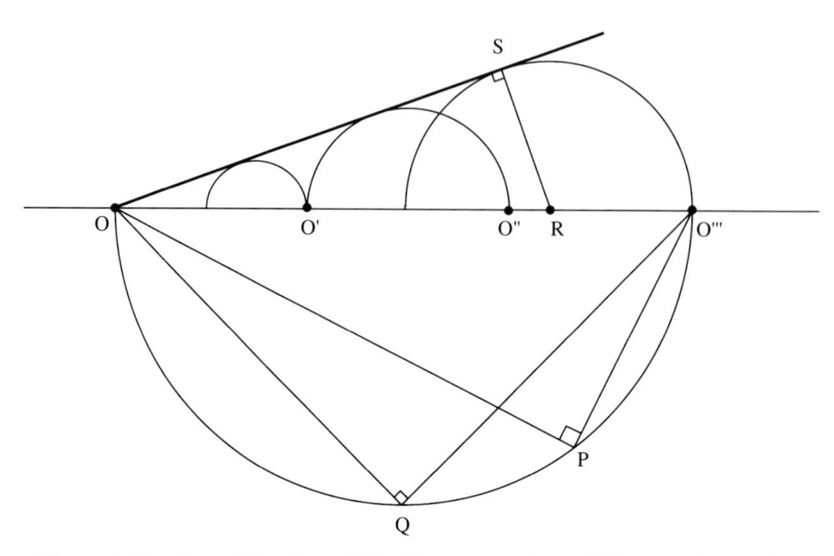

Figure 5.16 A modification of Mach's construction which correctly predicts the angle of the Kelvin wedge.

together with similar points associated with other segments of the hypothetical dispersion curve, lie on a semicircle with O‴O as diameter.

Now let us recognise that in our universe the dispersion curve is in fact continuous and that c_g is only half as big as c, and let us suppose that the only correction we need to make on that account is to redraw the semicircle with a diameter of $\frac{1}{2}$O‴O, as in the top half of fig. 5.16. Similar semicircles originating from O″ and O′ are included in the top half of the figure. They, and semicircles originating from points between O‴ and O which are not drawn, reinforce one another on the common tangent OS, and the inclination of this is

$$\sin^{-1}\left(\frac{\mathrm{RS}}{\mathrm{OS}}\right) = \sin^{-1}\left(\frac{1}{3}\right) = 19{\cdot}5^{\circ}$$

as required.

This argument is aesthetically attractive, like most arguments which rely upon geometry, but the extent to which it illuminates the physics of the Kelvin wedge is open to question.

Further reading

Chapters dealing in some detail with surface waves will be found in *Hydrodynamics*, Cambridge University Press, 1932, by H. Lamb. A more up-to-date reference on this topic is *Waves in Fluids*, Cambridge University Press, 1978, by J. Lighthill.

6

Viscosity

6.1 Shear stresses in Newtonian fluids

Throughout the last four chapters, the fact that real fluids possess viscosity has been almost completely ignored. We have supposed shear stress to be negligible and normal stress to be isotropic, and we have found that in so far as isotropic normal stress – the pressure p – depends upon fluid velocity $\boldsymbol{u}$, it does so through formulae in which only the local magnitude of $\boldsymbol{u}$ and its rate of change with time appear. We cannot proceed much further on that simple, Eulerian, basis. The principal aims of the present chapter are firstly to establish the Newtonian formulae which relate the components of stress in viscous fluids to *gradients* of $\boldsymbol{u}$, secondly to use these formulae to establish a more general equation of motion for fluids than Euler's equation, and thirdly to discuss a variety of relatively simple problems in which the effects of viscosity are dominant – so dominant in most cases that the fluid's inertia is negligible instead. The motion of fluids in such circumstances is sometimes referred to as *creeping flow*.

Newton himself may have considered only the simple situation illustrated by fig. 6.1, where planar laminae of fluid lying normal to the x_2 axis are moving

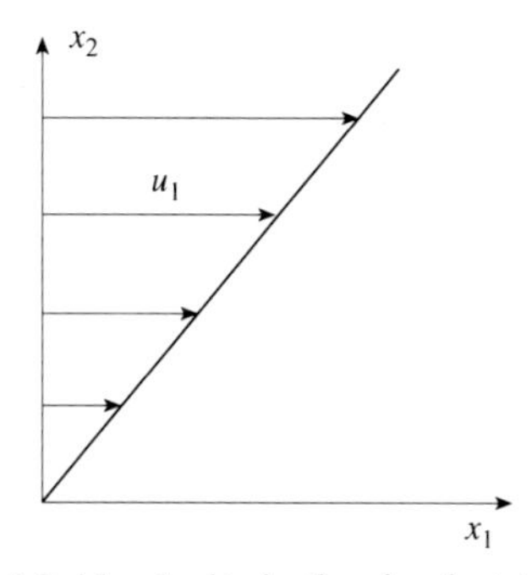

Figure 6.1 Profile of fluid velocity in the simplest type of laminar flow.

steadily in the x_1 direction and sliding over one another, so that there exists a uniform velocity gradient $\partial u_1/\partial x_2$. He postulated that in such circumstances a frictional shear stress arises between adjacent laminae, of magnitude

$$s_{12} = \eta \frac{\partial u_1}{\partial x_2}, \tag{6.1}$$

where η is a coefficient – the *viscosity* or *shear viscosity* – which depends upon the nature of the fluid. [See fig. 1.1 for a reminder as to how s_{12} is defined.] The assumption that the relation between shear stress and velocity gradient takes this linear form evidently satisfies the basic requirement that when the gradient vanishes everywhere, in which case there exists an inertial frame in which the fluid is at rest in equilibrium, s_{12} must vanish. It also satisfies the symmetry requirement that s_{12} must change sign when the motion is reversed. These two requirements would be satisfied equally well by more complicated formulae including non-linear terms in $(\partial u_1/\partial x_2)^3$, $(\partial u_1/\partial x_2)^5$ etc., but it is plausible to suppose that such higher-order terms, if they occur, are relevant only at rates of shear which are larger than those normally encountered.

Equation (6.1) cannot suffice, however, in situations which are more complicated than the one illustrated by fig. 6.1. Suppose that the fluid is moving not only along x_1 but also along x_2, and that there exist two non-zero velocity gradients, $\partial u_1/\partial x_2$ and $\partial u_2/\partial x_1$. It would follow from (6.1) that in that case a shear stress would exist on planes normal to the x_1 axis of magnitude

$$s_{21} = \eta \frac{\partial u_2}{\partial x_1}, \tag{6.2}$$

the coefficient being identical to the coefficient in (6.1) if the properties of the fluid are isotropic, as we assume to be the case. But (6.1) and (6.2) do not satisfy the requirement that s_{12} and s_{21} are always equal [§1.3]. Evidently, we must symmetrise them by writing

$$s_3 \ (= s_{12} = s_{21}) = \eta \left(\frac{\partial u_1}{\partial x_2} + \frac{\partial u_2}{\partial x_1} \right). \tag{6.3}$$

Now consider an infinitesimal element of fluid having a square cross-section ABCD of side d, as sketched in fig. 6.2. Relative to its centre O, the corner B has a velocity whose component along the line OB is

$$(u_{1,\mathrm{B}} - u_{1,\mathrm{O}}) \cos \frac{\pi}{4} + (u_{2,\mathrm{B}} - u_{2,\mathrm{O}}) \cos \frac{\pi}{4} = \frac{d}{2\sqrt{2}} \left(\frac{\partial u_1}{\partial x_1} + \frac{\partial u_1}{\partial x_2} + \frac{\partial u_2}{\partial x_1} + \frac{\partial u_2}{\partial x_2} \right).$$

This means that in a different frame of reference – say S′ as opposed to S – such that the x_1' axis coincides with OB we have

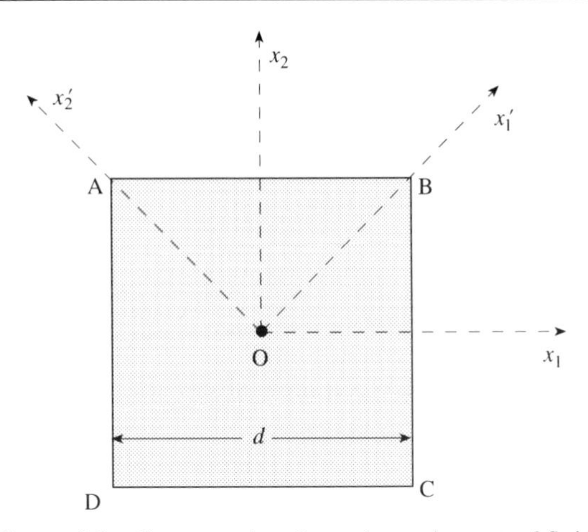

Figure 6.2 Cross-section through an element of fluid.

$$\frac{\partial u_1'}{\partial x_1'} = \frac{d}{2\sqrt{2}\mathrm{OB}} \left(\frac{\partial u_1}{\partial x_1} + \frac{\partial u_1}{\partial x_2} + \frac{\partial u_2}{\partial x_1} + \frac{\partial u_2}{\partial x_2} \right),$$

or, if we make use of (6.3) and of the fact that d and $\sqrt{2}$ OB are the same,

$$2\frac{\partial u_1'}{\partial x_1'} = \frac{\partial u_1}{\partial x_1} + \frac{\partial u_2}{\partial x_2} + \frac{s_3}{\eta}. \tag{6.4}$$

A similar argument applied to the rate of change of OA shows that

$$2\frac{\partial u_2'}{\partial x_2'} = \frac{\partial u_1}{\partial x_1} + \frac{\partial u_2}{\partial x_2} - \frac{s_3}{\eta}, \tag{6.5}$$

and it is apparent from (6.4) and (6.5) that

$$2s_3 = 2\eta \left(\frac{\partial u_1'}{\partial x_1'} - \frac{\partial u_2'}{\partial x_2'} \right). \tag{6.6}$$

Since an elementary modification of the argument used in §1.3 to prove (1.2) tells us that

$$2s_3 = -p_1' + p_2',$$

we may deduce from (6.6) that

$$p_1' + 2\eta \frac{\partial u_1'}{\partial x_1'} = p_2' + 2\eta \frac{\partial u_2'}{\partial x_2'}. \tag{6.7}$$

This result, like (6.3) on which it is based, remains valid when a Newtonian fluid undergoes three-dimensional shear flow, i.e. when u_3 varies with position as well as u_1 and u_2. In the general case we have

$$p_1' + 2\eta \frac{\partial u_1'}{\partial x_2'} = p_2' + 2\eta \frac{\partial u_2'}{\partial x_2'} = p_3' + 2\eta \frac{\partial u_3'}{\partial x_3'}, \tag{6.8}$$

and corresponding equations necessarily hold for the normal stresses in frame S. From these we may readily deduce that

$$p_1 = p - \frac{2}{3}\eta \left(2\frac{\partial u_1}{\partial x_1} - \frac{\partial u_2}{\partial x_2} - \frac{\partial u_3}{\partial x_3}\right), \tag{6.9}$$

where

$$3p = p_1' + p_2' + p_3' = p_1 + p_2 + p_3. \tag{6.10}$$

Equations (6.3) and (6.9), and four other equations which may be obtained from these by permutation of the suffices 1, 2 and 3, all stem from Newton's initial postulate that the relation between shear stress and velocity gradient is linear, coupled with (a) the requirement that shear stresses are symmetric in the sense that $s_{12} = s_{21}$ etc. and (b) our assumption that the fluid is isotropic. They provide the starting point for any discussion of viscous flow. They turn out to predict results which are in agreement with experiment, at any rate for fluids whose molecular structure is relatively simple, and therein lies their ultimate justification.

A reader who has previously met viscosity but only in the context of fig. 6.1, where its effects are adequately expressed by (6.1) alone, may be taken aback by the complexity of these results. A moment's reflection shows, however, that (6.9) is eminently reasonable. It tells us that in order to draw out an element of fluid in the x_1 direction, while allowing it to contract in the other two principal directions, we need to make p_1 less than p_2 and p_3. The degree of anisotropy which is required in these normal stresses depends upon the rate of deformation and, of course, upon the fluid's viscosity. Equation (6.9) can, incidentally, be written in other ways in circumstances where, for one reason or another, the fluid is effectively incompressible. The continuity condition which then applies, (2.2), means that

$$p_1 = p - 2\eta \frac{\partial u_1}{\partial x_1}, \tag{6.11}$$

or else that

$$p_1 = p + 2\eta \left(\frac{\partial u_2}{\partial x_2} + \frac{\partial u_3}{\partial x_3}\right). \tag{6.12}$$

Both forms will prove useful below.

6.2 Stress and rate of deformation as tensors

The results derived so laboriously in the previous section are almost obvious to anyone who is well grounded in elementary tensor analysis. The stress at each point in a continuous medium has nine components (though not all of them are different), which transform under rotation of the reference frame in the same way that the nine components of the generalised product of two vectors transform. Thus stress is a tensor of the second rank. It is usually represented, using conventional suffix notation, by σ_{ij}; the components for which $i \neq j$ are the shear stresses represented hitherto in this book by s_{ij}, but the normal components are defined as positive when they act outwards and tend to decrease the local density, so that σ_{ii} is equivalent to $-p_i$. In a solid medium, stress normally generates elastic strain, and this too has nine components which transform in such a way that they constitute a second-rank tensor; in terms of a local displacement vector, say ξ, it may be represented by $\partial\xi_i/\partial x_j$. In a fluid medium what matters is the rate of change of strain, following the motion of the fluid, or *rate of deformation*. This is yet another second-rank tensor,

$$\frac{\mathrm{D}(\partial\xi_i/\partial x_j)}{\mathrm{D}t} = \frac{\partial(\mathrm{D}\xi_i/\mathrm{D}t)}{\partial x_j} = \frac{\partial u_i}{\partial x_j}. \tag{6.13}$$

Now any second-rank tensor may be expressed as the sum of three parts, one of which is isotropic in character and the other two anisotropic. The two anisotropic parts are traceless tensors (the word 'traceless' means in this context that when the tensors' components are written out in matrix form the diagonal ones sum to zero), one of them symmetric and the other antisymmetric; the components of the antisymmetric part change sign when the reference axes are reflected (i.e. when they are labelled according to the left-handed convention instead of the right-handed one, or *vice versa*), but the components of the symmetric part are unaffected by reflection. The stress tensor, for example, may be divided thus:

$$\sigma_{ij} = \frac{1}{3}\,\delta_{ij}\sigma_{\mathrm{mm}} + \frac{1}{2}\left(\sigma_{ij} + \sigma_{ji} - \frac{2}{3}\,\delta_{ij}\sigma_{\mathrm{mm}}\right) + \frac{1}{2}\left(\sigma_{ij} - \sigma_{ji}\right), \tag{6.14}$$

where, according to the standard summation convention for repeated dummy suffices,

$$\sigma_{\mathrm{mm}} = \sigma_{11} + \sigma_{22} + \sigma_{33} = -3p; \tag{6.15}$$

σ_{mm} is necessarily a *scalar* quantity, unaffected by changes in the reference frame. Using the fact that s_{ij} and s_{ji} (and therefore σ_{ij} and σ_{ji}) are always equal, for reasons given in §1.3, we may write the symmetric anisotropic part of the stress tensor – the second term on the right-hand side of (6.14) – as

$$\sigma_{ij} + \delta_{ij}p = q_{ij} \text{ (say)}, \tag{6.16}$$

while the antisymmetric anisotropic part – the third term – evidently vanishes.

If the rate of deformation tensor is divided in this way its isotropic part turns out to be

$$\frac{1}{3}\,\delta_{ij}\,\frac{\partial u_{\mathrm{m}}}{\partial x_{\mathrm{m}}}\quad\left[\equiv\frac{1}{3}\,\delta_{ij}\boldsymbol{\nabla}\cdot\boldsymbol{u}\right],\tag{6.17}$$

while its symmetric and antisymmetric anisotropic parts are respectively

$$\frac{1}{2}\left(\frac{\partial u_i}{\partial x_j}+\frac{\partial u_j}{\partial x_i}-\frac{2}{3}\,\delta_{ij}\,\frac{\partial u_{\mathrm{m}}}{\partial x_{\mathrm{m}}}\right)=\zeta_{ij}\ \text{(say)}\tag{6.18}$$

and

$$\frac{1}{2}\left(\frac{\partial u_i}{\partial x_j}-\frac{\partial u_j}{\partial x_i}\right)=\omega_{ij}\ \text{(say)}.\tag{6.19}$$

The symbol ω is appropriate in (6.19) because what ω_{ij} describes on its own is the local rate of rotation of the medium; its six non-zero components are the components of the vectors $+\frac{1}{2}\boldsymbol{\Omega}$ and $-\frac{1}{2}\boldsymbol{\Omega}$, where $\boldsymbol{\Omega}$ is the vorticity. What ζ_{ij} describes on its own is a type of shear flow which is vorticity-free. Note that in the situation illustrated by fig. 6.1, where $\partial u_2/\partial x_1$ and $\partial u_{\mathrm{m}}/\partial x_{\mathrm{m}}$ (i.e. $\boldsymbol{\nabla}\cdot\boldsymbol{u}$) are zero while $\partial u_1/\partial x_2$ is not, we have $\zeta_{12}=\omega_{12}$. The velocity profile in that figure represents a combination in equal parts of clockwise rotation and vorticity-free shear [fig. 6.3].

Where one second-rank tensor depends upon another in a linear fashion, the coefficient is a fourth-rank tensor which in general may have up to 81 independent components. However, the fourth-rank tensor which relates stress to rate of deformation in a Newtonian fluid must be isotropic if the fluid itself is isotropic, and this greatly reduces its complexity. It turns out that each of the three parts of the stress tensor must then be separately related in a linear fashion to the corresponding part of the rate of deformation tensor, and that the coefficient is in each case a scalar. For example, we must expect

$$\frac{1}{2}\,(\sigma_{ij}-\sigma_{ji})\propto\omega_{ij}.$$

In this case the scalar coefficient of proportionality must be zero because the antisymmetric part of the stress is always zero, and this is no surprise; local rotation does not change the separation between any two points embedded in the fluid an infinitesimal distance apart, so there is no reason to expect it to give rise to stress. More significantly, we must expect q_{ij} to be proportional to ζ_{ij}, and by choosing the constant of proportionality to be 2η we arrive at once, using (6.16) and (6.18), at

$$s_{ij}=\eta\left(\frac{\partial u_i}{\partial x_j}+\frac{\partial u_j}{\partial x_i}\right);\quad -p_i+p=\eta\left(2\,\frac{\partial u_i}{\partial x_i}-\frac{2}{3}\,\frac{\partial u_{\mathrm{m}}}{\partial x_{\mathrm{m}}}\right).$$

These results correspond exactly to (6.3) and (6.9).

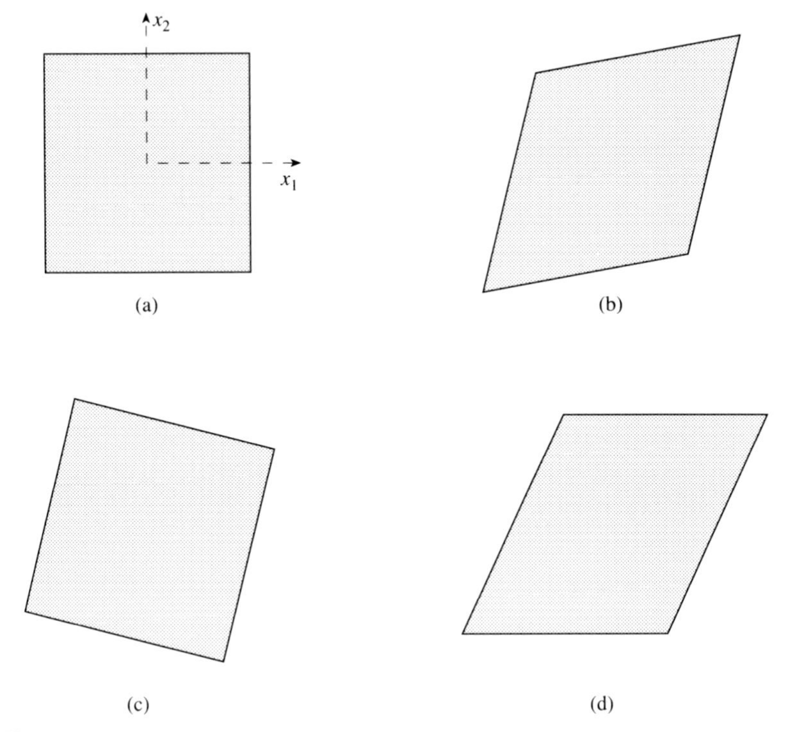

Figure 6.3 The effect on the square fluid element shown in (a) of (b) vorticity-free shear ($\zeta_{12} > 0, \omega_{12} = 0$), (c) pure rotation ($\zeta_{12} = 0, \omega_{12} > 0$), and (d) an equal combination of the two ($\zeta_{12} = \omega_{12} > 0$).

6.3 Bulk viscosity

The results of the previous section suggest that in a Newtonian fluid a linear relation is likely to exist between the isotropic part of the stress tensor, $\delta_{ij}p$, and the isotropic part of the rate of deformation tensor, $\frac{1}{3}\delta_{ij}(\partial u_{\rm m}/\partial x_{\rm m})$ or $\frac{1}{3}\delta_{ij}\boldsymbol{\nabla}\cdot\boldsymbol{u}$. The two cannot be simply proportional to one another, however, because p does not vanish when $\boldsymbol{\nabla}\cdot\boldsymbol{u}$ vanishes. Instead, it is related in that limit to the local density ρ and temperature T through an equation of state

$$p = p_{\rm equ}\{\rho, T\},$$

where the suffix stands for 'equilibrium'. Hence we should presumably write

$$-p = -p_{\rm equ} + \eta_{\rm b}\boldsymbol{\nabla}\cdot\boldsymbol{u}, \tag{6.20}$$

on the understanding that, if the rate of deformation is so rapid that the medium is no longer in local thermal equilibrium at a well-defined temperature, $p_{\rm equ}$ can in principle be calculated from the instantaneous energy density of the fluid instead. The scalar constant $\eta_{\rm b}$ is known as the *bulk viscosity* of the fluid, or sometimes as its *second viscosity*, to distinguish it from the shear viscosity η. Since $\nabla \cdot \boldsymbol{u}$ is related to rate of change of density through the continuity condition [(2.6)], (6.20) can be expressed as

$$p = p_{\text{equ}} + \frac{\eta_{\text{b}}}{\rho}\frac{\text{D}\rho}{\text{D}t} \tag{6.21}$$

if preferred.

Bulk viscosity is almost always irrelevant in fluid dynamics, either because the Mach Number is small enough for changes of density to be negligible, or because, in compressible flow, the Reynolds Number is large enough for the shear viscosity to be negligible as well. It does play a part, however, in the theory of the attenuation of sound waves. The velocity field which characterises a planar longitudinal sound wave of small amplitude, propagating along the x_1 direction say, is such that u_1 varies sinusoidally with x_1 but is independent of x_2 and x_3, while u_2 and u_3 are everywhere zero. It follows from (6.9) and (6.20) that in such circumstances

$$p_1 = p_{\text{equ}} - \left(\frac{4}{3}\eta + \eta_{\text{b}}\right)\frac{\partial u_1}{\partial x_1}, \tag{6.22}$$

and this result provides the basis for a discussion of viscous attenuation in the appendix at the end of this book.

The point is made there that in so far as experimental values are available for the bulk viscosity of fluids they are derived from ultrasonic measurements at high frequencies, where 'anomalous' attenuation is frequently observed which cannot be explained in terms of thermal conduction and shear viscosity alone. At these high frequencies bulk viscosity is normally comparable in magnitude with shear viscosity, except in the case of monatomic gases for which it is virtually zero. It is, however, a frequency-dependent quantity, and in polyatomic gases at low frequencies η_{b} may greatly exceed η. Nevertheless, bulk viscosity is normally irrelevant. Except in the appendix, and briefly in §6.12, it does not reappear in this book.

6.4 The Navier–Stokes equation

For an Euler fluid in which no shear stress exists and the pressure is isotropic, the force per unit mass which acts on the fluid due to pressure gradients and to its own weight has a component in the x_1 direction given by

$$f_1 = -\frac{1}{\rho}\frac{\partial p}{\partial x_1} - g\frac{\partial z}{\partial x_1}$$

[§§1.7 and 2.3]. If the fluid has viscosity, we clearly need to replace p in this expression by the pressure which acts in the x_1 direction, p_1. In addition, we need

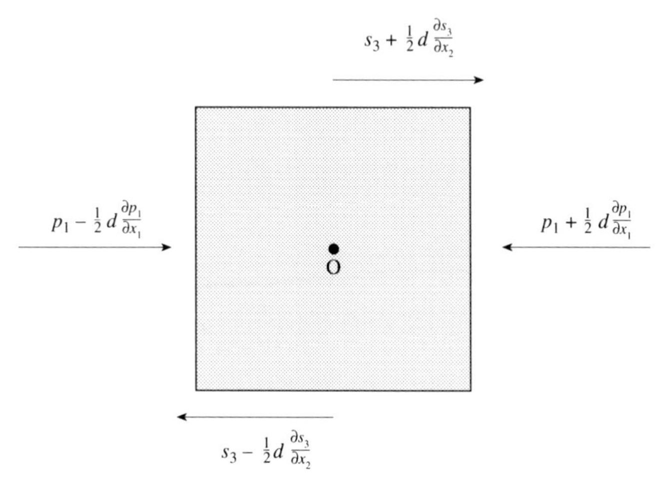

Figure 6.4 Two of the three stress components acting on the fluid element shown in figs. 6.2 and 6.3(a) which may cause it to accelerate in the x_1 direction; the shear stress s_2, which acts on faces parallel to the plane of the diagram and is not represented, is also relevant in this connection.

to allow for forces associated with gradients of shear stress, and the full expression for f_1 becomes [fig. 6.4]

$$f_1 = \frac{1}{\rho}\left(-\frac{\partial p_1}{\partial x_1} + \frac{\partial s_3}{\partial x_2} + \frac{\partial s_2}{\partial x_3} - g\frac{\partial z}{\partial x_1}\right). \tag{6.23}$$

Supposing the fluid to be Newtonian and effectively incompressible, we may conveniently use (6.12) for p_1, while permutations of (6.3) may be used for s_2 and s_3. If the coefficient of viscosity is uniform, a condition which in practice may limit the validity of the results that follow to situations in which the temperature of the fluid is uniform as well as its density, (6.23) may then be expressed in terms of second derivatives of the velocity as follows:

$$f_1 = -\frac{\partial}{\partial x_1}\left(\frac{p}{\rho} + gz\right) + \frac{\eta}{\rho}\left(-2\frac{\partial^2 u_2}{\partial x_1 \partial x_2} - 2\frac{\partial^2 u_3}{\partial x_1 \partial x_3} + \frac{\partial^2 u_1}{\partial x_2^2} + \frac{\partial^2 u_2}{\partial x_2 \partial x_1} + \frac{\partial^2 u_3}{\partial x_3 \partial x_1} + \frac{\partial^2 u_1}{\partial x_3^2}\right).$$

After rearrangement of terms this becomes

$$f_1 = -\frac{\partial}{\partial x_1}\left(\frac{p}{\rho} + gz\right) - \frac{\eta}{\rho}\left\{\frac{\partial}{\partial x_2}\left(\frac{\partial u_2}{\partial x_1} - \frac{\partial u_1}{\partial x_2}\right) - \frac{\partial}{\partial x_3}\left(\frac{\partial u_1}{\partial x_3} - \frac{\partial u_3}{\partial x_1}\right)\right\},$$

and the complicated expression enclosed by curly brackets on the right-hand side is just the x_1 component of $\boldsymbol{\nabla} \wedge (\boldsymbol{\nabla} \wedge \boldsymbol{u})$, i.e. of $\boldsymbol{\nabla} \wedge \boldsymbol{\Omega}$, where $\boldsymbol{\Omega}$ is the local vorticity. Since similar expressions hold for f_2 and f_3, the total force per unit mass which acts upon the fluid may be expressed in vector form as

$$\boldsymbol{f} = -\nabla\left(\frac{p}{\rho} + gz\right) - \frac{\eta}{\rho}\nabla \wedge \boldsymbol{\Omega}. \tag{6.24}$$

By equating this to the local acceleration [(2.7)] we arrive at the simplest version of the celebrated *Navier–Stokes equation*,

$$-\nabla p^* - \eta\nabla \wedge \boldsymbol{\Omega} = \rho\frac{\partial \boldsymbol{u}}{\partial t} + \rho(\boldsymbol{u}\cdot\nabla)\boldsymbol{u}, \tag{6.25}$$

where p^* is the local *excess mean pressure* defined by (2.21). Equation (6.25) is the equation of motion which replaces Euler's equation for a fluid which has viscosity but which is still effectively incompressible and also, to be on the safe side, isothermal. It differs from Euler's equation only, of course, in so far as it includes a viscous term.

Readers who find the Navier–Stokes equation indigestible should not lose heart, for little use is made of it in what follows. Few exact analytical solutions of the equation are known, and most of these are trivial enough to be reachable from simpler starting points. There is a growing branch of fluid dynamics in which relatively complex flow problems are tackled by computer simulation, and the Navier–Stokes equation provides the basis for this approach, but a discussion of the techniques of computer simulation would be out of place in this book. For our purposes, the chief interest of the equation is that it proves the assertion first made in §1.10, which was justified there by reference to the simple example of §1.9, that viscosity *has no effect upon the motion of a fluid which is vorticity-free throughout*. Obviously, the term involving η drops out when $\boldsymbol{\Omega}$ is uniformly equal to zero. We have relied extensively on this result throughout chapters 4 and 5; potential theory would have little relevance to the behaviour of real fluids without it.

The term involving η also drops out, of course, when $\boldsymbol{\Omega}$ is uniformly equal to some constant other than zero and, more generally still, whenever $\boldsymbol{\Omega}$, although non-uniform, is expressible as the gradient of some scalar potential (not to be confused with the velocity potential ϕ). Trivial examples of steady states of flow in which the vorticity satisfies one or other of these conditions, and which owe their steadiness to the fact that the viscous forces acting on any element of fluid exactly cancel, are discussed in §6.9.

6.5 Viscous dissipation

A formula for the rate at which energy is dissipated as heat in an incompressible fluid through the action of viscosity may be derived by reference to fig. 6.4. The rate at which work is being done by the normal stress on the right-hand and left-hand faces of the cubic element of side d sketched there are clearly given, to order d^3, by

$$- d^2 \left(p_1 + \frac{1}{2} d \frac{\partial p_1}{\partial x_1}\right)_O \left(u_1 + \frac{1}{2} d \frac{\partial u_1}{\partial x_1}\right)_O$$

and

$$+ d^2 \left(p_1 + \frac{1}{2} d \frac{\partial p_1}{\partial x_1}\right)_O \left(u_1 + \frac{1}{2} d \frac{\partial u_1}{\partial x_1}\right)_O$$

respectively, where the suffix O indicates values taken at the centre of the cube. Taking both of these into account the rate of work done, per unit volume of the element, is

$$- \left(p_1 \frac{\partial u_1}{\partial x_1} + u_1 \frac{\partial p_1}{\partial x_1}\right)_O .$$

Taking the shear stress components into account in a similar way and also the other four faces of the cube, and remembering that $\nabla \cdot \boldsymbol{u} = 0$ for an incompressible fluid, we may arrive at the following expression for the total rate of work per unit volume:

$$\begin{aligned}
&- (p_1 - p) \frac{\partial u_1}{\partial x_1} - (p_2 - p) \frac{\partial u_2}{\partial x_2} - (p_3 - p) \frac{\partial u_3}{\partial x_3} \\
&+ s_1 \left(\frac{\partial u_3}{\partial x_2} + \frac{\partial u_2}{\partial x_3}\right) + s_2 \left(\frac{\partial u_1}{\partial x_3} + \frac{\partial u_3}{\partial x_1}\right) + s_3 \left(\frac{\partial u_2}{\partial x_1} + \frac{\partial u_1}{\partial x_2}\right) \\
&- u_1 \frac{\partial p_1}{\partial x_1} - u_2 \frac{\partial p_2}{\partial x_2} - u_3 \frac{\partial p_3}{\partial x_3} \\
&+ u_1 \left(\frac{\partial s_3}{\partial x_2} + \frac{\partial s_2}{\partial x_3}\right) + u_2 \left(\frac{\partial s_1}{\partial x_3} + \frac{\partial s_3}{\partial x_1}\right) + u_3 \left(\frac{\partial s_2}{\partial x_1} + \frac{\partial s_1}{\partial x_2}\right).
\end{aligned}$$

However, only the first two lines of this expression represent work which is dissipated as heat; comparison with (6.23) shows that the terms in the third and fourth lines together amount to

$$\rho \boldsymbol{u} \cdot \{\boldsymbol{f} + \nabla(gz)\} = \frac{\mathrm{D}}{\mathrm{D}t} \left(\frac{1}{2} \rho u^2 + \rho g z\right)$$

and therefore represent work which is expended in increasing the kinetic or gravitational potential energy of the fluid. When (6.3) and (6.11) (and permutations of these) are used to describe the stress components, the dissipation per unit volume turns out to be expressible in terms of the local rate of shear [(6.18)] as

$$\frac{1}{2}\eta\left(\frac{\partial u_i}{\partial x_j}+\frac{\partial u_j}{\partial x_i}\right)^2 = 2\eta\zeta_{ij}\zeta_{ij}, \tag{6.26}$$

where the suffices i and j are each to be set equal to 1, 2 or 3 in turn, and where the nine terms thus obtained are to be summed in the usual way.

In deriving this result we have not had to assume that η is independent of position, and it can therefore be trusted, unlike the simple version of the Navier–Stokes equation quoted as (6.25), even when the fluid is not isothermal. Variations of temperature T from place to place are almost inevitable, of course, where dissipation is occurring, but they are rarely significant except perhaps in highly viscous fluids which are undergoing large rates of shear. Consider, for example, a stream of water which is moving tangentially to a plane solid surface with velocity u in the x_1 direction, when there is a boundary layer of thickness δ adjacent to the surface. The dissipation within the boundary layer is of order $\eta(u/\delta)^2\delta$ per unit area. The corresponding amount of heat per unit area must either be conducted sideways into the solid or the bulk fluid, or else carried downstream. In the first case we may expect transverse temperature gradients of order $\eta u^2/\delta\kappa$ to arise within the boundary layer, and hence temperature differences ΔT of order $\eta u^2/\kappa$, where κ is thermal conductivity. In the second case we may expect longitudinal temperature gradients governed by the equation

$$\eta\left(\frac{u}{\delta}\right)^2 \approx \rho c_V\frac{\mathrm{D}T}{\mathrm{D}t} \approx \rho c_V u\frac{\partial T}{\partial x_1},$$

and therefore temperature differences over a length L of order $\eta uL/\rho c_V\delta^2$, which if δ^2 is roughly $\eta L/u\rho$ [(1.29)] is equivalent to u^2/c_V. The viscosity of water at room temperature falls by about 2% for every 1 K of temperature rise, so one can hope to get away with treating it as uniform provided that $\Delta T/T$ is less than say 10^{-2}. That suggests, in the boundary layer situation, that both $\eta u^2/\kappa T$ and $u^2/c_V T$ need to be less than 10^{-2}. For gases η/κ and $1/c_V$ are comparable quantities [(3.17) and (A.35)], but for liquids (other than liquid metals) the former is the larger of the two. The condition that matters for water, therefore, is $\eta u^2/\kappa T < 10^{-2}$, and at room temperature this corresponds to $u < 40\ \mathrm{m\ s^{-1}}$. In the flow situations to which this chapter is devoted, the velocity is normally well below that sort of limit.

Note that if the temperature variations are too small to affect the viscosity significantly they are certainly too small for thermal expansion to be relevant; we are entitled to treat ρ as uniform as well.

6.6 Laminar viscous flow, 1 (planar laminae)

We are now equipped to examine a range of relatively simple *laminar flow* problems in which viscosity plays a dominant role. The term 'laminar' is often used to mean 'non-turbulent', but here it is used in a rather more restricted sense, to imply that the fluid can be treated as an assembly of laminae of uniform thickness, whose boundaries remain fixed as the fluid moves between them. In the simplest case these boundaries are parallel planes, the flow is *unidirectional*, say in the x_1 direction, and within any one lamina of infinitesimal thickness u_1 is uniform, though it may vary in the direction, say x_2, which is normal to the planes. The assumption that $\partial u_1/\partial x_1$ is zero means, since u_1 and u_2 are everywhere zero, that $\nabla \cdot \boldsymbol{u}$ and therefore $\mathrm{D}\rho/\mathrm{D}t$ are zero; it is justified only when the fluid is effectively incompressible. On this account, some of the results obtained below are valid only for liquids. The essential difference between liquids and gases, in so far as laminar viscous flow is concerned, is discussed in subsection (vi) below.

In this simplest case, it is clear from (6.9) that p_1 does not differ from the mean pressure p. Likewise, p_2 and p_3 are equal to p, i.e. the pressure is isotropic. The shear stress s_2 evidently vanishes, while (6.3) for s_3 reduces to (6.1). Hence we may simplify (6.23) to

$$\rho f_1 = -\frac{\partial p^*}{\partial x_1} + \eta \frac{\partial^2 u_1}{\partial x_2^2}.$$

But, since $\partial u_1/\partial x_1$ is zero, the acceleration $\mathrm{D}u_1/\mathrm{D}t$ does not differ from $\partial u_1/\partial t$, so the appropriate equation of motion is

$$-\frac{\partial p^*}{\partial x_1} + \eta \frac{\partial^2 u_1}{\partial x_2^2} = \rho \frac{\partial u_1}{\partial t}. \tag{6.27}$$

Various elementary applications of this equation are discussed in the subsections below. Note that when the the fluid is not accelerating in the x_2 and x_3 directions there are no transverse gradients of p^*. The longitudinal gradient of excess pressure which appears as the first term on the left-hand side of (6.27) is therefore independent of x_2 in each case except the one discussed in subsection (v), where the flow is not unidirectional. It is expressed where abbreviation seems appropriate by $\nabla_1 p^*$.

(i) *Steady flow between stationary parallel plates*

When the flow is steady $\partial u_1/\partial t$ vanishes, so

$$\eta \frac{\partial^2 u_1}{\partial x_2^2} = \frac{\partial p^*}{\partial x_1},$$

which may readily be integrated twice over x_2 to give, as one possible solution,

$$u_1 = \frac{1}{2\eta}\left(x_2^2 - \frac{1}{4}d^2\right)\frac{\partial p^*}{\partial x_1} = \frac{1}{2\eta}\left(x_2^2 - \frac{1}{4}d^2\right)\nabla_1 p^*. \tag{6.28}$$

Here the constants of integration have been chosen to ensure that u_1 is zero where $x_2 = \pm\frac{1}{2}d$, so the solution satisfies the no-slip boundary conditions applicable to steady flow between stationary parallel plates a distance d apart. The *discharge rate per unit width*, a quantity denoted by q which is related to the mean value of u_1 by the equation

$$q = d\langle u_1 \rangle, \tag{6.29}$$

is

$$q = \int_{-d/2}^{+d/2} u_1 \mathrm{d}x_2 = -\frac{d_3}{12\eta} \nabla_1 p^*. \tag{6.30}$$

For it to be positive the excess pressure must *decrease* with increasing x_1, and since q is independent of x_1 the rate of decrease must be uniform.

If the plates are horizontal $\partial(\rho g z)/\partial x_1$ is zero and the flow has to be driven by a gradient of pressure p. If they are vertical, however, a gradient of pressure is not essential; liquid may flow downwards between two plates under its own weight alone, with

$$p = p_o, \quad \nabla_1 p^* = -\frac{\partial(\rho g z)}{\partial z} = -\rho g, \quad q = \frac{\rho g d^3}{12\eta}. \tag{6.31}$$

(ii) *Steady flow between parallel plates with one plate moving*

Suppose that the plate at $x_2 = +\frac{1}{2}d$ is moving in the x_1 direction at constant speed U, while the plate at $x_2 = -\frac{1}{2}d$ is stationary. A different choice of integration constants is now required to satisfy the no-slip boundary condition. Instead of (6.28) we have

$$u_1 = \frac{1}{2\eta}\left(x_2^2 - \frac{1}{4}d^2\right)\nabla_1 p^* + U\left(\frac{x_2}{d} + \frac{1}{2}\right), \tag{6.32}$$

so

$$q = -\frac{d^3}{12\eta}\nabla_1 p^* + \frac{1}{2}dU. \tag{6.33}$$

The shear stresses acting on the two plates in the x_1 direction are

$$\pm\eta\left(\frac{\partial u_1}{\partial x_2}\right)_{x_2 = \mp d/2} = -\frac{1}{2}d\nabla_1 p^* \pm \frac{\eta U}{d}. \tag{6.34}$$

Figure 6.5 illustrates this problem. The results are used in §6.7.

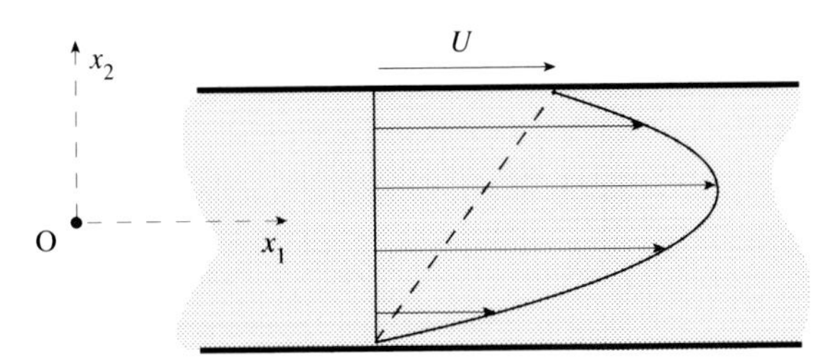

Figure 6.5 Velocity profile between parallel plates; the fluid is being pushed to the right by a longitudinal pressure gradient, and it is also being dragged to the right by motion of the upper plate.

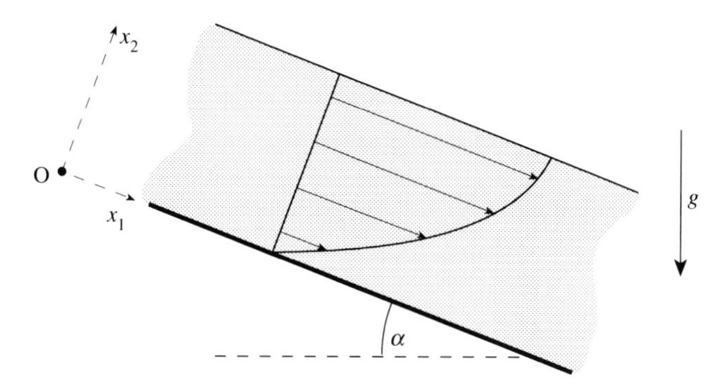

Figure 6.6 Velocity profile in open-channel flow over an inclined plate.

(iii) *Open-channel flow*

Suppose that a single flat plate inclined at an angle α to the horizontal is covered with a layer of liquid having uniform thickness d, within which p^* is equal to the pressure of the surrounding atmosphere (perhaps consisting only of the liquid's saturated vapour) and is therefore necessarily uniform [fig. 6.6]. Then, if the x_2 direction is chosen as the normal to the plate and if the fluid is flowing steadily downhill with velocity u_1, (6.27) becomes

$$\frac{\partial^2 u_1}{\partial x_2^2} = -\frac{\rho g \sin \alpha}{\eta}.$$

If the planes $x_2 = 0$ and $x_2 = d$ are chosen to describe the upper surface of the plate and the free surface of the liquid, the boundary conditions on these planes are respectively $u_1 = 0$ and $\partial u_1/\partial x_2 = 0$. The latter condition reflects the fact that shear stresses are always continuous at a boundary between two fluids; this means that, at any rate if the viscosity of the atmosphere is negligible, s_3 in the liquid layer must vanish at its free surface. The solution which satisfies these boundary conditions is

$$u_1 = \frac{\rho g \sin \alpha}{2\eta}(2dx_2 - x_2^2),$$

which leads to

$$q = \int_0^d u_1 \mathrm{d}x_2 = \frac{\rho g d^3 \sin \alpha}{3\eta}. \tag{6.35}$$

We may use this result to estimate the rate at which an inclined plate of finite extent, which at time $t = 0$ is covered by a thick layer of liquid, loses this coverage as the liquid drains off it. An exact analysis is difficult, because while the plate is draining d and $u_1\{x_2\}$ are not independent of x_1; the flow is not strictly steady, u_2 is not strictly zero, and the pressure is not strictly isotropic. Let us set these complications on one side for the moment and assume that (6.35) provides us with a reasonably reliable description of how the local discharge rate depends upon the local thickness. The local rate of change of d with time can then be related to the variation of d with x_1, by means of the continuity condition

$$\frac{\partial d}{\partial t} = -\frac{\partial q}{\partial x_1} = -\frac{\rho g d^2 \sin \alpha}{\eta}\frac{\partial d}{\partial x_1}.$$

A solution of this partial differential equation which is appropriate if the top edge of the plate – where d presumably vanishes except in the limit $t \to 0$ – lies in the plane $x_1 = 0$ is

$$d^2 = \frac{\eta x_1}{\rho g t \sin \alpha}. \tag{6.36}$$

Granted this solution, the mean velocity is

$$\langle u_1 \rangle = \frac{q}{d} = \frac{x_1 \sin \alpha}{3t}.$$

We are justified in treating the flow as steady if, and only if, the term $\partial u_1/\partial t$ on the right-hand side of (6.27) is much smaller than each of the terms on the left-hand side, and this condition amounts to

$$-\frac{1}{g \sin \alpha}\frac{\partial \langle u_1 \rangle}{\partial t} = \frac{x_1}{3gt^2} \ll 1. \tag{6.37}$$

We are justified in ignoring u_2 compared with u_1 if and only if

$$\frac{\partial d}{\partial x_1} = \frac{d}{2x_1} \ll 1. \tag{6.38}$$

Finally, we are justified in treating the pressure as isotropic because the mean value of the longitudinal gradient of the anisotropic component $(p_1 - p)$ is proportional to

$$\frac{\partial}{\partial x_1}\left(\eta \frac{\partial \langle u_1 \rangle}{\partial x_1}\right),$$

and this is zero if $\langle u_1 \rangle$ is proportional to x_1/t.

Try inverting an uncorked bottle which contains water and holding it upside down for a further 10 seconds, say, after it appears to have emptied. If you then turn it the right way up again and leave it to drain for a further minute or two, you will find that it still contains a millilitre of water or more. During most of the drainage period the two conditions expressed by (6.37) and (6.38) should have been adequately satisfied, so the thickness of the water layer at the end of that period may legitimately be estimated from (6.36), with $t = 10\,\text{s}$, $\eta/\rho = 10^{-6}\,\text{m}^2\,\text{s}^{-1}$, and $\alpha = \pi/2$. The equation turns out to account for the residual volume in a convincing manner.

(iv) *Oscillatory flow near a flat plate*

It is possible to establish unidirectional laminar flow in a large body of fluid which is initially at rest in hydrostatic equilibrium, even though no gradients exist in p^*, by moving a solid body immersed it. If the solid body is a flat plate whose surface occupies the plane $x_2 = 0$, with fluid occupying the region $x_2 > 0$, and if it is moved with velocity U in the x_1 direction, then the appropriate differential equation for the fluid's velocity is

$$\eta \frac{\partial^2 u_1}{\partial x_2^2} = \rho \frac{\partial u_1}{\partial t},$$

[(6.27)], which was quoted in chapter 1 as (1.25). If U is oscillatory with angular frequency ω we may treat it as a complex quantity proportional to $\exp(-\mathrm{i}\omega t)$, on the understanding that only the real part is of physical significance. In that case

$$\eta \frac{\partial^2 u_1}{\partial x_2^2} = -\,\mathrm{i}\omega\rho u_1,$$

which has travelling-wave solutions of the form $u_1 \propto \exp\{\mathrm{i}(kx_2 - \omega t)\}$ with

$$k = \pm(1 + \mathrm{i})\sqrt{\frac{\omega\rho}{2\eta}} = \pm\frac{1 + \mathrm{i}}{\delta},$$

where $\delta = \sqrt{2\eta/\omega\rho}$ is the oscillatory boundary layer thickness previously defined by (1.31). The particular solution which satisfies the no-slip boundary condition at $x_2 = 0$ and the condition that u_1 tends to zero when x_2 becomes large is

$$u_1 = U\,\mathrm{e}^{+\mathrm{i}(1+\mathrm{i})x_2/\delta} = U\,\mathrm{e}^{(-1+\mathrm{i})x_2/\delta}. \tag{6.39}$$

The waves which it describes are transverse ones, and they are heavily attenuated. Transverse electromagnetic waves are attenuated by conducting materials in a similar fashion, of course, and the oscillatory boundary layer thickness corresponds to what, in the electromagnetic case, is called the *skin depth*.

If the fluid does not extend to large values of x_2 but is bounded on the plane $x_2 = d$ by a second plate which is stationary, the solution is a mixture of attenuated transverse waves travelling in both directions,

$$u_1 = U\,\frac{\mathrm{e}^{(1-\mathrm{i})(d-x_2)/\delta} - \mathrm{e}^{(-1+\mathrm{i})(d-x_2)/\delta}}{\mathrm{e}^{(1-\mathrm{i})d/\delta} - \mathrm{e}^{(-1+\mathrm{i})d/\delta}},$$

which when $d \ll \delta$ may be expanded to second order thus:

$$u_1 \approx U\left(1 - \frac{x_2}{d}\right)\left(1 + \mathrm{i}\,\frac{2dx_2 - x_2^2}{3\delta^2} + \ldots\right).$$

The force per unit area which shear stress exerts on the moving plate in the x_1 direction is $\eta(\partial u_1/\partial x_2)_{x_2=0}$. When $d \gg \delta$, so that (6.39) is a good approximation, this may be written as

$$(-1 + \mathrm{i})\,\frac{\eta U}{\delta} = -\frac{\eta U}{\delta} - \frac{1}{2}\,\rho\delta\,\frac{\partial U}{\partial t},$$

while when $d \ll \delta$ it may be written as

$$\frac{\eta U}{d}\left(-1 + \mathrm{i}\,\frac{2d^2}{3\delta^2}\right) = -\frac{\eta U}{d} - \frac{1}{3}\,\rho d\,\frac{\partial U}{\partial t}. \tag{6.40}$$

In each case, part of the force is in antiphase with the velocity of the plate and tends to damp its oscillations. The other part is in antiphase with the acceleration; it increases the effective inertia of the plate [see §4.8 for a discussion of virtual mass in potential flow] by the inertia of a layer of fluid of thickness $\frac{1}{2}\delta$ in one limit and $\frac{1}{3}d$ in the other. It therefore tends to slow the oscillations down.

(v) *Oscillating disc viscometers*

The damping effect referred to in subsection (iv) and – less commonly – the increase in effective inertia have both been exploited by experimenters seeking to measure the viscosity of fluids. In 1866, for example, Maxwell observed the logarithmic decrement of a system of horizontal glass discs undergoing torsional oscillations in air, and he thereby determined the viscosity of air as a function of pressure and temperature. His apparatus is preserved in Cambridge in the Cavendish Laboratory. Each of his oscillating discs lies midway between a pair of stationary discs of larger diameter, and the separation between moving discs and stationary ones can be varied between 1 inch and about $\frac{1}{5}$ inch. Under the conditions of Maxwell's experiments, such separations were not quite small enough compared with δ for (6.40) to suffice, i.e. the damping force per unit area experienced by the oscillating discs at each of their surfaces, at a radius r from their axis of rotation, was not given exactly in terms of their instantaneous angular velocity $\mathrm{d}\theta/\mathrm{d}t$ by $(\eta r/d)(\mathrm{d}\theta/\mathrm{d}t)$. He therefore took account of higher-order correction terms, and he also corrected for edge effects.

Later experimenters have followed Maxwell by studying the torsional oscillations of discs and have also followed him in assuming the induced flow to be laminar. In fact it cannot be strictly laminar, in the sense in which that word is used here. A fluid lamina cannot undergo torsional oscillations about a vertical axis without an oscillating pressure gradient in the radial direction to provide the fluid with its centripetal acceleration [§2.5]. The pressure gradient which is needed depends upon the square of the instantaneous angular velocity of the lamina, and in an oscillating disc viscometer this varies with x_2. Since it is not possible for the needs of all laminae to be met simultaneously, a complicated circulation pattern is established, whereby the fluid in the laminae adjacent to the disc, which oscillate with a relatively large mean square angular velocity, moves slowly away from the axis of rotation, while the fluid in more distant laminae, whose mean square angular velocity is relatively small, moves slowly inwards towards this axis. In Maxwell's experiments, however, the circulation was almost certainly too slow to have affected his results.[1]

(vi) *Gas flow through narrow channels*

Our assumption that compressibility is negligible for a fluid undergoing steady viscous flow between, say, two parallel plates is almost always justified for liquids; in water, for example, a pressure difference Δp of 1 atmosphere produces a density difference of only about 0·005%. When the fluid is gaseous, however, it can be treated as incompressible only when Δp is everywhere small compared with the ambient pressure p. In circumstances where Bernoulli's theorem is applicable that condition is satisfied provided u is everywhere much less than the velocity of sound, but Bernoulli's theorem is certainly *not* applicable in situations where viscosity is dominant. It is easy to imagine a gas-filled channel so narrow that even when the pressure difference between its two ends is much more than one atmosphere the gas creeps through it with a mean velocity which is tiny compared with c_s. It would be absurd to treat the gas as incompressible in such a case.

The continuity condition for steady flow of an incompressible fluid through a channel whose dimensions do not change with time is, of course, that the volume

[1] Other examples of this type of secondary circulation are discussed in §7.13.

discharge rate Q must be the same at all points along the channel. Thus if the channel lies between parallel plates, as in subsection (i) above, the discharge rate per unit width, q, must be independent of x_1, and we may deduce from (6.30) that the gradient $\nabla_1 p^*$ is also independent of x_1. This means that we can rewrite (6.30) in terms of finite differences, as

$$q = -\frac{d^3}{12\eta}\frac{\Delta p^*}{\Delta x_1}; \tag{6.41}$$

here Δx_1 could be the total length of the channel, provided that this is so much larger than d that the inlet length discussed in §1.13 is negligible by comparison, in which case Δp^* would be the difference in excess pressure between its two ends. If the fluid is compressible, however, it is the *mass* discharge rate ρQ which is uniform in steady flow; if it is an ideal gas, maintained at uniform temperature by thermal contact with the walls of the channel (in which case we may safely assume that not only p/ρ but also η is constant), then pQ should be uniform. In that case, on the assumption that (6.30) remains locally valid, $p(\partial p^*/\partial x_1)$ is independent of x_1 for an ideal gas undergoing steady flow between parallel plates. The distinction between $\partial p^*/\partial x_1$ and $\partial p/\partial x_1$ may here be ignored, because even if the channel is vertical the weight of the gas which it contains is too small to be significant, so

$$p\frac{\partial p^*}{\partial x_1} \approx \frac{1}{2}\frac{\partial p^2}{\partial x_1}.$$

Hence instead of (6.41) we should use

$$\rho q = -\frac{\rho_o d^3}{24\eta p_o}\frac{\Delta p^2}{\Delta x_1}, \tag{6.42}$$

where ρ_o is the density at some point where the pressure is p_o.

Strictly speaking, (6.30) is *not* locally valid for a fluid which is compressible, because $\partial u_1/\partial x_1$ is not zero; this means that the fluid is accelerating even when its flow is steady, and that the pressure is not completely isotropic. Similar complications arose in subsection (iii), though from the variation with x_1 of d rather than ρ. They proved negligible there, and they are equally negligible in the context of gas flow through narrow channels.

6.7 Hydrodynamic lubrication

The best way to reduce friction between two solid surfaces is to separate them by a layer of fluid, normally oil, but when the two surfaces are pressed together by a load of some sort the lubricant is liable to get squeezed out. The idea of *hydrodynamic lubrication* is to use the relative motion of the two solid surfaces to generate an excess pressure in the fluid layer, so that it can withstand a load. In

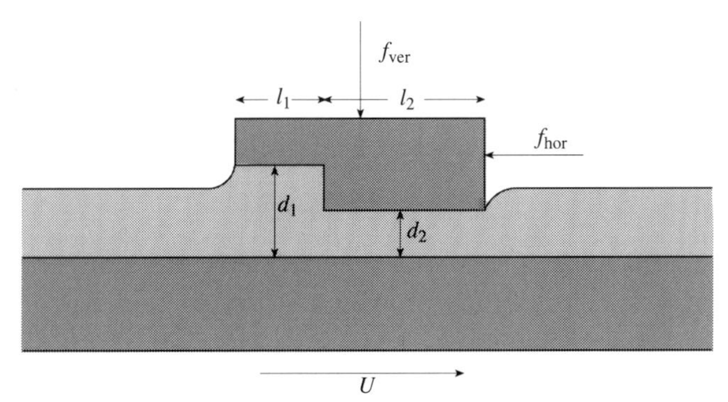

Figure 6.7 A pad with a step in it, separated from a moving plate by a layer of lubricating liquid. The forces f_{ver} and f_{hor} are needed to keep the pad stationary.

hydrostatic lubrication an excess pressure is maintained continuously, even when the surfaces are at rest, by supplying the lubricant from a pressurised source.

Relative motion is liable to generate an excess pressure whenever the fluid layer is thicker at its front edge, where the two solids are moving towards one another, than at its back edge where they are moving apart. A somewhat artificial example to illustrate this general principle is sketched in cross section in fig. 6.7; the bearing in this case takes the form of a stationary solid pad with a stepped profile, adjacent to a horizontal plane surface which is moving to the right in the x_1 direction with velocity U. The separation between pad and plane is d_1 over a length l_1, and d_2 ($< d_1$) over a length l_2. The interspace is entirely filled with oil of viscosity η, and oil extends over the plane to left and right of the pad. To simplify the analysis, we may suppose that the thickness of the oil layer on the left is exactly q/U, where q is the volume of fluid which flows past the pad per unit time and per unit width in the x_3 direction; if the thickness were less than this the oil would not fill the length l_1 completely, while if it were greater excess oil would accumulate steadily to the left of the pad. Both pad and plane are wide enough for the flow pattern to be regarded as two-dimensional. The pattern is a steady one in the frame of reference in which the pad is stationary, and the problem is more easily handled in this frame than in any other. The results obtained below are equally valid, however, if the plane is stationary and the pad moves with velocity $-U$.

According to (6.33) the discharge rates per unit width over the lengths l_1 and l_2 respectively are in this frame given by

$$q_1 = \frac{1}{2} d_1 U - \frac{d_1^3}{12\eta} \frac{p_s^*}{l_1},$$

$$q_2 = \frac{1}{2} d_2 U - \frac{d_2^3}{12\eta} \frac{p_s^*}{l_2},$$

where p_s^* is the excess pressure at the step itself and p^* on either side of the pad is taken to be zero; over the length l_1 the uniform pressure gradient is positive and tends to reduce the discharge rate generated by the relative motion, whereas over the length l_2 it is negative and enhances the discharge rate. The continuity condition for this problem is evidently

$$q_1 = q_2 \,(= q),$$

from which we may deduce that

$$p_s^* = \frac{6\eta U(d_1 - d_2)}{\dfrac{d_1^3}{l_1} + \dfrac{d_2^3}{l_2}}. \tag{6.43}$$

It follows that the pad can withstand a vertical load per unit width of

$$\begin{aligned} f_{\text{ver}} &= \frac{1}{2} p_s^*(l_1 + l_2) \\ &= \frac{3\eta U(d_1 - d_2)(l_1 + l_2)l_1 l_2}{l_2 d_1^3 + l_1 d_2^3}, \end{aligned} \tag{6.44}$$

and the pad moves up or down slightly until this equation is satisfied.

We may also deduce, using (6.34), that the horizontal force per unit width which acts on the plane and opposes its motion relative to the pad is

$$f_{\text{hor}} = -\frac{1}{2} p_s^*(d_1 - d_2) - \eta U\left(\frac{l_1}{d_1} + \frac{l_2}{d_2}\right). \tag{6.45}$$

The horizontal force per unit width on the pad, including a contribution due to the excess pressure acting on the vertical face of the step, is

$$-\frac{1}{2} p_s^*(d_1 - d_2) + \eta U\left(\frac{l_1}{d_1} + \frac{l_2}{d_2}\right) + p_s^*(d_1 - d_2),$$

which naturally differs from (6.45) only in sign. With the aid of (6.44), (6.45) may be expressed in the form

$$|f_{\text{hor}}| = F\left\{\frac{d_1}{d_2}, \frac{l_1}{l_2}\right\} \sqrt{\eta U f_{\text{ver}}}. \tag{6.46}$$

The coefficient $F\{(d_1/d_2), (l_1/l_2)\}$ is a complicated function of its two dimensionless arguments, but computation shows that it can be reduced to a minimum value which is close to $\sqrt{3}$ by designing the bearing in such a way that, under normal operating conditions, l_1/l_2 is about 13 and d_1/d_2 about 2·5.

In practice, bearings are more often designed so that the layer of oil trapped between the two moving parts is wedge-shaped, rather than stepped, but such

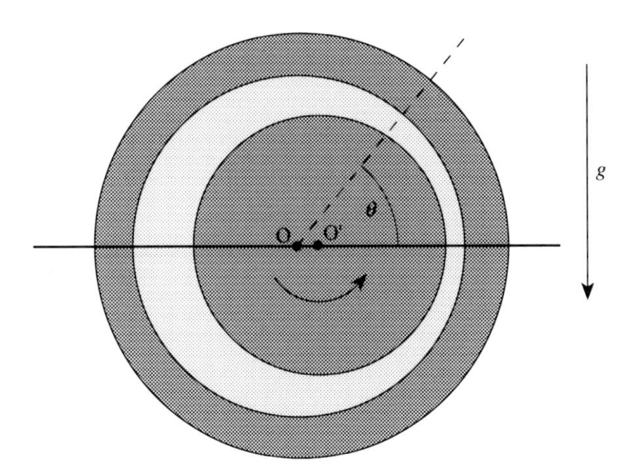

Figure 6.8 Cross-section through a lubricated journal bearing. Anticlockwise rotation of the shaft has caused its centre to move from O to O′; the pressure in the lubricating liquid is now greater below the line OO′ than at a corresponding point above this line, and the pressure difference supports the weight of the shaft. The difference in radius between the shaft and its bush is greatly exaggerated.

bearings depend on the same principles and may be analysed in the same way as the bearing illustrated in fig. 6.7. In one ingenious design, which permits motion in either direction in a way that the design of fig. 6.7 does not, the pad is pivoted; it switches from one stable inclination to another when U is reversed, thereby reversing the orientation of the wedge of oil.

Very similar principles govern the operation of lubricated *journal bearings*, like the one sketched in fig. 6.8. This figure shows a cross-section through a horizontal cylindrical shaft, or *journal*, which is rotating anticlockwise in a fixed cylindrical sleeve, or *bush*, the interspace being entirely filled with oil; the axis of the shaft (O′) is slightly to the right of the axis (O) of the sleeve. The rotation drives the oil round in an anticlockwise direction, and the discharge rate q per unit length along the axis of the sleeve must be independent of the angle θ to satisfy the continuity condition. That is only possible if the flow rate is enhanced by a pressure gradient in the region near $\theta = 0$, where the thickness of the oil film is relatively small, and reduced by a contrary pressure gradient in the region near $\theta = \pi$ where the thickness is relatively large; thus if the excess pressure p^* is chosen to be zero at $\theta = 0$ it is negative in the range $0 < \theta < \pi$ and positive in the range $0 > \theta > -\pi$, and the symmetry is such that $p^*\{\theta\} = -p^*\{-\theta\}$. Clearly, the net force which the excess pressure exerts on the shaft is vertically upwards, and it is this force which supports the weight of the shaft and any load which the shaft carries. When the shaft is *not* rotating it rests directly on the sleeve, with its axis vertically below the point labelled O in the figure; when it *is* rotating it loses contact with the sleeve and reaches a new stable equilibrium with its axis displaced horizontally from O, as the figure suggests. For a given load, the displacement OO′ diminishes as the rate of rotation increases.

6.8 Hele Shaw flow

Suppose that a layer of liquid between two plane parallel plates a distance d apart is interrupted by an impermeable obstacle which occupies, over an area normal to x_2 whose boundary is fixed, the whole region between $x_2 = -\frac{1}{2}d$ and $x_2 = +\frac{1}{2}d$. If this liquid is set into motion by gradients of p^*, the motion may well be laminar in the sense that u_2 is zero everywhere, but because the liquid is obliged to avoid the obstacle it will not be unidirectional, and the acceleration will not be zero everywhere, even though the flow is steady. Provided that d is small enough, however, we may ignore these complications and assume that the local discharge rate per unit width is related to the local gradient of p^* by a two-dimensional version of (6.30), namely

$$\boldsymbol{q} = -\frac{d^3}{12\eta}\boldsymbol{\nabla}p^*;$$

the vector $\boldsymbol{q}$ is confined to the q_1q_3 plane, and so, of course, is the vector on the right-hand side. In that case the mean velocity defined by (6.29) is

$$\langle \boldsymbol{u} \rangle = \boldsymbol{\nabla}\left(-\frac{d^2p^*}{12\eta}\right), \tag{6.47}$$

which implies that the mean velocity is derivable from a single-valued flow potential, $\phi = -d^2p^*/12\eta$. Moreover, it is obvious from the continuity condition

$$\boldsymbol{\nabla}\cdot\boldsymbol{q} = \frac{\partial q_1}{\partial x_1} + \frac{\partial q_2}{\partial x_2} = 0$$

that the potential obeys Laplace's equation in two dimensions,

$$\frac{\partial^2\phi}{\partial x_1^2} + \frac{\partial^2\phi}{\partial x_2^2} = 0.$$

As Hele Shaw pointed out in 1898, one can therefore reproduce within the narrow confines of a *Hele Shaw cell* the lines of flow to be expected for an incompressible bulk fluid undergoing two-dimensional potential flow past, say, a transverse cylinder; all that is necessary is to include in the cell an obstacle with a circular cross-section and to mark the fluid with dye so as to render the lines of flow visible. However, one cannot use a Hele Shaw cell to mimic potential flow round a rotating cylinder [§4.12], because potentials which describe flow with non-zero

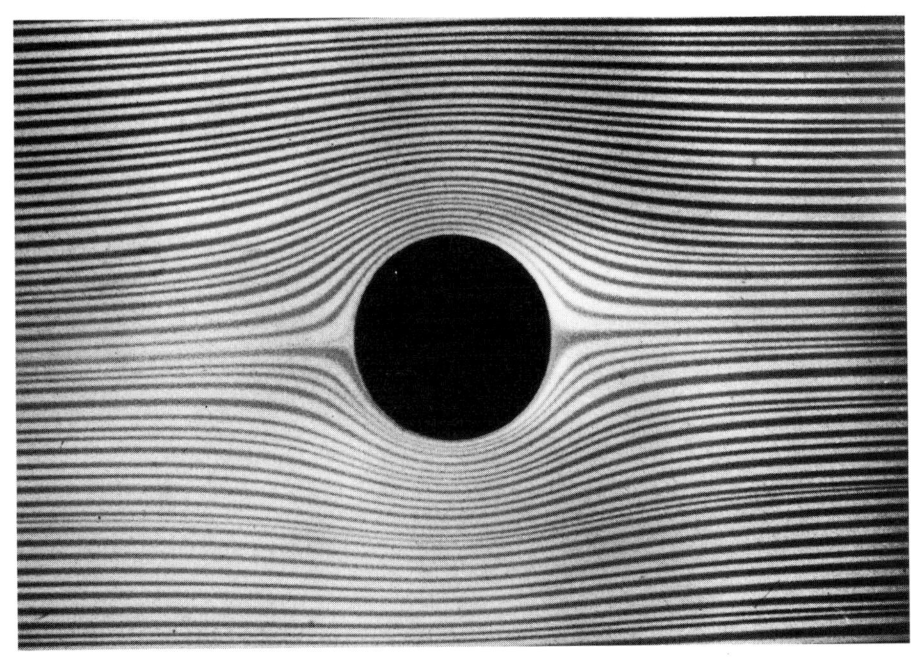

Figure 6.9 Lines of flow, revealed by streaks of dye, around a circular spacer between two glass plates separated by 1 mm.
[Photograph by D. H. Peregrine.]

circulation are necessarily multi-valued. A photograph of a Hele Shaw cell in action is reproduced in fig. 6.9.

Just how narrow must the cell be? Equation (6.47) is only to be trusted when (a) $\partial^2 u_1/\partial x_3^2$ is negligible compared with $\partial^2 u_1/\partial x_2^2$ (the x_1 direction being chosen to coincide with the local direction of $\boldsymbol{u}$), and (b) the local acceleration $u_1(\partial u_1/\partial x_1)$ is negligible compared with $(\eta/\rho)(\partial^2 u_1/\partial x_2^2)$. If L is a scale length describing the lateral extent of the obstacle – e.g. its diameter, if it is circular – then condition (a) should be satisfied when $(d/L)^2 \ll 1$, and condition (b) should be satisfied when

$$\frac{\langle u \rangle^2}{L} \ll \frac{\eta \langle u \rangle}{\rho d^2},$$

i.e. when

$$\left(\frac{d}{L}\right)^2 \ll \frac{\eta}{\rho \langle u \rangle L} = \frac{1}{Re}, \tag{6.48}$$

where Re is the Reynolds Number for the problem in hand. There is no difficulty about satisfying both conditions in practice. Incidentally, the time taken for $\langle u \rangle$ to adjust to any sudden change in pressure gradient is of order $\rho d^2/\eta$. When the inequality labelled (6.48) is satisfied, the distance through which the fluid travels during this time is less than d and therefore much less than L. This means that for practical purposes the flow *is* always steady; its response to changes of pressure gradient is effectively instantaneous.

6.9 Laminar viscous flow, 2 (cylindrical laminae)

The cases of laminar flow discussed in §6.6 have analogues in which the laminae are concentric cylindrical shells rather than plane sheets. In these analogues, which are best discussed using cylindrical coordinates (r, θ, x_3), the fluid may move unidirectionally, parallel to the axis of the cylinders, with

$$u_3 = u\{r\}, \quad u_r = u_\theta = 0, \tag{6.49}$$

or it may circulate about the axis with

$$u_\theta = u\{r\}, \quad u_r = u_3 = 0. \tag{6.50}$$

Subsection (i) below refers to longitudinal flow of the type described by (6.49), while subsections (ii) and (iii) refer to circulating flow of the type described by (6.50).

(i) *Longitudinal flow*

When liquid moves in concentric cylindrical lamina with a longitudinal velocity which depends on radius r but not on θ or x_3, the longitudinal shear stress components s_{31} and s_{32} do not depend on the choice of orientation for the x_1 and x_2 axes, and they are both equal to $\eta(\partial u/\partial r)$. Hence the force per unit length which shear stress exerts over a cylindrical surface of radius r on the liquid inside this surface is $2\pi r\eta(\partial u/\partial r)$. An element of liquid in the form of a cylindrical shell with inner radius r and outer radius $r + \mathrm{d}r$ therefore experiences a net longitudinal force per unit length due to shear of

$$\left(2\pi r\eta\,\frac{\partial u}{\partial r}\right)_{r+\mathrm{d}r} - \left(2\pi r\eta\,\frac{\partial u}{\partial r}\right)_r = 2\pi\eta\,\frac{\partial}{\partial r}\left(r\,\frac{\partial u}{\partial r}\right)\mathrm{d}r,$$

while the longitudinal force on it due to pressure gradients and its own weight is

$$-\,2\pi r\mathrm{d}r\,\frac{\partial p^*}{\partial x_3} = -\,2\pi r\mathrm{d}r\nabla_3 p^*.$$

Hence the equation of motion which replaces (6.27) when the laminae are cylindrical rather than planar is

$$-\,\nabla_3 p^* + \frac{\eta}{r}\frac{\partial}{\partial r}\left(r\,\frac{\partial u}{\partial r}\right) = \rho\,\frac{\partial u}{\partial t}. \tag{6.51}$$

The solution equivalent to (6.28) which describes steady laminar flow within a cylindrical pipe of internal radius a has been quoted already, as (1.32). It may be derived by integrating the left-hand side of (6.51) twice over r with the right-hand side set equal to zero or, equivalently, by the argument in §1.13 which requires only one integration. It leads to the *Poiseuille*, or *Hagen–Poiseuille*, equation

$$Q = -\frac{\pi a^4}{8\eta}\nabla_3 p^*$$

for the discharge rate Q in laminar pipe flow, already quoted as (1.33). This equation provides the basis for what is probably the most popular of all methods for the measurement of liquid viscosity.

Poiseuille's law applies locally to laminar pipe flow of gases as well as liquids. However, whereas in cases of liquid flow through pipes of uniform bore the local gradient $\nabla_3 p^*$ is uniform – and therefore the same as $\Delta p^*/\Delta x_3$, where Δp^* is the finite difference of excess pressure between two points a finite distance Δx_3 apart – this is not true of gas flow. Section 6.6(vi) should be consulted in this connection.

Suppose that the pipe has an internal radius b and that a cylindrical rod of radius a ($< b$) lies along its axis, with liquid filling the interspace, and suppose that the liquid is set into motion by movement of the rod with some uniform longitudinal velocity U rather than by the application of a pressure gradient. The equation of motion which describes steady flow in such circumstances is

$$\frac{\partial}{\partial r}\left(r\frac{\partial u}{\partial r}\right) = 0,$$

and the solution which satisfies the no-slip boundary condition at $r = a$ and $r = b$ is

$$u = U\frac{\ln(r/b)}{\ln(a/b)}. \tag{6.52}$$

This solution is wholly unremarkable until one looks at the corresponding vorticity. This has components

$$\Omega_\theta = -\frac{\mathrm{d}u}{\mathrm{d}r} = \frac{U}{r\ln(a/b)}, \quad \Omega_r = \Omega_3 = 0.$$

In its direction and in its variation with the coordinates (r, θ, x_3), $\boldsymbol{\Omega}$ evidently resembles the magnetic field around a current-carrying wire. It can clearly be derived from a potential, though the potential is not single-valued, and it therefore satisfies – as expected – one of the conditions which are sufficient to ensure that the viscous term in the Navier–Stokes equation is irrelevant [see the closing remarks in §6.4]. When liquid is in the steady state of motion described by (6.52) there is continuous dissipation of energy for which viscosity is responsible, and viscosity naturally determines the external force which is required to keep the

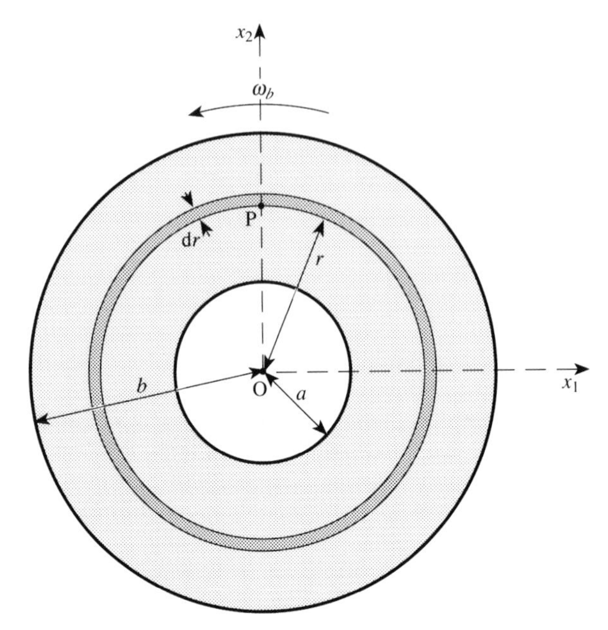

Figure 6.10 A Couette viscometer consisting of coaxial cylinders, the outer one rotating with uniform angular velocity ω_b. The fluid between the cylinders circulates with a velocity u_q which, in the steady state, depends only on radius r.

rod moving. Each cylindrical shell of liquid experiences equal and opposite forces on its inner and outer surfaces, however, and the viscous stresses cause no acceleration for that reason.

(ii) *Circulating (Couette) flow*

When fluid, whether liquid or gas, is circulating round the x_3 axis in the manner described by equations (6.50), the only non-zero component of shear stress is s_3. In cartesian coordinates this is of course determined, according to (6.3), by the two velocity gradients $\partial u_1/\partial x_2$ and $\partial u_2/\partial x_1$. At a point such as P in fig. 6.10, which lies on the x_2 axis through the origin O, these gradients are given, for reasons explained in §2.7 [fig. 2.5 in particular], by:

$$\frac{\partial u_1}{\partial x_2} = -\frac{\partial u}{\partial r}; \quad \frac{\partial u_2}{\partial x_1} = \frac{u}{r} = \omega.$$

Thus in cylindrical coordinates we have

$$s_3 = \eta\left(\frac{u}{r} - \frac{\partial u}{\partial r}\right) = -\eta r\frac{\partial}{\partial r}\left(\frac{u}{r}\right) = -\eta r\frac{\partial \omega}{\partial r} \tag{6.53}$$

at P, and the symmetry is such that s_3 must take the same value at all points around the circle on which P lies. Shear stress is to be expected, of course, only when adjacent layers of fluid slide over one another; it is *not* to be expected if a fluid rotates about an axis like a rigid body, with an angular velocity ω which is the same for all r, and (6.53) is therefore an unsurprising result.

Due to this shear stress, the liquid inside a cylinder of radius r experiences a viscous torque about the x_3 axis (positive when it acts in an anticlockwise sense in fig. 6.10) given per unit length by

$$g_3 = -r(2\pi r s_3) = 2\pi\eta r^3 \frac{\partial}{\partial r}\left(\frac{u}{r}\right). \tag{6.54}$$

Hence the cylindrical shell of liquid of thickness dr which is indicated by darker shading in fig. 6.10 experiences, in circulating flow, a net torque (the sum of torques exerted on its inner and outer surfaces) of magnitude

$$\mathrm{d}g_3 = 2\pi\eta \frac{\partial}{\partial r}\left\{r^3 \frac{\partial}{\partial r}\left(\frac{u}{r}\right)\right\} \mathrm{d}r$$

per unit length, and by equating this to its rate of change of angular momentum per unit length we may arrive at the equation

$$\eta \frac{\partial}{\partial r}\left\{r^3 \frac{\partial}{\partial r}\left(\frac{u}{r}\right)\right\} = \rho r^2 \frac{\partial u}{\partial t}. \tag{6.55}$$

There is no pressure gradient term in this equation because p^* is a single-valued quantity and $\partial p^*/\partial\theta$ must vanish when there is cylindrical symmetry. There must of course be a radial pressure gradient to provide the centripetal acceleration characteristic of circulating flow [§2.5], and the full Navier–Stokes equation describes this as well as the content of (6.55). However, the radial pressure gradient is of no special interest here.

There are two basic patterns of circulation which correspond to steady flow because for each of them the left-hand side of (6.55) vanishes everywhere. They are both discussed in earlier chapters [§§2.7 and 4.11 in particular]. For the first, u is proportional to r; for the second, u is inversely proportional to r. The first corresponds to rigid-body rotation with uniform angular velocity, and the second to circulation which is vorticity-free. In both cases the viscous term in the Navier–Stokes equation vanishes, though for different reasons.

A *Couette viscometer* consists of two concentric solid cylinders of radii a and b ($b > a$), the space between them being filled with fluid, which are arranged to rotate with different angular velocities, ω_a and ω_b; in practice, the inner cylinder is normally stationary, i.e. $\omega_a = 0$. The steady flow pattern is a hybrid of the two basic patterns referred to above, and the solution which satisfies the no-slip boundary conditions ($u = a\omega_a$ at $r = a$, $u = b\omega_b$ at $r = b$) is

$$u = \frac{(b^2\omega_b - a^2\omega_a)r - b^2a^2(\omega_b - \omega_a)r^{-1}}{b^2 - a^2}.$$

The fluid viscosity may be determined from measurements of the torque transmitted from one cylinder to the other once a steady state has been achieved, and per unit length this torque amounts to

$$g_3 = \frac{4\pi\eta b^2 a^2}{b^2 - a^2}(\omega_b - \omega_a). \tag{6.56}$$

Circulating flow of the sort described here, because it occurs inside a Couette viscometer, is often referred to as *Couette flow*.

(iii) *Vortex lines*

It may be verified by substitution that one possible solution of the time-dependent equation (6.55) is

$$u = \frac{K}{2\pi r}(1 - \mathrm{e}^{-\rho r^2/4\eta t}), \tag{6.57}$$

where K is any constant. This describes rigid-body rotation ($u \propto r$) for small values of r, and vorticity-free circulation ($u \propto r^{-1}$) for large values of r, with the change from one regime to the other occurring over a range of values of r in the neighbourhood of $\sqrt{2\eta t/\rho}$. It therefore corresponds in every particular to the flow pattern about a free vortex line, as predicted in §4.13. The constant K is, of course, the *circulation* round the line, a convenient measure of its strength.

Although (6.57) may be said to constitute the ideal model of a vortex line in a viscous fluid, it is only one of many possible mathematical models, all describing circulating flow around a core which is rotating more or less like a rigid body and growing in radius like $\sqrt{t}$. An equally valid solution of (6.55), associated with the name of *Taylor* whereas (6.57) is associated with the names of *Oseen* and *Lamb*, is

$$u \propto \frac{\rho r}{16\pi\eta^2 t^2}\mathrm{e}^{-\rho r^2/4\eta t}. \tag{6.58}$$

According to the Taylor model the vorticity changes sign at $r = \sqrt{2\eta t/\rho}$, and the circulation $K = 2\pi r u$ decreases to zero at large values of r. Consequently, the total angular momentum of the fluid about the axis $r = 0$ is a finite and well-defined quantity for a Taylor vortex; per unit length, it is just the constant of proportionality in (6.58). For an ideal Oseen/Lamb vortex, on the other hand, this quantity diverges at the upper limit on integration over r. Vortices created by localised stirring of a fluid for a limited period of time, rather than by one of the specialised procedures suggested in §4.11, may well conform rather more closely to the Taylor model than to that of Oseen and Lamb.

6.10 The Ekman layer

The winds that blow over the Earth's oceans excite waves, of course, but in addition they give rise to surface currents, especially when they blow steadily in a fixed direction. In that case the ocean may approach a steady state of motion, such that the shear force exerted on it by the wind is balanced, in a frame of reference which rotates with the Earth, by Coriolis forces in the opposite direction. A theory due originally to Ekman, which is outlined below, suggests that in this steady state both the magnitude and the direction of the velocity $\boldsymbol{u}$ vary with depth below the surface. Suitably defined, however, the *mean* direction of the surface current must be at right angles to the direction of the wind. In the northern hemisphere, where Coriolis forces cause moving objects to veer to their right, a wind which blows steadily from the north should give rise to a mean surface current which flows towards the west. Just such a surface current occurs off the coast of California, where the prevailing wind is northerly. Here the surface water is replaced, as it moves out to sea in a westerly direction, by water brought up from the depths below, and it is because this water is so cold that the coast of California is notorious for fog; the upwelling water is rich in nutrients, incidentally, so fish flourish in it in and birds are attracted by the fish. The best known example of a wind-driven surface current in the Atlantic Ocean is the Gulf Stream.

Per unit volume, the Coriolis force experienced by a fluid of density ρ which is moving relative to the rotating Earth with velocity $\boldsymbol{u}$ is $2\rho\boldsymbol{u} \wedge \boldsymbol{\omega}_E$, where $\boldsymbol{\omega}_E$ is the angular velocity with which the Earth spins about its axis. Thus in a right-handed frame of reference such that the z axis points vertically upwards, and in conditions such that p^* does not vary with x or y, the equations of motion which describe steady laminar flow in a horizontal direction are

$$\begin{aligned} \eta\frac{\partial^2 u_x}{\partial z^2} + (2\rho\omega_E \sin\lambda)u_y &= 0, \\ \eta\frac{\partial^2 u_y}{\partial z^2} - (2\rho\omega_E \sin\lambda)u_x &= 0, \end{aligned} \tag{6.59}$$

where λ is the angle of latitude measured from the equator in a northerly direction (i.e. $\lambda = +\pi/2$ at the north pole, $\lambda = -\pi/2$ at the south pole). Elimination of u_y from equations (6.59) yields for u_x the fourth-order differential equation

$$\frac{\partial^4 u_x}{\partial z^4} = -\left(\frac{2\rho\omega_E \sin\lambda}{\eta}\right)^2 u_x.$$

This has four forms of solution,

$$u_x \propto e^{\pm z/\varepsilon} e^{\pm iz/\varepsilon},$$

where

$$\varepsilon = \sqrt{\frac{\eta}{\rho\omega_E \sin|\lambda|}}, \tag{6.60}$$

but only those forms which attenuate to zero as z tends to $-\infty$ are of interest here. If the velocity of the water is $\boldsymbol{u}_o$ at its free surface (where $z = 0$) and zero well below the surface, and if we choose the x axis to coincide with the direction of $\boldsymbol{u}_o$, then an appropriate solution is

$$u_x = u_o e^{z/\varepsilon} \cos(z/\varepsilon), \tag{6.61}$$

while substitution back into (6.59) shows that then

$$u_y = \frac{\sin|\lambda|}{\sin\lambda} u_o e^{z/\varepsilon} \sin(z/\varepsilon). \tag{6.62}$$

According to this theory, therefore, the magnitude of $\boldsymbol{u}$ attenuates by a factor e^{-1} over a depth ε, and the direction of $\boldsymbol{u}$ makes an angle θ with the x axis such that

$$\tan\theta = \frac{u_y}{u_x} = \frac{\sin|\lambda|}{\sin\lambda}\tan(z/\varepsilon);$$

hence $\theta = \pm(z/\varepsilon)$, the sign depending upon which hemisphere we are in. If we define the mean velocity of the surface current as

$$\langle \boldsymbol{u} \rangle = \frac{1}{\varepsilon}\int_{-\infty}^{0} \boldsymbol{u} dz,$$

then

$$\langle u \rangle_x = -\frac{\sin\lambda}{\sin|\lambda|}\langle u \rangle_y = \frac{1}{2}u_o.$$

which means that the angle which $\langle \boldsymbol{u} \rangle$ makes with the x axis is $\mp\pi/4$. To establish this steady state, a shear stress with components

$$s_{xz} = \eta\left(\frac{\partial u_x}{\partial z}\right)_{z=0} = \eta\varepsilon u_o,$$

$$s_{yz} = \eta\left(\frac{\partial u_y}{\partial z}\right)_{z=0} = \eta\varepsilon\frac{\sin|\lambda|}{\sin\lambda}u_o,$$

must be applied to the free surface, so the angle which the direction of the wind must make with the x axis is $\pm\pi/4$. The relation between the directions of the

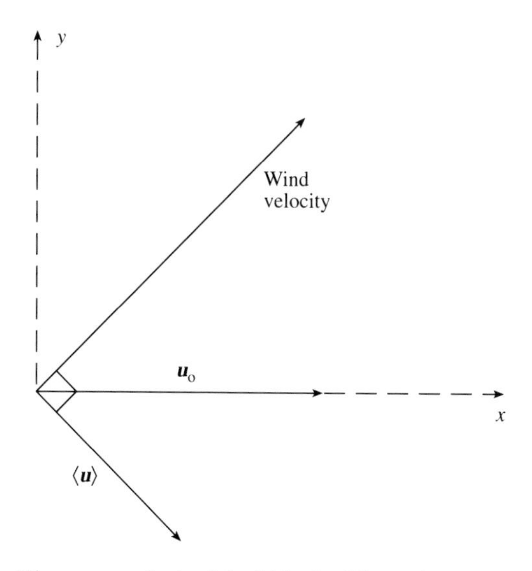

Figure 6.11 The mean velocity $\langle \boldsymbol{u} \rangle$ within the Ekman layer is at right angles to the wind velocity, while the velocity $\boldsymbol{u}_o$ at the surface (where $z = 0$) is at an angle of $\pi/4$ to both. This diagram represents the situation in the northern hemisphere, looking down on the ocean from above.

wind, of the surface velocity $\boldsymbol{u}_o$, and of $\langle \boldsymbol{u} \rangle$ is illustrated for the northern hemisphere in fig. 6.11.

The above theory is not wholly realistic, because it predicts unacceptably small values for the thickness of the so-called *Ekman layer* – the layer to which the wind-driven surface currents are confined. In practice this thickness seems to be of the order of 100 m at moderate latitudes, whereas ε is only about 15 cm according to (6.60). The discrepancy can be explained using the concept of *eddy viscosity*, which is discussed in §9.5.

In circumstances where the deep water of an ocean is not stationary, as assumed above, but is drifting relative to the ocean floor, friction between the water and the floor may excite transverse currents in an Ekman layer at the bottom, resembling the currents which winds are liable to excite at the free surface. They are, however, of relatively little significance.

6.11 Creeping flow past a sphere: Stokes's law

When fluid is draining from a plate or moving through a Hele Shaw cell, its inertia is irrelevant because [§§6.6(iii) and 6.8 respectively] the term in the Navier–Stokes equation which involves density, i.e. the term $\rho \, \mathrm{D}\boldsymbol{u}/\mathrm{D}t$, otherwise

$$\rho \frac{\partial \boldsymbol{u}}{\partial t} + \rho(\boldsymbol{u} \cdot \boldsymbol{\nabla})\boldsymbol{u},$$

although not equal to zero, is negligible; $\partial \boldsymbol{u}/\partial t$ is negligible because the flow is normally steady or quasi-steady, and $(\boldsymbol{u}\cdot\boldsymbol{\nabla})\boldsymbol{u}$ is negligible because u is so small. These are both examples of *creeping flow*. Here we consider a more complicated creeping flow problem, first solved by Stokes in 1851: what is the drag force exerted on a stationary solid sphere of radius a, by a fluid stream whose velocity at large distances from the sphere is uniformly equal to $\boldsymbol{U}$, when the Reynolds Number, $Re = 2\rho a U/\eta$, is very much less than unity? Under such conditions the analysis in §4.7 is useless; in so far as there exists a boundary layer around the sphere it is so thick that even at distances large compared with a we are not entitled to rely upon potential theory. An alternative analysis is outlined below. Readers who are disinclined to follow it step by step may jump to the results.

When its inertial term is neglected, the Navier–Stokes equation becomes

$$-\boldsymbol{\nabla}p^* - \eta\boldsymbol{\nabla}\wedge(\boldsymbol{\nabla}\wedge\boldsymbol{u}) = 0, \tag{6.63}$$

which, since

$$\boldsymbol{\nabla}\wedge(\boldsymbol{\nabla}\wedge\boldsymbol{u}) = \boldsymbol{\nabla}(\boldsymbol{\nabla}\cdot\boldsymbol{u}) - \nabla^2\boldsymbol{u},$$

is equivalent for an effectively incompressible fluid such that $\boldsymbol{\nabla}\cdot\boldsymbol{u}$ is zero to

$$\nabla^2\boldsymbol{u} = \frac{1}{\eta}\boldsymbol{\nabla}p^*. \tag{6.64}$$

This is the basic equation of motion for creeping flow. Its solutions for $\boldsymbol{u}$ consist in general of a *particular integral*, $\boldsymbol{u}_{\mathrm{PI}}$, and a *complementary function*, $\boldsymbol{u}_{\mathrm{CF}}$. The latter is a solution of $\nabla^2\boldsymbol{u} = 0$, which means that it is normally a solution of $\boldsymbol{\nabla}\wedge\boldsymbol{u} = 0$ and can therefore be described by a potential ϕ_{CF}. In the present problem the complementary function has to be chosen in such a way that it corresponds to uniform flow in the x_1 direction at large distances from the sphere, so in the spherical polar coordinates defined in fig. 4.6 we may expect [§4.7]

$$\phi_{\mathrm{CF}} = UR\cos\theta + AR^{-2}\cos\theta,$$

or

$$u_{R,\mathrm{CF}} = (U - 2AR^{-3})\cos\theta,$$
$$u_{\theta,\mathrm{CF}} = (-U - AR^{-3})\sin\theta,$$

where the coefficient A remains to be determined.

We cannot hope to match the boundary condition that $\boldsymbol{u} = 0$ at $R = a$ for all values of θ unless $u_{R,\mathrm{PI}}$ and $u_{\theta,\mathrm{PI}}$ are likewise proportional to $\cos\theta$ and $\sin\theta$ respectively. But application of the divergence operator $(\boldsymbol{\nabla}\cdot)$ to (6.63) shows at once that p^* obeys Laplace's equation,

$$\nabla^2 p^* = 0. \tag{6.65}$$

Where the flow is axially symmetric, as it is here, p^* must therefore be expressible, like ϕ_{CF}, in solid harmonic functions. If it is defined to be zero at large values of R where $\boldsymbol{u} = \boldsymbol{U}$, then the only credible possibility is that

$$p^* = BR^{-2} \cos \theta, \tag{6.66}$$

where the coefficient B is independent of θ and R. In that case $\boldsymbol{\nabla} p^*$ is proportional to R^{-3}, and $\boldsymbol{u}_{\text{PI}}$ must therefore be proportional to R^{-1}. Let us try

$$u_{R,\text{PI}} = CR^{-1} \cos \theta.$$

Then in order to satisfy the condition

$$\boldsymbol{\nabla} \cdot \boldsymbol{u}_{\text{PI}} = \frac{1}{R^2} \frac{\partial (R^2 u_{R,\text{PI}})}{\partial R} + \frac{1}{R \sin \theta} \frac{\partial (\sin \theta \, u_{\theta,\text{PI}})}{\partial \theta} = 0$$

we must set

$$u_{\theta,\text{PI}} = -\frac{1}{2} CR^{-1} \sin \theta.$$

These guesses have now to be checked by substitution into (6.64). Both sides of that equation are, of course, vectors, but to simplify the analysis we shall consider only their components in the longitudinal x_1 direction; it can easily be verified that when these are equal to one another the transverse components are equal to one another also. On the left-hand side we have

$$\nabla^2 u_{1,\text{PI}} = \left\{ \frac{1}{R^2} \frac{\partial}{\partial R} \left(R^2 \frac{\partial}{\partial R} \right) + \frac{1}{R^2 \sin \theta} \frac{\partial}{\partial \theta} \left(\sin \theta \frac{\partial}{\partial \theta} \right) \right\} (u_R \cos \theta - u_\theta \sin \theta),$$

which simplifies to

$$\begin{aligned} \nabla^2 u_{1,\text{PI}} &= \frac{1}{2} C \frac{1}{R^3 \sin \theta} \frac{\partial}{\partial \theta} \left\{ \sin \theta \frac{\partial}{\partial \theta} (2 \cos^2 \theta + \sin^2 \theta) \right\} \\ &= -\frac{C}{R^3} (2 \cos^2 \theta - \sin^2 \theta). \end{aligned}$$

On the right-hand side we have

$$\begin{aligned} \frac{1}{\eta} \frac{\partial p^*}{\partial x_1} &= \frac{1}{\eta} \left(\cos \theta \frac{\partial p^*}{\partial R} - \frac{1}{R} \sin \theta \frac{\partial p^*}{\partial \theta} \right) \\ &= -\frac{B}{\eta R^3} (2 \cos^2 \theta - \sin^2 \theta). \end{aligned}$$

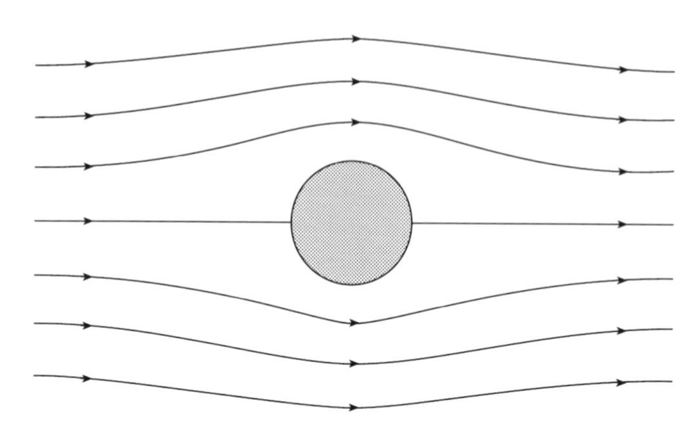

Figure 6.12 Lines of flow past a sphere according to Stokes's solution.

These expressions can indeed be made equal to one another, by choosing $C = B/\eta$. Finally, to ensure that both u_R and u_θ vanish at $R = a$ we need to let $A = -\frac{1}{4}a^3U$, $C = -\frac{3}{2}aU$.

The full solution, which is the only solution which satisfies the given boundary conditions, is therefore

$$\begin{aligned} u_R &= u_{R,\mathrm{CF}} + u_{R,\mathrm{PI}} = U\cos\theta\left(1 - \frac{3a}{2R} + \frac{a^3}{2R^3}\right), \\ u_\theta &= u_{\theta,\mathrm{CF}} + u_{\theta,\mathrm{PI}} = -U\sin\theta\left(1 - \frac{3a}{4R} - \frac{a^3}{4R^3}\right). \end{aligned} \tag{6.67}$$

The principal respects in which in which it differs from the solution of Euler's equation worked out in §4.7, on the basis of potential theory alone, are:

(i) it satisfies the no-slip boundary condition at the sphere's surface;
(ii) it describes a velocity u_θ in the equatorial ($\theta = \pi/2$) plane which increases monotonically towards U with increasing R instead of decreasing;
(iii) the terms in a/R which it contains represent a perturbation of the flow field which is of a *long-range* nature.

Lines of flow for Stokes's solution are sketched in fig. 6.12.

According to this solution, the excess stress which acts upon the surface of the sphere has a normal component given by

$$p_R^* = p^* - 2\eta\left(\frac{\partial u_R}{\partial R}\right)_{R=a} = -\frac{3\eta U\cos\theta}{2a},$$

[(6.11)] and a shear component acting in the direction of increasing θ given by

$$s_{\theta R} = \eta a \left\{ \frac{\partial}{\partial R}\left(\frac{u_\theta}{R}\right) + \frac{1}{a^2}\frac{\partial u_R}{\partial \theta} \right\}_{R=a} = -\frac{3\eta U \sin\theta}{2a}$$

[(6.3) and (6.53)]. Taken together, these components are equivalent to a uniform force per unit area in the direction of $\boldsymbol{U}$ of magnitude $3\eta U/2a$. The total drag force in the direction of $\boldsymbol{U}$ is therefore

$$F_{\mathrm{D}} = 4\pi a^2 \frac{3\eta U}{2a} = 6\pi\eta a U. \tag{6.68}$$

This expression constitutes *Stokes's law*.

It is only in the limit when velocity U and Reynolds Number Re $(= 2\rho a U/\eta)$ tend to zero that the assumption on which Stokes's law is based is fully consistent with the details of his solution. Since the leading term in $\boldsymbol{u}$ is $\boldsymbol{U}$, while the next terms in (6.67) are proportional to aU/R, the inertial term in the Navier–Stokes equation, $\rho(\boldsymbol{u}\cdot\boldsymbol{\nabla})\boldsymbol{u}$, is of order $\rho U^2 a/R^2$ at large values of R according to Stokes, while the viscous term $\eta\boldsymbol{\nabla}\wedge(\boldsymbol{\nabla}\wedge\boldsymbol{u})$ is of order $\eta a U/R^3$. Far from being negligible, the inertial term is clearly liable to exceed the viscous term at distances such that

$$R > \frac{\eta}{\rho U} = \frac{2a}{\mathit{Re}}.$$

The inconsistency may suggest to the reader that we cannot trust equations (6.67) to describe the velocity distribution in the immediate vicinity of the sphere, and that we therefore cannot trust Stokes's law, unless Re is really quite small compared with unity. It is only when Re reaches about 0·5, however, that deviations from the law become detectable experimentally.

Needless to say, if Stokes's law applies in a frame of reference such that the sphere is stationary then it applies also in the frame in which the distant fluid is stationary and the sphere is moving instead. Thus a solid sphere of radius a and density ρ_{sol}, falling down the axis of a vertical cylinder of sufficiently large radius which is filled with liquid of density ρ_{liq}, may be expected to reach a terminal velocity U such that

$$6\pi\eta a U = \frac{4}{3}\pi a^3(\rho_{\mathrm{sol}} - \rho_{\mathrm{liq}})g, \tag{6.69}$$

provided that

$$\mathit{Re} = \frac{4a^3\rho_{\mathrm{liq}}(\rho_{\mathrm{sol}} - \rho_{\mathrm{liq}})g}{9\eta^2} < 0{\cdot}5. \tag{6.70}$$

The *falling sphere viscometer*, a simple device commonly used for the determination of the viscosity of rather viscous liquids, is based upon (6.69). In this device the sphere is usually a steel ball. If the liquid is something like olive oil, which at room temperature has a viscosity about 300 times greater than that of water, then the radius of the ball must be less than about 1 mm for (6.70) to be satisfied, while the radius of the cylinder should be greater by a factor of at least 100.

If the falling sphere is itself liquid, with viscosity η', circulating currents arise within it as it falls which modify the flow pattern outside the sphere. The modified form of Stokes's law which applies in these circumstances is

$$F_D = \frac{4\pi\eta aU(\eta + \frac{3}{2}\eta')}{\eta + \eta'}. \qquad (6.71)$$

This evidently reduces to (6.68) when $\eta' \gg \eta$, e.g. under the conditions of Millikan's celebrated experiment, where the spheres were oil drops moving through air. At the opposite extreme where $\eta' \ll \eta$, however, e.g. where the spheres are very small bubbles of gas rising (rather than falling) through soda water or champagne, it reduces to $F_D = 4\pi\eta aU$, so the terminal velocity of such bubbles should be

$$U = \frac{a^2 \rho_{liq} g}{3\eta} \qquad (6.72)$$

[(6.69), but with 6 replaced by 9 and with ρ_{sol} replaced by ρ_{gas}; ρ_{gas} is negligible compared with ρ_{liq}]. In fact, (6.72) does not describe the terminal velocity of rising soda water bubbles at all accurately. That is partly because the Reynolds Number normally exceeds 0·5 but also, it seems, because impurities adsorbed on the gas–liquid interface endow this interface with some measure of rigidity. It can be shown, incidentally, that the stresses which act on a gas bubble which is rising steadily with $Re \ll 1$ do not tend to distort it; it should – and does – remain spherical.

6.12 The viscosity of suspensions

The route which led us, in §6.11, to Stokes's solution for the creeping flow equation may be followed to find a solution to describe flow in the neighbourhood of a stationary solid sphere of radius a, with its centre at the origin where $R = 0$, when the incompressible liquid surrounding it is being subjected at large values of R to continuous extension along say the x_1 axis. The complementary function is derivable in this case from a potential of the form

$$\phi_{CF} = \frac{1}{4}\zeta R^2(3\cos^2\theta - 1) + AR^{-2}(3\cos^2\theta - 1)$$

in spherical polar coordinates [§4.6(iv)], or

$$\phi_{CF} = \frac{1}{4}\zeta(2x_1^2 - x_2^2 - x_3^2) + A\,\frac{(2x_1^2 - x_2^2 - x_3^2)}{(x_1^2 + x_2^2 + x_3^2)^2}$$

in cartesian coordinates. The corresponding velocity components are

$$\begin{aligned} u_{\mathrm{CF},R} &= \left(\frac{1}{2}\,\zeta R - 2AR^{-3}\right)(3\cos^2\theta - 1), \\ u_{\mathrm{CF},\theta} &= \left(-\frac{3}{2}\,\zeta R - 6AR^{-3}\right)\sin\theta\cos\theta, \end{aligned} \tag{6.73}$$

or, at such large distances that the terms in A are negligible,

$$\begin{aligned} &u_{\mathrm{CF},1} \approx \zeta x_1, \\ &u_{\mathrm{CF},2} \approx -\frac{1}{2}\,\zeta x_2, \quad u_{\mathrm{CF},3} \approx -\frac{1}{2}\,\zeta x_3. \end{aligned} \tag{6.74}$$

We shall stick to spherical polars in what follows, but equations (6.74) are nevertheless helpful because they reveal with particular clarity the nature of the flow pattern at large distances, where the rate of extension or vorticity-free shear in the fluid is uniformly equal to ζ. By combining (6.74) with (6.26) we may show at once that the rate at which energy is being dissipated as heat per unit volume asymptotically approaches

$$2\eta\left\{(\zeta)^2 + \left(\frac{1}{2}\,\zeta\right)^2 + \left(\frac{1}{2}\,\zeta\right)^2\right\} = 3\eta\zeta^2. \tag{6.75}$$

This result is needed later.

Including the particular integral, the full solution turns out to be

$$\begin{aligned} u_R &= \frac{1}{2}\,\zeta R(3\cos^2\theta - 1)\left\{1 - \frac{5}{2}\left(\frac{a}{R}\right)^3 + \frac{3}{2}\left(\frac{a}{R}\right)^5\right\} \\ u_\theta &= -\frac{3}{2}\,\zeta R\sin\theta\cos\theta\left\{1 - \left(\frac{a}{R}\right)^5\right\}, \end{aligned} \tag{6.76}$$

[(6.67)] and the associated excess mean pressure distribution is

$$p^* = -\frac{5}{2}\,\eta\zeta\left(\frac{a}{R}\right)^3(3\cos^2\theta - 1) \tag{6.77}$$

[(6.66)]. Let us use these results to calculate what difference the sphere makes to the total dissipation. It turns out to be laborious to rely upon (6.26) in this context, i.e. to integrate $2\eta\zeta_{ij}\zeta_{ij}$ per unit volume from $R = a$ up to say $R = R_o$ ($\gg a$). The expression for $\zeta_{ij}\zeta_{ij}$ includes terms of order 1, $(a/R)^3$, $(a/R)^5$, $(a/R)^6$, $(a/R)^8$ and $(a/R)^{10}$, and all of them except the first and second (which turns out to vanish on integration over θ) contribute something proportional to a^3 to the final answer. Instead, therefore, let us calculate the rate at which work is being done on the fluid which instantaneously occupies the sphere of radius R_o by fluid outside it; all of this must be dissipated as heat when the creeping flow approximation is justified, because changes in the kinetic energy of the fluid are negligible in comparison and changes in its gravitational potential energy, if any, are taken care of by the use of p^* rather than p.

The rate of working which we need is

$$4\pi R_o^2 \langle - p_R^* u_R + s_{\theta R} u_\theta \rangle_{R=R_o},$$

where

$$p_R^* = p^* - 2\eta \frac{\partial u_R}{\partial R}$$

$$s_{\theta R} = \eta \left\{ R \frac{\partial}{\partial R}\left(\frac{u_\theta}{R}\right) + \frac{1}{R}\frac{\partial u_R}{\partial \theta} \right\}$$

as in §6.11, and where the angle brackets $\langle\ \rangle$ here imply an average over θ. It proves a relatively simple matter to show that to third order in (a/R_o) this amounts to

$$3\eta\zeta^3 \left(\frac{4}{3}\pi R_o^3\right)\left\{1 + \frac{5}{2}\left(\frac{a}{R_o}\right)^3 + \dots\right\}. \tag{6.78}$$

When a is zero, (6.78) describes the dissipation to be expected from (6.75). For non-zero a, however, the total dissipation within a radius $R_o \gg a$ is evidently equal to the total dissipation within the same radius for a *homogeneous* incompressible fluid endowed at all values of R with the velocity distribution described by (6.73), the viscosity of which is

$$\eta_{\text{eff}} \approx \eta \left\{1 + \frac{5}{2}\left(\frac{a}{R_o}\right)^3\right\}.$$

This is the basis for a result due originally to Einstein, that the effective viscosity of a fluid which contains a large number of spherical solid particles in suspension is

$$\eta_{\text{eff}} \approx \eta \left(1 + \frac{5}{2} f\right), \tag{6.79}$$

where f is the fraction of the total volume of the suspension which is occupied by solid matter. The result is valid only when the spheres are far enough apart from one another not to interact (in practice this requires $f < 0{\cdot}02$, or thereabouts), and at such low rates of shear that the Reynolds Number defined by

$$Re = \frac{\rho a(\zeta a)}{\eta}$$

is small compared with unity; otherwise, the creeping flow approximation is unreliable.

The above argument can be modified to describe suspensions in which the spherical inclusions are fluid rather than solid, and in the particular case where they are bubbles containing gas of negligible viscosity the end result is

$$\eta_{\text{eff}} = \eta(1 + f). \tag{6.80}$$

An effectively incompressible liquid which contains spherical inclusions of compressible material – in particular, one which contains bubbles of gas – behaves like a homogeneous liquid which possesses not only compressibility but also an effective bulk viscosity. The latter can be estimated from the energy dissipated when the suspension is compressed, or more directly by the following argument. Imagine a sphere of the liquid of radius R_{o} with a single spherical bubble of radius a at its centre, containing gas whose compressibility is β_{gas}; the shear viscosity of the liquid is η_{liq}. This system is initially at rest in equilibrium, with p^* uniformly equal to zero throughout. The pressure outside the spherical surface of radius R_{o} is now altered uniformly by a small amount Δp, which may subsequently vary with time. This causes a change ΔV in the volume $V = \frac{4}{3}\pi_{\text{o}}^3$ of the whole system, and the *same* volume change (because the liquid is incompressible) in the volume of the bubble. The flow which occurs while ΔV is changing is radial, and such that at radius R

$$4\pi R^2 u_R = \frac{\partial(\Delta V)}{\partial t},$$

and hence

$$\frac{\partial u_R}{\partial R} = -\frac{1}{2\pi R^3}\frac{\partial(\Delta V)}{\partial t}. \tag{6.81}$$

The flow is vorticity-free, so we can use Bernoulli's theorem in its generalised form [(4.14)] to relate the mean excess pressure p^* where R is just greater than a to the same quantity at $R = R_{\text{o}}$. At small values of the Reynolds Number, however, i.e. when

$$Re = 2\rho\left(\frac{Ru_R}{\eta_{\text{liq}}}\right)_{R=a} \ll 1, \tag{6.82}$$

the inertia of the liquid is negligible, and the terms $\frac{1}{2}u^2$ and $\partial\phi/\partial t$ in (4.14) are negligible too; in this limit, the increase of p^* or p in the liquid must be uniformly equal to Δp throughout. The increase of pressure in the gas is different, however, because the quantity which is continuous across the liquid–gas interface is the radial component of normal stress, p_R rather than p. Since we may safely assume that $\eta_{\text{gas}} \ll \eta_{\text{liq}}$, we have

$$\Delta p_{\text{gas}} = \Delta p - 2\eta_{\text{liq}}\left(\frac{\partial u_R}{\partial R}\right)_{R=a},$$

or, in view of (6.81),

$$\Delta p_{\text{gas}} = \Delta p + \frac{\eta_{\text{liq}}}{\pi a^3}\frac{\partial(\Delta V)}{\partial t}.$$

With

$$\beta_{\text{gas}}\Delta p_{\text{gas}} = -\frac{\Delta V}{\frac{4}{3}\pi a^3} = -\frac{\Delta V}{fV},$$

this reduces to

$$\Delta p = -\frac{1}{f\beta_{\text{gas}}}\frac{\Delta V}{V} - \frac{4\eta_{\text{liq}}}{3f}\frac{1}{V}\frac{\partial(\Delta V)}{\partial t}.$$

The effective compressibility of the liquid sphere with a gas bubble at its centre is therefore

$$\beta_{\text{eff}} = f\beta_{\text{gas}}, \tag{6.83}$$

while comparison with (6.21) shows that the effective bulk viscosity is

$$\eta_{\text{b,eff}} = \frac{4\eta_{\text{liq}}}{3f}. \tag{6.84}$$

The smaller the bubble is, the greater its effect on $\eta_{\text{b,eff}}$. This seems strange, until one remembers that reducing the size of the bubble reduces β_{eff}. If raising the pressure has very little effect on the volume of the system, the fact that $\eta_{\text{b,eff}}$ is very large is scarcely material.

Equations (6.83) and (6.84) should, like (6.80), hold for suspensions in which there are many gas bubbles, provided that $f < 0{\cdot}02$ (say), that (6.82) is satisfied, and that the compressibility of the liquid is indeed negligible. The third of these conditions corresponds to $\beta_{\text{liq}} \ll \beta_{\text{eff}}$, which is equivalent in view of (6.83) to $f \gg \beta_{\text{gas}}/\beta_{\text{liq}}$.

6.13 Percolation

Suppose that many small solid spheres of radius a, collectively occupying a fraction f of the total volume, are falling through an otherwise stationary liquid of viscosity η, what is their terminal velocity likely to be? An answer to this question which is plausible when f is small may be obtained by replacing the η which appears in (6.69) by the η_{eff} of (6.79); it is

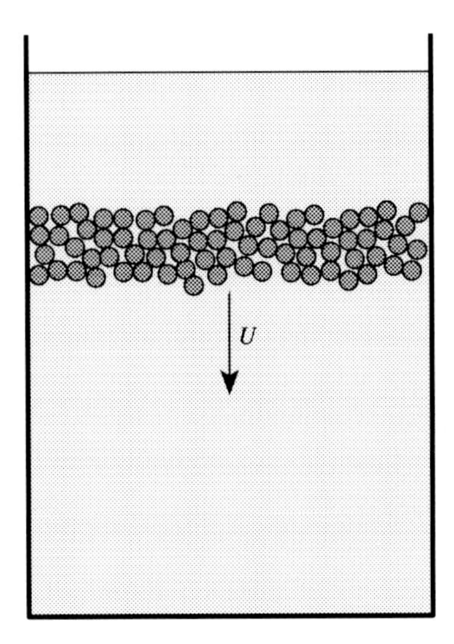

Figure 6.13 A layer of solid spheres in contact with their neighbours, falling through a liquid which is stationary above and below them. In a different frame of reference the spheres are stationary and the liquid is percolating upwards through the interstices between them.

$$U \approx \frac{2a^2g}{9\eta}(\rho_{\text{sol}} - \rho_{\text{liq}})\left(1 - \frac{5}{2}f\right).$$

If we cannot trust this unless f is less than about 0·02, however, it is of rather little interest. It certainly will not tell us what to expect when f is large enough (greater than 0·5, say) for the spheres to be jammed together in contact with their neighbours and to fall as a coherent, but permeable, solid block [fig. 6.13]. Dimensional analysis suggests an answer in the form

$$U = \frac{2a^2g}{9\eta}(\rho_{\text{sol}} - \rho_{\text{liq}})\,F\{f\}, \tag{6.85}$$

and a semi-empirical relation employed by engineers (the *Kozeny–Carman* relation) corresponds to this, with

$$F\{f\} = \frac{(1-f)^3}{10f}.$$

It is likely, however, that the coefficient $F\{f\}$ depends not only on the filling factor f but also on the geometry of the packing. It may not be the same for packings which are regular rather than random. It may not be the same for cubic close-packed arrays of spheres as for hexagonal close-packed ones, though f is identical in these two cases (about 0·74). It certainly does not remain the same if the spheres are deformed, without changing f, into objects of different shape.

This unsolved problem in creeping flow is of interest because of its relevance to practical situations where liquids are percolating under a pressure gradient through some porous medium, e.g. water through soil, or oil through shale. From one point of view, the layer of compacted spheres which is sketched in fig. 6.13 is falling downwards with velocity U through liquid which is stationary below it. From another point of view, in a different reference frame, the spheres are stationary and the liquid is creeping upwards through the interstices between them; U then represents the volume of liquid which crosses a stationary horizontal plane, per unit area and per unit time. If d is the vertical thickness of the layer of spheres, the excess pressure below this layer must be greater than it is above by $gdf(\rho_{\text{sol}} - \rho_{\text{liq}})$, to support the spheres' weight. Thus in the frame of reference in which the spheres are stationary the volume flux rate U in the $+z$ direction is associated with a gradient of excess pressure

$$\frac{\mathrm{d}p^*}{\mathrm{d}z} = -\, gf(\rho_{\text{sol}} - \rho_{\text{liq}}).$$

From this point of view, therefore, (6.85) is an example of *Darcy's law* of percolation through porous media, which states that the local volume flux rate of the fluid is related to the local pressure gradient by the equation

$$\boldsymbol{U} = -\,\frac{k}{\eta}\,\boldsymbol{\nabla} p^*, \tag{6.86}$$

where k is a constant, known as the *permeability*, which depends upon the properties of the medium but not on the properties of the fluid; (6.85) is consistent with (6.86), of course, if

$$k = \frac{2a^2F\{f\}}{9f}.$$

Darcy's law is useful to soil engineers because (6.86) may be solved, once k has been determined by experiment, to predict the movements of water or oil in situations where $\boldsymbol{\nabla} p^*$ varies in direction and magnitude from place to place. It is valid only when U is small enough to make the inertia of the fluid irrelevant.

Further reading

An Introduction to Fluid Mechanics, Cambridge University Press, 1967, by G. K. Batchelor, covers nearly all the topics discussed in this chapter, at a higher level of mathematical sophistication. Readers who wish to see a more elegant demonstration of Stokes's law than the one outlined in §6.11, for example, should consult Batchelor's book, or else *Fluid Dynamics*, Pergamon Press, Oxford, 2nd edition

1987, by L. D. Landau & E. F. Lifshitz. *Cartesian Tensors*, Cambridge University Press, 1931, by H. Jeffreys, provides proofs for certain unsupported statements contained in §6.2. For more information concerning the theory and practice of lubrication consult *Mechanics of Fluids*, Van Nostrand Reinhold (UK), Wokingham, 4th edition 1979, by B. S. Massey.

7

Vorticity

7.1 Lines of vorticity

Most of this chapter concerns incompressible flow past solid obstacles, and the drag and lift forces which they experience, at values of the Reynolds Number which are too large compared with unity for the approximations employed in chapter 6, e.g. in the derivation of Stokes's law for the drag force on a solid sphere, to be valid. The effects to be discussed depend critically on the behaviour of boundary layers, and boundary layers, as we have seen, are layers within which the fluid is contaminated by vorticity. To understand these effects properly we need to understand how vorticity behaves, and that is why the chapter has 'Vorticity' as its heading.

The properties of free vortex lines, set in otherwise vorticity-free fluid, have already been described in §§4.13 and 4.14, but we can explore the subject of vorticity dynamics in a more general fashion now that we have the Navier–Stokes equation to use as a starting point. The first point to note is that because $\boldsymbol{\Omega}$ is defined as the curl of another vector its divergence is necessarily zero everywhere; vorticity, like the electromagnetic fields $\boldsymbol{E}$ and $\boldsymbol{B}$ in free space and like the velocity $\boldsymbol{u}$ of an incompressible fluid, is what is called a *solenoidal* vector. This means that its spatial variation can be described by continuous field lines whose direction coincides everywhere with the local direction of $\boldsymbol{\Omega}$ and whose density is proportional to the magnitude of $\boldsymbol{\Omega}$. The relevant field lines are called *vortex lines* in some textbooks, where the vortex lines discussed in chapter 4 are rechristened *line vortices* in an effort to prevent confusion, but to emphasise the distinction a little more strongly the field lines will here be referred to as *lines of vorticity* (cf. *lines of flow*). Using this language, we may say that a bundle of lines of vorticity runs through the core of every free vortex line.

In any Euler fluid, lines of vorticity may be regarded as embedded in the fluid and obliged to move with it, and they are conserved in number.[1] A proof of that assertion may readily be constructed on the basis of Kelvin's circulation theorem, along the lines of the proof offered in §4.14 for the fact that vortex lines are embedded and conserved in strength. Alternatively, one may prove it by applying the curl operator, $\nabla \wedge$, to both sides of the Euler equation

$$-\frac{1}{\rho}\nabla p^* = \frac{\partial \boldsymbol{u}}{\partial t} + (\boldsymbol{u}\cdot\nabla)\boldsymbol{u}.$$

This operation leads, when ρ is uniform and $\nabla\cdot\boldsymbol{u} = 0$, to

$$0 = \frac{\partial \boldsymbol{\Omega}}{\partial t} + (\boldsymbol{u}\cdot\nabla)\boldsymbol{\Omega} - (\boldsymbol{\Omega}\cdot\nabla)\boldsymbol{u}$$

and hence to

$$\frac{\mathrm{D}\boldsymbol{\Omega}}{\mathrm{D}t} = (\boldsymbol{\Omega}\cdot\nabla)\boldsymbol{u}.$$

To see what this result means, let us write out its components, choosing the x_3 direction to coincide locally with the direction of $\boldsymbol{\Omega}$, in which case $\Omega_1 = \Omega_2 = 0$. We have

$$\frac{\mathrm{D}\Omega_1}{\mathrm{D}t} = \Omega_3\frac{\partial u_1}{\partial x_3}, \tag{7.1}$$

a similar equation to describe $\mathrm{D}\Omega_2/\mathrm{D}t$, and

$$\frac{\mathrm{D}\Omega_3}{\mathrm{D}t} = \Omega_3\frac{\partial u_3}{\partial x_3}. \tag{7.2}$$

Now the quantity $\partial u_1/\partial x_3$ which appears in (7.1) is the angular velocity with which a short line embedded in the fluid, initially lying along the x_3 axis and therefore coinciding in direction with $\boldsymbol{\Omega}$, is rotating about the x_2 axis; (7.1) tells us that within the element of fluid which contains this line the vorticity vector precesses about x_2 at the same rate. Since it also precesses about x_1 at the same rate as the embedded line, the embedded line and the local line of vorticity must remain coincident at all times. As for (7.2), this tells us that where the fluid is being elongated in the direction of $\boldsymbol{\Omega}$ the magnitude of $\boldsymbol{\Omega}$ increases. We have met this phenomenon of the *intensification of vorticity by stretching* in §4.13, where we noted that the angular velocity ω_o within the core of a vortex line is proportional to

[1] Note the cautious phrase 'may be regarded'. A pedant might argue that lines of vorticity are figments of our imagination, that we have no way of labelling them, and no way of telling whether they are moving or not. The notion that they are embedded and conserved is fully consistent with the behaviour of Euler fluids, however, and it is a notion which proves helpful in practice.

its length L and therefore inversely proportional to its cross-sectional area. Equation (7.2) shows that within any cylindrical element of fluid whose axis coincides with the direction of $\boldsymbol{\Omega}$, whether it forms part of the core of a vortex line or not, the product of Ω and the cross-sectional area of the element is constant. Both equations are clearly consistent with the notion that lines of vorticity are embedded and conserved.

To find what difference viscosity makes, we need to repeat the above analysis using the Navier–Stokes equation as our starting point, rather than the Euler equation. The viscous term on the left-hand side of (6.25) is $-\eta\boldsymbol{\nabla} \wedge \boldsymbol{\Omega}$, and the curl of this, since $\boldsymbol{\nabla} \cdot \boldsymbol{\Omega} = 0$, is $\eta\nabla^2\boldsymbol{\Omega}$. Hence we now have

$$\frac{\mathrm{D}\boldsymbol{\Omega}}{\mathrm{D}t} = (\boldsymbol{\Omega} \cdot \boldsymbol{\nabla})\boldsymbol{u} + \frac{\eta}{\rho}\nabla^2\boldsymbol{\Omega}. \tag{7.3}$$

Apart from the $(\boldsymbol{\Omega} \cdot \boldsymbol{\nabla})\boldsymbol{u}$ term, the effects of which are as described above, this is just a three-dimensional diffusion equation for each of the components of $\boldsymbol{\Omega}$; to be more precise, it becomes a three-dimensional equation in the co-moving frame for which $\mathrm{D}\boldsymbol{\Omega}/\mathrm{D}t$ and $\partial\boldsymbol{\Omega}/\partial t$ are the same. Thus vorticity is not permanently embedded if the fluid has viscosity; where $\nabla^2\boldsymbol{\Omega}$ is non-zero it spreads by diffusion, and its *diffusivity* is the kinematic viscosity, $\nu = \eta/\rho$. Since the process described by the diffusion equation always conserves the thing which is diffusing, whether it be dye or heat or whatever, the fact that vorticity is liable to diffuse does not affect our conclusion that lines of vorticity are conserved.

If, however, the vorticity is positive in region A and negative in an adjoining region B, diffusion from A to B and *vice versa* is bound to result in some degree of cancellation. The lines of vorticity in such situations tend to form closed loops which disappear by collapsing to a point. For example, consider the simple case of a fluid undergoing Poiseuille flow along a straight cylindrical pipe whose axis is the x_3 axis. In the plane $x_2 = 0$, say, Ω_1 and Ω_3 both vanish, while

$$\Omega_2 = -\frac{\partial u_3}{\partial x_1} = -\frac{x_1}{2\eta}\nabla_3 p$$

[(1.32)]. Thus $\boldsymbol{\Omega}$ changes sign on the axis in the plane $x_2 = 0$, and it also does so in the plane $x_1 = 0$ where Ω_1 is the non-vanishing component; evidently the lines of vorticity are closed circular loops coaxial with the pipe. Now the direction in which the lines of vorticity diffuse is determined by the sign of $\partial\Omega/\partial r$. Because this is positive we should picture the loops as diffusing inwards, to smaller values of radius r, and ultimately collapsing on the axis. We should therefore picture the surface of the fluid, where it is in contact with the solid wall of the pipe, as a vorticity source at which new loops are continuously created to replace those which collapse.

In other situations lines of vorticity may diffuse *towards* surfaces rather than away from them and may disappear in that way. Whether a surface is to be thought

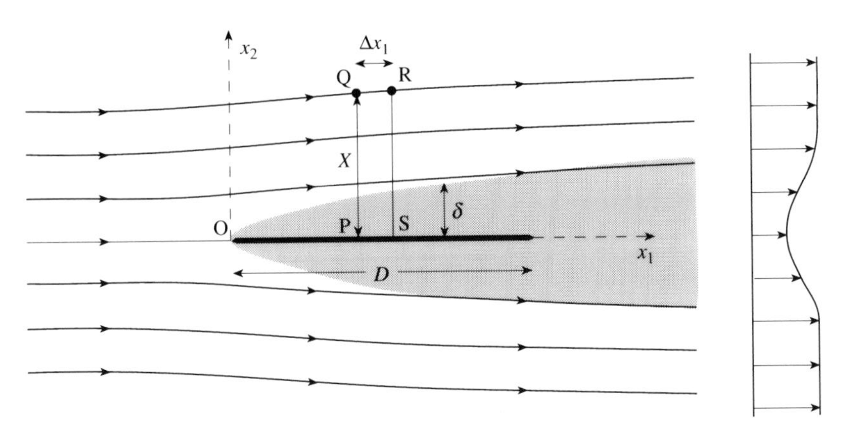

Figure 7.1 Two-dimensional flow past a flat plate with a straight leading edge through O; the boundary layer of thickness $\delta\{x_1\}$ is indicated by shading. PQRS is a cross-section through a control volume, whose length X in the x_2 direction is $1/\Delta x_1$. The profile on the right suggests how the velocity u_1 is likely to vary with x_2 behind the plate.

of as a vorticity source, a vorticity sink, or neither depends entirely upon the gradient of vorticity adjacent to it, as we shall see in the following section. It may also depend upon one's point of view, however, since a surface which can be described as absorbing vorticity can equally well be described as generating vorticity with the opposite sign.

7.2 Boundary layers on plates

A simple analysis outlined in §1.12 shows that in contact with an infinite flat plate which is moving in its own plane relative to otherwise stationary fluid there is a boundary layer whose uniform thickness is of order $\sqrt{2\eta t/\rho}$ or $\sqrt{2\eta/\omega\rho}$, depending upon whether the plate has been in steady motion for a time t or is oscillating with angular frequency ω. Here we shall discuss the boundary layer on a plate of finite extent which has been in uniform motion relative to the surrounding fluid for an effectively infinite time. This is a problem of greater practical importance but also of greater complexity, principally because the vorticity is subject to downstream convection as well as to diffusion outwards from the plate. It is illustrated by figs. 1.11 and 7.1; as the latter suggests, the plate is to be regarded as a rectangular strip, having breadth D in the x_1 direction, negligible thickness in the x_2 direction, and a length L in the x_3 direction which is much greater than D, and it is aligned *edge-on* to the flow. In the reference frame of these figures, where the plate is treated as stationary, the fluid velocity at large distances from it has components $(U, 0, 0)$.

The full equations of motion for the problem are the continuity condition

$$\frac{\partial u_1}{\partial x_1} + \frac{\partial u_2}{\partial x_2} = 0 \tag{7.4}$$

and two components of the Navier–Stokes equation [(6.25)] which, since the flow is two-dimensional and steady, may in view of (7.4) be written as

$$u_1 \frac{\partial u_1}{\partial x_1} + u_2 \frac{\partial u_1}{\partial x_2} = -\frac{1}{\rho}\frac{\partial p^*}{\partial x_1} - \frac{\eta}{\rho}\frac{\partial}{\partial x_2}\left(\frac{\partial u_2}{\partial x_1} - \frac{\partial u_1}{\partial x_2}\right),$$

$$u_1 \frac{\partial u_2}{\partial x_1} + u_2 \frac{\partial u_1}{\partial x_1} = -\frac{1}{\rho}\frac{\partial p^*}{\partial x_2} + \frac{\eta}{\rho}\frac{\partial}{\partial x_1}\left(\frac{\partial u_2}{\partial x_1} - \frac{\partial u_1}{\partial x_2}\right).$$

However, the Navier–Stokes equation may be significantly simplified if we are prepared to assume that the thickness of the boundary layer, $\delta\{x_1\}$, is much less than x_1; it can readily be shown, using the results stated below, that this assumption is likely to be justified over most of the plate's surface (though it fails near the leading edge where x_1 is small) provided that the Reynolds Number $Re = \rho DU/\eta$ is much greater than unity. In that case the lines of flow are almost parallel to the plate and almost straight (though not completely so, because where the fluid is being slowed down by the effects of viscosity they must move apart from one another, and in doing so they must displace other lines of flow). Granted that their curvature is negligible we may ignore $\partial p^*/\partial x_2$, and since p^* is effectively uniform in the vorticity-free fluid outside the boundary layer where $u \approx U$ this means that we may ignore $\partial p^*/\partial x_1$ too. Moreover, we may ignore u_2 in comparison with u_1 and $\partial u_2/\partial x_1$ in comparison with $\partial u_1/\partial x_2$, though we cannot safely ignore $u_2(\partial u_1/\partial x_2)$ in comparison with $u_1(\partial u_1/\partial x_1)$. Thus all the terms in the third of the three equations above are very small, and the second may by simplified to

$$u_1 \frac{\partial u_1}{\partial x_1} + u_2 \frac{\partial u_1}{\partial x_2} \approx \frac{\eta}{\rho}\frac{\partial^2 u_1}{\partial x_2^2}. \tag{7.5}$$

Equations (7.4) and (7.5) constitute the *boundary layer equations* for constant U, and together they suffice to define the two unknowns u_1 and u_2.

A solution in series form which matches the appropriate boundary conditions was first obtained by Blasius in 1908, and it has the property of *self-similarity*. That is to say, the velocity profiles which it describes are the same for all values of x_1, apart from a scaling length l which, as could have been anticipated on dimensional grounds alone, is proportional to $\sqrt{\eta x_1/\rho U}$. We may choose the coefficient of proportionality to be unity if we wish and write

$$\frac{u_1}{U} = f_o\left\{\frac{x_2}{l_o}\right\}, \quad l_o = \sqrt{\frac{\eta x_1}{\rho U}}; \tag{7.6}$$

the suffices attached to both l and f in this expression are meant as reminders of the choice we have made, and of the fact that other choices are possible [see

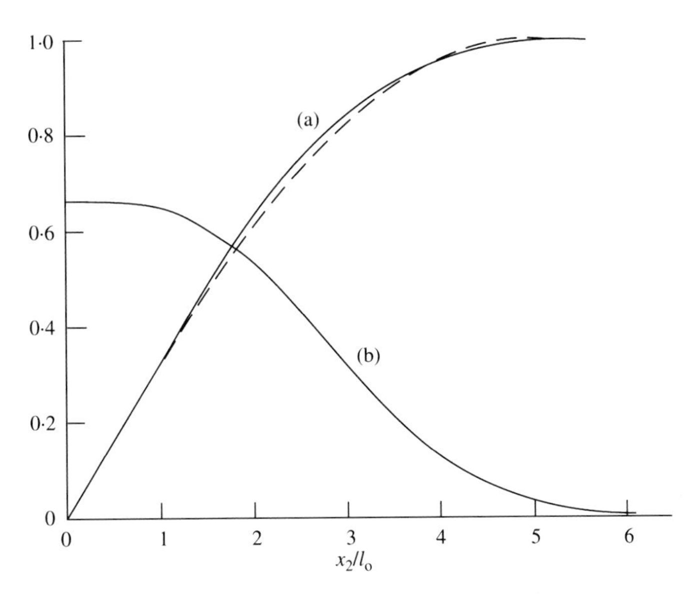

Figure 7.2 Variation with x_2 of (a) velocity u_1 and (b) vorticity $\partial u_1/\partial x_2$ near a stationary flat plate. The quantities plotted are dimensionless, namely x_2/l_o along the abscissa scale, with $l_o = \sqrt{(\eta x_1/\rho U)}$, and (a) u_1/U, (b) $2(l_o/U)(\partial u_1/\partial x_2)$, along the ordinate scale. The broken curve corresponds to (7.13) with $\alpha = 0{\cdot}326$.

below]. The way in which f_o depends upon its argument, according to Blasius, is shown by the full curve labelled (a) in fig. 7.2.

For two-dimensional flow of the sort we are considering the only non-zero component of vorticity is Ω_3, and when $u_2 \ll u_1$ we have

$$\Omega_3 \approx -\frac{\partial u_1}{\partial x_2} = \frac{U}{l_o} f_o'.$$

A second curve in fig. 7.2 shows how this varies with x_2/l_o in a Blasius boundary layer. In a boundary layer attached to an infinite plate which has been in motion for a limited time the vorticity profile is a Gaussian [(1.27)], but the corresponding profile in fig. 7.2 cuts off more sharply. Nevertheless, there is no value of x_2/l_o beyond which Ω_3 is precisely zero, and there is therefore room for disagreement as to how δ, the thickness of the boundary layer, should be defined. In §1.12 we set δ equal to the r.m.s. half-width of the Gaussian, namely $\sqrt{2\eta t/\rho}$, thereby including within it about 68% of the vorticity present in the fluid at time t, but in the present context it is more usual to define δ so that it includes 99% of the vorticity instead. However, for a reason which appears later in this section a third convention is adopted here: we define the boundary layer thickness to be

$$\delta_{90\%} = 3{\cdot}4 l_o, \tag{7.7}$$

which is sufficient, as the suffix implies, to include about 90% of the vorticity.[2]

Given that $f_o'\{0\} = 0{\cdot}3320\ldots$ [this is the initial slope of the curve in fig. 7.2] it is a simple matter to calculate, as a function of x_1, the viscous shear stress to which the surfaces of the plate are subject and hence, by integration over x_1, to find an expression for the drag force F_D on the plate.[3] Taking both sides of the plate into consideration, it is given by

$$F_D \approx 1{\cdot}33 L\sqrt{\eta\rho D U^3}. \tag{7.8}$$

We shall later discover that Blasius's solution is not always stable and that (7.8) breaks down on that account when *Re* exceeds about 3×10^5, but we may leave this complication on one side for further discussion in §9.9.

It would not be appropriate in a book of this sort to go through the mathematics of Blasius's solution in detail, but it is worth noting that the physical principles of conservation of vorticity, conservation of momentum, and conservation of energy impose certain conditions upon the the function f_o which are almost sufficient on their own to determine its behaviour. When δ is small enough compared with x_1 for the boundary layer equations to hold, the rate at which lines of vorticity are carried by convection across the fixed line labelled PQ in fig. 7.1 is proportional to

$$\int_0^\infty \frac{\partial u_1}{\partial x_2} u_1 \mathrm{d}x_2 = \int_0^U u_1 \mathrm{d}u_1 = \frac{1}{2} U^2,$$

and the corresponding rate due to diffusion in the x_1 direction may be shown to be smaller by a factor of order $(\delta/x_1)^2$ and therefore irrelevant. In this limit, therefore, the total vorticity flux across lines such as PQ is independent of x_1. We may at once infer that no vorticity enters or leaves the fluid by diffusion at the plate's top and bottom surfaces; all the vorticity present in the boundary layer must have entered it abruptly at the plate's leading edge, where the boundary layer equations fail. We may therefore infer that $\partial^2 u_1/\partial x_2^2$, which determines the diffusive flux of vorticity in the x_2 direction, is zero at $x_2 = 0$, and hence that $f_o''\{0\} = 0$. Blasius's solution confirms this, as fig. 7.2 makes clear.[4] The creation of vorticity at the plate's leading edge and its downstream convection are continuous processes, of course, so vorticity must be present in the wake immediately behind the trailing edge. In the absence of complications to be discussed in other contexts

[2] Equation (7.7) provides adequate justification for the statement in §1.12 that $\delta\{x_1\}$ can be roughly estimated from the formula $\delta = \sqrt{2\eta t/\rho}$ by using x_1/U for the time t.

[3] Equation (7.8) describes only the drag due to what is called *skin friction*, but the *form drag* discussed in §7.8 is irrelevant if the plate is genuinely thin.

[4] In fact it follows from the boundary layer equations that $f_o'''\{0\}$ vanishes as well as $f_o''\{0\}$. From the boundary condition that $u_1 = u_2 = 0$ at $x_2 = 0$ for all x_1 we may deduce: (i) that $\partial u_1/\partial x_1 = 0$ at $x_2 = 0$; (ii) that, as a consequence of (i) and of (7.4), $\partial u_2/\partial x_2 = 0$ at $x_2 = 0$; (iii) that both the terms on the left-hand side of (7.5) are therefore proportional, for small x_2, to x_2^2 or to some higher power of x_2.

later in this chapter it may be expected to show up as a dip in the velocity profile there, as fig. 7.1 suggests.

The dip is also a manifestation of the fact that, in passing the plate and in exerting a drag force upon it, the fluid loses momentum. Let us now look at the rate at which momentum is lost by the fluid *per unit area of the plate's surface*. One way to evaluate this is to integrate $-\rho\{u_1(\partial u_1/\partial x_1) + u_2(\partial u_1/\partial x_2)\}$ over x_2. It can be evaluated more simply, however, as the net flux of momentum into a fixed control volume such as the one outlined by PQRS in fig. 7.1, which has unit cross-sectional area in any plane parallel to the plate and which is bounded at $x_2 = X\{x_1\} \gg \delta\{x_1\}$ by a surface which is almost but not quite parallel to the plate because it follows the local lines of flow. The momentum flux into this volume is

$$-\frac{\mathrm{d}}{\mathrm{d}x_1}\int_0^X \rho u_1^2 \mathrm{d}x_2 = -\frac{\mathrm{d}}{\mathrm{d}x_1}\int_0^X \rho(u_1^2 - Uu_1)\mathrm{d}x_2$$
$$= -\frac{\mathrm{d}}{\mathrm{d}x_1}\int_0^\infty \rho(u_1^2 - Uu_1)\mathrm{d}x_2,$$

where use has been made of the continuity condition which determines $\mathrm{d}X/\mathrm{d}x_1$, in the form

$$\frac{\mathrm{d}}{\mathrm{d}x_1}\int_0^X \rho u_1 \mathrm{d}x_2 = 0.$$

It can be equated to the force per unit area which the plate exerts in the $-x_1$ direction on the fluid in contact with it, and it thereby follows that the function f_o and the scaling length l_o of (7.6) must obey the condition

$$\frac{\eta U}{l_o} f_o'\{0\} = \rho U^2 \frac{\mathrm{d}l_o}{\mathrm{d}x_1}\int_0^\infty (f_o - f_o^2)\,\mathrm{d}\left(\frac{x_2}{l_o}\right), \tag{7.9}$$

which becomes

$$2f_o'\{0\} = \int_0^\infty (f_o - f_o^2)\,\mathrm{d}\left(\frac{x_2}{l_o}\right) \tag{7.10}$$

on integration over x_1.

Finally, consider the flux of kinetic energy into the same control volume, which can be expressed as

$$-\frac{\mathrm{d}}{\mathrm{d}x_1}\int_0^X \frac{1}{2}\rho u_1^3 \mathrm{d}x_2 = -\frac{\mathrm{d}}{\mathrm{d}x_1}\int_0^\infty \frac{1}{2}\rho(u_1^3 - U^2 u_1)\mathrm{d}x_2. \tag{7.11}$$

This may be equated to the rate at which energy is dissipated within the volume, which can be evaluated, since $\partial u_2/\partial u_1$ is negligible, by integrating $\eta(\partial u_1/\partial x_2)^2$ over

x_2 [(6.26)]. The result is another condition on f_o and l_o which may, like (7.9), be integrated over x_1 to yield the following result equivalent to (7.10):

$$4\int_0^\infty f_o'^2\,\mathrm{d}\left(\frac{x_2}{l_o}\right) = \int_0^\infty (f_o - f_o^3)\,\mathrm{d}\left(\frac{x_2}{l_o}\right). \tag{7.12}$$

These various conditions based on conservation laws are insufficient to determine f_o completely, but they provide a useful base for some intelligent guesswork. Suppose, for example, we guess that

$$\begin{aligned} f_o\left\{\frac{x_2}{l_o}\right\} &= \sin\left(\frac{\alpha x_2}{l_o}\right) \quad [\alpha x_2/l_o < \pi/2] \\ &= 1 \quad\quad\quad\quad\quad [\alpha x_2/l_o > \pi/2], \end{aligned} \tag{7.13}$$

which does at least satisfy the prime requirements that $f_o\{0\} = 0$, $f_o\{\infty\} = 1$, and also the vorticity-conservation condition, $f''_o\{0\} = 0$. What values of α need to be chosen in order to satisfy (7.10) and (7.12)? The answers are, respectively, $\sqrt{(4-\pi)/8} \approx 0{\cdot}328$ and $\sqrt{1/3\pi} \approx 0{\cdot}326$. They are not exactly the same, but the agreement between them is sufficiently close to suggest that (7.13), with either value for α, constitutes an adequate first approximation to the true form of f_o. That this is indeed the case is shown by the broken curve in fig. 7.2.

By way of preparation for what is to follow in the next section, let us now consider how the thickness of the boundary layer in fig. 7.1 is likely to vary with x_1 when the velocity U of the fluid outside the boundary layer is *not* uniform but itself varies with x_1. In these circumstances the fluid is subject, inside the boundary layer as well as outside it, to a longitudinal pressure gradient

$$\frac{\partial p^*}{\partial x_1} = -\frac{\partial}{\partial x_1}\left(\frac{1}{2}\rho U^2\right) \approx -\rho U\frac{\partial U}{\partial x_1},$$

and the boundary layer equations need to be modified by the addition of $U(\partial U/\partial x_1)$ to the right-hand side of (7.5). It turns out that the modified equations possess self-similar solutions if, but only if, U is proportional to some power of x_1, say x_1^m. The existence of the pressure gradient affects the momentum balance argument, of course, and it can be shown that in place of (7.9) we now have

$$\frac{\eta U}{\rho l_o} f'_o\{0\} - U\frac{\mathrm{d}U}{\partial x_1} l_o\int_0^\infty (1-f_o)\,\mathrm{d}\left(\frac{x_2}{l_o}\right) = \frac{\mathrm{d}}{\mathrm{d}x_1}\left\{U^2 l_o\int_0^\infty (f_o - f_o^2)\,\mathrm{d}\left(\frac{x_2}{l_o}\right)\right\}. \tag{7.14}$$

The energy balance argument is less affected because the term in $U^2 u_1$, which we included in (7.11) for other reasons, takes care of work done by the excess pressure; from that equation we now have

$$\frac{2\eta U^2}{\rho l_o}\int_0^\infty f_o'^2\,\mathrm{d}\left(\frac{x_2}{l_o}\right) = \frac{\mathrm{d}}{\mathrm{d}x_1}\left\{U^3 l_o\int_0^\infty (f_o - f_o^3)\,\mathrm{d}\left(\frac{x_2}{l_o}\right)\right\}. \tag{7.15}$$

When $U \propto x_1^m$ and $l_o = \sqrt{\eta x_1/\rho U} \propto x_1^{(1-m)/2}$, in which case the self-similarity of the flow ensures that the integrals in equations (7.14) and (7.15) are independent of x_1, these results integrate over x_1 to

$$2f_o'\{0\} = (3m + 1)\int_0^\infty (f_o - f_o^2)\,\mathrm{d}\left(\frac{x_2}{l_o}\right) + 2m\int_0^\infty (1 - f_o)\,\mathrm{d}\left(\frac{x_2}{l_o}\right) \tag{7.16}$$

[cf. (7.10)] and

$$4\int_0^\infty f_o'^2\,\mathrm{d}\left(\frac{x_2}{l_o}\right) = (5m + 1)\int_0^\infty (f_o - f_o^3)\,\mathrm{d}\left(\frac{x_2}{l_o}\right) \tag{7.17}$$

[cf. (7.12)].

It is clear from (7.17) that the function $f_o\{x_2/l_o\}$ must be affected by m, and that indeed is why no self-similar solutions can be found when the dependence of U on x_1 is not such that a simple power law can describe it. However, by introducing a different scale length, namely

$$l_m = \sqrt{\frac{\eta x_1}{(5m + 1)\rho U}} = \frac{l_o}{\sqrt{5m + 1}},$$

which involves replacing f_o by a different function f_m such that $f_o'/l_o = f_m'/l_m$, one may arrive at a version of (7.17) in which the factor $(5m + 1)$ does not appear. This suggests that the best hope of bringing results for different m into coincidence lies in plotting u_1/U against x_2/l_m rather than against x_2/l_o. The suggestion is confirmed by three of the four curves in fig. 7.3, where some computed results which extend to non-zero m the original results of Blasius are plotted in this way. Note in particular how the curves labelled (a), (b) and (c) in this figure come together where u_1/U is about 0·9. Evidently (7.7) is adequately satisfied over the range of m between +1 and say −0·06, provided that we generalise it to $\delta_{90\%} \approx 3{\cdot}4 l_m$.

That result can be further generalised for use in situations where U cannot be expressed in power-law form, though the argument is speculative in places. Note first that (7.15) is always satisfied, and that it can be re-expressed without the suffices in terms of a scaling length l, and a corresponding function f, which varies with x_1 in any way we choose. Let us choose $l\{x_1\}$ so that the integral on the right-hand side of (7.15) is independent of x_1. If the integral on the left-hand side is then almost independent of x_1 also (which is probably the case, at any rate when $U^{-1}(\mathrm{d}U/\mathrm{d}x_1)$ lies in the range between +1 and −0·06), we may adjust the magnitude of l until four times this integral is equal to the integral on the right-hand side. When $U \propto x_1^m$ the scaling length arrived at in this way is just l_m, and it differs from $\delta_{90\%}$ by a factor of about 1/3·4. If the same factor relates l to $\delta_{90\%}$ in other circumstances, then we may estimate $\delta_{90\%}$ from $U\{x_1\}$ by integrating the following equation based upon (7.15):

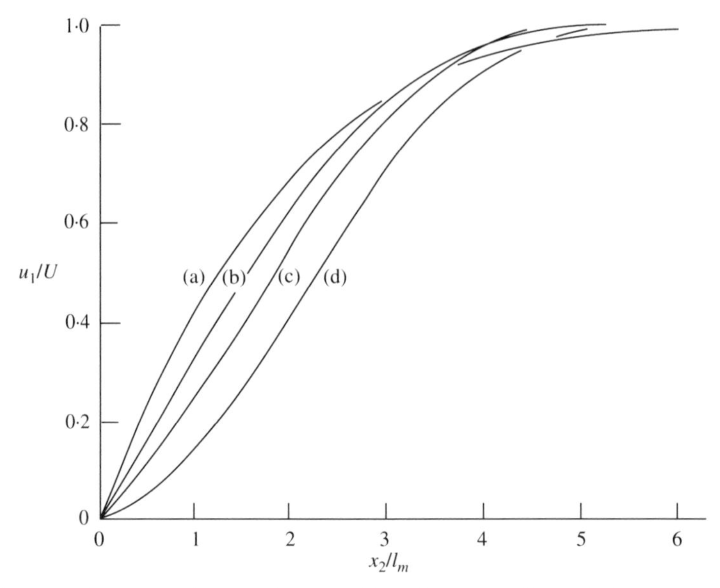

Figure 7.3 The variation of u_1/U with x_2/l_m when $U \propto x_1^m$, for (a) $m = 1$ (b) $m = 0$, (c) $m = -0{\cdot}0654$, and (d) m $= -0{\cdot}0868$. Curve (b) is Blasius's; the other three have been brought into some sort of coincidence with it, especially near $u_1/U = 0{\cdot}9$, by choosing $l_m = \sqrt{(\eta x_1/(5m + 1)\rho U)}$.
[Based on tabulated results in D. R. Hartree, *Proc. Camb. Phil. Soc.*, **33**, 223,1937.]

$$(3{\cdot}4)^2 \frac{\eta U^2}{2\rho\delta_{90\%}} \approx \frac{\mathrm{d}(U^3\delta_{90\%})}{\mathrm{d}x_1},$$

or equivalently

$$\frac{\mathrm{d}(U^6\delta_{90\%}^2)}{\mathrm{d}x_1} \approx 11{\cdot}6\,\frac{\eta U^5}{\rho}. \tag{7.18}$$

In order to make (7.18) seem plausible it has been necessary to gloss over the differences in shape between the curves in fig. 7.3, but they are in fact significant, particularly where x_2/l_m is small: for $m > 0$, the initial slope of the curve, $f'_m\{0\}$, is greater than it is for $m = 0$, while $f''_m\{0\}$ is not zero but is negative in sign; for $m < 0$ the initial slope is reduced rather than enhanced and the initial curvature is positive. There are straightforward physical reasons for these differences. Positive m corresponds to *convergent* flow outside the boundary layer and to a gradient of excess pressure which tends to accelerate the fluid. The almost stagnant fluid at the heart of the boundary layer is not free to accelerate, however, so the forward force per unit volume due to the pressure gradient must in this region be almost balanced by a viscous force per unit volume which is negative

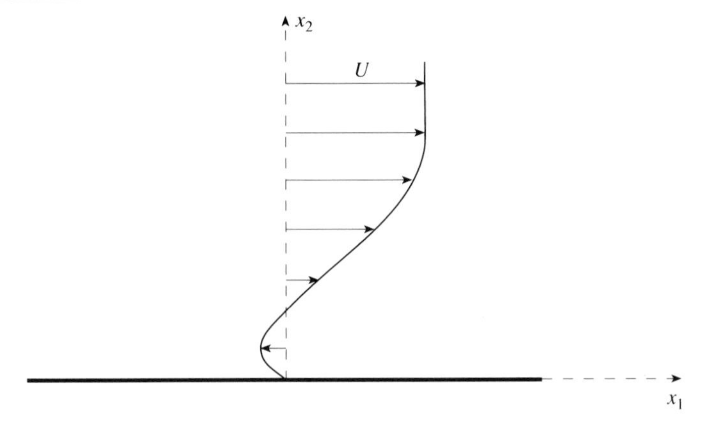

Figure 7.4 Sketch of the velocity profile to be expected near a flat plate in circumstances where the vorticity-free flow outside the boundary layer is diverging ($\partial U/\partial x_1 < 0$) at a rate corresponding to $m < -0{\cdot}0904$; near the plate the flow has reversed.

(implying $f''_m\{0\} < 0$), and any excess force per unit volume due to the pressure gradient must be transmitted to, and taken up by, the plate (implying an increase in $f'_m\{0\}$). Moreover, the increase of U with x_1 in cases of convergent flow means that the vorticity in the boundary layer must be continuously augmented by diffusion from the surface of the plate, and that too implies $f''_m\{0\} < 0$. When the flow outside the boundary layer is *divergent*, corresponding to negative m, the excess pressure gradient tends to decelerate the fluid and the surface of the plate must be taking vorticity out of the boundary layer rather than feeding it in, with the converse consequences described above.

There is a critical value of m at which $f'_m\{0\}$ and $f'_o\{0\}$ are reduced to zero, and it may be roughly estimated from (7.16) by setting the right-hand side equal to zero and by using for f_o the sine function which has been shown to constitute an adequate approximation when $m = 0$. That procedure suggests a critical value of about $-0{\cdot}12$. Considering that the sine approximation cannot be remotely adequate once $f'_m\{0\}$ is approaching zero, the estimate is in surprisingly close agreement with the results of numerical solution of the modified boundary layer equations: the critical value is in fact $-0{\cdot}0904$. What happens when m is less than that? Presumably the modified boundary layer equations have self-similar solutions for which $f'_m\{0\}$ is negative rather than positive, such that the direction of flow near the surface of the plate is reversed in the manner suggested by fig. 7.4. No one seems to have computed the form of these solutions in detail, however, partly because of technical difficulties and partly because the flow patterns which they describe cannot be realised in practice; when a thin plate is exposed edge-on to a fluid stream which is diverging at a rate corresponding to $m < -0{\cdot}0904$ the boundary layer in practice *separates* from the plate near its leading edge [§7.3], and the boundary layer equations then cease to apply.

7.3 Boundary layer separation and eddy formation

Although the problem considered in the previous section, of a thin strip set edge-on to the flow, is more complicated than the case of an infinite plate, it is a lot simpler than the problem of flow at high values of the Reynolds Number past a bluff body, e.g. a strip whose thickness is comparable with its breadth D, a cylinder with its axis transverse to the flow, or – as an example of a body past which the flow is three-dimensional rather than two-dimensional – a sphere. If one needs to know in detail how the boundary layer behaves in such situations one had no doubt better, in these days of fast computers, embark upon a numerical simulation than hunt for analytical solutions.

We have in (7.18), however, an analytical expression which should be fairly reliable for flat plates in convergent or weakly divergent fluid streams, and application of this expression to the two-dimensional problem of steady flow past a transverse cylinder of radius a yields results which, at any rate over the front half of the cylinder where the flow is convergent, are correct in essence as long as $\delta_{90\%}$ remains small compared with a. In such circumstances the fact that the boundary layer is curved rather than planar is irrelevant, and the velocity of the vorticity-free fluid just outside the boundary layer may be taken from (4.56): it is

$$-u_{\theta,r=a} = 2U \sin \theta$$

where U is the velocity of the fluid relative to the cylinder at large distances from it, or in terms of the angle $\chi\ (= \pi - \theta)$ defined in fig. 7.5 it is

$$u_{\chi,r=a} = 2U \sin \chi \quad [0 < \chi < \pi].$$

Here U is constant and it is $u_{\chi,r=a}$ (or u_a for short) which plays the role of the variable U in (7.18). Note that u_a is zero where $\chi = 0$ and that its initial rate of increase with χ is linear. For small values of χ, therefore, the boundary layer on a transverse cylinder is analogous to the boundary layer on an edge-on plate in circumstances (not easily realisable in practice) such that U is proportional to x_1^m with $m = 1$. In those circumstances the boundary layer thickness, being proportional to $x_1^{(1-m)/2}$, would be independent of x_1, and the boundary layer thickness on a cylinder is likewise independent of χ for small χ.

With U replaced by u_a and x_1 replaced by $a\chi$, (7.18) becomes

$$\frac{\mathrm{d}(\sin^6 \chi\, \delta_{90\%}^2)}{\mathrm{d}\chi} \approx 11{\cdot}6\, \frac{\eta a}{2\rho U} \sin^5 \chi,$$

and integration over χ yields

$$\delta_{90\%}^2\{\chi\} \approx 11{\cdot}6\, \frac{a^2}{Re} X\{\chi\}, \tag{7.19}$$

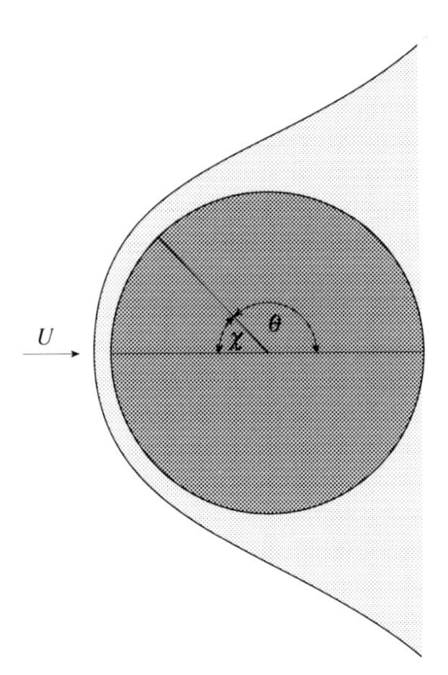

Figure 7.5 Steady flow past a cylinder set transverse to a fluid stream; the lightly shaded area represents a boundary layer with a thickness δ described by (7.19). The diagram is principally intended to show how δ varies with the angle $\chi\ (=\pi - \theta)$; its magnitude depends upon the Reynolds Number, and the magnitude depicted here, which corresponds to $Re \approx 200$, is too large in relation to the cylinder's radius for (7.19) to be reliable.

where Re is the Reynolds Number ($= 2\rho a U/\eta$) and where

$$X\{\chi\} = \frac{(\frac{8}{15} - \cos\chi + \frac{2}{3}\cos^3\chi - \frac{1}{5}\cos^5\chi)}{\sin^6\chi}.$$

Evidently the assumption that $\delta_{90\%}$ is small compared with a is justified when $Re \gg 1$ but not otherwise. The function X increases rather slowly with χ over the front half of the cylinder, but over the rear half, where the exterior flow is divergent, it increases rapidly. Figure 7.5 gives an idea of its behaviour up to about $\chi = 115°$. Beyond that its value is scarcely relevant because, normally before χ reaches 115°, the boundary layer is found in practice to *separate* from the cylinder.

The process which leads to separation begins when the velocity profile in the boundary layer, over some part of the cylinder's surface where the exterior flow is divergent, acquires the form indicated in fig. 7.4, i.e. *when the velocity adjacent to the surface reverses in sign*. In practice it normally begins while the fluid is still accelerating to its final velocity and while the instantaneous Reynold's Number is still quite small, i.e. less than 10. Because the conditions are far from steady at that stage, (7.18) does not apply and fig. 7.5 gives a misleading impression of what the boundary layer looks like; its thickness is presumably of order $\sqrt{\eta t/\rho}$, where t is

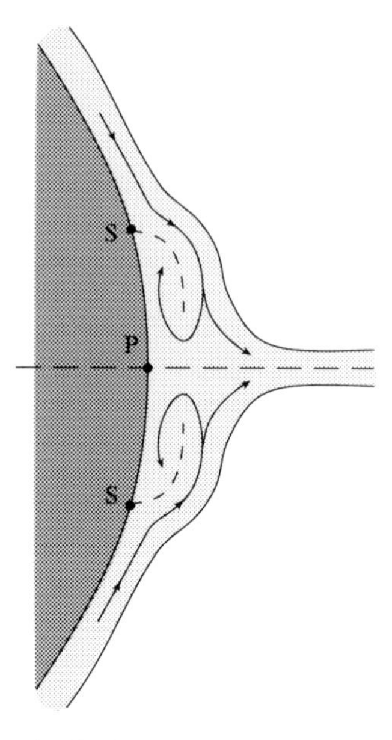

Figure 7.6 An early stage in the formation of two eddies behind a cylinder.

the time that has elapsed since the fluid was first set in motion, and is therefore small compared with a because t is small, besides being independent of χ. In such circumstances the steady-state theory of the previous section is of little help to us in predicting just where and when the flow reverses first. However, the physical reason why it does so is the one outlined above: it reverses because the fluid adjacent to the surface is subject to the same retarding pressure gradient as the vorticity-free fluid outside the boundary layer, and because it lacks the necessary momentum to continue moving forward despite this. The retarding pressure gradient is proportional to $-\sin\chi\cos\chi$ and reaches a maximum value where $\chi \approx \pm\frac{3}{4}\pi$, but the first signs of flow reversal are seen, during rapid acceleration, at values of χ which are quite close to π; the pressure gradient is smaller there but the forward momentum of the boundary layer fluid is smaller too.

Where does the fluid which is moving backwards come from, once flow reversal has occurred, and where does it go to? The answers would scarcely be obvious if we had to rely on imagination alone, but experimental observation and numerical simulation suggest answers which are embodied in the stylised sketch of fig. 7.6. This represents a detailed view of the rear surface of the cylinder near $\chi = \pm\pi$, covered by a boundary layer (the lightly shaded region in the figure) which a moment earlier was unseparated and of more or less uniform thickness; it continues in the wake downstream. The fluid which is moving backwards in each half of the figure, from P to S, has been drawn from this boundary layer and is accumulating near S, thereby forcing the oncoming boundary layer fluid to move away from the surface. The broken lines through S lie on surfaces which divide fluid which has doubled back from fluid which has not yet done so. These lines would be streamlines if they were stationary but they are not: they, and the *points of separation* S at which they originate, are moving away from P as more fluid accumulates within them. The arrowed lines are not necessarily streamlines either, but they give an idea of where the oncoming boundary layer fluid is going to.

The boundary layer fluid which accumulates behind the cylinder carries clockwise vorticity in the top half of the figure and anticlockwise vorticity in the bottom half, though some of this is lost to the surface of the cylinder between S and P. The presence of this vorticity imposes clockwise and anticlockwise rotation on the fluid in the neighbourhood, so it is scarcely surprising that as more time passes the boundary layer rolls up like a rolled carpet, so as to form two circulating *eddies* behind the cylinder, which increase steadily in size as more and more of the oncoming boundary layer fluid is trapped inside them.

The eddies do not grow indefinitely once U has reached its final value. Depending on the magnitude of U they may (a) settle down at a size which is consistent with steady flow, (b) cease to grow but start to oscillate instead, in which case the flow is never steady, or (c) oscillate with such an amplitude that some of the circulating fluid which constitutes each eddy is periodically *shed* by the cylinder. These different types of behaviour are further discussed in later sections, where we shall discover that at large enough values of the Reynolds Number the flow behind a cylinder is so turbulent that it is hard to say whether eddies exist or not.

Eddies are a common feature of flow past bluff obstacles of all sorts. The more rapidly the lines of flow diverge in the vorticity-free region outside the boundary layer, the more likely it is that separation will occur, so eddy formation is favoured by sharp edges or corners, where the radius of curvature of the obstacle's surface is small. Eddies form easily, for example, behind a thin plate which is set broadside-on to the flow, as the reader may verify by drawing a broad knife blade, held vertically, through a basin filled with water which has been allowed to come to rest. The water should have some dust in it or on its surface to make the eddies visible at low speeds, but at higher speeds the dimple in the surface of the water for which each eddy is responsible is readily detectable without dust. Comparable eddies are often detectable in rivers, behind groynes or rocks.

Although eddies are most pronounced *behind* obstacles, they may also occur in front in certain circumstances. Some lines of flow which might, in the absence of boundary layer separation, describe a stream of air past a vertical wall (though in practice winds are rarely, if ever, steady and vorticity-free) are sketched in fig. 7.7(a). They diverge from one another in the neighbourhood of P as well as beyond the top of the wall, near Q. Separation is therefore liable to occur in both places, with the result that two eddies form as shown in fig. 7.7(b). The separated

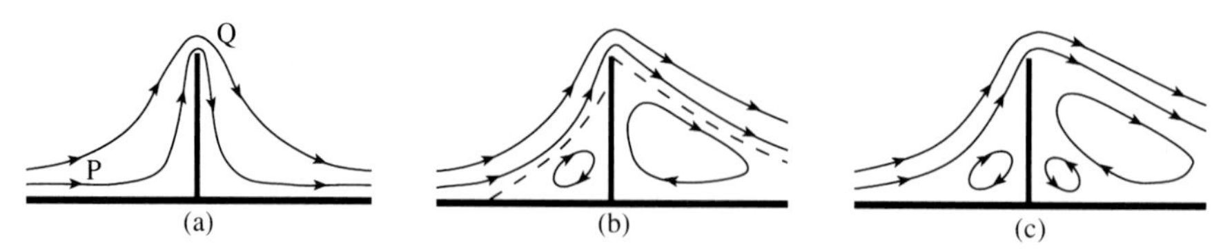

Figure 7.7 Lines of flow for a steady wind passing over a wall; the symmetrical pattern shown in (a) is unlikely to occur, because eddies tend to form as in (b) or (c). The broken lines in (b) represent separated boundary layers.

boundary layers are suggested by broken lines in this figure, and it will be seen that the one that leaves the ground near P becomes *reattached* at the wall. Further separation may occur on the downstream side, leading to the formation of a secondary eddy as in fig. 7.7(c).

7.4 Steady eddies behind cylinders and spheres

Provided that the flow of an effectively incompressible fluid past a transverse cylinder is steady, it must be *dynamically similar* [see §1.5] to the flow of another incompressible fluid past another transverse cylinder when the Reynolds Number Re $(=2\rho aU/\eta)$ is the same for both. The truth of this important principle is essentially obvious on dimensional grounds. As long as the flow is steady, the radial component of velocity u_r at a point with coordinates (r, χ) can depend only upon r, χ, U, a, η and ρ. Among the dimensionless groups which may be built up from these variables are clearly u_r/U, r/a, χ, and Re, and it is easily verified that there are no other independent possibilities. Hence [§1.5] we may write

$$\frac{u_r}{U} = f_r\left\{\frac{r}{a}, \chi, Re\right\},$$

where $f_r\{r/a, \chi, Re\}$ is any function of its arguments. Similarly

$$\frac{u_\chi}{U} = f_\chi\left\{\frac{r}{a}, \chi, Re\right\}.$$

These expressions immediately imply that round two cylinders for which Re is the same the flow patterns are scaled versions of one another – that the magnitude of Re uniquely determines the nature of any stable eddies which arise.

The photograph reproduced in fig. 7.8 was taken under conditions corresponding to $Re = 41{\cdot}0$, by the simple method of setting into motion relative to water a cylinder which had previously been coated with condensed milk; we may think of the cylinder as stationary and the water at large distances as moving from left to right with uniform velocity, though in fact the cylinder was moved, from right to left. The exposure was a short one, and the photograph shows an effectively

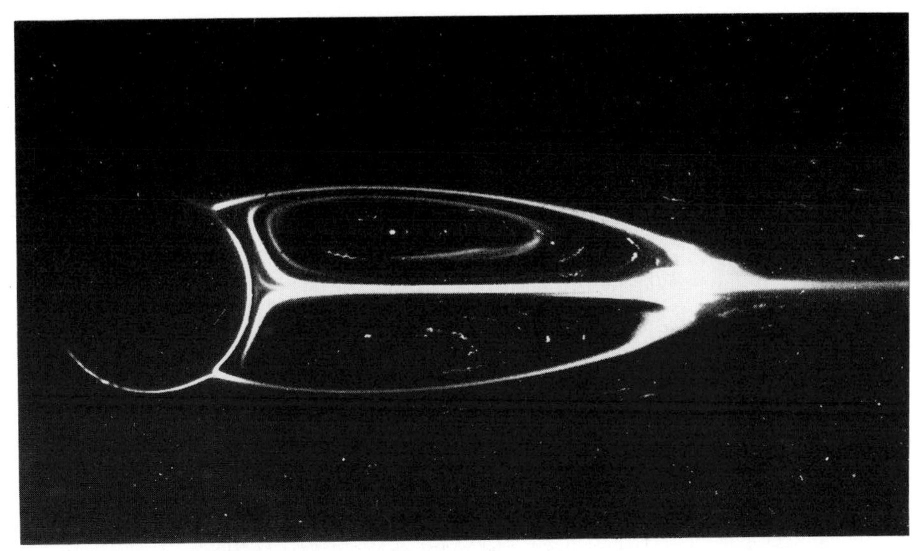

Figure 7.8 Eddies behind a cylinder at $Re = 41$; in the frame of reference in which the cylinder is stationary, the fluid – water in this case – is moving from left to right.
[Photograph by S. Taneda. [*Rep. Res. Inst. Appl. Mech., Kyushu Univ.*, **4**, 29, 1955.]]

instantaneous view of the *streaklines* [§1.6] of milk particles which have entered the fluid and been carried with it, past points of separation which have evidently reached $\chi \approx \pm 125°$; the streaklines are spirals, providing a vivid record of the way in which the eddies have formed. Note that spiralling streamlines are not possible in two-dimensional flow when $\nabla \cdot \boldsymbol{u}$ is zero everywhere, so here we have a case where streaklines and streamlines do not exactly coincide. That may suggest that the flow pattern has not had time to become (in the frame of reference in which the cylinder is stationary) completely steady; the points of separation may still be moving to smaller values of $|\chi|$ and the eddies may be increasing still in size. The steady-state pattern at $Re = 41{\cdot}0$ is unlikely to differ in any essential particular, however, from the pattern which the figure reveals.

The diffusivity of milk in water is much smaller than the diffusivity of vorticity, so the width of the white streaks in fig. 7.8 is much less than the thickness of the boundary layer. Indeed, when $Re = 41{\cdot}0$ the steady-state boundary layer thickness at $\chi = \pi/2$ as predicted by (7.19) is about $0{\cdot}3a$. This is far too big in relation to a to justify the approximations implicit in (7.19) and, more significantly, it is too big for the concept of a boundary layer to be really meaningful; most of the water behind the cylinder in the volume covered by fig. 7.8 must, in the steady state, be contaminated by vorticity. In particular, vorticity must be present in the water which is trapped in the eddies themselves; lines of vorticity are fed

Figure 7.9 Cross-sectional view of the ring-shaped eddy behind a sphere at $Re = 118$.
[Photograph by S. Taneda. [*J. Phys. Soc. Jap.*, **11**, 1104, 1956.]]

continuously into this water by diffusion from the vorticity-laden water which is passing by outside the eddies, only to be removed at the same rate over the rear half of the cylinder's surface, which can be viewed in this context as a vorticity sink [§7.1]. In the interior of each eddy $\boldsymbol{\Omega}$ is probably almost uniform in the steady state, so one may picture the fluid in these regions as rotating about axes which are stationary with respect to the cylinder with almost uniform angular velocity.

Experiments show that the flow pattern behind a transverse cylinder is characterised by steady eddies which resemble those shown in fig. 7.8 throughout the range $6 < Re < 41$ (both limits are approximate). They dwindle in size as Re is reduced and are vanishingly small at the lower limit. In the range $1 < Re < 6$ the lines of flow around a cylinder are not completely symmetrical about the equatorial ($\chi = \pm\pi/2$) plane, as they become at still smaller values of Re where the creeping flow approximation is valid [(6.64)], but the asymmetry is modest compared with the asymmetry which fig. 7.8 reveals. At values of Re which exceed 41 the eddies start to oscillate, and they are then prevented from growing much larger by the fact that some of the fluid trapped in each of them is periodically shed [§7.5].

Most of the above remarks apply qualitatively to flow past a sphere. The critical value of the Reynolds Number at which eddies first form is close to 24 for spheres, and they are steady up to $Re \approx 130$. The photograph reproduced as fig. 7.9 was taken using a sphere under conditions corresponding to $Re = 118$; the water contained particles of aluminium dust, uniformly distributed in suspension but made visible in one plane only, by illumination through a narrow slit. In this case [cf. fig. 7.8] the flow was in all probability sufficiently steady for the streaklines, streamlines and lines of flow to coincide. The extent to which the velocity of the water varies from place to place may be judged by the lengths of traces left by the individual particles, and it vanishes at the centre of the eddy. The eddy is, of course, a single toroidal one, whereas there are two distinct eddies in fig. 7.8, extending indefinitely in the x_3 direction. The toroidal eddy starts to oscillate when Re exceeds the upper limit of about 130, and shedding of trapped fluid occurs as Re increases further.

7.5 Eddy shedding by cylinders and spheres

The easiest way of detaching an eddy from the obstacle behind which it has formed is simply to remove the obstacle. Try drawing a knife blade through water, as described in §7.3, and lifting it out in the process; two free eddies are left behind, which behave in the manner to be expected of two free line vortices of opposite circulation, as described in §4.14. The eddies may also be detached by suddenly increasing the speed of the knife blade or by shaking it, though this is less easy to demonstrate.

Here, however, we are concerned with spontaneous shedding, due to the fluid stream rather than to an external disturbance. In the case of flow past transverse cylinders the process is a remarkably regular one, which has been carefully studied. It appear to starts with an instability in the wake, discussed in §8.12, which feeds back to the eddies and stimulates the oscillations briefly referred to above. They increase in amplitude as Re increases, and periodic shedding starts to occur when Re reaches about 60. To begin with, the amount of fluid which is shed in one period constitutes only a small part of the eddies' contents. It seems that by the time Re is about 100, however, the eddies are being shed almost completely, only to reform before the next shedding. The eddies on the two sides of the cylinder are shed alternately, at a well-defined frequency which depends slightly upon Re but is close to $0{\cdot}1\ U/a$, and in the wake of the cylinder they form a beautifully regular array known as a *Kármán vortex street* [fig. 7.10]. Von Kármán analysed the behaviour of such arrays of ideal free vortex lines and found that they are stable with respect to small perturbations in the position of the vortices if and only if the ratio D/L, where D is the distance between the row of positive vortices and the row of negative vortices and L is the repeat distance along each row, is 0·283. The eddies shown in fig. 7.10 are indeed spaced in more or less this way.

According to potential theory a cylinder which carries a bound vortex [§4.11] experiences a transverse Magnus force [§4.12], and the same is true of a cylinder to which one, but only one, eddy is attached. Hence cylinders which are shedding clockwise and anticlockwise eddies in regular succession experience a transverse

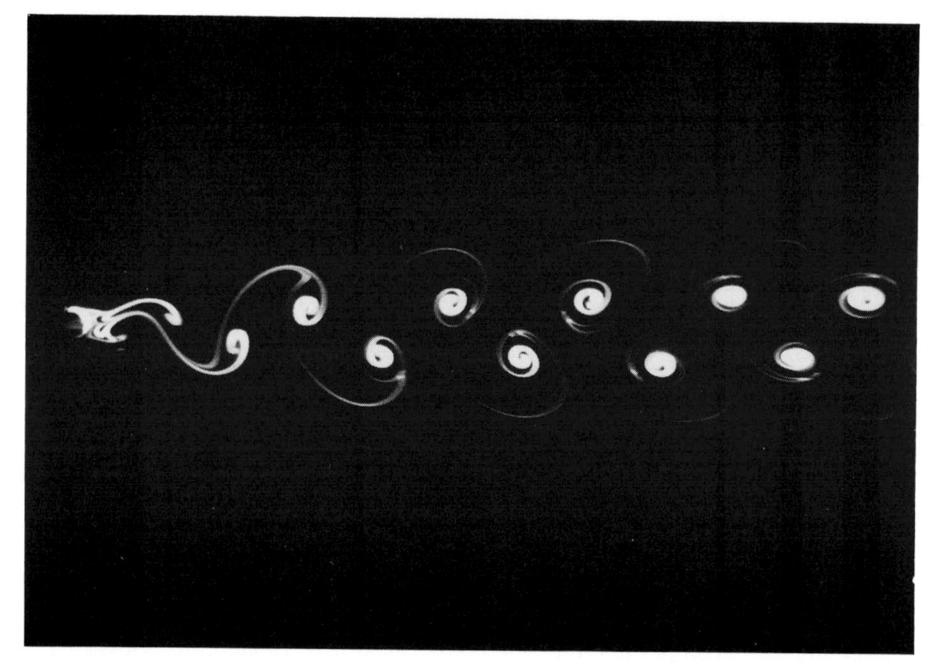

Figure 7.10 A Kármán vortex street in water flowing from left to right past a cylinder at $Re = 105$.
[Photograph by S. Taneda.]

force which oscillates in sign with the shedding frequency, and if the cylinder has some degree of flexibility it oscillates in sympathy. The shedding frequency is liable to shift slightly as a result, by an amount that depends upon the amplitude of the motion. The singing sound emitted by taut wires when a strong wind is blowing, and the howling sound emitted by a cane when it is swished through the air, are to be explained in this way; sound is generated by periodic vortex shedding even if the cylinder is completely rigid, but it is louder if the cylinder is free to oscillate. In many cases mechanical resonance is probably involved, and it is certainly involved in the production of sound by the so-called *Aeolian harp* discussed by Lord Rayleigh. Rayleigh's harp consisted of a taut wire set into motion by the draught generated by a steady fire, and he recounts how the slightest change in velocity, due to his adding a small piece of paper to the fire, was enough to cut off the sound. In such circumstances the wire presumably oscillates at its natural frequency, with an amplitude such that the shedding frequency locks on to this.

The toroidal eddy which forms behind a sphere is subject to periodic shedding too. It seems to peel away from the sphere by degrees rather than all at once, starting from one side in one half-cycle and from the opposite side in the next.

Thus the wake is thought to contain, at any rate when Re exceeds about 400, a train of elliptical vortex rings which are inclined to the axis of flow and linked to one another, instead of circular rings which are coaxial and separate.

7.6 Turbulent wakes behind cylinders and spheres

When flow past an obstacle is unsteady it is not uniquely determined by the magnitude of Re. In principle, the list of variables on which $\boldsymbol{u}$ may depend then includes one or more to describe the initial acceleration of the flow, e.g. if the acceleration itself was infinitely rapid, the time t which has since elapsed. In that case, dimensional analysis implies

$$\frac{u_r}{U} = f_r\left\{\frac{r}{a},\ \chi,\ Re,\ \frac{Ut}{a}\right\},$$

which in turn implies

$$\frac{\overline{u}_r}{U} = f_r\left\{\frac{r}{a},\ \chi,\ Re\right\}, \tag{7.20}$$

where $\overline{u}_r$ is an average of u_r from which all time dependence has been eliminated. In practice, unsteady fluid motion which is *turbulent* is not deterministic in the way that these equations suggest, but (7.20) and a similar equation for $\overline{u}_\theta$ apply to it nevertheless. Time-averaged flow patterns are uniquely determined by the Reynolds Number in cases of turbulent flow, even though instantaneous flow patterns are not.

Figure 7.11 is a time exposure of water flowing past a stationary sphere at $Re \approx 15\,000$. The white streaks are images left by small air bubbles in the water, and collectively they provide a good impression of the nature of the time-averaged flow under such conditions; a photograph taken with a cylinder rather than a sphere would look very similar. At such values of the Reynolds Number the boundary layer is at last genuinely thin. It evidently separates quite close to the equator (in fact at points slightly upstream of this), and it continues behind the equatorial plane as a surface separating flow which is still steady – or at any rate looks steady after time averaging – from flow which is clearly turbulent. The turbulence, which in fact results from an instability in the separated boundary layer to be discussed in §8.12, is so marked that large-scale eddies never have time to form, let alone to be shed in a regular fashion. Undoubtedly, however, turbulent fluid is continuously leaking into the wake and disappearing downstream; none of it is trapped in the neighbourhood of the obstacle for ever. The vorticity which is fed into the fluid over the front half of the sphere or cylinder must be convected downstream in the separated boundary layer, so this separated layer is a layer of high vorticity, across which the difference of velocity is large.

Figure 7.11 Flow past a sphere at $Re = 15\,000$.
[Photograph by H. Werlé, by courtesy of ONERA (Office National d'Etudes et de Recherches Aerospatiales), Châtillon, France.]

Hence the turbulent fluid enclosed by the separated boundary layer moves at speeds which are distinctly less than U.

At values of Re which are higher still – of the order of 2×10^5 – there is a sudden and dramatic change: the point of separation shifts downstream, from values of the angle χ which are by this stage slightly less than $\pi/2$ to values which are considerably greater than $\pi/2$. The size of the turbulent region diminishes sharply as a result; so does the rate of energy dissipation in the wake, and so too, as we shall see in §7.8, does the drag force. This effect is associated with the sudden onset of turbulence in the boundary layer itself, where it is still attached to the obstacle rather than downstream from the point of separation. At a qualitative level, the explanation is as follows. The process of flow reversal which leads to separation is inhibited by the inertia of the boundary layer fluid [§7.3]. When the boundary layer fluid becomes turbulent, the rate at which neighbouring elements of it exchange forward momentum increases to a marked degree. Consequently, any tendency of the fluid adjacent to the obstacle's surface to slow down and reverse its direction of flow is resisted by the inertia of fluid further out. [5]

[5] See also §8.13 and the remarks concerning Reynolds stress and eddy viscosity in §9.5.

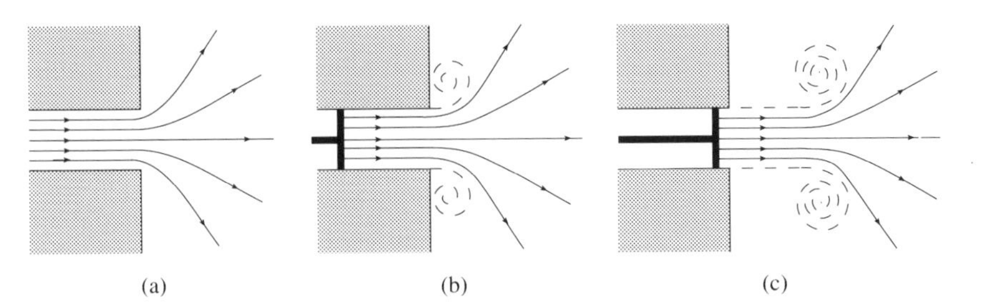

Figure 7.12 Fluid emerging from a cylindrical tube into a half-space full of fluid: (a) lines of potential flow for an ideal Euler fluid (sketch only); (b) instantaneous streamlines for a real fluid, shortly after an impulsive start; (c) as in (b) but a moment later. In (b) the boundary layer (broken curve) has separated and is rolling up into a ring-shaped eddy; in (c) the eddy has been shed and is moving to the right.

7.7 Submerged jets revisited

Figure 7.12 illustrates stages in a hypothetical experiment in which fluid inside a cylindrical tube of radius a is rapidly accelerated, by motion of a piston, to a constant velocity U and driven out through the open end of the tube into a large vessel full of the same fluid, initially stationary. What flow pattern is to be expected, if the value of $Re\,(=2\rho aU/\eta)$ is too large for the inertia of the fluid to be negligible, i.e. too large for the creeping flow approximation to apply? Potential theory implies lines of flow diverging as in fig. 7.12(a), and immediately after the initial acceleration the flow pattern may indeed look like this. There must be a boundary layer inside the tube, however, and it will surely separate at the sharp corners where tube and vessel meet. The separated boundary layer will then roll up into a toroidal eddy, not unlike the eddy behind a sphere, as in fig. 7.12(b). What happens next may depend to some extent upon the magnitude of Re, but when Re exceeds 1000 the eddy is undoubtedly shed from the opening, as in fig. 7.12(c); it becomes a more or less free ring vortex which moves forward as described under the heading of 'Smoke rings' in §4.15. Indeed, this is just how smokers generate smoke rings, by a sharp exhalation through pursed lips or through the larynx.

The boundary layer continues to receive fresh supplies of vorticity by convection from the open end of the tube, and behind the first ring vortex it forms a tube itself, which increases in length as the ring vortex moves forward [fig. 7.12(c)]. Enclosed by this tubular *vortex sheet* there is a *submerged jet*, i.e. a column of fluid with velocity U surrounded by fluid which is still almost stationary. The vortex sheet is unstable, however, and when a submerged jet has been in existence for some time it is almost invariably turbulent at distances greater than a few diameters from the orifice through which it emerges [§§8.12 and 9.7].

The problem of why inflow and outflow of fluids, to and from orifices, are not symmetrical processes was addressed in general terms at the end of §2.13. In the light of what we now know about the asymmetry of flow past obstacles, and of what we know about boundary layer separation in particular, the problem is no longer so mysterious.

7.8 Drag forces

In so far as the flow round an obstacle which is moving through otherwise stationary fluid with a uniform velocity $\boldsymbol{U}$ is vorticity-free and can be represented by a potential which obeys Laplace's equation, the instantaneous flow pattern is uniquely determined by $\boldsymbol{U}$. It must therefore remain the same as time passes, except in so far as the whole pattern is subject to translation with the obstacle. In that case the kinetic energy stored in the fluid must be constant, so the obstacle does no work on the fluid and the drag force must be precisely zero. This is *d'Alembert's paradox*, already referred to in §4.7, where the case of potential flow round a sphere was discussed in some detail. There it was shown that the *skin friction* which is inevitable if the fluid has viscosity is exactly cancelled, provided that the flow is everywhere vorticity-free, by the so-called *form drag*; the latter term is used to describe drag exerted by normal stresses as opposed to shear stresses, and it is because normal stress is in general anisotropic that the cancellation occurs. It must occur for obstacles of any shape, not just for spheres. However, to ensure that the fluid is genuinely vorticity-free the surface of the obstacle must at all points move with the fluid that is in contact with it [§4.7] and in practice it never does so. Hence d'Alembert's paradox is of little relevance to the real world.

The non-zero drag force F_{D} which is in practice experienced by a sphere in uniform motion through fluid is determined by the velocity distribution $\boldsymbol{u}\{\boldsymbol{x}\}$, and provided that the relative velocity U is small enough for compressibility to be irrelevant it depends only on the parameters U, a, ρ and η. The gravitational acceleration g plays a part in the static Archimedean upthrust experienced by the sphere, but drag force is always defined so that it vanishes when U vanishes and does not include this upthrust. A dimensionless group which includes F_{D} is the so-called *drag coefficient*, C_{D}, conventionally defined by

$$C_{\mathrm{D}} = \frac{F_{\mathrm{D}}}{\frac{1}{2}\rho U^2 A}, \tag{7.21}$$

where $\frac{1}{2}\rho U^2$ is the excess pressure at a stagnation point where the fluid is stationary relative to the obstacle and A is a cross-sectional area of the obstacle; for bluff obstacles, A is normally taken to be the maximum cross-sectional area in a plane normal to $\boldsymbol{U}$, e.g. πa^2 in the case of a sphere of radius a and $2aL$ in the case of a transverse cylinder of radius a and axial length L ($\gg a$). We know at once from dimensional analysis, therefore, that it must be possible to express drag forces by an equation of the form

$$\overline{C}_{\mathrm{D}} = f\{Re\};\tag{7.22}$$

a bar is included on the left-hand side in case the flow is turbulent, in which case the equation will only describe the time-averaged drag. So far so good, but what is the function $f\{Re\}$?

It is *only* in the region $Re < 0{\cdot}5$, where Stokes's law applies, that theory provides us with an answer to this question. In this region we have

$$C_{\mathrm{D}} = C_{\text{skin friction}} + C_{\text{form drag}} = \frac{24}{Re}\tag{7.23}$$

[(6.68) and (7.21)], which does of course conform to (7.22), and it is readily shown that skin friction accounts for two-thirds of the whole and form drag for the remaining one-third. Outside this region we are obliged to rely either upon numerical simulation or upon experiments for determination of the function f, and at large values of the Reynolds Number, such that turbulence is liable to occur, experiments provide the more reliable guide.

One may attempt to estimate the drag coefficient of a sphere for values of Re in the range 10^4–10^5 on the following assumptions, justified to a limited extent by the photograph reproduced as fig. 7.11: (a) that separation occurs fairly near the equator in this range, (b) that over the front half of the sphere the fluid velocity is little affected by separation, and (c) that the turbulent fluid enclosed by the separated boundary layer is moving with respect to the sphere at speeds which are genuinely small compared with U. To the extent that (b) is correct, the excess pressure p^* should vary over the front half of the sphere in the way that (4.36) describes, between $+\frac{1}{2}\rho U^2$ at $\chi = 0$ and $-\frac{5}{8}\rho U^2$ at $\chi = \pi/2$. To the extent that (c) is correct, p^* should be essentially uniform within the turbulent region. Furthermore, since the boundary layer is thin at these large values of Re, p^* within the turbulent region should be almost equal to p^* in the vorticity-free fluid just outside the separated boundary layer; it should perhaps be something like $-\frac{5}{8}\rho U^2$. These postulates (all of them involve guesswork and none of them is fully borne out, in fact, by measurements of pressure distribution) can be shown to imply that

$$C_{\text{form drag}} \approx \frac{9}{8}.\tag{7.24}$$

As for skin friction alone, the local shear stress exerted over the front half of the sphere should, according to (b), be of order

$$\frac{\eta u_{\mathrm{o}}}{\delta} \approx \frac{3\eta U \sin\chi}{2\delta},$$

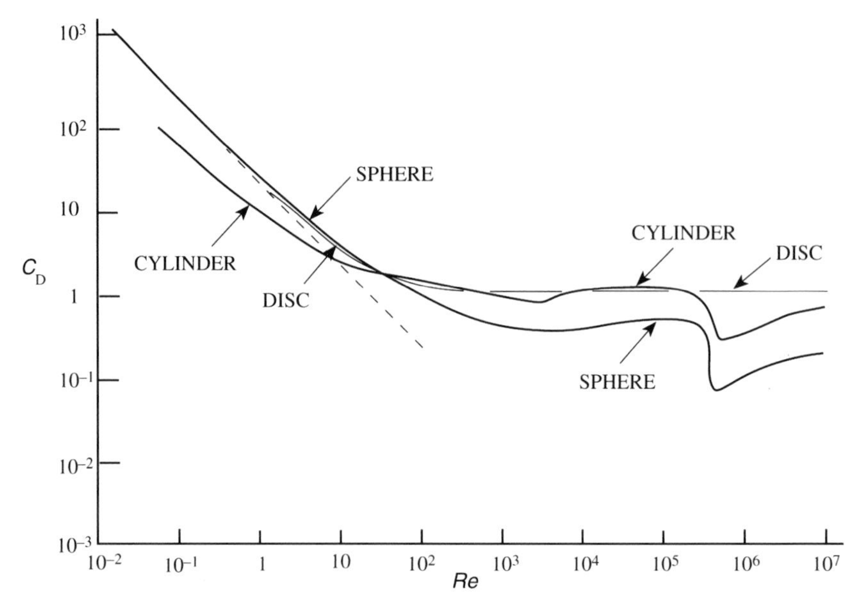

Figure 7.13 Log–log plot of drag coefficient C_D as a function of Reynolds Number Re for spheres, transverse cylinders, and face-on discs. The broken straight line represents Stokes's law.

while over the rear half it should, according to (c), be negligible. An integration over the front half, using (7.19) for δ, yields

$$C_{\text{skin friction}} \approx \frac{5}{\sqrt{Re}}. \tag{7.25}$$

The argument suggests that $C_{\text{skin friction}}$ at these large values of Re is small enough compared with $C_{\text{form drag}}$ to be neglected. That general conclusion is undoubtedly correct, and the argument is also correct in so far as it suggests that the drag force is largely attributable to the fact that p^* is negative near the rear pole of the sphere, where according to potential theory it should be just as positive as p^* near the front pole. But (7.24) and (7.25) are not to be trusted in detail.

What experiments tell us about the variation of C_D with Re is shown in fig. 7.13. As is customary when one dimensionless group of variables is plotted against another, the scales are logarithmic; this mode of presentation enables data collected over a very wide range of variables to be compressed into a single diagram. Between $Re \approx 10^3$ and $Re \approx 2 \times 10^5$ the curve applicable to spheres is indeed reasonably flat, but C_D is about $\frac{1}{2}$ rather than $\frac{9}{8}$. Throughout this range one may estimate the drag force to an accuracy which is adequate for many purposes using the simple equation

$$F_D \approx a^2 \rho U^2, \tag{7.26}$$

in which the viscosity of the fluid does not occur. Between $Re \approx 2 \times 10^5$ and $Re \approx 3 \times 10^5$, however, there is an abrupt fall in the drag coefficient from about 0·5 to about 0·1, so rapid that not only C_D but also F_D is decreased by acceleration of the sphere. The fall, sometimes referred to as the *drag crisis*, is clearly associated with the sudden onset of turbulence in the boundary layer. This increases the skin friction but decreases the form drag, because of the way in which the point of separation moves to higher values of χ [§7.6]. The exact value of Re at which the fall occurs is very sensitive, not surprisingly, to irregularities of any sort in the sphere's surface or else in the motion of the fluid through which the sphere is moving.

Figure 7.13 includes a curve for long cylinders with their axes transverse to $\boldsymbol{U}$, and a curve for thin circular discs with their axes parallel to $\boldsymbol{U}$. The former is similar to the curve for spheres (except in so far as there is no precise equivalence between the flow round spheres and cylinders in the region where $Re < 1$) and it needs no separate discussion. The latter is distinctly flatter over a wider range of Re, and it shows no sudden fall. Those features are attributable to the fact that separation occurs very readily at the edges of a thin disc whose plane is transverse to the flow – so readily that no amount of turbulence in the boundary layer can cause the point of separation to move backwards.

The forces on bluff bodies of more complex shape – automobiles, for example – are liable to include *induced drag* as well as form drag [§7.11], but it is still true that for a given shape the dimensionless drag coefficient is completely determined by the Reynolds Number, provided only that the Mach Number $M = U/c_s$ is small enough compared with unity for the compressibility of the fluid to be irrelevant.[6] Although designers are nowadays making increasing use of computer simulation, the curve relating C_D to Re is still most often determined experimentally, using scale models in wind tunnels. An alternative experimental procedure involves towing scale models through tanks of water; because the kinematic viscosity $\nu = \eta/\rho$ is smaller for water than for air by a factor of about 15, the value of U which is needed to produce a given value of Re for a given model size is 15 times smaller too.

The drag force on an obstacle which is *not* bluff, namely a thin plate set edge-on to the flow whose breadth in the direction of flow is D and whose length in a transverse direction is L, is entirely due to skin friction and is described according to Blasius's solution by (7.8). It is worth noting that when this is expressed in dimensionless terms the area which is used in defining the drag coefficient is normally LD rather than the much smaller cross-sectional area of the plate in a plane perpendicular to $\boldsymbol{U}$. Thus

$$C_D = \frac{F_D}{\frac{1}{2}\rho U^2 LD},$$

[6] A brief discussion of drag forces when $M \approx 1$ is to be found in §3.12.

and since the Reynolds Number is defined in a corresponding fashion, by

$$Re = \frac{\rho D U}{\eta},$$

(7.8) becomes

$$C_{\mathrm{D}} = 2{\cdot}66\sqrt{\frac{1}{Re}}.$$

Reynolds Numbers and drag coefficients for aircraft wings, which are comparable to edge-on plates in some respects, are normally defined in a similar fashion [§7.11].

7.9 Techniques for drag reduction

Values of Re which exceed 10^4 are frequently achieved in practice. When a golf ball leaves the tee after a powerful drive, for example, it is travelling at about 150 m.p.h., and that corresponds to $Re \approx 2 \times 10^5$. Fast bowlers and professional pitchers deliver cricket balls and baseballs respectively at speeds of about 85 m.p.h., which likewise correspond to $Re \approx 2 \times 10^5$. Sprinters and free-style swimmers of Olympic standard cover about 10 metres through air and 2 metres through water per second respectively, and for them Re is up to 10^6. For submarines at top speed Re may well reach 10^8. It is in such contexts that the reduction of drag is particularly desirable, and because form drag is much stronger than skin friction at large Re the strategies to be described below are aimed at the reduction of form drag, often at the cost of a considerable increase in skin friction.

The most familiar strategy is that of *streamlining* the solid body, either by rounding off its edges in some way or by adding to it an otherwise redundant shield or *fairing*, which increases the superficial area of the body but diminishes the form drag nevertheless. The so-called *Rankine fairing*, designed to reduce almost to zero the form drag exerted on the front surface of a bluff obstacle, has been described in passing in §4.4. It is only at speeds where $M \approx 1$ or $M > 1$, however, that it is essential to streamline the front of a body [§3.12]. At lower speeds it is far more important to streamline the rear. Figure 7.14(a) shows a cross-section through a transverse cylinder which has been enlarged by the addition of a rear fairing so that it now has the shape associated with aerofoils [§7.10]. Such an addition, properly designed, can reduce the drag force at a value of U which corresponds to $Re \approx 10^5$ for the unmodified cylinder by a factor of about 50. The rear fairing is effective in this case because it decreases the rate at which the lines of flow diverge in the vorticity-free fluid behind the obstacle, and thereby decreases the pressure gradient which is the root cause of flow reversal in the boundary layer and of boundary layer separation. It is because separation occurs

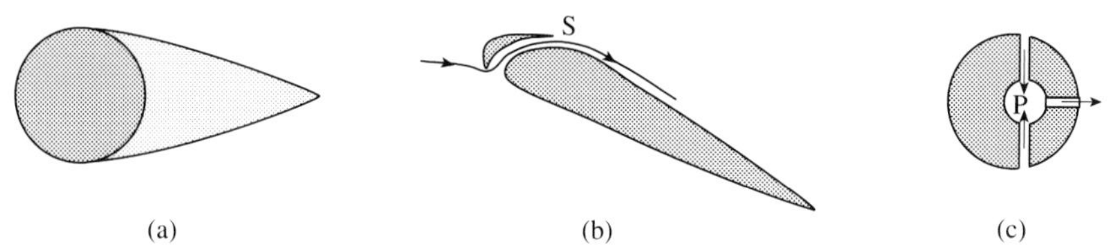

Figure 7.14 Some strategies for drag reduction: (a) a cylinder fitted with a fairing at the rear; (b) an inclined aerofoil with a slot in it; (c) a sphere containing an internal pump P, which sucks fluid in through apertures near the equatorial plane and discharges it through an aperture at the rear pole. The external flow, not represented in these diagrams, is in each case from left to right.

that the excess pressure is smaller behind a moving obstacle than in front; it is because separation occurs that the rate of energy dissipation behind a moving obstacle is as large as it is. It is clear from either of these points of view that separation increases the drag force, and the aim in streamlining the rear of obstacles, not always fully achievable, is to prevent it from occurring at all.

A second strategy for the prevention of separation involves what is sometimes referred to as the *injection of forward momentum* into the boundary layer. It is applied particularly in the design of slotted wings for aircraft, where the object is not just to reduce drag but also to prevent the sudden loss of lift which accompanies a stall [§7.11]. Figure 7.14(b) shows a cross-section through a wing whose leading edge is slotted. Where a current of air emerges through the slot, at S, the boundary layer which is advancing from the upper surface of the wing to the left of S is pushed upwards, and a fresh boundary layer is formed over the upper surface to the right of S. This fresh boundary layer has fresh momentum to the right, and it is the more able on that account to survive without flow reversal.

A third strategy, restricted to rather specialised applications, involves preventing separation, but *not* flow reversal, by a suction process which suppresses potential eddies at birth. The boundary layer fluid which is moving backwards is drawn into the body through small apertures in its surface and is discharged where it will do no harm, as fig. 7.14(c) suggests.

Finally, it is possible in some contexts to stimulate turbulence in the boundary layer, and thereby to exploit the drop in C_D which is associated with the onset of such turbulence. The dimples on the surface of a golf ball have this purpose and, though reliable data are hard to come by, they appear to reduce the drag coefficient for a golf ball which is leaving the tee with $Re \approx 2 \times 10^5$ from about 0·5 to about 0·2. Tennis players may benefit from the drag crisis, at any rate if they are serving rather than receiving serve, tennis balls being naturally rough. There have been claims that swimmers can swim faster if they are clad in rough material, but they seem rather improbable. If swimmers are close enough to the drag crisis for the roughness of their swimsuits to trigger boundary layer turbulence it is likely to

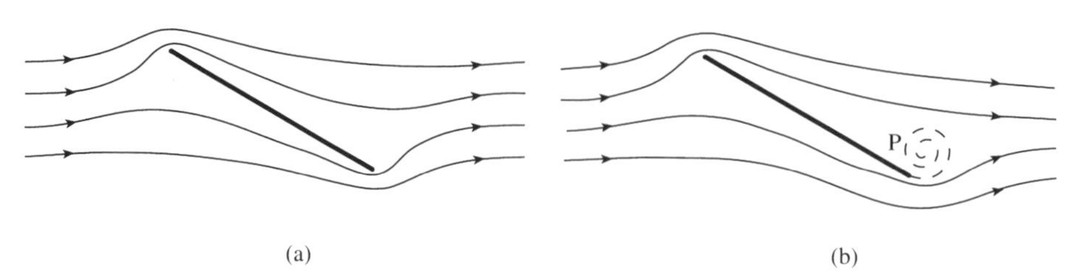

Figure 7.15 Flow past an inclined plate: (a) lines of potential flow for an ideal Euler fluid (sketch only); (b) instantaneous streamlines for a real fluid, shortly after an impulsive start. In (b) the boundary layer (broken curve) has separated and is rolling up into an eddy.

be triggered in any case, e.g. by turbulence in the water through which they are moving; moreover, the drag forces which swimmers have to cope with are due largely to wave resistance [§5.10], to which the behaviour of the boundary layer is almost wholly irrelevant.

7.10 Starting, stopping and trailing vortices

Readers who attempted the experiment with a knife blade described in §7.3 may care to try it again, this time with the blade held neither edge-on nor broadside-on but at an orientation half way between those two. They are likely to observe that the first eddy to form behind the trailing edge, though not the less conspicuous one behind the leading edge, is rapidly shed, to form an isolated *starting vortex* which is left behind in the fluid as the knife blade moves on; if the motion of the knife blade is slow and steady, the shedding process is not repeated [cf. the periodic shedding of eddies by cylinders, §7.5]. The effect may be understood by reference to fig. 4.10 and fig. 7.15(a), which show the lines of flow around an inclined plate as predicted by potential theory, in a frame of reference such that the plate is at rest. They diverge behind both edges, but the divergence is more rapid behind the trailing edge than behind the leading edge, so it is there that flow reversal first occurs in the boundary layer when the velocity of the fluid relative to the plate is increased from zero. Formation of an eddy behind the trailing edge, as suggested by fig. 7.15(b), rapidly ensues, and its subsequent shedding is presumably the result of a build-up of excess pressure near P, where the fluid velocity is small.

Now consider fig. 7.16, which shows the starting vortex being carried away from the stationary inclined plate by the current from left to right and a loop PQRSTUP which is sufficiently far from both plate and vortex to be embedded in fluid which is wholly vorticity-free. Because the viscous term in the Navier–Stokes equation vanishes in vorticity-free fluid [(6.25)], Kelvin's circulation theorem is valid for this loop: the circulation around it was undoubtedly zero when all the fluid was stationary so it must be zero still. Clearly the circulation round the sub-loop

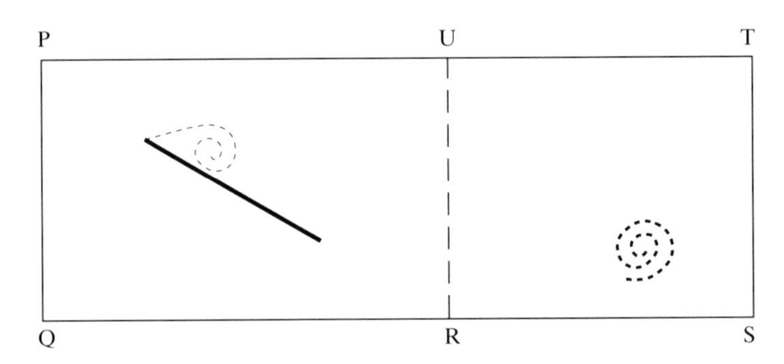

Figure 7.16 The inclined plate of fig. 7.15(b) a moment later; the eddy behind the trailing edge has been shed as a starting vortex and is moving to the right. A fainter spiral suggests that the residual bound vortex may become visible behind the leading edge, if separation occurs there also.

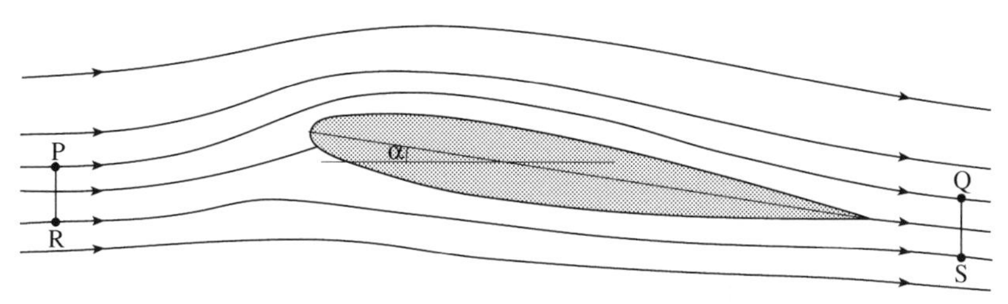

Figure 7.17 Lines of potential flow past an aerofoil inclined at an angle α to the direction of the stream, when the aerofoil carries a bound vortex.

PQRUP which encloses only the plate must be equal and opposite to the circulation round the sub-loop URSTU which encloses only the vortex. The latter is of course the strength of the starting vortex, say K_0. Evidently the plate now carries a *bound vortex* of strength $-K_0$.

If the plate has two sharp edges, separation is almost certain to occur behind the leading edge, and the bound vortex becomes visible as a more or less steady eddy there [see fig. 7.16]. Separation behind the leading edge can be discouraged, however, by giving the plate a finite thickness and by rounding the front of it while preserving the sharpness of its trailing edge; the plate is then an *aerofoil*. Figure 7.17 shows the nature of the lines of flow at large *Re* past an aerofoil which carries a bound vortex, when it is inclined to the flow at a sufficently small angle for no eddy to form behind the leading edge; the frame of reference corresponds to that of fig. 7.15, i.e. the aerofoil is stationary. The existence of the bound vortex is not immediately obvious from this figure. If, however, one were to evaluate the integral of $\boldsymbol{u}\cdot d\boldsymbol{l}$ from say P to Q, along a flowline that lies above the aerofoil and just outside the boundary layer, and then to evaluate the same integral from R to S, along a flowline that lies below the aerofoil, the first integral would turn out to be somewhat larger than the second; the length PQ is larger than RS for one thing

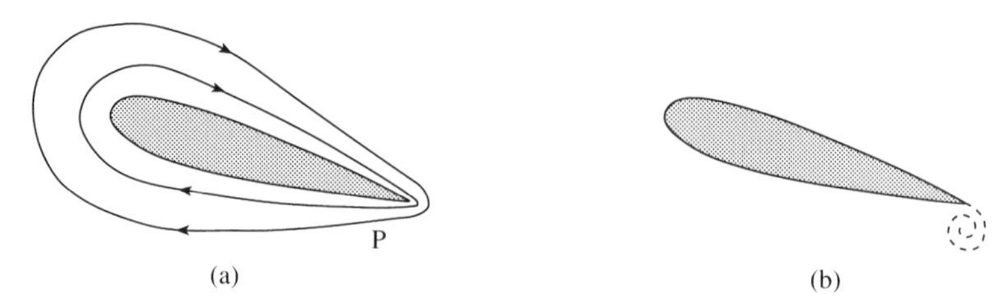

Figure 7.18 (a) Instantaneous streamlines (sketch only) round an aerofoil carrying a bound vortex which has been suddenly brought to rest relative to the surrounding fluid. The boundary layer separates near P, with subsequent formation of a stopping vortex as in (b).

and u tends to be greater above the aerofoil than it is below. Hence the circulation round the aerofoil, evaluated anticlockwise round the closed loop PRSQP, is certainly non-zero and negative.

The flowlines in fig. 7.17 are those appropriate when the ratio K_0/U (U being the velocity of the fluid at large distances), takes a particular value which characterises the steady state. In this state the fluid velocity just outside the boundary layer attached to the aerofoil is the same immediately above and below the trailing edge. Once these two velocities have become equal, there is no longer any divergence of the lines of flow in the neighbourhood of this edge, of the sort that caused the starting vortex to form in the first place. Moreover, the rate at which anticlockwise vorticity is carried into the wake from the bottom surface of the aerofoil is then equal, since the flux of vorticity through any boundary layer is proportional to the square of the velocity outside it [§7.2], to the rate at which clockwise vorticity is carried into the wake from the top surface; the two vorticity streams are free to annihilate completely by diffusing into one another, just as they do in the wake behind a plate which is not inclined [§7.2 and fig. 7.1]. The steady value of K_0/U can be calculated from potential theory for aerofoils of special shapes which permit analysis by the method of conformal mapping [§4.6(iii)]; for the *symmetrical Zhukovskii aerofoil* sketched in fig. 7.17, for example, it is

$$\frac{K_0}{U} = \pi \sin \alpha(D + 0{\cdot}77d), \tag{7.27}$$

where D is the dimension previously referred to as a breadth – though in the context of aerofoil theory it is called the *chord* – and d is the much smaller maximum thickness.

Experiments suggest that when a fluid stream is rapidly accelerated past a stationary aerofoil to a value of U which is subsequently held steady, or when the aerofoil is accelerated in the reverse direction through stationary fluid, the strength of the bound vortex reaches its steady value in a single step, with the shedding of a single starting vortex. On further acceleration, however, a second

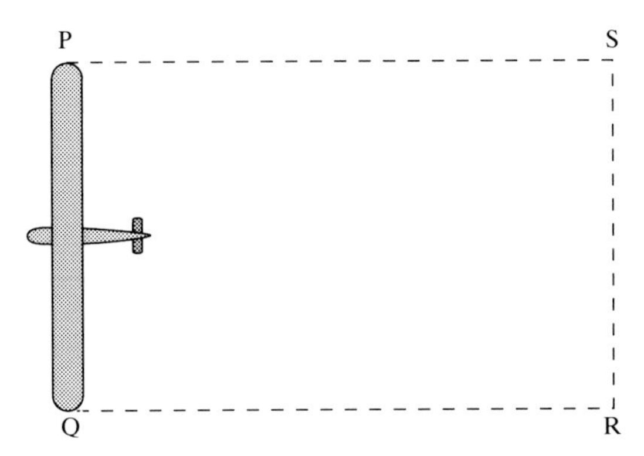

Figure 7.19 View from above of an aerofoil of uniform cross-section, as it were the wing of an aeroplane, which has been in motion relative to the surrounding fluid for a limited time. The bound vortex which runs between Q and P is linked to the starting vortex SR by trailing vortices PS and RQ.

vortex must be shed, with the same sense of rotation as the original starting vortex, if the strength of the bound vortex is to increase to match the new value of U. By the same token, any deceleration results in the shedding of a vortex with the opposite sense of rotation, and when the fluid stream or the aerofoil, whichever is moving, is brought to rest a so-called *stopping vortex* is shed which reduces the strength of the bound vortex to zero. The mechanism of that process is illuminated by the sketch in fig. 7.18(a) of the streamlines round an aerofoil which still carries a bound vortex but for which U is zero. Because these streamlines diverge below the trailing edge, near P, flow reversal occurs in the boundary layer there, the boundary layer separates, and a clockwise eddy – the stopping vortex – appears.

If the aerofoil has been moving in a straight line for a limited time the starting and stopping vortices are parallel to one another, and each of them has the same length as the aerofoil (supposed here to have a cross-section which is uniform all the way along its length). Neither vortex can suddenly terminate in the fluid, however, for reasons given in §4.11, so together they must form two sides of a rectangular vortex ring. They are linked along the other two sides of the rectangle by *trailing vortices* [fig. 7.19], which contribute significantly, as we shall see, to the drag force which the aerofoil experiences.

7.11 Wing theory

The wings which support aircraft in subsonic flight are of course aerofoils, with sharp trailing edges to promote the shedding of a starting vortex and rounded leading edges, and in the simplest version of wing theory they are treated as long aerofoils of uniform cross-section. In that version the flow past a wing is essentially two-dimensional. At reasonably large values of Re the wing carries a starting vortex of uniform strength $-K_0$, where, even if the cross-section is not

Figure 7.20 Asymmetric aerofoil cross-sections which offer increased lift at the expense of increased drag.

such that (7.27) applies, K_o is proportional to the flight speed U and, for small angles of inclination with respect to the orientation for which K_o vanishes, to $\sin\alpha$ or α. Interaction between the bound vortex and the 'external' velocity field U generates an upwards *lift force* F_L, at right angles to both $\boldsymbol{U}$ and $\boldsymbol{K}_o$, of magnitude

$$F_L = \rho U K_o L \tag{7.28}$$

[(4.55)], where L is the length or *span* of the wing; for reasons outlined in §4.12 it is independent of the cross-section of the aerofoil except in so far as this determines K_o. It may be described in terms of a dimensionless *lift coefficient*, normally defined in terms of the chord D as

$$C_L = \frac{F_L}{\frac{1}{2}\rho U^2 DL} \tag{7.29}$$

[§7.8, final paragraph] and when (7.27) applies we have

$$C_L \approx 2\pi\alpha\left(1 + 0{\cdot}77\,\frac{d}{D}\right) \tag{7.30}$$

for small α.

Equation (7.30) for the Zhukovskii aerofoil holds *only* for small α, because when α reaches a critical angle of about 8° the boundary layer separates behind the rounded leading edge. An eddy then forms there, and when this is shed the strength of the residual bound vortex suddenly falls. The aircraft is liable to fall in a dramatic fashion too, as a large part of the force which was supporting it suddenly disappears; it is said to have *stalled*. The maximum value of C_L achievable before stalling occurs may be increased by varying the cross-section of the aerofoil, e.g. by giving it the asymmetrical *cambered* shape of fig. 7.20(a), or by giving it rear flaps as in fig. 7.20(b). Such modifications increase the drag coefficient

$$C_D = \frac{F_D}{\frac{1}{2}\rho U^2 DL}, \tag{7.31}$$

however. For this reason, the rear flaps which are regular features on the wings of commercial aircraft are hinged; they are lowered during take-off, when lift is of prime importance, and raised during steady flight, when drag needs to be minimised so as to conserve fuel.

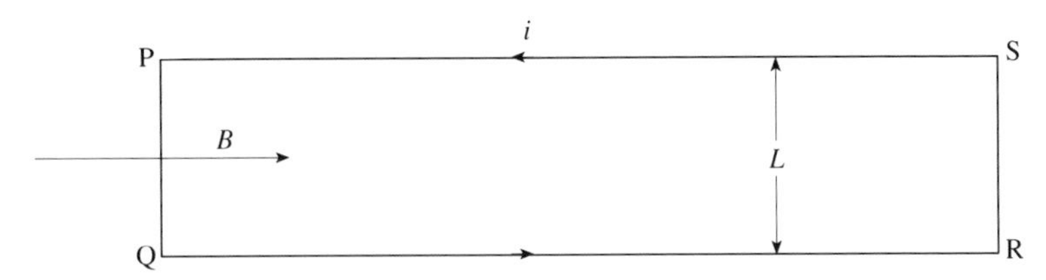

Figure 7.21 A rectangular loop of wire carrying a current; the length PQ experiences a force to the left, in addition to a force perpendicular to the plane of the diagram due to the applied magnetic field.

Of the two types of drag force discussed earlier in this chapter, namely form drag and skin friction, the latter is the more significant for unstalled aerofoils at large values of Re, even though $C_{\text{skin friction}}$ varies like $1/\sqrt{Re}$ [(7.25)] and is therefore small. Aircraft wings are not infinitely long, however, and the flow around them is never exactly two-dimensional. On this account they experience a third type of drag force known as *induced drag*, the magnitude of which may be estimated using an electromagnetic analogue.

Consider a rectangular loop of wire carrying a current i, in the presence of a uniform external magnetic field B in the plane of the loop, as shown in fig. 7.21. The short side PQ, of length L, experiences a force normal to the plane of the diagram of magnitude BiL. Even in the absence of B, however, PQ experiences a force which tends to expand the loop, i.e. which acts in the plane of the diagram and in a leftwards direction. It is due to an interaction between the current in PQ and the magnetic field perpendicular to the plane of the diagram, say b, produced by the current in the long side arms, SP and QR; the magnetic field due to the current in PQ itself exerts no force upon PQ and may be ignored. It is readily shown that when SP $\gg L$ the field b is given, at a point on PQ which is a distance x from its mid point, by

$$b = \frac{\mu_o i}{4\pi}\left(\frac{1}{\frac{1}{2}L + x} + \frac{1}{\frac{1}{2}L - x}\right),$$

provided that $\frac{1}{2}L \pm x$ is less than the radius of the wire a. Hence the additional force to the left is approximately

$$\int_{-\frac{1}{2}L+a}^{+\frac{1}{2}L-a} ib\,\mathrm{d}x \approx \frac{\mu_o i^2}{2\pi}\ln\left(\frac{L}{a}\right).$$

This result may be translated into terms that apply to an aerofoil of length L which carries a bound vortex of uniform strength K_o, by replacing μ_o by $-1/\rho$ and i by $-\rho K_o$ [§4.5]. It implies, in addition to the lift force $\rho U K_o L$ (the analogue of BiL), an induced drag force of magnitude

$$\frac{\rho K_o^2}{2\pi}\ln\left(\frac{L}{a}\right),$$

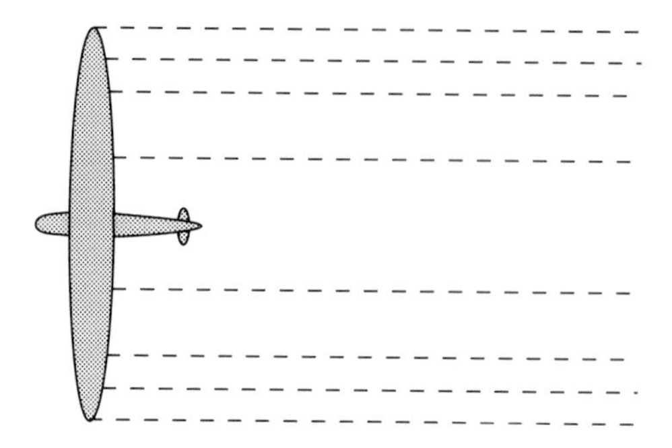

Figure 7.22 View from above of a tapered aerofoil [cf. fig. 7.19]. The broken lines represent trailing vortices, or rather lines of vorticity, which form part of a continuous vortex sheet.

and therefore a third contribution to the drag coefficient,

$$C_{\text{induced drag}} = \frac{K_o^2}{\pi U^2 DL} \ln\left(\frac{L}{a}\right). \tag{7.32}$$

The length a is in this context determined by the core radius of the trailing vortices.

The induced drag force may be interpreted as the result of tension exerted on the aerofoil by the trailing vortices [compare (7.32) with (4.64)], and the work that is done against induced drag in moving an aerofoil forwards evidently supplies these vortices with their kinetic energy. The electromagnetic analogue reveals, however, the way in which the force is exerted. Just as the external magnetic field seen by the current in the wire PQ is tilted away from $\boldsymbol{B}$ by the additional field $\boldsymbol{b}$, so the external flow field seen by the aerofoil is tilted by the flow associated with the trailing vortices. The lift force is thereby tilted backwards, and it is the small component of the lift force which acts in the direction of $\boldsymbol{U}$ rather than perpendicular to $\boldsymbol{U}$ which constitutes the induced drag.

It is obviously desirable in practice to minimise C_D/C_L, and therefore to minimise $C_{\text{induced drag}}/C_L$, and since the latter ratio is proportional, according to (7.32) and (7.30), to $L^{-1} \ln(L/a)$ it looks as though aircraft wings should be made as long as is possible, given that they must be reasonably robust. That conclusion is by and large correct, but there is more to the design of an efficient wing than the simple model we have so far relied on reveals. The fact that the tension exerted by a line vortex is proportional to K^2 shows that it can be reduced by splitting the line: $2 \times (\frac{1}{2}K)^2$ is less than K^2. Not surprisingly, therefore, the induced drag on a wing can be reduced by arranging things so that the trailing vorticity is not all concentrated in two line vortices attached to the wing tips but is shared by an array of weaker vortices, attached to points all along the wing in the manner that fig. 7.22 suggests. To achieve this end the wings are tapered in cross section, so that

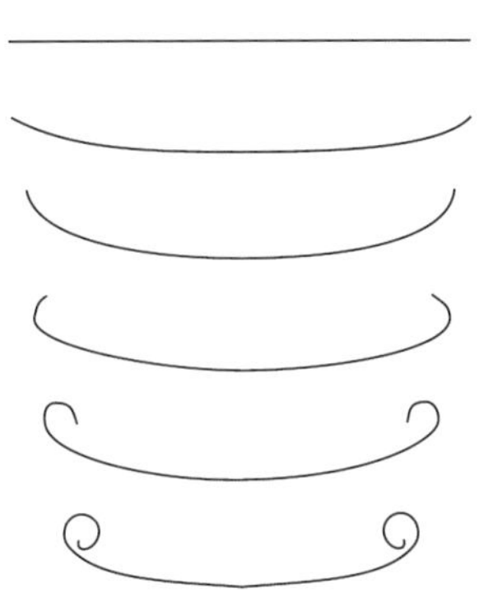

Figure 7.23 Cross-sections, in vertical planes, of the vortex sheet behind a tapered aerofoil, at various distances from it. As distance from the aerofoil increases, the sheet rolls up and moves downwards.
[From J. Lighthill, *An Informal Introduction to Fluid Mechanics*, Clarendon Press, Oxford, 1986.]

the strength of the bound vortex falls from K_o at the centre ($x = 0$) to zero at the tips ($x = \pm\frac{1}{2}L$). The tapering is normally continuous, of course, so that the trailing vortices are not really discrete; they form a continuous *vortex sheet* behind the wings, the sign of the vorticity in this sheet being positive behind one wing and negative behind the other. As time passes and the aircraft moves on, however, the flow field associated with this sheet causes it to roll up at its ends [fig. 7.23]; this rolling-up effect is related to a phenomenon illustrated by fig. 4.19, where two line vortices of the same sign are shown as circling around a common centre. Two discrete trailing vortices are eventually generated by the rolling-up process, but they have relatively large cores. Note how the vortex sheet drifts downwards as it rolls up. This is a consequence of the downwards velocity of the air behind the wings in the region near $x = 0$, which is itself a consequence of the existence of the trailing vortex sheet. The fact that the wings experience a lift force implies, of course, that they impart downwards momentum to the airstream, and it is in the downwards drift that this momentum shows up. It can be shown that in order to minimise $C_{\text{induced drag}}/C_{\text{L}}$ one should arrange things so that the tilt angle (the equivalent in flow terms of b/B) is independent of x, and that this requires the strength of the bound vortex to vary with x according to the equation

$$K = K_o\left(1 - \frac{4x^2}{L^2}\right).$$

The curves sketched in fig. 7.23 were calculated for this optimum arrangement.

Incidentally, the induced drag on an aerofoil which is moving horizontally through air is significantly less when the aerofoil is close to the ground beneath it than when it is at some height, an effect which is said to be exploited by hang gliders and others. The effect of the ground on the flow pattern of the air above it

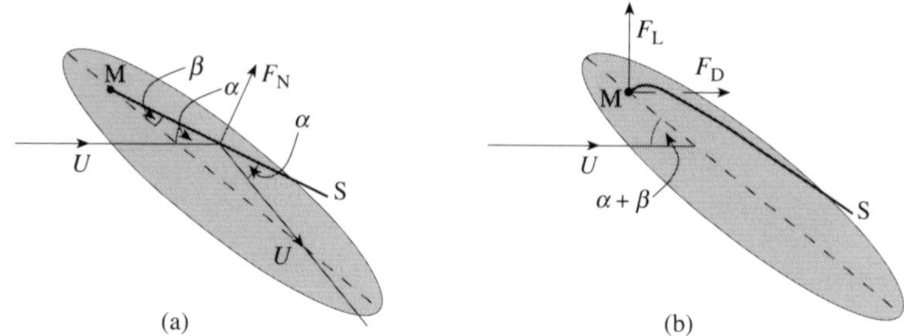

Figure 7.24 A boat sailing into the wind: (a) an unrealistic picture, which suggests that the sail MS remains straight and experiences a normal force F_N as the wind is reflected by it, which is transferred to the boat through the mast M; (b) the sail in fact becomes an aerofoil and experiences a lift force F_L and a drag force F_D.

can be described by postulating the existence of a system of vortices which are images, with reversed circulation, of the bound and trailing vortices associated with the aerofoil [§4.14]. The images of the two trailing vortices are responsible for reducing the induced drag; they reduce, for obvious reasons, the flow equivalent of the vertical field b. The image of the bound vortex repels the bound vortex itself and therefore causes some reduction in the lift force experienced by the aerofoil, but a hang glider can probably compensate for that in practice by slightly increasing the angle of inclination of his aerofoil.

The principle of the aerofoil is exploited by nature and by engineers in many different ways. The flight of birds depends upon it, of course, the only essential difference between a bird's wing and an aircraft wing being that the former is designed to *flap*, and thereby to provide not only lift but also the propulsive force that is needed to overcome drag. The sails of windmills and the blades of turbines operate as (cambered) aerofoils, as do the blades of compressors and propellers, and the tails and fins of fishes.

Finally, yachtsmen rely upon the aerofoil principle when they sail into the wind. An elementary explanation of why it is possible to do this, to which many physics students are exposed in their early years of training, is based upon the diagram in fig. 7.24(a). This shows a sail of area A, inclined at angles α to the wind velocity $\boldsymbol{U}$ (this being the wind velocity relative to the boat rather than the water) and β to the boat's keel, which is intercepting a stream of air in which the momentum flux is $\rho U^2 A \sin\alpha$ and deflecting it through an angle 2α. The implied rate of change of momentum of the air in a direction normal to the plane of the sail, and hence the implied reaction F_N on the sail, is $2\rho U^2 A \sin^2\alpha$, and the component of this which drives the boat forwards is

$$2\rho U^2 A \sin^2 \alpha \sin \beta. \tag{7.33}$$

The picture is nonsensical, of course; if a stream of air were to be deflected in this manner it would get in the way of adjacent streams which the sail does not intercept. Instead, one should recognise that the sail is bellied by the wind into a shape which corresponds to the top half of an aerofoil [fig. 7.24(b)]. The lift force F_L on the aerofoil, at right angles to $\boldsymbol{U}$, is perhaps something like

$$\pi\rho U^2 L\langle D\rangle \sin\alpha \approx \pi\rho U^2 A \sin\alpha$$

[(7.27)], where L is the height of the mast and $\langle D\rangle$ is the mean breadth of the sail. If so, the force driving the boat forward may be something like

$$\pi\rho U^2 A \sin\alpha \sin(\alpha+\beta), \tag{7.34}$$

though in so far as that answer does not take into account the drag force F_D it must be grossly over-simplified. The drag force on the sail is partly induced drag associated with trailing vortices. For the same reason that it pays to make the wing of an aircraft long – i.e. because this tends to reduce the ratio of induced drag to lift – it therefore pays to provide sailing boats with tall masts.

7.12 The Magnus effect revisited

When a cylinder possesses not only a uniform velocity $\boldsymbol{U}$ relative to otherwise stationary fluid, the direction of $\boldsymbol{U}$ being transverse to the cylinder's axis, but also a steady angular velocity ω_0 about its axis, the time-averaged flow pattern must depend upon *two* dimensionless quantities, e.g. upon

$$Re = \frac{2\eta a U}{\rho}$$

and

$$P = \frac{\omega_0 a}{U}, \tag{7.35}$$

where Re is the familiar Reynolds Number and P has been termed the *roll parameter*. When a cardboard tube is launched through the air with a rubber band, as described in §4.12, Re is usually in the range 10^3–10^4, and P is close to 1.

The early stages of the development of flow round a rotating cylinder which, like the cardboard tube, is launched impulsively, i.e. in which U and ω_0 rise simultaneously and almost instantaneously to their final values, have been intensively studied for values of Re up to 10^3, both experimentally and by computer simulation. The general effect of the rotation when $P < 1$ is, as one would surely expect, to encourage the formation and growth behind the cylinder of eddies which rotate in the opposite sense (i.e. clockwise if the cylinder rotates anticlockwise), and to discourage the formation and growth of eddies which rotate with the same sense. For values of Re of 60 or more, such that in the absence of rotation eddies of both sorts would be shed periodically, there is still periodic

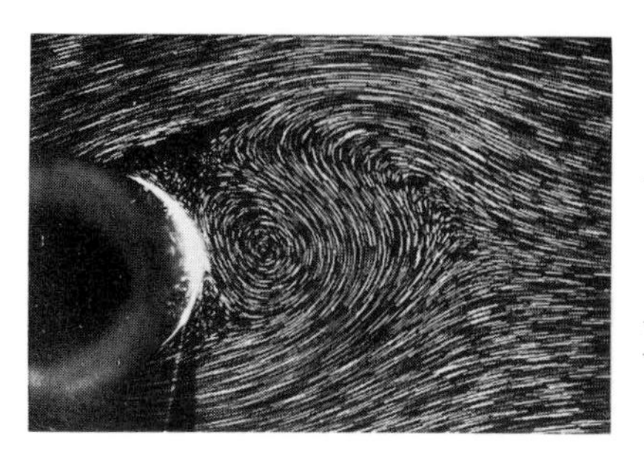

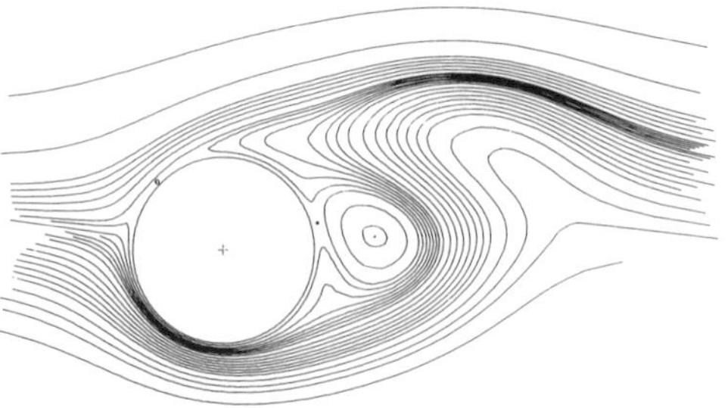

Figure 7.25 Flow past a rotating cylinder; computer simulation of streamlines compared with a photograph of a real system.
[From M. Coutanceau & C. Ménard, *J. Fluid Mech.*, **158**, 399, 1985 (photograph) and H. M. Badr & S. C. R. Dennis, *J. Fluid Mech.*, **158**, 447, 1985 (computer simulation).]

shedding for $P < 1$, but as P increases from zero the even eddies (supposing them to be numbered from 1 upwards in the order of their shedding) become less conspicuous than the odd ones, and because their formation is delayed the shedding frequency falls. The very first eddy differs from all its successors, and the shedding of this one establishes a transverse lift, or Magnus, force which never subsequently disappears, though the lift force does of course oscillate about its non-zero mean value as later eddies are shed. If the mean lift force is expressed in terms of an effective circulation $K_{\text{eff}} = \overline{F_{\text{L}}})/\rho UL$, K_{eff} is found to be significantly less than the steady-state circulation around a rotating cylinder in the limit $U \to 0$, namely $K_{\text{o}} = 2\pi a^2\omega_{\text{o}}$ [see §4.11].

Figure 7.25 shows sets of observed and computed streamlines for the case $Re = 200$, $P = 0{\cdot}5$, at a time which corresponds to translation of the cylinder through 3·5 diameters since its impulsive start. Apart from its intrinsic interest, the figure provides a nice example of what can nowadays be achieved by computer simulation; the agreement with experiment is evidently close. The rotation of the cylinder is anticlockwise in this figure, and the rotation of the substantial eddy attached to the rear of the cylinder is anticlockwise too. This is the second eddy, in fact; the first one has been shed and is moving downstream in the wake. If the first eddy looks less impressive than the second here, that is only because it is moving relative to the cylinder (and, in the experimental case, relative to the camera). It is only on that account that the streamlines in the neighbourhood of the first eddy are not closed loops.

Less is known about the flow when $Re > 10^4$, $P < 1$. At such large values of the Reynolds Number there is, of course, a turbulent wake in the absence of any rotation, enclosed within separated boundary layers [§7.6], and a turbulent wake no doubt continues to exist when P is non-zero but still small. It seems likely, however, that the points of separation are displaced in the direction of the

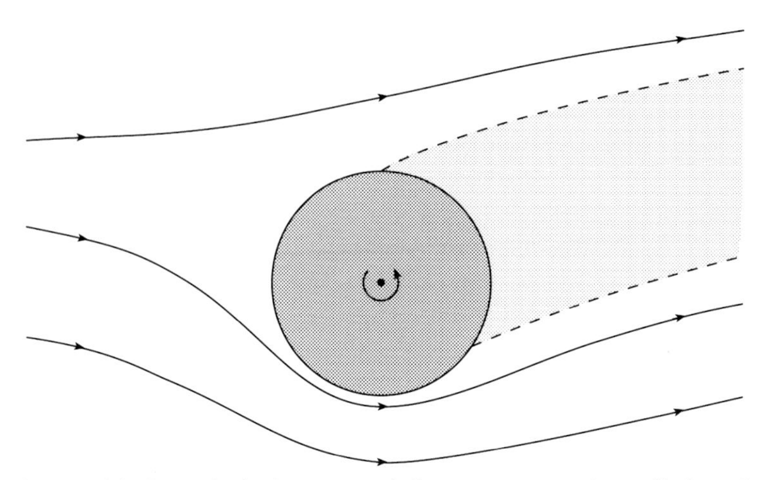

Figure 7.26 Hypothetical pattern of flow past a rotating cylinder when $Re > 10^4$ and $P < 1$. The broken lines represent separated boundary layers, between which the wake is turbulent.

cylinder's rotation, and if at the same time the vorticity-free fluid outside the boundary layers contains some circulation around the cylinder (acquired as a result of vorticity shedding during the early stages of the motion) the resultant mean flow pattern may resemble the one sketched in fig. 7.26. In the vorticity-free fluid outside the boundary layers the excess pressure p^* will then be smaller below the cylinder than above it, and the Magnus force arising on this account will act in the usual direction – downwards in the diagram in fig. 7.26. However, p^* is relatively small within the turbulent wake [§7.8], and if the points of separation are shifted in the manner suggested by the figure the cylinder is liable to be pulled upwards by this pressure deficit. This may be the reason why, at values of Re and P in the region of 3×10^5 and 0·3 respectively, the transverse force experienced by a rotating cylinder is in the opposite direction to the normal Magnus force – an effect known as the *inverse Magnus effect*.

If the shedding frequency for $Re < 10^3$ decreases as P increases, does it at some stage go to zero? The answer seems to be yes it does, at $P \approx 2{\cdot}5$ or thereabouts; beyond that only the first eddy is shed, and when P exceeds about 4 no second eddy is detectable. For $P > 4$, in fact, the steady state flow pattern bears a strong resemblance to the vorticity-free flow pattern discussed in §4.12, for which (if K is the strength of the bound vortex)

$$u_\theta = \frac{K}{2\pi\rho} - U \sin\theta \left(1 + \frac{a^2}{r^2}\right) \tag{7.36}$$

[(4.57)]. There is reason to suppose that, almost independent of the magnitude of Re, the resemblance gets steadily stronger as P increases to infinity, and that at the same time K approaches the limiting value of $K_o = 2\pi a^2 \omega_o$. There must always, of course, be a boundary layer around the cylinder, because (7.36) does not satisfy the no-slip boundary condition at $r = a$. If the boundary layer is sufficiently thin and if $K = K_o$, however, then the mismatch between the fluid velocity u_θ just outside the boundary layer and the velocity of the cylinder's surface is $-2U \sin\theta$ at a fixed value of θ, or, as a function of time t at a fixed point on the surface, $-2U \sin \omega_o t$. The displacement of the fluid just outside the boundary layer with respect to the surface underneath oscillates, in fact, with angular frequency ω_o and amplitude $2U/\omega_o = 2a/P$. It is surely true that if this amplitude is much less than a, i.e. if $P \gg 1$, the boundary layer will not separate. Moreover, the amounts of vorticity fed into the boundary layer from the surface during successive half-periods should then be equal in magnitude but opposite in sign, in which case the thickness of the boundary layer should be

$$\delta \approx \sqrt{\frac{2\eta}{\omega_o \rho}} = 2a\sqrt{\frac{1}{ReP}}, \tag{7.37}$$

independent of time [(1.31)]. Now the vorticity-free solution implies the existence, for sufficiently large values of K/aU or P, of a surface Σ which divides fluid which is trapped near the cylinder and rotates with it from fluid which passes by with the main stream [fig. 4.17]. Under conditions such that the boundary layer thickness predicted by (7.37) is small compared with the minimum distance separating Σ from the surface of the cylinder, which is of order

$$\frac{K}{U} = 2\pi a P,$$

there is surely no chance of vorticity diffusing across Σ and no other way in which the strength of the bound vortex can change. The argument implies that $K \approx K_o$ when

$$\sqrt{\frac{1}{ReP}} \ll P,$$

or

$$P \gg Re^{-1/3}.$$

Since we are in any case assuming that $P \gg 1$, this condition is almost automatically satisfied; it certainly does not suggest the existence of any upper limit to Re, beyond which the 'classical' formula for the Magnus force,

$$|F_L| = \rho K_o UL = 2\pi\rho\omega_o a^2 UL,$$

is liable to fail.

If classical potential theory is to be trusted for rotating cylinders in the limit of large P, then the drag coefficient should presumably tend to zero in this limit. That C_D does at least fall as P increases is confirmed by much experimental evidence.

Any reader who plays tennis or golf or cricket or baseball is likely to be more interested in spinning spheres than spinning cylinders, but though the story for cylinders is complicated it is more complicated still for spheres. Transverse forces do, of course, act on spinning balls for which the Reynolds Number is up to say 2×10^5 and the roll parameter less than say 1, and they generally have the sign to be expected of classical Magnus forces; a sliced golf ball swerves to the right, an undercut golf ball appears to ignore gravity during the early part of its trajectory, a tennis ball with top spin dips sharply after crossing the net, and so on. But the origin of these forces is not understood in detail. There is evidence that the swing of a cricket ball is due largely to an asymmetry in the flow for which the raised seam is responsible, and that swing bowlers impart spin to the ball in order to keep the plane of the seam at a fixed angle to its direction of translation rather than to exploit the Magnus effect; the seam on a baseball seems to be less important, no doubt because it meanders over the surface of the ball in an elaborate curve instead of following a great circle. All cricketers, and perhaps baseball players too, are convinced that balls swing more in some weather conditions than in others – particularly when it is humid. Physicists are unable explain this correlation to their satisfaction, and some have claimed that it therefore cannot exist.

7.13 Tealeaves and suchlike

When flow reversal occurs in a boundary layer it does so, according to one way of looking at the phenomenon, because fluid close to the solid surface has less forward momentum to lose than fluid further out and is less able on that account to withstand an adverse gradient of excess pressure. From that point of view, flow reversal is one example of a class of phenomena in which a *secondary circulation* arises due to a mismatch between the pressure gradients needed to maintain steady motion without this circulation in adjacent bodies of fluid.

Perhaps the most familar example of a secondary circulation is the one which carries the tealeaves to the centre of a teacup when the tea is set into rotation by stirring it with a spoon. Most of the liquid rotates about the vertical axis as a solid body would, with a uniform angular velocity ω, and this steady motion implies the existence of a radial pressure gradient, $\mathrm{d}p^*/\mathrm{d}r = \rho\omega^2 r$ [§2.5]. A layer of fluid at the bottom is slowed down, however, by contact with the cup; it is a sort of boundary layer, but a boundary layer within which the vorticity is less, rather than greater, than it is in the adjacent fluid. To maintain that boundary layer in steady rotation requires radial pressure gradients which are less than $\rho\omega^2 r$, and the mismatch generates a slow circulation in which fluid moves inwards towards the axis in the boundary layer and returns to the perimeter higher up.

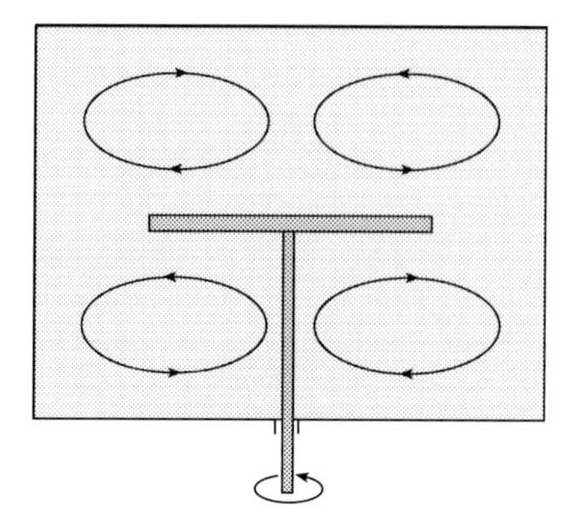

Figure 7.27 Secondary circulation in an enclosed body of fluid, brought about by rotation of a submerged disc

Much the same thing, only in reverse, happens when a solid disc or short cylinder is set into rotation in otherwise stationary fluid, as in fig. 7.27. In this case $\mathrm{d}p^*/\mathrm{d}r$ is positive in the boundary layer fluid which rotates with the solid object but would like to be zero in stationary fluid outside the boundary layer. The mismatch imposes a secondary circulation of the sort suggested by the figure. Since its speed is proportional to ω^2, it does not average to zero if the rotation is oscillatory rather than steady. It must therefore occur in rotating disc viscometers of the sort used by Maxwell [§6.6(v)], though it may well be insignificant in practice.

When water in a river flows round a bend it undergoes centripetal acceleration towards the inside of the bend, and on that account there must be a positive excess pressure gradient $\mathrm{d}p^*/\mathrm{d}r$; the free surface of the water must be slightly higher on the outside of the bend than on the inside. As in a teacup when the tea is stirred, the more slowly moving water near the bottom cannot withstand the same pressure gradient, and a secondary circulation is established which carries this slow moving water towards the inside of the bend. Like the tealeaves in the teacup, particles of soil at the bottom of the river are slowly carried in the same direction, with the result that the inside of the bend silts up, while the outside is scoured out further. It is in this way that the loops of meandering rivers develop.

One last example of a secondary circulation concerns the horizontal flow of air or water past a vertical cylinder, as it were a tree or post, projecting from the ground. The patterns of flow past non-rotating cylinders discussed earlier in this chapter, which are two-dimensional and apply exactly only when the cylinders are very long, are associated with values of p^* which are positive at the front stagnation point where the fluid divides and negative at the sides of the cylinder and normally to its rear as well. Near the ground, however, the fluid moves more

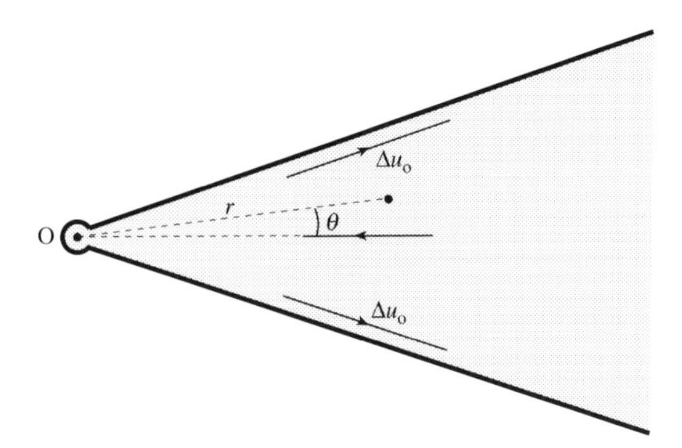

Figure 7.28 Acoustic streaming in a wedge of fluid between two plates, when there is a pulsating line source at its apex O; near each plate, though outside the boundary layers attached to them, the fluid drifts away from O with velocity Δu_o, while other fluid near the middle of the wedge drifts back towards O. The phenomenon is discussed in the text in terms of a polar coordinate system such that θ represents angle of rotation in an anticlockwise direction.

slowly and cannot withstand the same gradients of p^*. A secondary circulation therefore arises, in which oncoming fluid flows down the front surface of the tree or post and round its sides to the rear. The existence of this circulation is apparent under snowy conditions, from the way in which the snow in front of trees is scooped out when a wind blows. It is equally apparent from the scooping out of sand or gravel in front of posts submerged in rivers.

7.14 Acoustic streaming

The final topic to be discussed in this chapter is a more subtle type of secondary circulation, associated with the behaviour of oscillatory boundary layers in circumstances where the amplitude of oscillation of the fluid velocity just outside the boundary layer varies in a longitudinal direction. Consider, for example, the flow induced in a wedge between two inclined plates [fig. 7.28] by a line source along the apex of the wedge which is pulsating, i.e. which is alternately emitting and drawing in fluid with angular freqency ω. In so far as the fluid is vorticity-free, its motion is radial and its velocity varies like $r^{-1}\cos\omega t$. In that case, we may expect vorticity to exist within boundary layers adjacent to each plate, of uniform thickness $\delta \approx \sqrt{2\eta/\omega\rho}$ [(1.31)]. But per unit length normal to the plane of the diagram in fig. 7.28 the instantaneous volume flux of fluid through each boundary layer in a radial direction is about $\frac{1}{2}\delta u_o$, where u_o denotes the radial velocity outside the boundary layer, and if u_o varies with r the flux does too. A continuity condition then requires that at the interface between the boundary layer and the vorticity-free fluid $\boldsymbol{u}$ must instantaneously have a component u_θ normal to this interface; in the vicinity of the lower plate in the figure

$$u_\theta \approx -\frac{\partial(\frac{1}{2}\delta u_o)}{\partial r} = \frac{\delta u_o}{2r},$$

while in the vicinity of the upper plate the sign of u_θ is reversed. Now any fluid which crosses this interface carries longitudinal (i.e. radial) momentum with it, and the transfer of longitudinal momentum per unit area per unit time corresponds to an instantaneous shear stress in a radial direction, exerted *by* the fluid in the boundary layer *on* the fluid outside the boundary layer, of magnitude

$$\rho u_r u_\theta \approx \frac{\rho \delta u_o^2}{2r}.$$

Because this stress is proportional to u_o^2 its time average is not zero, in the way that the time averages of u_o and u_θ are. Thus it continuously feeds momentum (and incidentally vorticity) into the fluid outside the boundary layer, and augments the radial velocity there by a non-oscillatory term, say Δu_o. This term goes on increasing until the shear stress estimated above is balanced by a viscous shear stress due to the velocity gradient in the boundary layer, i.e. until

$$\frac{\rho \delta \overline{u_o^2}}{2r} \approx \frac{\eta \Delta u_o}{\delta},$$

or until[7]

$$\Delta u_o \approx \frac{\rho \delta^2 \overline{u_o^2}}{2\eta r} = \frac{\overline{u_o^2}}{\omega r}. \tag{7.38}$$

Not all the wedge-shaped body of fluid between the two boundary layers can be drifting outwards with that velocity, however, for the source at the apex is a pulsating one and it does not supply the steady flux required. What happens is that fluid immediately adjacent to the two boundary layers acquires the radial velocity away from the apex of the wedge which (7.38) describes, and that fluid half way between the plates drifts back towards the apex. A particularly remarkable feature of this phenomenon is that the velocity of steady circulation is quite independent of the fluid's viscosity.

Secondary circulation of this sort occurs in many contexts and is known, at any rate when ω is in the audio region, as *acoustic streaming*. Figure 7.29 is reproduced from a paper by Andrade, who investigated the phenomenon using air contami-

[7] A more exact treatment leads to a result which differs from (7.38) only by a factor $\frac{3}{4}$ on the right-hand side.

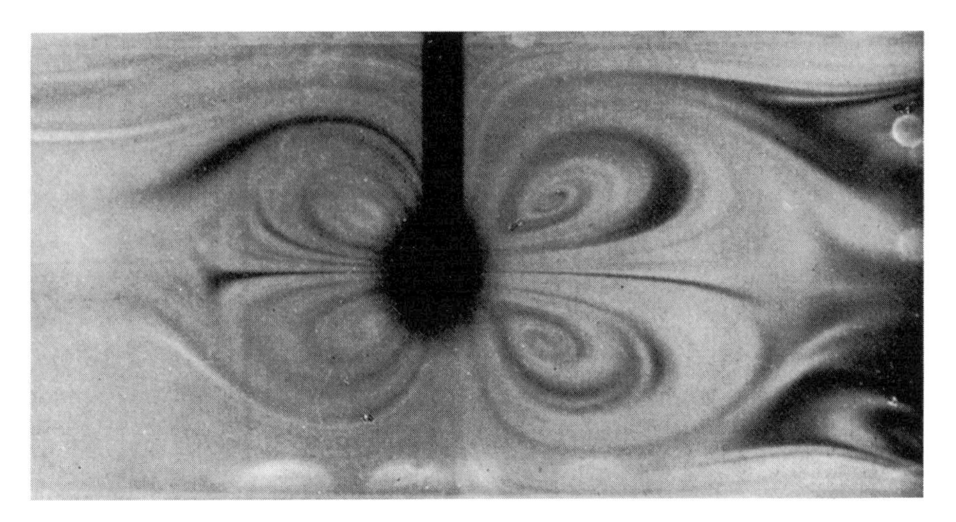

Figure 7.29 Secondary circulation pattern in air surrounding a cylinder of diameter 2·4 mm, stimulated by a standing sound wave of frequency 512 Hz. The displacements due to the sound wave are in the plane of this photograph and horizontal rather than vertical; the axis of the cylinder is normal to the plane and coincides with a displacement antinode.
[From E. N. da C. Andrade, *Proc. Roy. Soc. A*, **134**, 445, 1932.]

nated by smoke; he found that under the influence of the streaming motion the smoke particles tend to aggregate and thereby to exhibit streaklines. The streaklines in fig. 7.29 reveal the existence of a steady circulation, caused by interaction between a sound wave propagating in a horizontal direction and a cylindrical obstacle at the centre of the figure, the axis of the cylinder being normal to the plane of the diagram. The circulation is evidently anticlockwise in the top right and bottom left quadrants, and clockwise in the other two quadrants. Thus the air just outside the very thin boundary layer surrounding the cylinder is streaming towards the stagnation points ($\theta = 0, \pi$) where the oscillatory velocity associated with the sound wave, $u_o = 2U \sin\theta$ at $r = a$, vanishes; it is streaming away from the regions (near $\theta = \pm\pi/2$) where $|u_o|$ is relatively large. The phenomenon can be explained by the reasoning applied above to fig. 7.28.

Acoustic streaming accounts for the behaviour of two pieces of apparatus which used to be found in every establishment where physics was taught, and which are still useful for demonstration purposes: *Kundt's tube* and *Chladny's plate*. Kundt's tube is a horizontal glass cylinder containing air and a small amount of lycopodium powder. Standing sound waves are excited in the column of air, which then develops nodes and antinodes of longitudinal oscillatory velocity u_o. The powder at the bottom of the tube is carried by the streaming motion in the direction of decreasing $|u_o|$ and therefore collects at the nodes. Chladny's plate is a horizontal metal plate clamped at its centre and sprinkled with lycopodium powder, in which a variety of transverse standing waves may be excited. Vertical displacement of the plate's surface drives oscillatory air currents which are also vertical near the antinodes of the plate's displacement but which are horizontal in between. It is the horizontal component of the air's oscillatory motion which gives rise to acoustic streaming in the air, so it is towards the plate's antinodes that the powder tends to drift. This behaviour puzzled some nineteenth century physicists, who expected the powder to be bounced towards the nodes – as indeed it is when the plate oscillates *in vacuo*. It seems to have been Faraday who first appreciated the role played by the air.

Further reading

A lucid and readable account of several of the topics discussed in this chapter, and of wing theory in particular, is available in *An Informal Introduction to Theoretical Fluid Mechanics*, Clarendon Press, Oxford, 1986, by J. Lighthill. For a more comprehensive treatment consult, once again, *An Introduction to Fluid Mechanics*, Cambridge University Press, 1967, by G. K. Batchelor. Readers interested in how the dimples on a golfball affect its trajectory may like to look at a paper by J. M. Davies, *J. Appl. Phys.*, **20**, 821, 1949, while those who play cricket instead and are interested in the problem of swing may consult N. G. Barton, *Proc. Roy. Soc. Lond. A*, **379**, 109, 1982.

8

Instabilities

8.1 Stability, instability and overstability

As every physicist knows, a dynamical system which is in equilbrium may be *stable* or *unstable*. The simplest case of the distinction is that of a particle of mass m which can move only in one dimension, in circumstances where the particle's potential energy Φ varies with its position x in the manner suggested by fig. 8.1. The particle experiences no force when it is situated at the minimum, P, or at the maximum, Q, and in principle it can remain at rest indefinitely in either of these positions. However, if it is slightly displaced from P it accelerates towards P, whereas if it is slightly displaced from Q it accelerates away from Q; in the first position the particle is stable and in the second it is unstable. Near any minimum such as P the restoring force $\partial\Phi/\partial x$ can normally be expanded as a Taylor series in powers of displacement $\xi_{\mathrm{P}} = x - x_{\mathrm{P}}$. Since it is zero at P itself, an adequate approximation for small values of ξ_{P} is

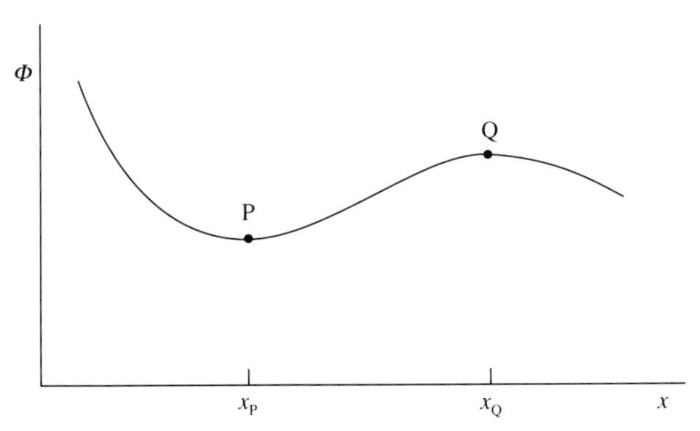

Figure 8.1 Positions of stable (P) and unstable (Q) equilibrium for a particle whose potential energy Φ varies with x in the manner shown.

$$\frac{\partial \Phi}{\partial x} \approx \left(\frac{\partial^2 \Phi}{\partial x^2}\right)_{\mathrm{P}} \xi_{\mathrm{P}},$$

in which case the equation of motion of the particle is linear in ξ_{P},

$$m \frac{\partial^2 \xi_{\mathrm{P}}}{\partial t^2} = -\left(\frac{\partial^2 \Phi}{\partial x^2}\right)_{\mathrm{P}} \xi_{\mathrm{P}}. \tag{8.1}$$

The oscillations which it describes are then simple harmonic, with angular frequency ω_{P} such that

$$\omega_{\mathrm{P}}^2 = \frac{1}{m}\left(\frac{\partial^2 \Phi}{\partial x^2}\right)_{\mathrm{P}}.$$

An equation of motion similar to (8.1) applies in the neighbourhood of Q, but since $(\partial^2 \Phi/\partial x^2)_{\mathrm{Q}}$ is negative the roots for ω are necessarily imaginary, $\omega_{\mathrm{Q}} = \pm \mathrm{i} s_{\mathrm{Q}}$ with s_{Q} real. Hence the displacement $\xi_{\mathrm{Q}} = x - x_{\mathrm{Q}}$ of a particle which starts at rest at $t = 0$ from a position such that $\xi_{\mathrm{Q}} = \xi_{\mathrm{o}}$ is given at later times by

$$\xi_{\mathrm{Q}} \approx \frac{1}{2} \xi_{\mathrm{o}} \left(\mathrm{e}^{s_{\mathrm{Q}} t} + \mathrm{e}^{-s_{\mathrm{Q}} t}\right),$$

as long as it remains small. If ξ_{o} is infinitesimal, then by the time the displacement becomes apparent $\exp(s_{\mathrm{Q}} t)$ must be very much greater than unity, in which case $\exp(-s_{\mathrm{Q}} t)$ must be negligible. When a particle leaves a position of unstable equilibrium, therefore, its displacement normally grows in an exponential fashion.

Suppose now that $(\partial \Phi/\partial x)_{\mathrm{P}}$ is necessarily always zero – perhaps because of some symmetry requirement – while $(\partial^2 \Phi/\partial x^2)_{\mathrm{P}}$ can be reduced in magnitude and ultimately reversed in sign by altering the external constraints which determine Φ. In that case P is always an equilibrium position, but the equilibrium is stable in one range of the constraints and unstable in an adjacent range. Where the changeover occurs one has

$$\omega_{\mathrm{P}}^2 = \frac{1}{m}\left(\frac{\partial^2 \Phi}{\partial x^2}\right)_{\mathrm{P}} = 0,$$

and this is the condition for what is called *marginal stability*. When it is satisfied, the force experienced by a particle near P is normally determined by $(\partial^3 \Phi/\partial x^3)_{\mathrm{P}}$ or, if $(\partial^3 \Phi/\partial x^3)_{\mathrm{P}}$ is zero for symmetry reasons, by $(\partial^4 \Phi/\partial x^4)_{\mathrm{P}}$; it is then proportional to ξ_{P}^2 or ξ_{P}^3 rather than to ξ_{P}.

Similar results apply, of course, to any mechanical system for which energy is conserved. If the system is a complicated one, a full description of its state requires specification of a great many different coordinates of position. There always exists a set of normal coordinates ζ_n, however, such that for small ζ_n the potential energy Φ and kinetic energy T of the system may be expressed in the form

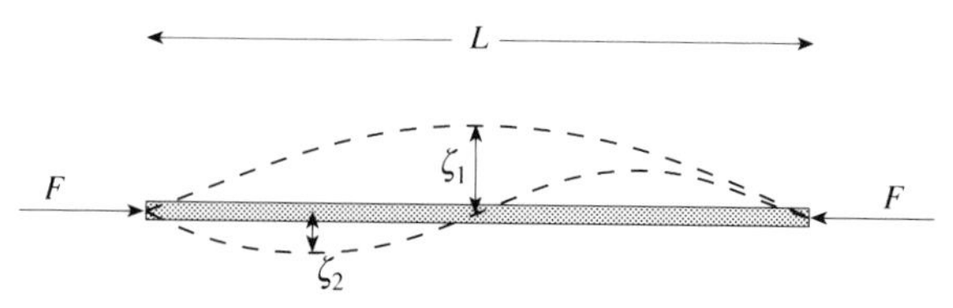

Figure 8.2 A strut under compression. The broken curves suggest two normal modes of flexural distortion, corresponding to $n = 1$ and $n = 2$.

$$\Phi = \Phi_o + \sum_n \frac{1}{2} m_n \omega_n^2 \zeta_n^2,$$

$$T = \sum_n \frac{1}{2} m_n \left(\frac{\partial \zeta_n}{\partial t}\right)^2,$$

where Φ_o is the potential energy of the equilibrium state for which all ζ_n are zero, and the equilibrium is stable if and only if $m_n\omega_n^2 > 0$ for *all* values of n. In continuous systems the normal coordinates often describe periodic modes of distortion of the system as a whole, rather than displacements of isolated parts of the system. Consider, as an example, the well-known problem first solved by Euler of a uniform elastic strut of length L which is straight in equilibrium but free to bend, and which is compressed by longitudinal forces $\pm F$ applied to its ends [fig. 8.2]. When the ends are pinned in such a way that they cannot move except towards one another, the normal coordinates describe the amplitude of flexural modes in which the transverse displacement varies sinusoidally, like $\sin(n\pi x/L)$ where x is distance measured along the strut from one end and where $n = 1, 2, 3$ etc. The potential energy stored in the elastic deformation of the strut associated with the nth mode is proportional for small ζ_n to $n^4\zeta_n^2$, while the corresponding potential energy stored in whatever system supplies the compressive force is proportional to $-Fn^2\zeta_n^2$. When $F = 0$ the strut in its undeformed state is completely stable, because the second derivatives of the total potential energy, $\partial^2\Phi/\partial\zeta_n^2$, are positive for all n. As F is increased from zero, the marginal stability condition, $\partial^2\Phi/\partial\zeta_n^2 = 0$, is reached first for $n = 1$. Once the load which corresponds to that condition is exceeded the strut becomes unstable and is bound to buckle. Because $\partial^4\Phi/\partial\zeta_1^4$ is positive it stabilises after buckling at a finite value of ζ_1 (which may have either sign), but that is an aspect of the problem which need not concern us here.

In so far as the above remarks apply to conservative systems they may seem to have little relevance to viscous fluids, which are inherently dissipative. If, however, a particle moving in the potential of fig. 8.1 is subject to a dissipative retarding force proportional to its velocity, the principal effect of this is merely to damp – and perhaps overdamp – oscillations in ξ_P and to slow down the exponential rate of growth of ξ_Q. That does not invalidate the conclusion that P and Q represent states of stable and unstable equilibrium respectively. Indeed, the fluctuations which always accompany dissipation in thermal equilibrium now make it impossible in principle, as well as in practice, for a particle to remain indefinitely at Q. Nor does the existence of dissipation invalidate the conclusion that when, as a result of a continuous change in the form of $\Phi\{x\}$, the equilibrium at P changes from being stable to being unstable, this equilibrium passes through a state of marginal stability.

The general procedure for investigating the stability or otherwise of patterns of fluid flow involves perturbing the pattern in various ways and calculating whether the amplitude – say $\zeta_n\{t\}$ – of each perturbation mode decreases or increases with time; the amplitude may well describe a velocity rather than a displacement, but that is a rather trivial distinction in this context. The modes must be consistent with the boundary conditions to which the fluid is subject, and they should form, like the periodic normal modes of the Euler strut, a complete set in terms of which any possible perturbation may be expanded. The exact equations of motion of the fluid are always non-linear in ζ_n, and one cannot achieve a detailed understanding of what happens once an instability has developed without taking non-linear terms into account. As a first step, however, it may suffice to establish the condition for a state of marginal stability to exist; having done that, one may confidently assert that true stability lies on one side of this condition and instability on the other. Since marginal stability requires

$$\frac{\partial \zeta_n}{\partial t} = 0 \tag{8.2}$$

to first order only in ζ_n, the condition for its existence may be established using approximate equations of motion from which all terms which are non-linear in ζ_n have been deleted. If, as is often the case, there are several competing modes of instability, the first to develop once the condition for marginal stability has been exceeded is normally the one for which s_n $(= \zeta_n^{-1}\partial\zeta_n/\partial t)$ is largest. Linearised equations of motion suffice to settle this question as well.

A characteristic of dissipative systems, which cannot readily be illustrated by reference to simple mechanical models of point masses moving in variable potentials but which is familiar in the context of electrical circuits, is that they may spontaneously oscillate. To do so they must incorporate a source of power, of course, and some feedback mechanism which selectively amplifies an oscillatory component in the thermal fluctuations of the system. At the onset of an oscillatory instability a system is said to become *overstable* (a potentially misleading term for which Eddington was responsible). If the amplitude $\zeta_n\{t\}$ of the overstable mode is taken to vary with time like $\exp\{-\mathrm{i}(\omega_n + \mathrm{i}s_n)t\}$ for small ζ_n, where ω_n $(\neq 0)$ and s_n are both real, overstability requires s_n to be positive whereas stability requires it to be negative. The condition for marginal overstability is $s_n = 0$, i.e.

$$\frac{\partial \zeta_n}{\partial t} = - \mathrm{i}\omega_n \zeta_n \tag{8.3}$$

to first order in ζ_n.

The discussion of flow instabilities below is limited in the main to linearised theory, and the reader must not expect detailed explanations for all the phenomena he will encounter. It falls into three main parts, with a very brief postscript: §§8.2–8.4 concern instabilities in which surface tension plays a key role, which illustrate the ideas outlined above in a relatively straightforward way; §§8.5–8.10 concern convection and convective instabilities; §§8.11–8.13 concern instabilities in vortex sheets; and §8.14 concerns an instability which affects Stokes waves.

The physicist who contributed most to an understanding of fluid instabilities during the nineteenth century was Lord Rayleigh, and his twentieth century counterpart was G. I. Taylor. Each of their names occurs three times in the section headings below, and further specification is obviously desirable to avoid confusion. Perhaps for that reason, fluid instabilities tend to be labelled by the names of *two* people who have studied them.

8.2 The Rayleigh–Taylor instability

The Rayleigh–Taylor instability arises when a vessel which contains two fluids separated by a horizontal interface – one at least of the fluids must of course be a liquid – is suddenly inverted so that the heavier fluid lies above the lighter one. The gravitational potential energy of the system, which was at its minimum value before inversion, is now at its maximum, and although the system is still in equilibrium while the interface remains horizontal the equilibrium is clearly liable to be unstable. Whether or not it is actually unstable with respect to any particular perturbation depends upon whether the gravitational energy which this releases is greater or less than the increase in surface free energy. The system is marginally stable with respect to the perturbation when the two are equal.

All possible small perturbations of the surface may be expressed in terms of their Fourier components, a typical Fourier component involving a vertical displacement of the interface

$$\zeta = \zeta_k\{t\}\mathrm{e}^{\mathrm{i}\boldsymbol{k}\cdot\boldsymbol{r}},$$

where $\boldsymbol{r}$ is a vector which lies in the $z = 0$ plane, i.e. the plane of the undisturbed interface. Per unit area of the interface, the reduction in gravitational potential energy associated with a single wave of this form, averaged over any integral number of wavelengths, is [(5.29)]

$$\frac{1}{4}(\rho' - \rho)g\zeta_k^2,$$

where ρ' and ρ are the densities of the heavier and lighter fluids respectively. The increase of the surface free energy, similarly averaged, is

$$\sigma \left\langle \left\{ 1 + \left(\frac{\partial \zeta}{\partial x}\right)^2 + \left(\frac{\partial \zeta}{\partial y}\right)^2 \right\}^{1/2} - 1 \right\rangle \approx \frac{1}{4} \sigma k^2 \zeta_k^2$$

to second order in ζ_k, where σ is the interfacial surface tension. Marginal stability is therefore only possible for one wavevector k_c, such that

$$(\rho' - \rho)g = \sigma k_c^2. \tag{8.4}$$

In order to find the rate at which modes for which $k < k_c$ grow in amplitude, one needs to know how the velocities of each fluid depend upon $\partial\zeta_k/\partial t$. With that information at one's disposal, one may follow the routine procedure of evaluating the mean kinetic energy per unit area and hence the total energy, a sum of gravitational and surface terms proportional to ζ_k^2 and kinetic terms proportional to $(\partial\zeta_k/\partial t)^2$; by equating the time derivative of the total energy to zero one may then obtain, after cancellation of a factor $\partial\zeta_k/\partial t$, a linear equation of motion relating ζ_k to $\partial^2\zeta_k/\partial t^2$ which provides the required answer. We, however, can make use of a result already available as (5.40), which tells us, in the notation of §8.1, that the dispersion relation for waves on the interface is

$$\omega_k^2 = -\frac{\rho' - \rho}{\rho' + \rho} gk + \frac{\sigma k^3}{\rho' + \rho}, \quad \text{with } s_k = 0,$$

as long as k is greater than the critical wavevector which (8.4) describes, which implies that when $k < k_c$ we have

$$s_k^2 = \frac{\rho' - \rho}{\rho' + \rho} gk - \frac{\sigma k^3}{\rho' + \rho}, \quad \text{with } \omega_k = 0.$$

The value of k, say k_{max}, which maximises s_k and hence the rate of growth is clearly such that

$$(\rho' - \rho)g = 3\sigma k_{max}^2,$$

so

$$k_{max} = \frac{k_c}{\sqrt{3}}. \tag{8.5}$$

We may infer from the above results that if it were possible to invert almost instantaneously a *large* vessel containing two fluids, so large that the boundary conditions imposed virtually no limitations on the allowed values of $\boldsymbol{k}$, the contents would be inherently unstable. The interface would inevitably develop corrugations whose periodicity would be the wavelength associated with k_{max}, i.e. $2\pi\sqrt{3\sigma/(\rho' - \rho)g}$, which amounts to about 3 cm when the heavier fluid is water

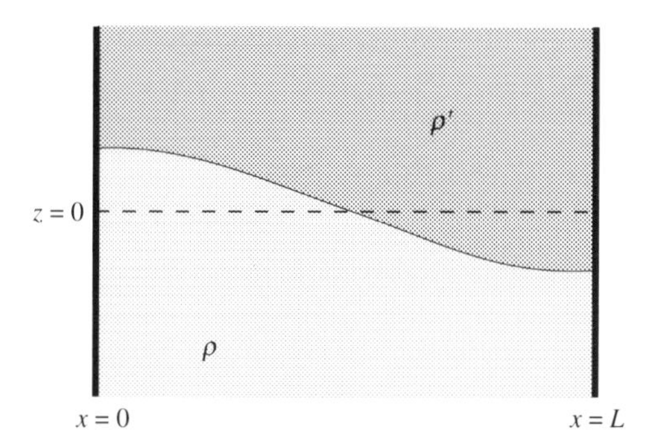

Figure 8.3 A layer of one fluid with a denser fluid above it, in a container of width L, is stabilised by surface tension against the perturbation suggested here provided that (8.6) is satisfied.

and the lighter one is air. In practice, however, rapid inversion is possible only with small vessels, and the fact that liquid inside an inverted bottle is stabilised by surface tension if the opening of the bottle is small enough must be familiar to every reader. For simplicity, suppose the vessel to be a rectangular one, with vertical sides and a cross-section in the $z = 0$ plane of which the larger dimension is L. The smallest non-zero value of k consistent with the boundary conditions [fig. 8.3 and some remarks about the boundary conditions applicable to water waves at the start of §5.8] is then π/L. In that case the inverted contents are stable provided that $\pi/L > k_c$, i.e. provided that

$$L < \pi\sqrt{\frac{\sigma}{(\rho' - \rho)g}}. \tag{8.6}$$

8.3 The Rayleigh–Plateau instability

A free jet of water, emerging from a circular orifice, is liable to break up into a regular succession of drops, and according to Plateau's analysis of some observations by Savart the drops are separated by a distance λ which is about 8·8 times the radius a of the jet before it disintegrates. If a stationary cylinder of water could be obtained it would break up in the same way, and indeed the droplets of water which are to be seen on spiders' webs after a damp cold night are probably formed by accretion from layers of dew which are cylindrical when first deposited. The explanation lies in the fact that, volume for volume, spheres have smaller surface areas than cylinders.

Suppose an initially uniform cylinder of liquid to be subject to a small *varicose deformation*, which preserves rotational symmetry about the x axis (the axis of the cylinder) but alters its radius in a periodic fashion from a to

$$b = \langle b\rangle + \zeta_k \cos kx \; (\zeta_k \ll a).$$

The volume of the cylinder per unit length, averaged over an integral number of wavelengths, is

$$V = \langle \pi b^2 \rangle = \pi \langle b \rangle^2 + \frac{1}{2}\pi\zeta_k^2,$$

and since this must equal the initial volume per unit length, πa^2, we have

$$\langle b \rangle = \sqrt{a^2 - \tfrac{1}{2}\zeta_k^2} \approx a - \frac{\zeta_k^2}{4a}.$$

Thus the surface area of the cylinder per unit length, similarly averaged, is

$$A = \left\langle 2\pi b \sqrt{1 + \left(\frac{\mathrm{d}b}{\mathrm{d}x}\right)^2} \right\rangle$$

$$\approx 2\pi a + \frac{\pi\zeta_k^2}{2a}\{(ka)^2 - 1\}.$$

In this problem there is no gravitational term to consider, and it is the surface free energy per unit length, σA, which plays the role of the potential energy Φ of §8.1. The condition for marginal stability is $\partial^2 A/\partial\zeta_k^2 = 0$, equivalent to

$$k = k_c = \frac{1}{a}.$$

The cylinder is inherently unstable, as Plateau was the first to note, to any periodic deformation for which k is less than k_c, i.e. for which the wavelength λ is greater than $2\pi a$.

To find the rate of growth of a mode for which $k < k_c$ one may follow the routine procedure outlined in §8.2. Provided that the viscosity of the liquid may be neglected, i.e. provided that potential theory may be employed, it is not difficult to calculate the fluid velocity $\boldsymbol{u}\{x, r\}$ associated with rate of change of ζ_k. It is described by a flow potential ϕ which is a solution of Laplace's equation proportional to $\cos(kx) f\{r\}(\partial\zeta_k/\partial t)$; the function $f\{r\}$ involves Bessel functions. Hence the constant of proportionality relating the fluid's mean kinetic energy per unit length to $(\partial\zeta_k/\partial t)^2$ may be found, and the equation of motion relating $\partial^2\zeta_k/\partial t^2$ to ζ_k follows immediately. According to Rayleigh, s_k, which is zero where $k = k_c$, reaches a maximum where $k = 0{\cdot}697k_c$ or where $\lambda = 9{\cdot}02a$, in reasonable agreement with Savart's observations. The 2% discrepancy, in the wrong direction to be due to viscosity, is attributable to experimental error.

When spiders spin their webs in the first place, and when glassblowers produce thin filaments by drawing out molten glass beads, viscosity is far from negligible. It reduces the maximum value of s_k and thereby prevents the threads from coalescing into droplets before they have had time to set.

8.4 The Saffman–Taylor instability

The Saffman–Taylor instability arises, or may arise, when two fluids of different viscosity are pushed by a pressure gradient through a Hele Shaw cell [§6.8] or allowed to drain through such a cell under their own weight. It would be of little practical importance were it not for the fact that creeping flow in a Hele Shaw cell is the two-dimensional analogue of creeping flow through a porous medium [§6.13]. Something very like the Saffman–Taylor instability frustrates attempts to extract, by pushing it out with pressurised water, the last traces of oil from oil wells. Theoretically, the instability has features in common with the Rayleigh–Taylor instability discussed in §8.2; it differs in that the equibrium state is a dynamic one, in which the interface between the two fluids is moving rather than stationary, but the analysis required is nevertheless distinctly similar.

Suppose the cell to be horizontal, in which case the effects of gravity may be ignored. Suppose it to be bounded by straight edges at $y = \pm\frac{1}{2}L$, and suppose there to be pressure gradients which are driving the fluid contents in the $+x$ direction with some uniform velocity U. In the equilibrium state whose stability we are to investigate, the interface between the two fluids is the straight line $x = Ut$. Where $x < Ut$, the viscosity is η'; where $x > Ut$, the viscosity is η. According to (6.47), the pressure gradients needed to maintain this motion are given in the two regions by

$$\frac{\partial p'}{\partial x} = -\frac{12\eta' U}{d^2}, \quad \frac{\partial p}{\partial x} = -\frac{12\eta U}{d^2},$$

where d is the thickness of the cell. The pressures p' and p are not necessarily equal at the interface, because the interface is liable to be curved in the vertical (z) direction. Provided that this curvature is constant, however, it does not affect the results of the analysis, so we may as well ignore it and write

$$p' = -\frac{12\eta' U}{d^2}(x - Ut) + p_o, \quad p = -\frac{12\eta U}{d^2}(x - Ut) + p_o,$$

for the equilibrium state, where p_o does not depend upon x.

Now suppose that the interface is perturbed, in such a way that at time t it lies at $x = X$, where

$$X = Ut + \zeta_k \, e^{iky}.$$

There must be some corresponding perturbation in p' and p, and it must have the same periodicity in the y direction. However, p' and p obey Laplace's equation in two dimensions [§6.8], so any perturbing term which varies like $\exp(iky)$ must vary like $\exp(\pm kx)$ [(5.12)]. Since the perturbation cannot affect the pressure at large distances from the interface, the perturbed pressures presumably have the form

$$p' = -\frac{12\eta' U}{d^2}(x - Ut) + p_o + A' e^{k(x-Ut)} e^{iky}$$

$$p = -\frac{12\eta U}{d^2}(x - Ut) + p_o + A e^{-k(x-Ut)} e^{iky}$$

when k is positive, where the coefficients A' and A are to be determined by reference to the boundary conditions at the interface.

These boundary conditions, applicable in each case at $x = X$, and linearised by omission of terms which are of higher than first order in A or ζ_k are as follows.

(i) $$\langle u' \rangle_x = \langle u \rangle_x = \frac{\partial X}{\partial t},$$

where $\langle \boldsymbol{u} \rangle$ is the mean velocity described by (6.47), or

$$-\frac{d^2}{12\eta'}\frac{\partial p'}{\partial x} = -\frac{d^2}{12\eta}\frac{\partial p}{\partial x} = U + \frac{\partial \zeta_k}{\partial t} e^{iky}.$$

To first order this corresponds to

$$-\frac{d^2 k}{12\eta'} A' = -\frac{d^2 k}{12\eta} A = \frac{\partial \zeta_k}{\partial t} e^{iky}. \tag{8.7}$$

(ii) $$p' - p = -\sigma \frac{\partial^2 X}{\partial y^2} = \sigma k^2 \zeta_k e^{iky},$$

where σ is the interfacial surface tension. To first order this corresponds to

$$A' - A = \left\{ \frac{12U}{d^2}(\eta' - \eta) + \sigma k^2 \right\} \zeta_k \exp(iky). \tag{8.8}$$

It is a trivial exercise to eliminate A' and A from (8.7) and (8.8), and so to obtain the result

$$s_k = \frac{1}{\zeta_k}\frac{\partial \zeta_k}{\partial t} = \frac{1}{\eta' + \eta}\left\{ -U(\eta' - \eta)k - \frac{\sigma d^2 k^3}{12} \right\}. \tag{8.9}$$

Thus if $\eta < \eta'$ the interface is stable for all k. When $\eta > \eta'$, however, i.e. when a viscous fluid is being displaced by a less viscous one, it is marginally stable with respect to a perturbation for which $k = k_c$, where

$$k_c^2 = \frac{12U(\eta - \eta')}{\sigma d^2},$$

and it is unstable with respect to perturbations for which $0 < k < k_c$. The perturbations which grow fastest (i.e. for which s_k is a maximum) have $k = k_c/\sqrt{3}$, i.e. a wavelength

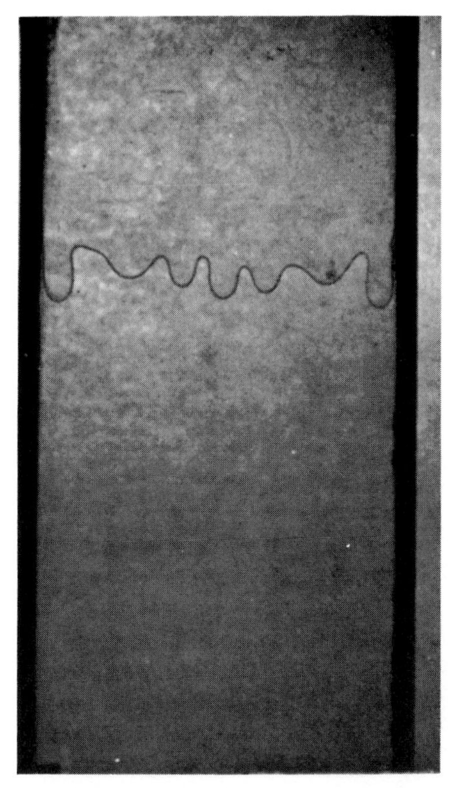

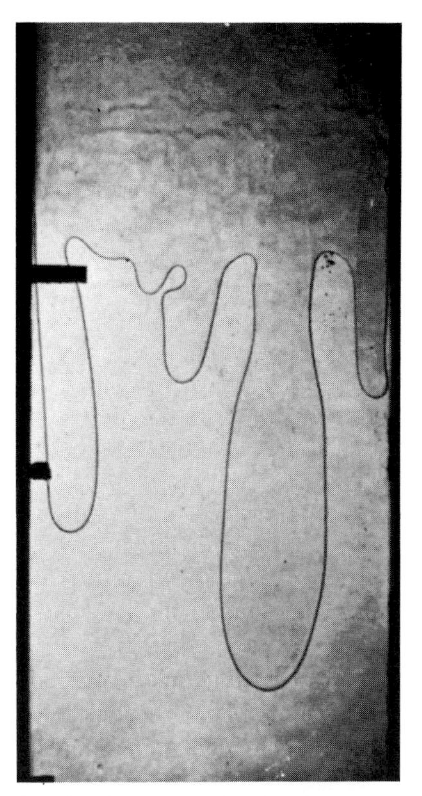

Figure 8.4 Air under pressure displacing glycerine in a vertical Hele Shaw cell; the mean downwards velocity of the two fluids is about 1 mm s^{-1}: (a) an early stage in development of the instability; (b) a later stage in a different experiment.
[From P. G. Saffman & G. I. Taylor, *Proc. Roy. Soc. A*, **245**, 312, 1958.]

$$\lambda = \pi d \sqrt{\frac{\sigma}{U(\eta - \eta')}}. \tag{8.10}$$

The smallest value of k which is consistent with the boundary conditions at the sides of the cell, where $y = \pm\frac{1}{2}L$, is π/L, and if the cell is so narrow, or if U is so small, that this exceeds k_c then no instabilities can be observed. In the experiments conducted by Saffman and Taylor, however, in which air was used to displace glycerine through a cell whose thickness was about 1 mm, L was 12 cm and the wavelength λ predicted by (8.10) was normally a bit less than 2 cm. Thus they expected to see, when the pressure gradient was first applied, six or seven corrugations develop in the interface over the full width of the cell, and so they did; one of their photographs is reproduced as fig. 8.4(a).

When the corrugations are no longer very small they do not all grow at the same rate, as is shown by fig. 8.4(b). One of the advancing *fingers* of the less viscous fluid tends to get ahead, whereupon it expands sideways and, by doing so, slows

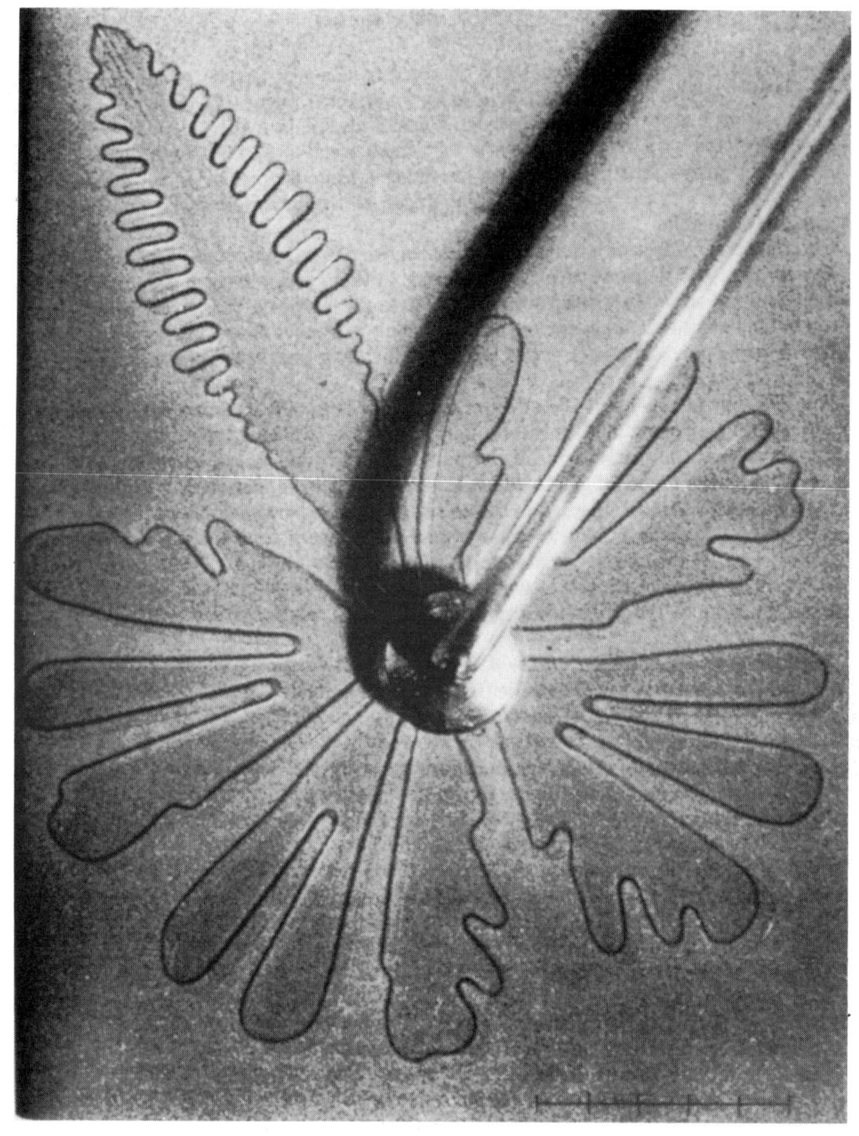

Figure 8.5 Fingers of less viscous fluid spreading outwards from a central source in a Hele Shaw cell; the dendritic finger is carrying with it at its tip a small bubble of gas.
[From Y. Couder, O. Cardoso, D. Dupuy, P. Tavernier & W. Thom, [*Europhys. Lett.*, **2**, 437, 1986.]]

down the advance of its competitors. In due course only a single finger survives. It continues to advance at its tip, but it appears to stop expanding sideways when its width reaches half the width of the cell. The tip has a characteristically rounded shape, which Saffman and Taylor were able to explain.

Figure 8.6 Fingers of less viscous fluid spreading outwards from a central source in a Hele Shaw cell; in this case a dendritic character is imposed on the fingers by the cell walls, which are glass plates etched along lines that form a triangular lattice.
[Photograph by E. Ben-Jacob, from D. A. Kessler, J. Koplik & H. Levine, *Adv. Phys.*, **37**, 255, 1988.]

Are the fingers stable and, if not, how do they split up? This question has proved in recent years to be of much greater complexity and interest than Saffman and Taylor could have guessed when their paper on this subject was published in 1958. A partial answer is provided by the two remarkable photographs of fingers spreading radially from a central source which are reproduced in figs. 8.5 and 8.6. The first one shows a number of fingers which are splitting in an irregular and unsurprising way, and one finger which has developed side branches of astonishing regularity; it differs from the others by having a defect at its tip, in the shape of a small gas bubble which has accidentally entered the apparatus and become entrained in the flow. The second photograph shows an even more regular pattern of dendritic growth; this one was taken using a Hele Shaw cell whose surfaces had grooves in them at angles of $\pi/6$ to one another, favouring flow in those directions. Considerable progress has been made towards a full understanding of the subtle effects which these photographs reveal, and the theoretical ideas involved contribute to our understanding of what is called *pattern formation* in many contexts that have little, if anything, to do with fluid dynamics. The resemblance between the pattern revealed by fig. 8.6 and the patterns associated with snowflakes will not escape the reader.

8.5 Thermal convection – an introduction

As every physics student knows, a warm body which is immersed in fluid is liable to lose heat by convection. The convection currents which carry the heat away may be *forced*: they may be driven by a fan, for example, or they may be draughts of air from an open window. Alternatively, the currents may have their origin in the temperature gradients which the warm body imposes on the fluid, in which case they are said to be *natural*. Forced convection is of some practical importance but is of rather little physical interest, and we shall consider only natural convection in this book; it is discussed at a qualitative level in the present section, and the relatively simple case of convection in a vertical slot which is heated from one side is treated analytically – though rather sketchily – in §8.6. Since both sections are concerned almost exclusively with the heat loss aspect of convection and since neither of them refers to instabilities they are perhaps out of place in the present chapter, but they will serve to introduce the reader to the case discussed in §8.7, of fluid in a horizontal slot which is heated from below; in that case instabilities breed in profusion.

Whatever the shape of the warm body, and whatever the boundary conditions to which the surrounding fluid is subject, three distinct regimes of natural convection may normally be distinguished. One passes from one regime to the next on increasing the temperature excess at the surface of the warm body, though a considerable increase may well be needed to effect each transition completely. This temperature excess, incidentally, is assumed in what follows to be independent of position and, except while it is being reset, to be independent of time as well. It is denoted by θ_o.

When θ_o is very small and the convection currents are very feeble it is a reasonable first approximation to ignore their effect upon the temperature distribution in the fluid, and because this distribution is effectively determined by thermal conduction alone one is then in what is called the *conduction regime*. As θ_o is increased the convection currents get stronger and the temperature distribution begins to change. The change tends to be such that the spatial domain within which the temperature excess of the fluid is comparable with θ_o is increasingly localised in the neighbourhood of the warm body [see §8.6 for an example]. The convection currents then become localised too, and at large enough values of θ_o one enters the *boundary layer regime*, where the currents are localised to such a degree that the presence of bodies other than the warm one is of little significance. At larger values still there is a *turbulent boundary layer regime*; the name needs no explanation.

Many elementary discussions of this topic convey the impression that convection and conduction are two separable mechanisms for heat loss, which can be thought of as functioning simultaneously but independently of one another. They are misleading in this respect, except perhaps in the conduction regime. In that regime it is legitimate [§8.6] to expand the total heat loss Q in powers of temperature excess and to ignore all but the first two terms, as follows

$$Q = c'\theta_o + c''\theta_o^2,$$

where c' and c'' are coefficients which depend upon the size and shape of the body and on the nature of the fluid. Since the first term in this expansion describes the heat loss due to conduction alone that would occur if convection could somehow be suppressed, the second term can safely be attributed to convection. But the approximations implicit in any analytical treatment of the conduction regime are valid only so long as the second term is much smaller than the first, and this regime is therefore of little interest from a practical point of view. Outside it, convection and conduction are certainly not separable. In a sense all the heat loss is due to conduction, for it is only by conduction that heat can pass from the warm body to the surrounding fluid; however, the convection currents affect the temperature gradients adjacent to the body and thereby affect – often very significantly – the rate of heat transfer. Series expansions for Q are then quite inappropriate. It turns out that in the boundary layer regime Q is normally proportional to $\theta_o^{5/4}$, while in the turbulent boundary layer regime it tends to vary like $\theta_o^{4/3}$.

Incidentally, elementary texts are misleading in another respect: they state *Newton's law of cooling* as though it were established truth. This celebrated law, which says that heat losses are proportional to temperature excess (θ_o), is adequately obeyed when there is no possibility of convection, even in circumstances where the warm body may lose heat by radiation as well as conduction. It may also be obeyed in a very draughty room where *forced* convection constitutes the principal mechanism for heat loss. But where natural convection is important it does not apply at all.

Exact analytical solution – and, indeed, exact numerical simulation – of even the simplest problems involving thermal convection is so prohibitively difficult that approximations are essential, and the approximations upon which most solutions are based are summed up in the so-called *Boussinesq equations*. The first of these is a version of the Navier–Stokes equation [(6.25)]. That equation applies as it stands to an effectively incompressible fluid of uniform density ρ, and the gradient of excess pressure which features in it is

$$\nabla p^* = \nabla p + \rho g \nabla z.$$

As long as the density is completely uniform, it is this gradient which vanishes if the fluid is at rest in equilibrium and which otherwise causes the fluid to accelerate. But suppose the fluid to be at some temperature $T + \theta\{\boldsymbol{x}\}$, where T is uniform but θ is not. In that case its local density is given to a good approximation, if ρ is the density characteristic of temperature T and if θ is reasonably small compared with T, by $\rho(1 - \alpha\theta)$, where

$$\alpha = -\frac{1}{\rho}\left(\frac{\partial \rho}{\partial T}\right)_p$$

is the thermal expansion coefficient of the fluid at this temperature. The gradient which must vanish everywhere if the fluid is to be at rest in equilibrium, and which must otherwise determine the fluid's acceleration, is now

$$\nabla p + \rho(1 - \alpha\theta) g \nabla z = \nabla p^* - \alpha\rho g \theta \nabla z.$$

That suggests that the Navier–Stokes equation be rewritten, on the assumption that convective flow is normally steady flow for which $\partial \boldsymbol{u}/\partial t$ is zero, as

$$-\nabla p^* + \alpha\rho g \theta \nabla z \approx \eta \nabla \wedge \boldsymbol{\Omega} + \rho(\boldsymbol{u}\cdot\nabla)\boldsymbol{u}, \tag{8.11a}$$

or, since ρ is a constant in this equation, as

$$-\nabla\left(\frac{p^*}{\rho}\right) + \alpha g \theta \nabla z \approx \nu \nabla \wedge \boldsymbol{\Omega} + (\boldsymbol{u}\cdot\nabla)\boldsymbol{u}, \tag{8.11b}$$

where $\nu = \eta/\rho$ is the kinematic viscosity or *viscous diffusivity*. This is the first Boussinesq equation. It is approximate on three grounds. Firstly, because it involves an assumption that θ is everywhere small; we cannot hope to express the variations of density in terms of a single expansion coefficient unless this is the case. Secondly, because the density which appears in the second term on the right-hand side of (8.11a) – the inertial as opposed to the viscous term – should really be the local density $\rho(1 - \alpha\theta)$ rather than ρ. And thirdly because the viscous term includes no allowance for the possibility that not only the density of the fluid but also its viscosity may vary from place to place.

The second Boussinesq equation is

$$\nabla\cdot\boldsymbol{u} \approx 0; \tag{8.12}$$

this is of course the continuity equation for a fluid of uniform density, and it is clearly approximate when temperature, and therefore density, are in fact non-uniform. The third is

$$\kappa \nabla^2 \theta \approx \rho c_p \frac{\mathrm{D}\theta}{\mathrm{D}t},$$

where c_p is specific heat per unit mass, which in terms of the *thermal diffusivity* $\chi = \kappa/\rho c_p$ becomes

$$\chi\nabla^2\theta \approx (\boldsymbol{u}\cdot\boldsymbol{\nabla})\theta \tag{8.13}$$

if the flow is steady. Its left-hand side describes, in circumstances where the fluid's thermal conductivity κ is uniform though not otherwise, the rate at which heat is conducted into an element of fluid per unit volume, and the reasons for relating this to $\mathrm{D}T/\mathrm{D}t$ are obvious enough [§3.2]. It is approximate not only because κ, like ρ and η, is liable to vary from place to place when temperature is non-uniform, but also for two more subtle reasons. Firstly, it includes no allowance for the heat which may be generated inside each fluid element by viscous dissipation. Secondly, since pressure is not strictly constant in a fluid element which is convecting, the specific heat appropriate for use in the equation is not necessarily c_p.

The three Boussinesq equations may in principle be solved to find how p^*, $\boldsymbol{u}$ and θ vary with position for any given set of boundary conditions. The heat loss Q may then be calculated, by integrating the product of κ and the local temperature gradient over the whole surface of the warm body. In some geophysical and astrophysical applications of convection theory the approximations implicit in the equations break down, but they are normally adequate under laboratory conditions. They may of course be justified in particular cases by order-of-magnitude estimates of the neglected terms, based upon solution of the Boussinesq equations, but no general justification will be attempted here.

Simplified though the Boussinesq equations are they are often hard to solve, so dimensional analysis has a valuable part to play in the discussion of thermal convection. A brief discussion of heat losses from long horizontal cylinders immersed in otherwise stationary fluid will illustrate the power of such analysis and will serve to introduce the reader to some dimensionless quantities which will recur in later sections. The problem is a relatively simple one because, at any rate if the length of the cylinder is large enough for the convective flow to be treated as two-dimensional and if the external boundaries of the region occupied by fluid are sufficiently far away, it involves only one characteristic length – the radius of the cylinder, a. If q is the heat loss of the cylinder *per unit length along its axis*, we may expect q/κ to be uniquely determined – provided that the Boussinesq equations are indeed sufficient – by θ_o, a, αg, ν and χ. There is no need to include the temperature T in this list because it is relevant only in so far as it determines the fluid properties α, ν etc. There is no need to include ρ (except to the extent that it is a component of both diffusivities, ν and χ), because it does not occur at all in (8.12) or (8.13), and in (8.11) it occurs only as a scaling factor for the excess pressure p^*; the magnitude of p^* affects the force exerted on the cylinder by the convecting fluid but it does not affect the heat loss. Now the dimensions of the six relevant quantities may readily be shown to be:

q/κ	$[\Theta]$,
θ_o	$[\Theta]$,
a	$[L]$,
αg	$[L][T]^{-2}[\Theta]^{-1}$,
ν	$[L]^2[T]^{-1}$,
χ	$[L]^2[T]^{-1}$.

Following principles enunciated in §1.5, we need to construct dimensionless combinations of these, and three such combinations are:

the *Nusselt Number*,

$$Nu = \frac{q}{\kappa\theta_o}, \tag{8.14}$$

the *Rayleigh Number*,

$$Ra = \frac{\alpha g\theta_o a^3}{\chi\nu}, \tag{8.15}$$

and the *Prandtl Number*,

$$Pr = \frac{\nu}{\chi} = \frac{\eta c_p}{\kappa}. \tag{8.16}$$

These three are clearly independent of one another, but no other independent possibilities exist; for example,

the *Grashof Number*,

$$Gr = \frac{\alpha g\theta_o a^3}{\nu^2}, \tag{8.17}$$

can be expressed as Ra/Pr. It follows that heat losses from horizontal cylinders must, to the extent that the Boussinesq equations are applicable, obey a functional relation of the form

$$Nu = f\{Ra, Pr\}. \tag{8.18}$$

The Prandtl Number differs from the Rayleigh and Nusselt Numbers in that it represents an intrinsic property of the fluid and has nothing to do with the geometry of the particular problem in hand. For air (and other divalent gases) it is close to 0·73 whatever the pressure and temperature [(3.17) and (A.35)]; for water it is about 6; and for more viscous non-conducting liquids it can be 10^3 or more. Only for liquid metals is it significantly less than unity, and for liquid mercury in particular it is about 0·025. It follows that one may test the validity of (8.18), and hence subject the Boussinesq equations to an indirect test, by measuring q for horizontal cylinders of various radii in air at various pressures, perhaps using various values of θ_o, and by subsequently plotting Nu *versus* Ra; the

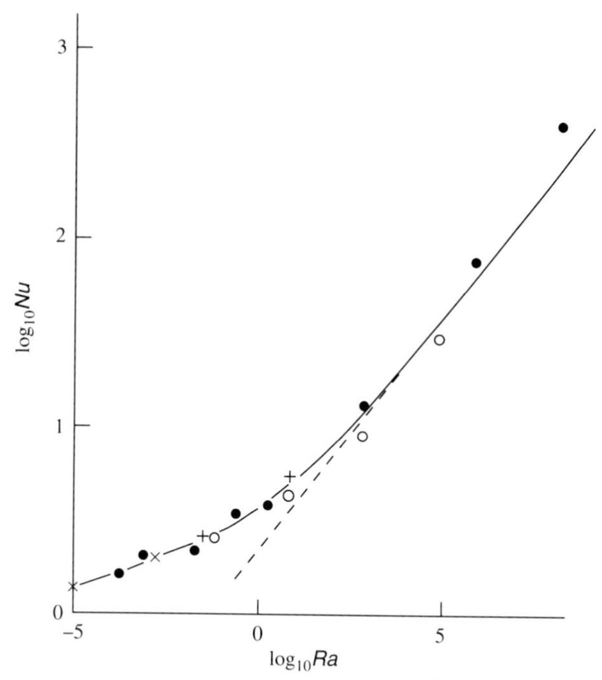

Figure 8.7 Heat loss due to natural convection from long horizontal cylinders: a log–log plot of Nusselt Number *versus* Rayleigh Number. Circles refer to experiments in which the fluid was air at room temperature with $\theta_o = 100$ K; some of these results (filled circles) were obtained at atmospheric pressure but using cylinders whose radii varied from about 0·015 to 150 mm, while others (empty circles) were obtained using a fixed value for a but pressures which varied between 0·1 and 100 atmospheres. The crosses refer to experiments using CCl_4 (+) or glycerine (×) rather than air. The curve drawn through the points extrapolates to a straight line which corresponds to $Nu \propto Ra^{1/4}$.
[Based on results measured by A. H. Davis or derived by him from earlier sources, *Phil. Mag.*, **43**, 329, 1922 and **44**, 920, 1922.]

points should all lie on a single curve. A selection of results referring to cylinders surrounded by air at room temperature, which between them demonstrate the effects of changing a by a factor of 10^4 and p by a factor of 10^3, are represented by filled and open circles in fig. 8.7. In each of the measurements which these points describe, θ_o was about 100 K, and that is scarcely small enough in relation to T (about 300 K) to make the Boussinesq approximations convincing. The author of the paper from which the data were taken went a step beyond them, therefore, by using for η, κ, ρ and α in his analysis the values appropriate for air at a temperature $T + \frac{1}{2}\theta_o$ rather than T. Considering the *ad hoc* nature of this correction and the many experimental errors which may have affected the raw results, the scatter of the points in fig. 8.7 is remarkably small. Note that the slope of the curve through them seems to be about $\frac{1}{4}$ for values of Ra in the range between 10^3 and 10^8 or more; that slope corresponds, since the graph is a logarithmic one, to $q \propto \theta_o^{5/4}$, the result expected in the boundary layer regime.

A few points representing results obtained using liquids instead of air are included in fig. 8.7; they are plotted as crosses rather than circles. Despite the fact that the Prandtl Numbers for both liquids – glycerine in particular – are distinctly larger than 0·73, these crosses seem to lie on very much the same curve as the circles. The implication – confirmed to some extent by fig. 8.10 below – is that at constant Ra the dependence of Nu on Pr, at any rate for $Pr > 0{\cdot}73$, is rather weak. However, the value of Ra at which boundary layer convection becomes turbulent depends upon Pr [§8.6], and it is arguable that in the turbulent boundary layer regime the Grashof Number has more physical significance than the Rayleigh Number and should be used in place of it.

The problem to be analysed in the following section involves *two* characteristic lengths, the height D and the width d of a vertical slot. A fourth independent dimensionless combination of variables can therefore be constructed, namely the *aspect ratio* of the slot,

$$H = \frac{D}{d}. \tag{8.19}$$

Moreover, it becomes necessary not only to distinguish two possible Rayleigh Numbers,

$$Ra_d = \frac{\alpha g \theta_o d^3}{\chi \nu} \tag{8.20}$$

and

$$Ra_D = \frac{\alpha g \theta_o D^3}{\chi \nu} = Ra_d H^3, \tag{8.21}$$

but also to think again about how the Nusselt Number is best defined. Some authors treat the Nusselt Number in contexts such as this as a local property, describing heat losses per unit area which are liable to vary from point to point over the surface of the warm body. In what follows, however, it describes an integrated heat loss Q, and it is normalised so that in the absence of convection it would be unity. The end result, as applied to a vertical slot, of the sort of dimensional analysis we have applied above to a horizontal cylinder would clearly be

$$Nu = f\{Ra_d, Pr, H\}, \tag{8.22}$$

or equivalent forms such as

$$Nu = f\{Ra_D, Pr, H\}, \quad Nu = f\{Gr_d, Pr, H\} \quad \text{etc.}$$

if preferred.

8.6 Convection in an open vertical slot

The slot we are now to consider may be thought of as the space between a vertical wall at uniform temperature T and a vertical panel radiator, of the sort used for central heating, at uniform temperature $T + \theta_o$. The slot is open in the sense that fluid – air, say – is free to enter it at the bottom from a large reservoir – the room – at temperature T and free to return to this reservoir at the top. The distance between the wall and the inner surface of the radiator is d, and the height of the radiator is D. The third dimension of the slot may be regarded as so large that the convective flow within it is essentially two-dimensional; that is to say, if we choose axes such that the wall occupies the whole of the $x = 0$ plane while the radiator occupies the $x = d$ plane between $z = 0$ and $z = D$, then u_y is zero while u_z and u_x do not depend upon y. On the assumption that $H (= D/d)$ is large compared with unity, we may surely ignore u_x in comparison with u_z except where fluid is converging upon the slot near $z = 0$, and likewise ignore $\partial u_x/\partial z$ in comparison with $\partial u_z/\partial x$. It does not follow, however, that we can ignore $u_x \partial u_z/\partial x$ in comparison with $u_z \partial u_z/\partial z$, or $u_x \partial T/\partial x$ in comparison with $u_z \partial T/\partial z$, since rates of change are likely to be larger with respect to x than with respect to z. For the same reason, however, we may ignore $\partial^2 u_z/\partial z^2$ and $\partial^2 T/\partial z^2$ in comparison with $\partial^2 u_z/\partial x^2$ and $\partial^2 T/\partial x^2$ respectively. Finally, we may treat $\partial p^*/\partial z$ as negligible within the slot, and likewise $\partial p^*/\partial x$. The justification for neglect of $\partial p^*/\partial z$ lies primarily in the fact that p^* is uniform in the reservoir and therefore the same above the radiator as it is below; a drop in p^* is needed to accelerate the fluid where it converges on the slot, but the end correction needed to take account of this drop is negligible provided $H \gg 1$. These further approximations allow one to reduce the Boussinesq equations to

$$\nu \frac{\partial^2 u_z}{\partial x^2} + \alpha g \theta \approx u_x \frac{\partial u_z}{\partial x} + u_z \frac{\partial u_z}{\partial z}, \tag{8.23}$$

$$\frac{\partial u_x}{\partial x} + \frac{\partial u_z}{\partial z} \approx 0, \tag{8.24}$$

$$\chi \frac{\partial^2 \theta}{\partial x^2} \approx u_x \frac{\partial \theta}{\partial x} + u_z \frac{\partial \theta}{\partial z}. \tag{8.25}$$

In the *conduction regime* [§8.5] we may further simplify (8.23) and (8.25) by ignoring all the terms on their right-hand sides, on the grounds that these are of second order in θ_o. The solution to (8.25) is then trivial, namely

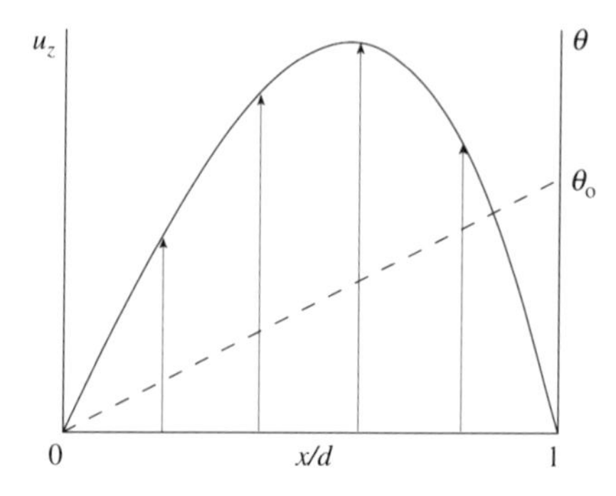

Figure 8.8 Natural convection at say half-height ($z/D \approx 0{\cdot}5$) in a vertical slot heated from one side, in the conduction regime. The excess temperature varies in a linear manner across the width of the slot (broken line) and (8.27) describes the velocity profile (full curve).

$$\theta \approx \frac{\theta_o x}{d}, \tag{8.26}$$

so that (8.23) becomes

$$\nu \frac{\partial^2 u_z}{\partial x^2} \approx -\frac{\alpha g \theta_o x}{d},$$

which integrates, in view of the no-slip boundary condition at $x = 0$ and $x = d$, to

$$u_z \approx \frac{\alpha g \theta_o}{6\nu d}(xd^2 - x^3). \tag{8.27}$$

The velocity profile has the form sketched in fig. 8.8.

The rising fluid transfers heat to the reservoir as it leaves the slot at the top, at a rate which is given per unit breadth in the y direction by

$$\int_0^d \rho c_p \theta u_z \mathrm{d}x = \frac{\rho c_p \theta_o}{d} \int_0^d x u_z \mathrm{d}x = \frac{\rho c_p \alpha g \theta_o^2 d^3}{45\nu}. \tag{8.28}$$

It acquires this heat soon after entering the slot at the bottom, where it warms up from the temperature of the reservoir T to $T + \theta_o x/d$. Thus over a range of z between 0 and l, say, where l must be very small compared with D to justify the approximations we have made in calculating u_z but may be large compared with d, θ cannot vary in a linear fashion with x; the rate per unit area at which the warm panel delivers heat to the fluid by conduction must be greater than the rate at which the wall extracts it, which implies the existence in this limited range of z of a *concave* temperature profile, such that

$$\left(\frac{\partial\theta}{\partial x}\right)_{x=d} > \frac{\theta_o}{d} > \left(\frac{\partial\theta}{\partial x}\right)_{x=0}.$$

For the purposes of a quick estimate we may plausibly assume that over much of this range

$$\left(\frac{\partial\theta}{\partial x}\right)_{x=d} - \frac{\theta_o}{d} \approx \frac{\theta_o}{d} - \left(\frac{\partial\theta}{\partial x}\right)_{x=0},$$

which implies that of the heat which is carried away by the rising fluid only about two-thirds represents an additional heat loss from the panel radiator; the fluid obtains the other one-third by way of commission, as it were, on the heat that, in the absence of convection, is transferred continuously by conduction from the radiator to the wall. The additional heat loss due to convection, $q - q_o$, is therefore something like

$$10^{-2}\,\frac{\rho c_p \alpha g \theta_o^2 d^3}{\nu}$$

per unit breadth in the y direction [(8.28)]. Since the heat loss from the inner surface of the radiator (its outer surface may be ignored for simplicity) is

$$q_o = \frac{\kappa\theta_o D}{d}$$

in the absence of convection, we may expect the Nusselt Number defined at the end of §8.5 to be given approximately in the conduction regime by

$$Nu = \frac{q}{q_o} \approx 1 + 10^{-2}\,\frac{\rho c_p \alpha g \theta_o d^4}{\kappa\nu D},$$

i.e. by

$$Nu \approx 1 + 10^{-2}\,\frac{Ra_d}{H}. \tag{8.29}$$

Corrections of order $1/H^2$ may well be needed to this expression, to take into account sundry complications in the entry zone, near $z = 0$, which have been passed over above on the grounds that $H \gg 1$, but with or without such corrections it conforms, as it must do, to the pattern of (8.22), predicted on the basis of dimensional analysis alone. Note, however, that it does not involve Pr. In so far as we expect the conduction regime to terminate when $Nu - 1$ is no longer small compared with unity, we may expect it to terminate where Ra_d is of order $1/H$. Domestic radiators operate well outside this range in practice, needless to say.

As θ_o and hence Ra_d increase in magnitude, the length l over which the temperature profile within the slot is concave inevitably increases, until in due

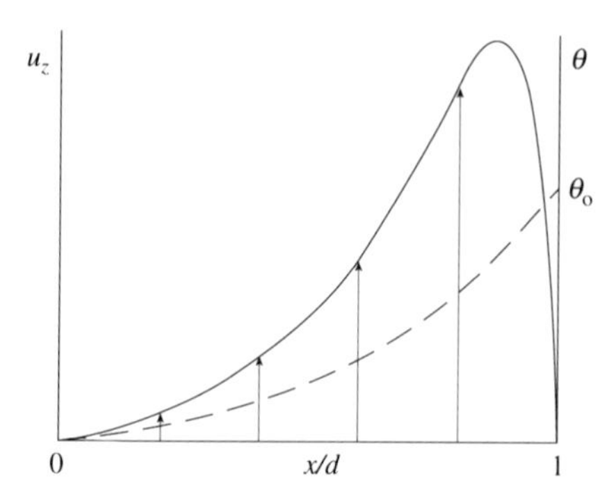

Figure 8.9 Natural convection at say half-height ($z/D \approx 0{\cdot}5$) in a vertical slot heated from one side, as the boundary layer regime is approached. The temperature profile is now concave [cf. fig. 8.8] and the peak in the velocity profile has shifted to the right. Both curves here are schematic only, and the vertical scales are different from the vertical scales of fig. 8.8; θ_o is larger, and so is the maximum value of u_z.

course it becomes comparable with the full height D. At that stage the temperature excess and vertical velocity of the fluid in the middle of the slot, i.e. near $z = \frac{1}{2}D$, vary with x in the manner suggested by fig. 8.9; in comparison with the velocity profile shown in fig. 8.8, the peak in u_z is closer to the warm surface. Further increases in θ_o enhance this tendency, until it is only in a thin layer that the convection current is appreciable. For large enough θ_o, the thickness δ of the convecting layer, even near $z = D$ where it is thickest, becomes small compared with d, and the magnitude of d becomes immaterial; one has then reached the *boundary layer regime*. In this limit the heat loss from the inner surface of the radiator may be analysed as though it were in contact with an effectively infinite body of fluid at temperature T.

It may be shown by substitution – though not without effort – that there exist solutions to equations (8.23)–(8.25) which describe this limit, such that

$$\theta = \theta_o F\{x^*\}, \quad u_z = \sqrt{\alpha g \theta_o z}\, G\{x^*\}, \tag{8.30}$$

where x^* is a dimensionless coordinate related to x by the equation

$$x^* = \sqrt[4]{\frac{\alpha g \theta_o}{\nu^2 z}}\,(x - d), \tag{8.31}$$

and it follows that δ increases with height z like $z^{1/4}$ and decreases with temperature excess like $\theta_o^{-1/4}$. The functions F and G obey two dimensionless non-linear ordinary differential equations which are coupled, and they are subject to the boundary conditions

$$F\{-\infty\} = G\{-\infty\} = 0, \quad F\{0\} = 1, \quad G\{0\} = 0.$$

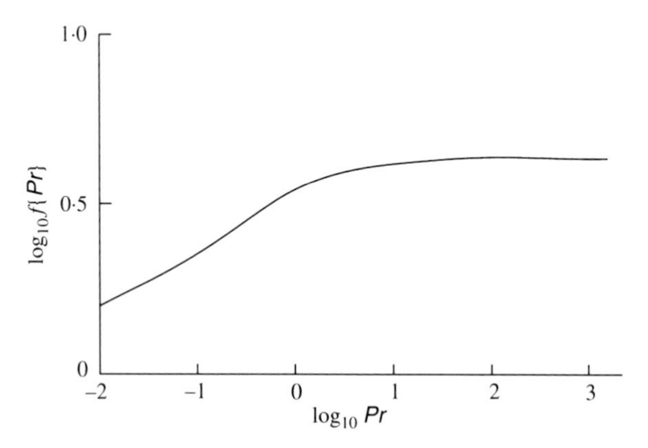

Figure 8.10 Natural convection near a heated vertical plate: dependence of *Nu* on *Pr* in the boundary layer regime.
[Based on an approximate formula quoted on p. 643 of *Modern Developments in Fluid Dynamics*, ed. S. Goldstein, Clarendon Press, Oxford, 1938.]

When (8.30) and (8.31) are applicable, the rate of loss of heat per unit breadth from the inner surface of the radiator may be expressed in terms of the derivative F' of F as

$$\kappa \int_0^D \left(\frac{\partial \theta}{\partial x}\right)_{x=d} \mathrm{d}z = \frac{4}{3}\,\kappa\theta_o \sqrt[4]{\frac{\alpha g \theta_o D^3}{\nu^2}}\, F'\{0\}.$$

Note its proportionality to $\theta_o^{5/4}$.

The equations for F and G may be solved, by series methods or numerically, to find $F'\{0\}$, and the answer turns out to depend to some extent upon the magnitude of the Prandtl Number, which occurs in one of the terms which couple the equations together. When the fluid surrounding the radiator is air, with $Pr \approx 0{\cdot}73$, the Nusselt Number as previously defined is found to be

$$Nu = \frac{4}{3}\frac{d}{D}\sqrt[4]{\frac{\alpha g \theta_o D^3}{\nu^2}}\, F'\{0\} \approx 0{\cdot}48\,\frac{1}{H}\sqrt[4]{\frac{Ra_D}{Pr}}$$

or

$$Nu \approx 0{\cdot}52 \sqrt[4]{\frac{Ra_D}{H^4}}, \tag{8.32}$$

but more generally one has

$$Nu = f\{Pr\} \sqrt[4]{\frac{Ra_D}{H^4}}. \tag{8.33}$$

The curve plotted in fig. 8.10 shows how $f\{Pr\}$ is thought to vary with Pr.

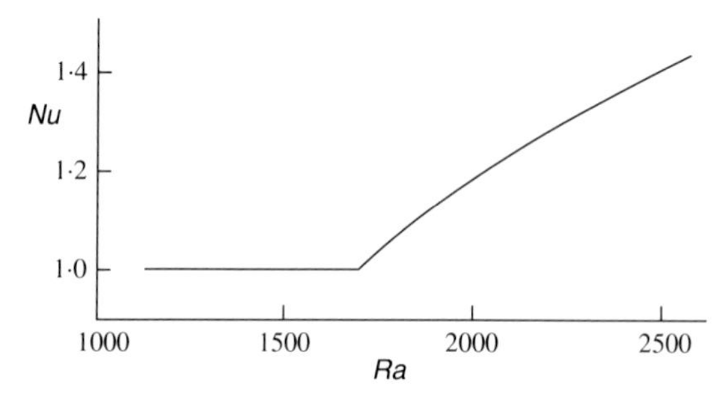

Figure 8.11 Natural convection in a horizontal slot heated from below: dependence of Nu on Ra near the onset of Bénard convection.

Turbulence in boundary layers normally begins when the Reynolds Number reaches a certain critical value, and in the context of thermal convection in the boundary layer regime it is appropriate to define the Reynolds Number as $\nu^{-1}(u_{z,\max}\delta)_{z=D}$, where $u_{z,\max}$ is the maximum velocity within the convecting boundary layer. It may be seen from (8.30) and (8.31) that the Reynolds Number thus defined is proportional to $\sqrt[4]{Ra_D/Pr}$. Hence the onset of turbulence is signalled by the magnitude of the Grashof Number, $Gr_D = Ra_D/Pr$, rather than of Ra_D. Experiments show that in the *turbulent boundary layer regime* convective heat losses tend to vary, as was stated in §8.5, like $\theta_o^{4/3}$ rather than like $\theta_o^{5/4}$.

8.7 The Rayleigh–Bénard instability

Fluid which lies between two parallel plates at different temperatures undergoes convection however small the temperature difference, as long as the plates are vertical. That is not true of fluid between horizontal plates. Let the two horizontal plates occupy the planes $z = \pm\frac{1}{2}d$ and be at uniform temperatures of $T \mp \frac{1}{2}\theta_o$ respectively. Then if the lower plate is the cooler one, i.e. if $\theta_o < 0$, the fluid may remain at rest in stable equilibrium for all values of θ_o; the temperature of the fluid in that case increases linearly with height, and the heat transfer between the plates, due to conduction alone, may be described by a Nusselt Number [defined as in §8.6] which is exactly unity. The fluid may likewise remain at rest if the lower plate is the hotter one, i.e. if $\theta_o > 0$, but only if the Rayleigh Number Ra_d or Ra (there is no need in the context of Bénard convection for a distinguishing suffix) is less than a certain critical value. At this critical Rayleigh Number, Ra_c, the so-called *Rayleigh–Bénard instability* arises, and above it a cellular pattern of convection currents develops, often associated with the name of Bénard alone. The effect of this on the rate of heat transfer from the lower plate to the upper one is shown in fig. 8.11.

The fluid near the bottom of a layer which is heated from below is inevitably hotter than the fluid near the top. It is therefore less dense than the fluid near the top, at any rate in laboratory experiments where the variation of density with height associated with the Earth's gravitational field is negligible. Thus any pattern of circulation which causes fluid from the bottom of the layer to change places with an equal volume of fluid from the top may be expected to release gravitational potential energy, and the circulation is liable to occur spontaneously provided that the energy available for release is more than sufficient to pay for the viscous dissipation which inevitably accompanies it. The cost of the process tends to zero as the time taken to effect it tends to infinity, but the available energy tends to zero too, because as the hotter fluid rises it loses heat by thermal conduction to the surrounding fluid and therefore gets more dense, while as the colder fluid falls it gains heat and gets less dense. Thus whether the process occurs, and if so at what rate, depends on the thermal conductivity of the fluid as well as on its viscosity; it also depends on the size of the two bodies of fluid relative to the thickness of the layer.

At $Ra = Ra_c$ the fluid layer is marginally stable, which means [§8.1] that there exists a mode of circulation which is independent of time, whose amplitude, expressed by some characteristic velocity u, is irrelevant as long as it is small. In attempting to analyse this mode and to calculate Ra_c, we may rely upon Boussinesq equations from which all terms which are time-dependent, or of higher order in u than the first, have been removed. Moreover, the simplifying assumption that the circulation pattern is two-dimensional involves no significant loss of generality at this stage. Granted that assumption, the relevant linearised equations for a fluid layer in which the temperature[1] gradient $\partial\theta/\partial z$ is $-\theta_0/d$ in the limit of zero u turn out to be

$$\nu\left(\frac{\partial^2 u_x}{\partial x^2}+\frac{\partial^2 u_x}{\partial z^2}\right)-\frac{1}{\rho}\frac{\partial p^*}{\partial x}\approx 0, \tag{8.34}$$

$$\nu\left(\frac{\partial^2 u_z}{\partial x^2}+\frac{\partial^2 u_z}{\partial z^2}\right)-\frac{1}{\rho}\frac{\partial p^*}{\partial z}+\alpha g\tau\approx 0, \tag{8.35}$$

$$\frac{\partial u_x}{\partial x}+\frac{\partial u_z}{\partial z}\approx 0, \tag{8.36}$$

$$\chi\left(\frac{\partial^2 \tau}{\partial x^2}+\frac{\partial^2 \tau}{\partial z^2}\right)+\frac{\theta_0 u_z}{d}\approx 0. \tag{8.37}$$

Equations (8.34) and (8.35) follow in a straightforward way from (8.11b); (8.36) is equivalent to (8.12); and (8.37) is the linearised version of (8.13). Note that in this problem [cf. the problem discussed in §8.6] p^* cannot be treated as uniform and

[1] Here θ represents, as in §§8.5 and 8.6, the difference between local temperature and mean temperature T. The τ which features in (8.37) is $\theta+\theta_0 z/d$, i.e. it represents the increase in local temperature for which, to first order in u, the circulation is responsible.

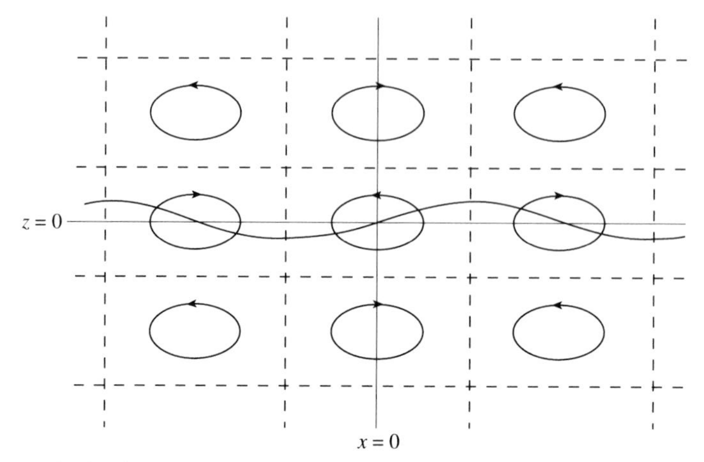

Figure 8.12 Cross-section through a body of fluid heated from below, showing a periodic mode of convection. The rectangular cells are each π/k_z in height and π/k_x in width, and in the case depicted here their aspect ratio is close to the favoured value of $1/\sqrt{2}$. The sinusoidal curve through the origin suggests how the upwards velocity u_z and the temperature enhancement τ vary with x in the plane $z = 0$.

rates of change with respect to x are not necessarily larger than rates of change with respect to z.

It is easy enough to find solutions such that all three of the variables in equations (8.34)–(8.37), u_x, u_z and τ, have the same periodic variation, described by a wavevector with components (k_x, k_z), and the circulation pattern associated with one of these is illustrated by fig. 8.12. It shows in cross-section an array of rectangular cells or *rolls*, each of indefinite extension in the y direction perpendicular to the plane of the diagram, having aspect ratio

$$H = \frac{(\pi/k_z)}{(\pi/k_x)} = \frac{k_x}{k_z}.$$

Within half of the rolls the fluid is rotating clockwise, and within the other half it is rotating anticlockwise. Because the temperature of every element of fluid reflects to some extent the ambient temperature of the region it has recently passed through, τ is positive – and the density of the fluid is in consequence reduced – on those boundaries between adjacent rolls where u_z is positive, and *vice versa*. For such solutions (8.35) and (8.37) reduce, after elimination of p^* and u_x with the aid of (8.34) and (8.36), to

$$\frac{(1 + H^2)^2}{H^2} k_z^2 u_z \approx \frac{\alpha g}{\nu} \tau,$$

$$(1 + H^2)k_z^2\,\tau \approx \frac{\theta_o}{\chi d}\,u_z.$$

It follows at once that a solution with given k_z and H is possible only if

$$Ra = \frac{\alpha g \theta_o d^3}{\chi \nu} = (k_z d)^4\,\frac{(1 + H^2)^3}{H^2}. \tag{8.38}$$

Not surprisingly, therefore, the Rayleigh Number at which a fluid layer heated from below is marginally stable with respect to the development of rolls depends upon their shape as well as on their size. If the rolls are too squat in cross-section, i.e. if H is too small, circulation is inhibited by enhanced viscous dissipation. If they are too tall for their width, on the other hand, it is inhibited by enhancement of the losses of available energy due to thermal conduction. The rolls which form first when Ra is increased steadily from zero are those for which

$$H = H_o = \frac{1}{\sqrt{2}},$$

since this value minimises the expression on the right-hand side of (8.38).

The above analysis is suggestive, but it cannot be applied without modification to a layer of liquid contained between solid surfaces, because no solution of the type sketched in fig. 8.12 is capable of satisfying the appropriate boundary conditions at $z = \pm\frac{1}{2}d$, where not only u_z but also u_x must vanish. The first of these boundary conditions can be satisfied with

$$k_z = \frac{n\pi}{d}\ [n = 1, 2 \text{ etc.}], \tag{8.39}$$

but then it is $\partial u_x/\partial z$ rather than u_x which vanishes at $z = \pm\frac{1}{2}d$. Thus the analysis is applicable to fluid layers which experience no shear stresses on their top and bottom surfaces, and for such layers it suggests

$$Ra_c = \frac{27\pi^4}{4} \approx 657, \tag{8.40}$$

since that is the smallest value which Ra can take according to (8.38) and (8.39); it corresponds, of course, to $n = 1$. But stress-free fluid layers with isothermal surfaces at different temperatures cannot be realised in the laboratory, so this result has little practical significance.

Solutions of equations (8.34)–(8.37) which satisfy realistic boundary conditions are not so amenable to analysis, but they have been exhaustively investigated nevertheless. It turns out that the lowest value of Ra at which a realistic solution is possible is 1707·762 . . ., and that the aspect ratio of the rolls to be expected at this value, H_o, is about 0·99 rather than $1/\sqrt{2}$. These results are in good agreement

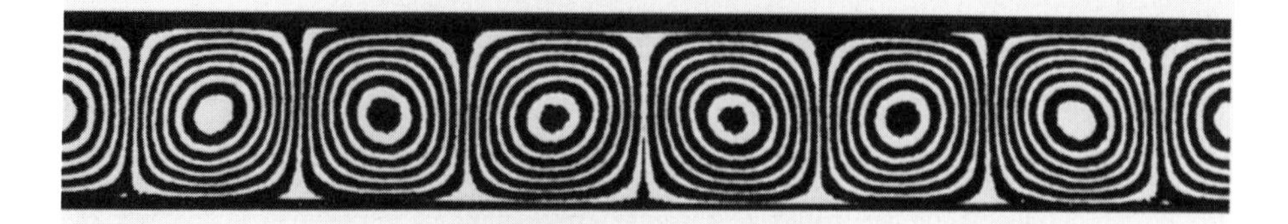

Figure 8.13 Interference fringes generated by Bénard convection in a layer of silicone oil, viewed along the horizontal axes of the convection rolls. Since the interference is in this case between coherent beams which have been split by a Wollaston prism and have travelled through the cell along paths at the same height but separated horizontally from one another by a small distance Δx, the fringes are contours of constant $\Delta n/\Delta x$, n being the refractive index of the oil. Hence they are contours of constant $\Delta T/\Delta x$, which in Bénard convection tend to follow lines of flow.
[From H. Oertel Jr [*Flow Visualization II*, ed. W. Merzkirch, Hemisphere, Washington, 1982.]]

with experimental observations. That Ra_c is close to 1700 is apparent from fig. 8.11, and the photograph reproduced as fig. 8.13 shows rolls whose cross-section is indeed virtually square. It should be noted, however, that many experimental studies of Bénard convection have been carried out using apparatus with lateral dimensions which are not very much greater than the thickness d of the fluid layer, and that in such apparatus the width of the rolls, and also their orientation, tends to be influenced by boundary conditions at the side walls of the fluid container which we have not discussed. The experiment of which fig. 8.13 provides a record was conducted on fluid in a container of breadth $4d$ and length $10d$. The rolls formed with their long axes parallel to the smaller of these dimensions and there were 10 of them, not all of which appear here.

The dependence of Nu on Ra just above the critical Rayleigh Number cannot be predicted without embarking upon a much more elaborate analysis than would be appropriate here, because the value at which u stabilises is necessarily determined by the non-linear terms in the Boussinesq equations. A non-linear treatment is also required if one wishes to examine the possibility that roll-like perturbations which differ in orientation may couple together in such a way as to favour a convection pattern which is three-dimensional rather than two-dimensional. It is in fact quite often observed that over a small range of Rayleigh Numbers which begins a little above Ra_c the convection cells are not roll-like but have a hexagonal cross-section when viewed along the z axis; together, they form a honeycomb structure [fig. 8.14], which can be regarded as a superposition of three sets of rolls with their axes at angles of $\pm\pi/3$ to one another. Two varieties of honeycomb are possible, one in which u_z is positive on the central axis of each hexagonal cell and negative on the planes that divide cells from their neighbours, and the reverse of this. The first variety tends to be observed with liquids – especially liquids whose

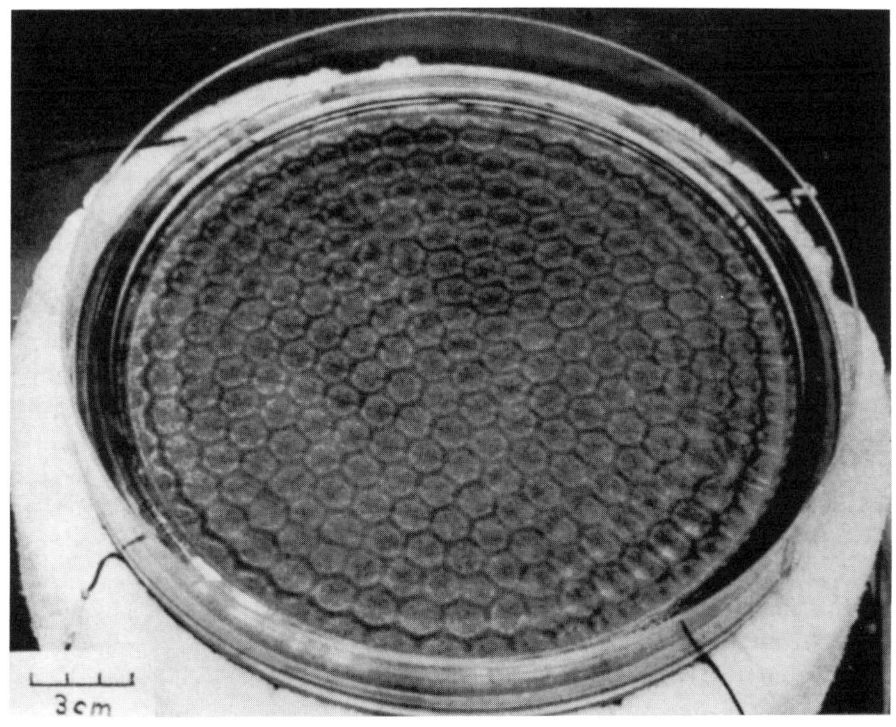

Figure 8.14 Hexagonal convection cells in a 4 mm layer of silicone oil heated from below; the oil contains aluminium powder in suspension. This photograph illustrates Marangoni rather than Bénard convection [see §8.8], but similar hexagonal cells may occur in Bénard convection at Rayleigh Numbers not far above the critical.
[From E. L. Koschmieder, *Beiträge zur Physik der Atmosphäre*, **39**, 1, 1966.]

viscosity is particularly sensitive to temperature – whereas the second is observed with gases. Since the temperature coefficient of viscosity is always negative for liquids but always positive for gases, it is perhaps not surprising that to explain the occurrence of honeycomb structures in Bénard convection it has proved necessary not only to include non-linear terms but also to go beyond the Boussinesq approximation by taking the temperature dependence of viscosity into account. The theory implies, incidentally, that once a honeycomb convection pattern has been established by suitable heating of the lower plate it may persist, when this plate is cooled again, at Rayleigh Numbers which are slightly below Ra_c.

Though honeycomb structures often occur not far above the critical Rayleigh Number, they give way to rolls again as Ra is increased further. The orientation of the rolls is not normally homogeneous throughout the convecting layer; there tend to be many 'grain boundaries' across which it changes abruptly, much as the orientation of the atomic planes changes from grain to grain in a polycrystalline solid. It is possible, however, to induce rolls which have the same orientation throughout the layer – the analogue of a single crystal – by an ingenious technique which involves supplementing the heat supplied to the bottom plate, during the warm-up phase of an experiment, by radiation delivered through a coarse grating. Moreover, it is possible to set the aspect ratio of the rolls to any desired value by choosing a grating of appropriate periodicity. Are the regular patterns of convection which may thus be established stable or unstable?

There are two values of H for which steady two-dimensional circulation should be possible when $Ra > Ra_c$ according to linearised theory [(8.38)], say H_1 ($< H_o$) and H_2 ($> H_o$), but neither of these is favoured in practice. Although there exists a range of H within which a homogeneous pattern of straight rolls can persist, it is centred on H_o and is not wide enough to include H_1 or H_2. If rolls are established with an aspect ratio which lies outside this range, they adjust themselves in an effort to achieve the preferred cross-section, which is almost square. One mode of adjustment which is sometimes observed if $H > H_o$ initially, i.e. if the rolls are narrower than they would like to be, involves the elimination of some rolls by a *pinch-off* process which starts at one end of a roll and runs along it; the rolls that remain can then expand. The converse process is possible if $H < H_o$ initially. But these are slow processes unless grain boundaries or other defects in the roll pattern happen to be present, and they tend to be anticipated by instabilities which introduce into the pattern a variation which is periodic along the length of the rolls.

The three principal instabilities of this type to which convection rolls are prone have been christened the *zigzag instability*, the *skewed varicose instability*, and the *cross roll instability*, and the sort of patterns to which they lead are illustrated by photographs in fig. 8.15. The first, which involves a periodic variation in the direction of the rolls, can occur only if H is initially less than H_o, i.e. only if the rolls are initially too wide; by developing zigzags they become longer and therefore narrower. The skewed varicose instability is more complicated, involving as it does a periodic variation in both orientation and width which is staggered in phase from one roll to the next. The cross roll instability leads ultimately to replacement of one set of rolls by another, oriented at right angles and having a different aspect ratio. The conditions under which straight rolls should be marginally stable with respect to these instabilities – conditions which involve not only the Rayleigh Number Ra and the initial value of H but also the Prandtl Number of the fluid Pr – have been worked out theoretically, and the results are in good agreement with observations of steady roll patterns up to a limiting value of Ra which is 2×10^4 or less, depending on the magnitude of Pr.

If the fluid is a liquid with a Prandtl Number of about 10 or more, then the pattern which is steady above $Ra = 2 \times 10^4$ is what is called *bimodal*; it is illustrated by a fourth photograph in fig. 8.15. At $Ra \approx 10^5$, however, bimodal patterns become overstable with respect to an oscillatory instability. If the fluid is

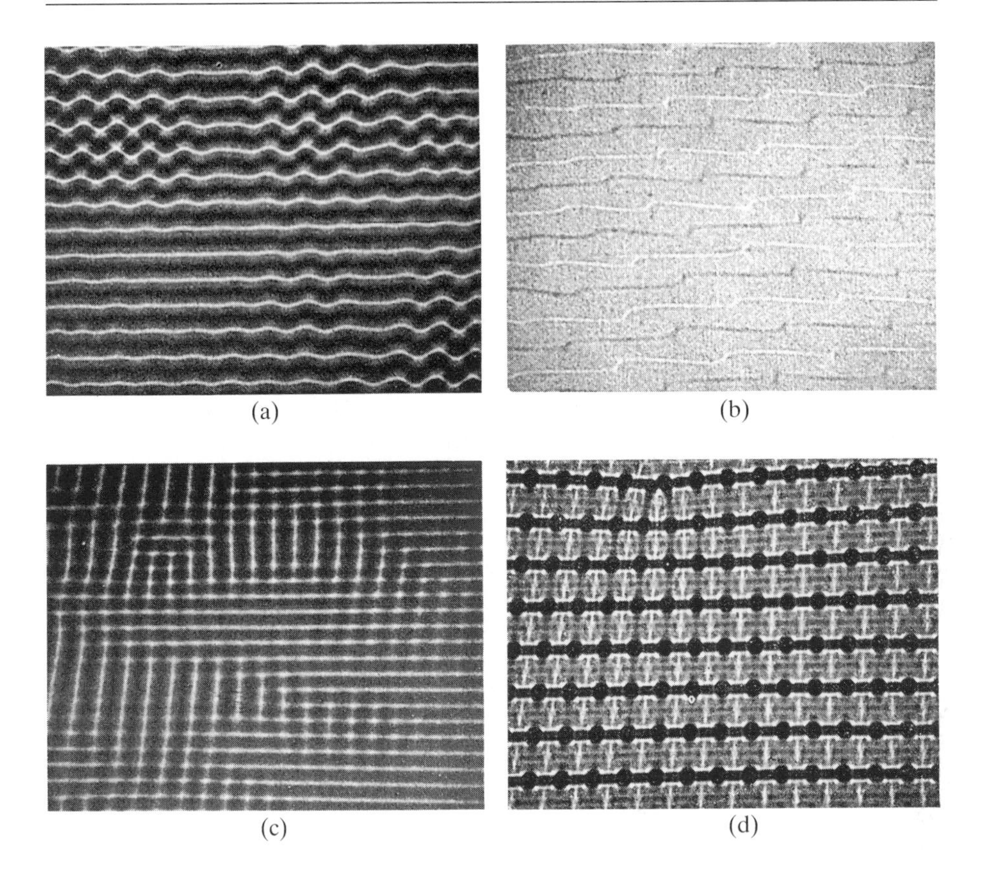

Figure 8.15 Three of these photographs are snapshots of patterns which are changing with time; they illustrate the effects on Bénard convection rolls which would otherwise run from left to right of (a) the zigzag instability, (b) the skewed varicose instability, and (c) the cross roll instability. Photograph (d) illustrates bimodal convection.
[From F. H. Busse & J. H. Whitehead, *J. Fluid Mech.*, **47**, 305, 1971, and F. H. Busse & R. M. Clever, *J. Fluid Mech.*, **91**, 319, 1979.]

a gas with $Pr \approx 1$, rolls start to oscillate at $Ra = 1 \times 10^4$ and bimodal patterns are not observed. In either case, further increases in the Rayleigh Number increase the amplitude of oscillation; the time-dependence of the pattern becomes more and more irregular, and the end result is *turbulent convection* [§9.10]. If the fluid is a liquid metal such as mercury, with a Prandtl Number in the range 10^{-1}–10^{-2}, the range of Rayleigh Numbers over which steady convection is possible is very limited, and the turbulent state is reached when Ra is only about 2500.

These details are not, in themselves, of particular significance. They have been included here for one reason only: to give the reader some impression, in the context of a problem which is well defined, well documented, and relatively well understood, of the enormous variety and complexity of the instabilities which fluid systems are liable to display.

8.8 Marangoni convection

Convection may often be observed in thin horizontal layers of liquid heated from below whose bottom surface is in contact with a solid but whose top surface is effectively free – in contact with air, perhaps, the viscosity and thermal conductivity of which are relatively small. In such circumstances the top surface is virtually stress-free, but it is not isothermal; it is somewhat hotter over regions where liquid is welling up from below than over regions where liquid is moving down again. The surface tension of liquids decreases on heating. Hence in so far as the convection process continuously removes liquid from the free surface and replaces it by warmer liquid, it continuously releases surface free energy. Like the gravitational potential energy which is released by convection, this helps to pay for losses due to viscous dissipation.

Convection which is driven primarily by the release of surface free energy rather than gravitational potential energy is invariably associated with hexagonal convection cells rather than rolls, and it is nowadays referred to as *Marangoni* rather than Bénard convection. It may be demonstrated by warming in a frying pan some silicone oil which contains, to render the cells visible, a little powdered aluminium in suspension. The effects of surface tension are likely to be dominant when

$$-\frac{\mathrm{d}\sigma}{\mathrm{d}T} > \alpha\rho g d^2,$$

a condition which is adequately satisfied for silicone oil if its depth d does not exceed say 3 mm. It seems that fig. 8.14 should perhaps be regarded as illustrating a process rather closer to Marangoni convection than to Bénard convection, since in that case the depth of oil was 4 mm.

In salt water, or brine, a type of convection may occur which is driven by the release of gravitational potential energy rather than of surface energy but which nevertheless has features in common with Marangoni convection. Imagine a layer of brine in a shallow lake which is drying up under a hot sun. Once convection is under way, elements of the brine are liable to lose water by evaporation during their periodic visits to the free surface and to increase in density in consequence. Hence the falling liquid is liable to be more dense than the rising liquid, even though the basic condition for Bénard convection – that mean temperature increases with depth – may not be fulfilled.

At some stage as the evaporating lake gets shallower the brine will start to precipitate salt crystals. Presumably they form at the free surface in the first

Figure 8.16 The bed of a salt lake which has dried out. [Photograph by G. Gerster.]

instance, and at boundaries between hexagonal convection cells where the brine is on its way down. By accumulating at the bottom of the lake in ridges which concide with these boundaries, the salt crystals may render the convection pattern visible when the lake has dried out completely. A spectacular pattern of ridges for which that may be the explanation (though other explanations have been suggested) is shown in fig. 8.16.

8.9 Bénard convection in binary fluids

Salt water is a mixture with two components which may diffuse through one another, and fig. 8.16 illustrates one example of a convective process which is determined by gradients of concentration as much as by gradients of temperature. In this section we shall be concerned with another such example: convection of the Bénard type in mixtures of water and alcohol.

In a horizontal layer of water–alcohol mixture which is heated from below, the state of stationary equilibrium is one in which the mass fraction c of the denser component (water) varies in a linear fashion with height z, being larger at the bottom than at the top by an amount

$$- c(1 - c)s_T\theta_o,$$

where θ_o, as in §8.6, is the temperature difference over the same range. The variation is a consequence of the *Soret effect*, and s_T, a negative quantity in this instance, is the *Soret coefficient* of the denser component. In equilibrium, therefore, the density of the mixture is smaller at the bottom than at the top by

$$\alpha\rho\theta_o + c(1 - c)s_T\theta_o\left(\frac{\partial\rho}{\partial c}\right)_T,$$

which, with

$$\alpha' = - c(1 - c)s_T\rho^{-1}\left(\frac{\partial\rho}{\partial c}\right)_T \tag{8.41}$$

(a positive quantity), may be expressed as

$$\alpha\rho\theta_o\left(1 - \frac{\alpha'}{\alpha}\right).$$

That may suggest that the Soret effect reduces the effective temperature difference across a water–alcohol layer by a factor $(1 - \alpha'/\alpha)$, and thereby enhances the critical Rayleigh Number by a factor $(1 - \alpha'/\alpha)^{-1}$, but makes rather little difference otherwise. The situation is not, however, so simple. Firstly, the condition for marginal stability, which already involves the viscous diffusivity ν and the thermal diffusivity χ as two ingredients in the Rayleigh Number, turns out to involve the molecular diffusivity D as well. Secondly, because D is less than χ in water–alcohol mixtures, a layer heated from below is liable to become *overstable*; it may support convective patterns which are oscillatory rather than steady.

These two points can both be illustrated in a relatively simple way by consideration of two-dimensional roll patterns of the type sketched in fig. 8.12 for which H is so large ($k_x \gg k_z$) that derivatives with respect to z may be ignored. The linearised Boussinesq equations which suffice in this limit, including derivatives with respect to time which were omitted from (8.34)–(8.37), may be written in the form

$$\nu \frac{\partial^2 u}{\partial x^2} - \frac{\partial u}{\partial t} + \alpha g(\tau - \tau') = 0, \tag{8.42}$$

$$\chi \frac{\partial^2 \tau}{\partial x^2} - \frac{\partial \tau}{\partial t} + \frac{\theta_o u}{d} = 0, \tag{8.43}$$

$$D \frac{\partial^2 \tau'}{\partial x^2} - \frac{\partial \tau'}{\partial t} + \frac{\alpha' \theta_o u}{\alpha d} = 0. \tag{8.44}$$

Here u is just u_z with its suffix suppressed for simplicity, and τ is the local temperature excess already defined in §8.7. The local excess mass fraction of water, similarly defined, is significant only in so far as it affects the local density of the fluid, and it may be fully represented by the temperature deficit, τ', which would affect the density to the same extent. Equations (8.42) and (8.43) are closely related to (8.35) and (8.37) respectively. Equation (8.44) resembles (8.43), but naturally it involves D in place of χ.

Following principles laid down in §8.1, we now assume that u, τ and τ' are all periodic in x with (real) wavevector k_x (or k), and that they all oscillate in time with the same (real) angular frequency ω_k. The equations then reduce to

$$- \nu k^2 u + \mathrm{i}\omega_k u + \alpha g(\tau - \tau') = 0,$$

$$- \chi k^2 \tau + \mathrm{i}\omega_k \tau + \frac{\theta_o u}{d} = 0,$$

$$- D k^2 \tau' + \mathrm{i}\omega_k \tau' + \frac{\alpha' \theta_o u}{\alpha d} = 0.$$

It then follows by elimination of u, τ and τ', and by equation of real and imaginary parts, that physically significant solutions of two kinds are possible.

(i) The frequency ω_k may be zero, in which case

$$\nu k^2 = \frac{g\theta_o}{d}\left(\frac{\alpha}{\chi k^2} - \frac{\alpha'}{D k^2}\right),$$

or

$$\frac{Ra}{(kd)^4} = \frac{1}{1 - \dfrac{\chi\alpha'}{D\alpha}}. \tag{8.45}$$

This is the condition for steady convection rolls to be marginally stable. In the absence of any Soret effect, i.e. when $\alpha' = 0$, it is completely equivalent to (8.38) when $k_x \gg k_z$, i.e. in the limit $H \to \infty$.

(ii) The frequency ω_k may be different from zero, in which case

$$\frac{Ra}{(kd)^4} = \frac{\nu\chi}{(\nu + D)(\chi + D)} \frac{1}{1 - \dfrac{(\nu + D)\alpha'}{(\nu + \chi)\alpha}} \tag{8.46}$$

and

$$\omega_k^2 = \frac{g\theta_o}{d(\chi + D)} \left\{ \frac{\alpha'\chi^2}{(\nu + \chi)} - \frac{\alpha D^2}{(\nu + D)} \right\}. \tag{8.47}$$

Equation (8.46) is the condition for oscillatory convection rolls to be marginally stable, and (8.47) tells us the angular frequency of these rolls. Since

$$\nu \approx 10\chi \approx 100D$$

for water–alcohol mixtures, an adequate approximation to (8.46) is

$$\frac{Ra}{(kd)^4} \approx \frac{0{\cdot}9}{1 - 0{\cdot}9\,\dfrac{\alpha'}{\alpha}}. \tag{8.48}$$

Curve (a) in fig. 8.17 represents the variation of Ra with α'/α according to (8.45). To the left of the origin, where α'/α is negative, the Soret effect enhances the tendency to instability of a layer heated from below, and the value of Ra associated with marginal stability of steady rolls is reduced in consequence. This part of curve (a) is not realisable with water–alcohol mixtures but could in principle be realised using other mixtures in which the Soret coefficient of the heavier component is positive rather than negative. To the right of the origin the Soret effect exerts a stabilising influence, and for $\alpha'/\alpha > D/\chi \approx 0{\cdot}1$ a layer which is heated from below is stabilised to such an extent that steady convection is impossible. Such convection may now occur in layers which are heated from above, however, and curve (a) switches to negative values of Ra for that reason.

Curve (b) represents the variation of Ra according to (8.48); it resembles curve (a) but switches to negative values at $\alpha'/\alpha \approx 1{\cdot}1$ instead. However, negative values

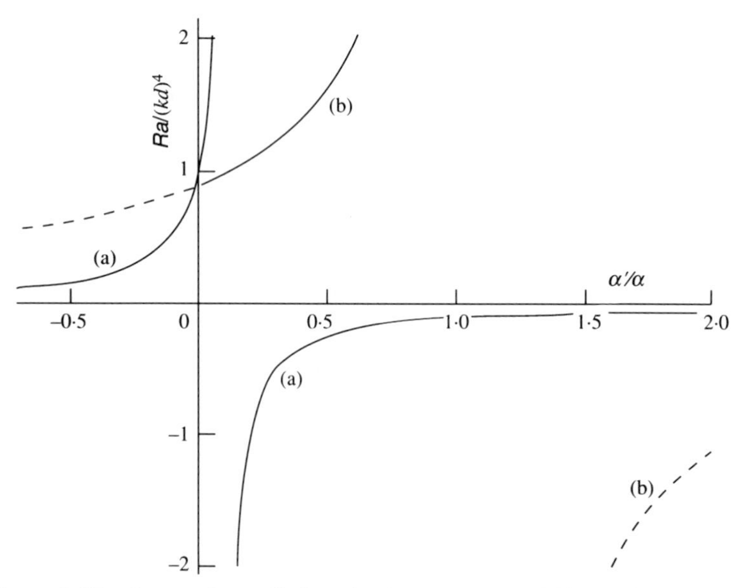

Figure 8.17 Convective rolls in a horizontal layer of binary fluid may be oscillatory rather than steady. Here the values of $Ra/(kd)^4$ associated with marginal stability (curve (a)) and marginal overstability (curve (b)) are plotted against the ratio α'/α, which is a measure of the strength of the Soret effect. Both curves refer to rolls whose aspect ratio $H = kd/\pi$ is large compared with unity, and to a binary fluid for which $\nu \approx 10\chi \approx 100D$.

of Ra and θ_o in this region of α'/α are inconsistent, according to (8.47), with the requirement that ω_k be real. That requirement implies, in fact, that only the section of curve (b) for which

$$\frac{D^2(\nu + \chi)}{\chi^2(\nu + D)} < \frac{\alpha'}{\alpha} < \frac{(\nu + \chi)}{(\nu + D)}$$

and which is drawn as a full line rather than a broken one is relevant in this discussion.

It will be noted that curve (a) and the full part of curve (b) do not quite intersect. However, these marginal stability curves apply only in the limit of large k ($\gg \pi/d$). More elaborate analyses are to be found in the literature, in which k is varied so as to minimise Ra for each value of α'/α, with proper attention to realistic boundary conditions at $z = 0$ and $z = d$. The curves which they predict for $Ra_c\{\alpha'/\alpha\}$ are superficially similar to the curves in fig. 8.17, apart from a vertical scaling factor which ensures that $Ra_c\{0\}$ on curve (a) is 1707·762 rather than $(kd)^4$, but they *do* intersect. At this intersection a fluid layer is marginally stable or overstable with respect to two distinct types of convective roll, which have different frequencies (one of them zero) and also different values of k. In the jargon of bifurcation theory, it is a *codimension-2 point*.

A convective roll pattern which has a non-zero frequency associated with it may have either a standing wave character or a travelling wave character. That is to say, the pattern may remain fixed while the flow velocities within it periodically change sign, or the flow velocities within each roll may remain constant while the rolls drift through the fluid to right or left with velocity ω_k/k. In practice the rolls prefer to travel, with velocities which in laboratory experiments are typically of order 1 centimetre per minute, though just what is observed depends upon the geometry of the cell in which the fluid is confined. Experiments have been done under a wide range of conditions – α'/α may be varied continuously between zero and at least 0·6 by adjustment of the mean concentration c or the mean temperature, and Ra can of course be raised well beyond Ra_c by increasing the temperature difference θ_0 – with results which are at least as varied as those described in §8.7. To explain them fully one is naturally obliged to take non-linear terms in the Boussinesq equations into account, and it may be necessary to take into account the possibility that fluid parameters such as α' are not really uniform but vary significantly with z.

Although most of the recent experimental work on oscillatory convection in binary fluids has been done on mixtures of water and alcohol, the phenomenon is certainly not confined to such mixtures, and indeed it does not have to be fuelled by the Soret effect. One can create a vertical gradient of salt concentration in water, for example, and at the same time a vertical gradient of temperature, by carefully pouring a relatively weak mixture which is relatively cold over a layer which is relatively strong and relatively warm. For a time, at any rate, the conditions for oscillatory convection may then exist. Convective mixing in layered systems of salty water, with particular reference to convective mixing in the Earth's oceans, is a fascinating subject in its own right, but it is not possible to do it justice here.

8.10 The Taylor–Couette instability

A type of convection which is closely related to Bénard convection may be observed in isothermal fluid which is undergoing Couette flow [§6.9(ii)] between rotating coaxial cylinders. The instability which marks its onset is known as the *Taylor–Couette instability*, G. I. Taylor having studied it both experimentally and theoretically during the 1920's.

Couette flow is vorticity-free when the angular velocity of the fluid about the axis of circulation, $\omega = u/r$, decreases with increasing radius r like r^{-2}, and as Rayleigh was the first to point out it is not prone to instability unless the rate of decrease of ω is more rapid than this. The reason may be understood by reference to fig. 8.18, which shows in cross-section two cylinders of radii a_1 and a_2, rotating in the same sense with fluid rotating steadily between them. Two toroidal elements of fluid, both of unit mass, are indicated on this diagram, one adjacent to

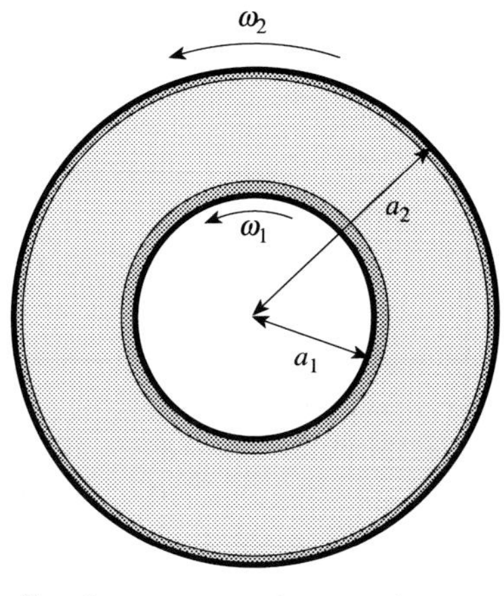

Figure 8.18 Couette flow between rotating coaxial cylinders. The cylinders are represented by black rings, and the rings of darker grey adjacent to them represent interchangeable elements of fluid of equal (unit) mass.

one cylinder and one adjacent to the other; they are rotating with the cylinders, i.e. with angular velocities ω_1 and ω_2. Suppose for a moment that the viscosity of the fluid is negligible, and suppose that a convection process occurs which has the effect of causing these two elements to change places. Provided that the fluid is effectively inviscid, we may infer from Kelvin's theorem [§4.2] that for each element

$$K = \oint \boldsymbol{u} \cdot \mathrm{d}\boldsymbol{l} = 2\pi\omega r^2$$

remains unchanged during the process, i.e. that its angular momentum remains unchanged. Hence after the exchange the angular velocities of the two elements are respectively $\omega_1 a_1^2/a_2^2$ and $\omega_2 a_2^2/a_1^2$, which means that the process has released an amount of kinetic energy given by

$$\frac{1}{2}\left(\omega_1^2 a_1^2 - \frac{\omega_1^2 a_1^4}{a_2^2}\right) + \frac{1}{2}\left(\omega_2^2 a_2^2 - \frac{\omega_2^2 a_2^4}{a_1^2}\right) = \frac{a_2^2 - a_1^2}{2a_1^2 a_2^2}(\omega_1 a_1^2 + \omega_2 a_2^2)(\omega_1 a_1^2 - \omega_2 a_2^2).$$

Only if $\omega_1 a_1^2 > \omega_2 a_2^2$ is this energy release positive; only then is the initial state potentially unstable.

The parallel with Bénard convection is most apparent in the particular case for which the gap between the two cylinders is so narrow ($d = a_2 - a_1 \ll a_1$) that the curvature of the fluid layer between them is unimportant, and for which the outer cylinder is stationary ($\omega_2 = 0$). In that case the energy release during the exchange process described above is just $\omega_1^2 ad$, and this quantity plays the role played by $\alpha g \theta_o d$ in the analysis of Bénard convection presented in §8.7. In that analysis, of

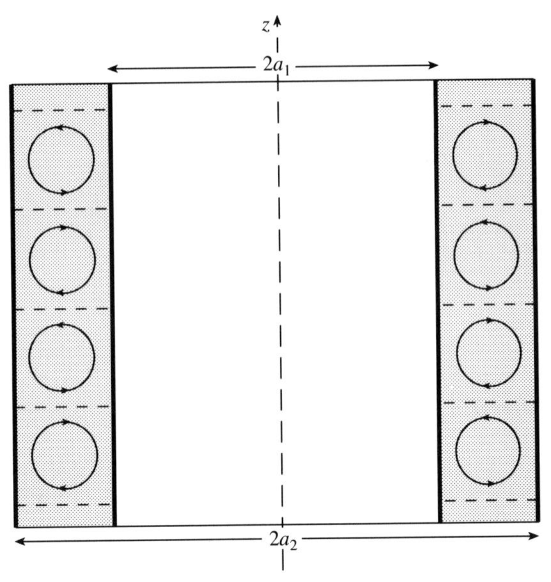

Figure 8.19 The secondary circulation associated with Couette flow between rotating coaxial cylinders when the Taylor Number exceeds its critical value. The cylinders resemble those shown in fig. 8.18, but here we see a lengthwise cross-section rather than a transverse one; it covers only part of the cylinders' full length. (The discussion in the text refers to the limiting case $a_2 - a_1 \ll a_2$, but this case does not lend itself to illustration.)

course, we took into account the viscous dissipation that occurs during convection, and also the effects of thermal conduction which prevent the rate at which energy is released to pay for viscous dissipation from being as high as it would be otherwise. Viscous dissipation is equally relevant in the present context, but it is the loss of angular momentum by elements of fluid which are moving from $r = a_1$ to $r = a_2$, rather than the loss of heat, which limits the rate at which energy is released. A plausible assumption, which is justified by a much more detailed analysis, is that the rate of loss of angular momentum is determined by the diffusion of vorticity, and that the role played in Bénard convection by the thermal diffusivity χ is therefore played by ν. In that case the role played by

$$Ra = \frac{\alpha g \theta_o d}{\chi \nu} d^2$$

is played by

$$\frac{\omega_1^2 a d}{\nu^2} d^2 = \frac{1}{2} T$$

(an equation which defines, in the simple case we are considering, the so-called

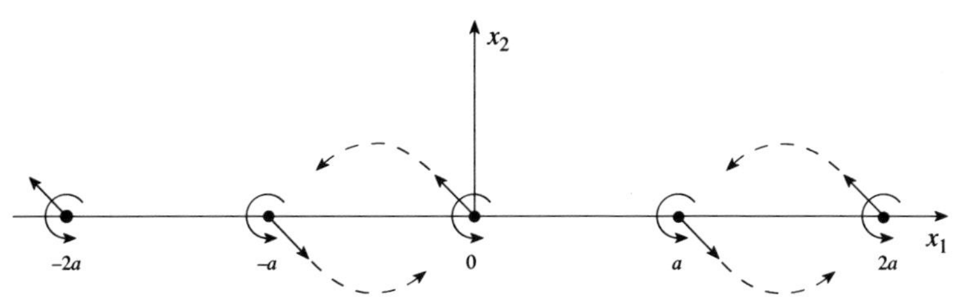

Figure 8.20 Cross-section through part of an infinite array of parallel vortex lines, all of strength K, situated on the lines $x_1 = na$, $(n = 0, \pm 1, \pm 2$ etc.). They are unstable [cf. fig. 8.21] to the set of small displacements suggested by inclined arrows. The continuation of these arrows by broken curves suggests that the vortex lines may subsequently circle around one another in pairs.

Taylor Number T), and instead of $Ra = 1707{\cdot}762$ the condition for marginal stability is likely to be $T \approx 3416$. In fact, the critical value of T which marks the onset of the Taylor–Couette instability in this case is 3390.

Not surprisingly, the convection rolls which form when this critical value is just exceeded have a height, measured in a direction parallel to the axis of circulation, which is very nearly equal to d, as indicated in fig. 8.19. At larger values of T they crinkle, as a result of a secondary instability related to the zigzag instability described in §8.7, and if ω_2 is varied as well as ω_1 a number of other secondary instabilities are liable to affect the flow.

8.11 The Kelvin–Helmholtz instability

By way of introduction to the subject matter of this section, it is instructive to consider the stability of the infinite array of vortex lines in otherwise stationary fluid which are sketched in cross-section in fig. 8.20; the vortices lie in the plane $x_2 = 0$ and are parallel to the x_3 axis, they are regularly spaced along the x_1 axis with separation a, and they all have the same strength K. In this condition the array is clearly stationary, because the symmetry is such that the components of velocity imparted to any one of the vortices by all the others are bound to cancel exactly. But if the vortices are given infinitesimal displacements, what then?

Rather than attempt to find a general answer to that question, let us consider the effect of displacements $(\pm\zeta_1, \pm\zeta_2)$ which are the same in magnitude for each vortex but which alternate in sign from one to the next, as suggested by the figure. Now the cancellation is not exact. It is a simple exercise to show that, as long as ζ_1 and ζ_2 remain small enough compared with a for terms of order $(\zeta/a)^2$ or higher to be discarded,

$$\frac{\mathrm{d}\zeta_1}{\mathrm{d}t} \approx -\frac{\pi K^2}{4a^2}\zeta_2, \quad \frac{\mathrm{d}\zeta_2}{\mathrm{d}t} - \frac{\pi K^2}{4a^2}\zeta_1,$$

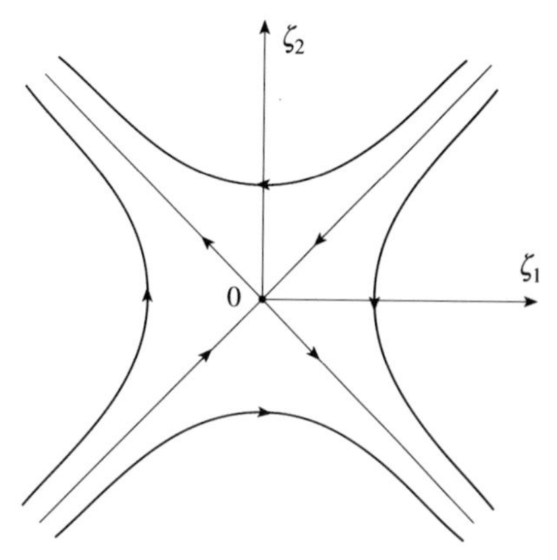

Figure 8.21 If all the vortex lines of even n in fig. 8.20 are simultaneously subjected to small displacements ζ_1 in the x_1 direction and ζ_2 in the x_2 direction, while all the lines of odd n are displaced by $(-\zeta_1, -\zeta_2)$, the lines are subsequently propelled along rectangular hyperbolae.

and it follows that the vortices are confined to trajectories such that

$$\zeta_1^2 - \zeta_2^2 = \text{constant},$$

i.e. the trajectories are rectangular hyperbolae, along which the vortices move in the directions indicated by arrows in fig. 8.21. Evidently, the unperturbed array of vortex lines is as it were at a saddle point with respect to the sort of periodic distortion we have considered. It is stable with respect to displacements such that ζ_1 and ζ_2 have the same sign, but unstable with respect to displacements such that ζ_1 and ζ_2 have opposite signs. Thus the vortex lines are liable to pair off with one another in the way that fig. 8.20 suggests, though there exists another mode of pairing which is equally probable. Broken lines in fig. 8.20 imply that the members of each pair subsequently circle round one another [fig. 4.19], but other instabilities are liable to complicate their long-term motion.

Let us now imagine that the separation a of the vortex lines in fig. 8.20 tends to zero, and that their strength tends to zero at the same rate. In the limit we have a continuous *vortex sheet*, which may also be visualised as a surface in the $x_2 = 0$ plane separating two large regions in which the fluid is moving uniformly in opposite directions, with velocities $u_1 = \pm U$ as in fig. 8.22.[2] Vorticity is present in the sheet itself because $\partial u_1/\partial x_2$ is infinite there, but it is not present in the fluid on either side of the sheet, and the stability of the flow pattern sketched in fig. 8.22 may therefore be investigated using potential theory. We may apply several of the results obtained in chapter 5 with little modification.

[2] The electromagnetic analogue of a vortex sheet is of course a current sheet, across which $\boldsymbol{B}$ rather than $\boldsymbol{u}$ changes discontinuously.

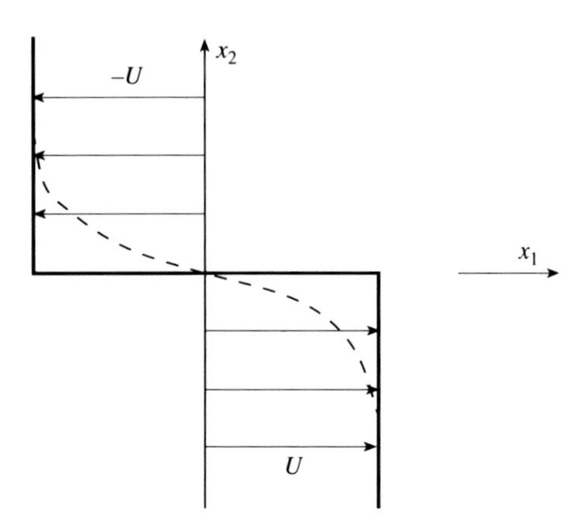

Figure 8.22 Velocity profile in the neighbourhood of an ideal vortex sheet. The broken curve, representing a vortex layer with finite thickness, is more realistic.

To be specific, suppose that the vortex sheet suffers a periodic disturbance of small amplitude which carries it from $x_2 = 0$ to $x_2 = \zeta$, where

$$\zeta = \zeta_o \, e^{ikx_1}, \tag{8.49}$$

with k positive and with $k|\zeta_o| \ll 1$; it is to be understood, as usual, that ζ_o may be complex and that only the real part of ζ is of physical significance. Appropriate potentials which have the same periodicity, and which satisfy not only Laplace's equation but also the boundary conditions that at large values of x_2 (negative or positive as the case may be) the fluid velocity $\boldsymbol{u}$ has components $(\pm U, 0, 0)$, are

$$\begin{aligned} \phi &= Ux_1 + ae^{kx_2} \, e^{ikx_1} \quad [x_2 < \zeta], \\ \phi' &= -\, Ux_1 + a'e^{-kx_2} \, e^{ikx_1} \quad [x_2 > \zeta] \end{aligned} \tag{8.50}$$

[(5.12)]. Two of the boundary conditions on the sheet itself, at $x_2 = \zeta$, are

$$\frac{\partial \phi}{\partial x_2} = \frac{\partial \zeta}{\partial t} + \frac{\partial \phi}{\partial \chi_1}\frac{\partial \zeta}{\partial x_1},$$

$$\frac{\partial \phi'}{\partial x_2} = \frac{\partial \zeta}{\partial t} + \frac{\partial \phi'}{\partial x_1}\frac{\partial \zeta}{\partial x_1},$$

[(5.16)] and to first order in $k\zeta$ these amount to

$$\begin{aligned} ka &\approx \frac{\partial \zeta_o}{\partial t} + ikU\zeta_o, \\ -\, ka' &\approx \frac{\partial \zeta_o}{\partial t} - ikU\zeta_o. \end{aligned} \tag{8.51}$$

A third boundary condition concerns the pressure in the fluid. We know from (4.13) that the pressure is given for $x_2 < \zeta$ by

$$p = p_o + \rho \left\{ \frac{1}{2}(u_1^2 + u_2^2) + g x_2 + \frac{\partial \phi}{\partial t} \right\},$$

in which case the pressure immediately adjacent to the vortex sheet on one side is given to first order in $k\zeta$ by

$$p \approx p_o + \rho \left\{ \frac{1}{2} U^2 + e^{ikx_1} \left(ikaU + g\zeta_o + \frac{\partial a}{\partial t} \right) \right\}.$$

A gravitational term has been included here as though the x_2 direction were vertically upwards, which is not necessarily the case. However, this term has been included only to demonstrate that gravity, and hence the orientation of the sheet with respect to the vertical, is irrelevant in this context – as it obviously must be, if the density of the fluid is the same on both sides. Immediately adjacent to the sheet on the other side, to the same order, the pressure is

$$p' \approx p_o' + \rho \left\{ \frac{1}{2} U^2 + e^{ikx_1} \left(-ika'U + g\zeta_o + \frac{\partial a'}{\partial t} \right) \right\},$$

and since the jump in pressure across the interface must be zero for all x_1 we therefore have $p_o' = p_o$ and

$$ikUa + \frac{\partial a}{\partial t} \approx -ikUa' + \frac{\partial a'}{\partial t}. \tag{8.52}$$

Equations (8.51) and (8.52) may be combined to yield

$$\frac{\partial^2 \zeta_o}{\partial t^2} \approx k^2 U^2 \zeta_o,$$

from which it follows that as long as the amplitude of the disturbance remains small it varies exponentially with time,

$$\zeta_o \propto e^{\pm kUt}. \tag{8.53}$$

From that analysis it follows that a plane vortex sheet can never be marginally stable with respect to periodic disturbances, except in two trivial limiting cases, $U = 0$ and $k = 0$. Such sheets are either stable or unstable, depending on the phases of a and a' [(8.51)]. Since we already know that arrays of discrete vortex lines are stable with respect to periodic displacements in one direction but unstable with respect to periodic displacements in another, the existence for a continuous vortex sheet of two distinct roots should be no surprise. In practice, as for the discrete array, the stable solution ($\zeta_o \propto \exp(-kUt)$) is irrelevant. The way

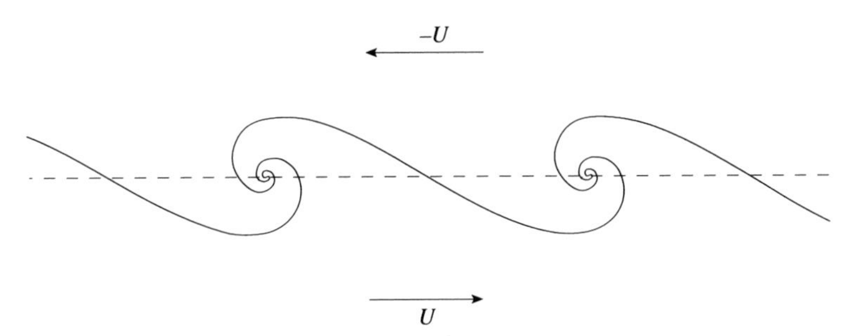

Figure 8.23 A stage in the development of the Kelvin–Helmholtz instability in an initially planar vortex sheet (sketch only).

in which the unstable solution develops when its amplitude is no longer small is indicated in fig. 8.23, which shows the sheet rolling up into spirals. The vorticity which initially was distributed uniformly in x_1 becomes concentrated in these spirals, and the continuous sheet thereby evolves into a structure which bears some resemblance to the array of discrete vortex lines in fig. 8.20.[3]

The perturbation which grows most rapidly, according to (8.53), is the one of largest k, i.e. of shortest repeat distance λ in the x_1 direction, and for the ideal case we have been considering, of a sheet of infinitesimal thickness in a fluid which has no viscosity, there is of course no lower limit to λ. In the real world, however, it is impossible to establish the velocity profile sketched in fig. 8.22; the change in velocity from $+U$ to $-U$ is bound to occur over a layer of finite thickness [fig. 8.22, broken curve]. The circumstances in which a vortex layer, as opposed to an ideal vortex sheet, is unstable in a fluid of uniform density and zero viscosity were first investigated by Lord Rayleigh. He showed that the occurrence of instability implies the existence of a point of inflexion in the velocity profile, at which $\partial^2 u_1/\partial x_2^2$ vanishes, and the complementary theorem that a point of inflexion guarantees instability for at least some values of k has since been shown to hold for profiles which are symmetrical, though not necessarily otherwise. One symmetrical profile with a point of inflexion in it which can be treated analytically is described by the equation

$$u_1 = U \tanh\left(\frac{z}{\delta}\right),$$

and this is known to be unstable only for values of k which lie within the range

[3] Like an array of discrete vortex lines, the structure sketched in fig. 8.23 must surely be unstable with respect to periodic displacements which cause spirals to spiral round one another in pairs; the structure which develops as a result is presumably itself unstable with respect to further displacements which cause adjacent pairs of spirals to spiral round one another; and so on. This remark may have some relevance to a phenomenon exhibited by turbulent vortex layers, i.e. mixing layers, which is illustrated by fig. 9.1 below.

$$0 < k\delta < 1.$$

In this case the perturbation which grows most rapidly is likely to be such that $k\delta \approx 0{\cdot}5$, which implies $\lambda \approx 4\pi\delta$. The repeat distances which arise in practice when the so-called *Kelvin–Helmholtz instability* occurs – some examples are discussed in the following section – are no doubt determined in a similiar fashion, by the magnitude of a layer thickness δ. In real fluids with viscosity, of course, δ tends to grow like $\sqrt{t}$ as the vorticity in the layer diffuses outwards. Viscosity undoubtedly affects observed repeat distances, and they are not easy to predict.

The theory outlined above can readily be extended to the case of a vortex sheet which separates two fluids of different density, in which case both gravity and the surface tension of the interface are, or may be, relevant. Suppose the sheet to be horizontal and the fluid below it to have density ρ and to be moving in the x direction with velocity U, while the fluid above it is moving with velocity $-U$ and has density ρ' ($\leq \rho$). The end result for $k > 0$ is then

$$\zeta_o \propto e^{-i\omega t} e^{st},$$

with

$$\omega + is = krU \pm ik\sqrt{(1 - r^2)U^2 - c_o^2}, \tag{8.54}$$

where

$$r = \frac{\rho - \rho'}{\rho + \rho'}$$

and where c_o is the velocity of waves on the interface, taking gravity and surface tension into account, when U is zero [(5.40)]. The fact that ω is now non-zero means that the disturbances which are potentially unstable have a travelling wave character in the frame of reference of fig. 8.22, though they are stationary in a frame of reference such that the velocities of the two bodies of fluid are $U(1 - r)$ and $-U(1 + r)$ rather than U and $-U$. However, they are not actually unstable unless the roots for s are real, i.e. unless

$$(1 - r^2)U^2 > c_o^2. \tag{8.55}$$

Otherwise, according to (8.54), the only effect of the vorticity which is present in the interface between the fluids is to modify the result expressed by (5.40): the velocity with which, in the alternative frame of reference just referred to, waves can propagate in either direction with constant amplitude is now

$$c = \sqrt{c_o^2 - (1 - r^2)U^2}. \tag{8.56}$$

8.12 Examples of the Kelvin–Helmholtz instability

A classic demonstration of the Kelvin–Helmholtz instability involves two layers of fluid of slightly different density in a long trough with transparent sides. When the

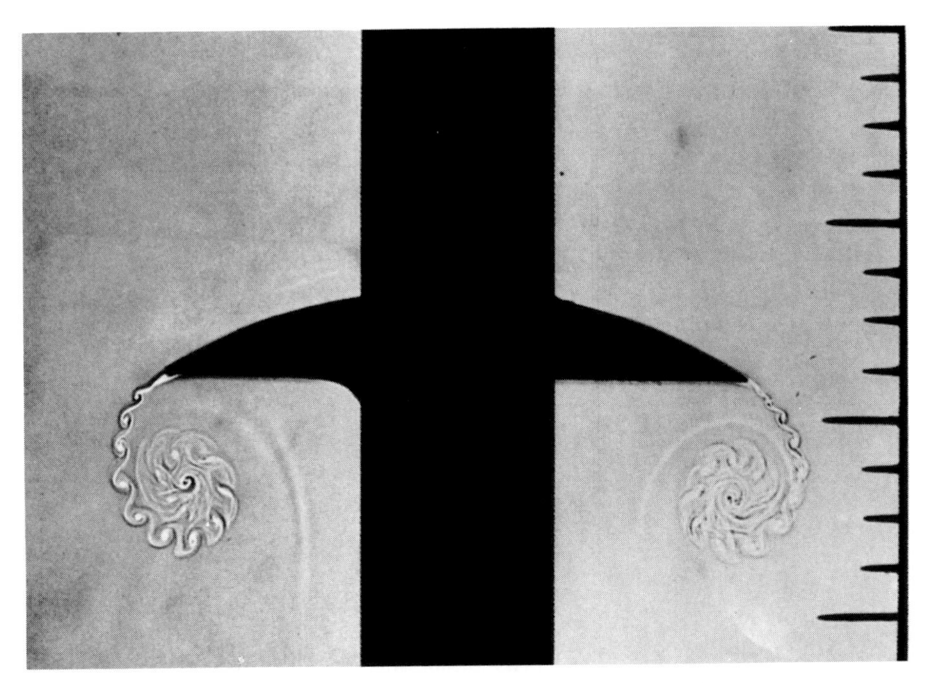

Figure 8.24 Shadowgraph showing a boundary layer which has separated from the edges of a square plate accelerating upwards through air. The separated boundary layer, which is rolling up to form an eddy in the usual way, has become corrugated as a result of the Kelvin–Helmholtz instability. (The optical contrast in shadowgraphs is due to scattering of light from regions where the refractive index of the fluid is non-uniform. In this case the top surface of the plate has been painted with benzene, and the contrast arises because benzene vapour is present in the boundary layer.)
[From D. Pierce, *J. Fluid Mech.*, **11**, 460, 1961.]

trough is tilted, one layer moves over the other and waves are seen to develop on the horizontal interface. The effect is manifested on a larger scale when the surface of a lake is ruffled by wind. It is a matter of common observation that no ripples appear until the wind speed exceeds a critical value, and according to (8.55) this critical speed, relative to water which is not moving, should be $2c_o^*/\sqrt{1 - r^2}$, where c_o^* is the minimum value of c_o as given by (5.36), i.e. about 23 cm s^{-1}. But $2/\sqrt{1 - r^2}$ is about 30 when one fluid is water and the other is air, so the predicted critical wind speed is about 7 m s^{-1}. That answer is much too high. Approximate theories which do not rely, as does the derivation of (8.54), upon potential methods, but which include viscosity from the start, suggest critical wind speeds which are only two or three times greater than c_o^*, and such predictions are in better agreement with what is normally observed.

Boundary layers contain vorticity, of course, and any boundary layer which has separated [§7.3] is also a vortex layer, though not necessarily a plane one. The

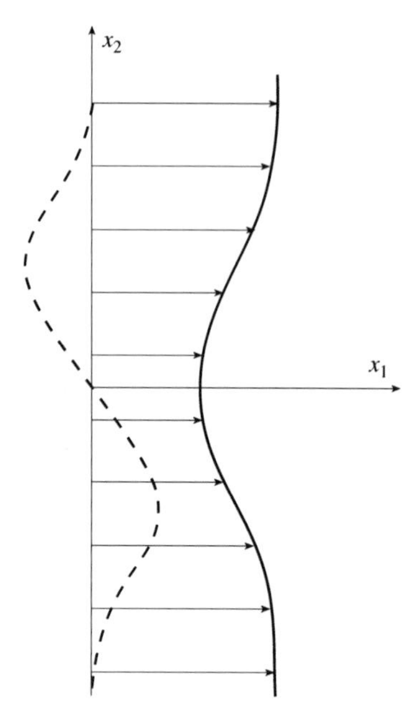

Figure 8.25 Hypothetical velocity profile in the wake behind a cylinder with its axis parallel to the x_3 axis. The broken curve refers to the vorticity, $\Omega_3 = -\partial x_1/\partial x_2$.

velocity profile across a separated boundary layer almost inevitably has a point of inflexion in it, and such layers are prone to the Kelvin–Helmholtz instability in consequence – see fig. 8.24 for an example. It is for this reason that the flow behind a spherical obstacle becomes turbulent at values of the Reynolds Number of order 10^4, as described in §7.6 and illustrated in fig. 7.11. It is for this reason that fluid emerging from a circular orifice as a submerged jet normally becomes turbulent at some distance from the orifice, as described in §7.7. It is arguably for this reason that, at moderate values of the Reynolds Number, a Kármán vortex street forms behind a cylindrical obstacle which is transverse to the flow, as shown in fig. 7.10. If instabilities could be suppressed, the velocity profile in the wake behind such an obstacle would presumably have the form sketched in fig. 8.25. The wake would consist, as the broken curve suggests, of a layer of positive vorticity adjacent to a layer of negative vorticity, both of them derived from separated boundary layers but both of them quite thick. Presumably on account of a Kelvin–Helmholtz-type instability the two sheets become corrugated, with an amplitude that increases downstream from the cylinder. As Re is increased, the region in which the corrugations are noticeable spreads upstream towards the cylindrical obstacle, and they seem to initiate the process of eddy-shedding discussed in §7.5.

The general circulation of the Earth's atmosphere is far too large and complex a subject to be discussed fully in a book of this nature, but some brief remarks may

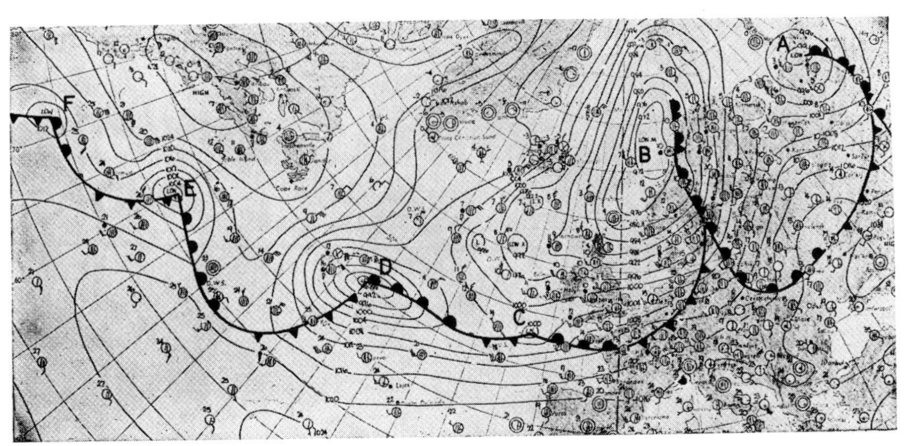

Figure 8.26 A family of depressions, labelled A–F, over the Atlantic Ocean and Europe at 1200 GMT on 15 October 1967; the polar front, cold in some places and warm in others, is shown as a heavy line.
[From D. H. McIntosh & A. S. Thom, *Essentials of Meteorology*, Wykeham Publications (London) Ltd, 1978.]

be appropriate here about the *polar front* in the northern hemisphere, and about the instability of this front which is responsible for the depressions which regularly afflict inhabitants of the British Isles. The front is a region of convergence, situated at a latitude of 40°–50° N, where relatively cold, dry air which has drifted south from polar regions meets relatively warm, moist air which has drifted north from sunnier parts. During its journey south or north the air is acted on by Coriolis forces, and in consequence the mean velocity of the cold air, near the surface of the Earth and relative to it, has a component from east to west, while the mean velocity of the warm air has a component from west to east; in this context, 'mean velocity' is velocity averaged over longitude as well as time. The front is therefore in some sense a vortex layer, though the associated flow pattern departs from the idealised model of fig. 8.22 in several particulars. Firstly, the mean velocity of the air on each side of the front has a north–south component as well as an east–west component. Secondly, the extent of the front in a vertical direction is limited; air which converges upon it at sea-level rises upwards and diverges again at high altitudes. Thirdly, pressure decreases with height more rapidly on the cold side of the front than on the hot side, because the air is denser there, which means that at high altitudes the air above the polar front is subject to a pressure gradient tending to displace it northwards; it responds by moving rapidly from west to east, in what is called the *jet stream*, at such a speed that forces due to the pressure gradient are balanced by Coriolis forces. Fourthly, the fluctuations about the mean are always large. These substantial complications apart, the polar front constitutes a vortex layer and as such is liable to develop periodic corrugations. The vorticity in the front is thereby concentrated into regions of spiralling winds, at the centre of which the pressure is low.

The pressure at the centre of a depression tends to become lower still as time passes, because the sea-level rate of convergence is less than the high-altitude rate of divergence, and the winds get stronger in consequence. Moreover, depressions are dragged eastward as they intensify, by the jet stream above them. The principal features of the weather map reproduced in fig. 8.26 – a map which is typical in the sense that it illustrates a regularly recurring pattern but untypical in the sense that it is relatively free of complicating features – can be explained in this way.

8.13 The Tollmien–Schlichting instability

The velocity profile in a boundary layer which is attached to a flat plate does not, according to Blasius's solution, have a point of inflexion in it [§7.2 and fig. 7.1], so a boundary layer of this sort should not be affected by the Kelvin–Helmholtz instability. If it is unstable, as indeed it proves to be at large enough values of the Reynolds Number, the instability must have a different origin; it must be due to the viscosity which real fluids possess. We cannot hope to explain it, therefore, using elementary arguments based upon potential theory.

More sophisticated, though nevertheless approximate, arguments suggest that boundary layers attached to flat plates become marginally stable with respect to periodic disturbances known as *Tollmien–Schlichting waves* when the Reynolds Number, defined in this context as $Re_\delta = \delta\rho U/\eta$, reaches about 10^3; here U is the velocity just outside the boundary layer and δ is the boundary layer thickness, defined in such a way that 90% of the vorticity is included within it [(7.7)]. Under very carefully controlled conditions the waves can be detected experimentally at values of Re_δ in the neighbourhood of 10^3, but once they have been excited the boundary layer tends to be unstable in other ways and rapidly becomes turbulent.

One manifestation of the transition to turbulence in attached boundary layers is the sudden shift in the point of boundary layer separation which is responsible, in cases of flow past spheres and transverse cylinders, for the so-called *drag crisis* [fig. 7.13]. An approximate formula for the thickness of the boundary layer attached to a transverse cylinder of radius a, in a fluid stream whose velocity at large distances is U, is derived in §7.3. In so far as we may trust that result [(7.19)], we may estimate that at an angle χ [fig. 7.5], where the velocity just outside the boundary layer is $2U \sin\chi$, we have

$$Re_\delta^2\{\chi\} = \left(\frac{2\rho U \sin\chi}{\eta}\right)^2 \delta^2\{\chi\} \approx 11{\cdot}6 \sin^2\chi X\{\chi\} Re,$$

Re being the Reynolds Number more conventionally defined, i.e. $2a\rho U/\eta$. Just before the onset of turbulence the boundary layer separates from a transverse

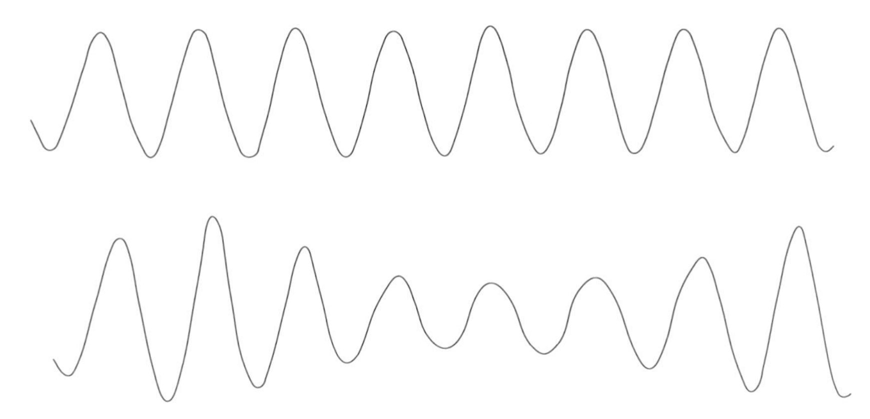

Figure 8.27 Experimental traces of wave height as a function of time: upper trace, 200 feet from the wavemaker; lower trace, 400 feet from the wavemaker. Fundamental frequency, 0·85 Hz.
[Based on a figure in T. B. Benjamin, *Proc. Roy. Soc. A*, **299**, 59, 1967.]

cylinder at a value of χ which is slightly greater than $\pi/2$, so where the boundary layer is still attached the coefficient $11{\cdot}6\sin^2\chi X\{\chi\}$ is everywhere less than about 6. A predicted critical value of about 10^3 for Re_δ is consistent, therefore, with the observed value of Re at the drag crisis, which is about 2×10^5.

8.14 The Benjamin–Feir instability

This final section harks back to Stokes waves of finite amplitude on the surface of deep water, a topic briefly discussed in §5.7. Until 1967 it was generally assumed that the state of motion associated with a single Stokes wave, i.e. a wave with straight crests, uniform amplitude and uniform wavelength, was stable. In that year, however, Benjamin and Feir published accounts of experiments on waves in wavetanks which showed the contrary to be true. One of their records is reproduced in fig. 8.27; the two curves show how the level of the water in the tank varied with time at distances of 200 feet (i.e. about 28 wavelengths) and 400 feet from the wave-making device at one end of it. At 200 feet the amplitude of the wave was virtually uniform, but at 400 feet it was not. Benjamin and Feir interpreted such results in terms of the spontaneous growth of two satellite waves, with wavevectors $k(1 + \kappa)$ and $k(1 - \kappa)$, where k is the wavevector of the initial disturbance. They showed how each of these satellites makes the amplitude of the other one grow, so that together they grow exponentially.

Since that pioneering work a number of other instabilities affecting Stokes waves have been discovered; one rather spectacular example was mentioned in §1.15 and is illustrated there by fig. 1.14. They are of too specialised interest to merit discussion, but the *Benjamin–Feir instability* is not confined to Stokes waves

– it may affect waves in any non-linear dispersive system – and physicists should know of its existence for that reason.

Further reading

The collected papers of Rayleigh and Taylor may be consulted in connection with the instabilities named after one or both of them (J. W. Strutt (Lord Rayleigh) *Scientific Papers*, *Vols I to IV*, Cambridge University Press, 1899 to 1920, and G. K. Batchelor, *The Scientific Papers of G. I. Taylor*, *Vols I to IV*, Cambridge University Press, 1958 to 1970. *Physical Fluid Dynamics*, by D. J. Tritton, Clarendon Press, Oxford, 2nd edition 1988, provides a good introduction to the complex field of thermal convection and convective instabilities, besides giving many references to the original literature; further details are to be found in *Modern Developments in Fluid Dynamics*, edited by S. Goldstein, Clarendon Press, Oxford, 1938, in *Hydrodynamic and Hydromagnetic Stability*, Clarendon Press, Oxford, 1961, by S. Chandrasekhar, and in a review article devoted to Bénard convection in particular by F. H. Busse, *Rep. Prog. Phys.*, **41**, 1929, 1978. The analysis presented in §8.9 is based primarily upon an early paper by D. T. J. Hurle & E. Jakeman, *J. Fluid Mech.*, **47**, 667, 1971; many more papers have since been published on the subject of Bénard convection in binary liquids, but no authoritative review of this recent work is yet available. In connection with §8.14, consult T. B. Benjamin, *Proc. Roy. Soc. A*, **299**, 59, 1967.

9

Turbulence

9.1 Introduction

It is not infrequently claimed that the subject of turbulence contains the last great unsolved problems that classical physics has to offer. What are these problems, and how important are they? These are not easy questions to answer in a short space, and the answers sketched under four headings below are partial in two senses of the word. They are partial in that they are incomplete, and they are also partial in that they reflect the prejudices of someone whose understanding of the subject derives at second hand from what others have written about it.

(i) *The development of turbulence*

Turbulence is often triggered by one of the instabilities discussed in chapter 8, and these have been exhaustively studied and seem well enough understood. Relatively little is known, however, about the processes which link trigger and explosion, i.e. which lead from an infinitesimal perturbation in one part of a fluid system to genuine turbulence downstream. Most fluid dynamicists probably believed until the 1970's that there were few general principles to be discovered in this area, apart from the essentially qualitative idea that once a state of laminar flow has been corrupted by one perturbation it tends to provide a breeding ground within which perturbations on a smaller scale may grow. It was recognised that turbulent flow is unpredictable: however precisely the initial conditions of potentially turbulent fluid are specified it is surely impossible to calculate the instantaneous velocity field $\boldsymbol{u}\{\boldsymbol{x}, t\}$ at some later time, though the time-average, $\overline{\boldsymbol{u}}\{\boldsymbol{x}\}$, may be calculable in principle, if not in practice. In so far as that unpredictability appeared a trifle paradoxical – after all, the classical laws which govern the motion of fluids are completely deterministic – the paradox was resolved using ideas due to Landau. Landau's philosophy was that each new instability which develops in a fluid introduces a new frequency into the spectrum of its motion, a frequency which is in general incommensurate with the frequencies already present; moreover, because each instability develops from noise it develops with random, i.e. unpredictable, phase.

During the 1970's it became apparent, partly as result of the increasing availablity of computers, that dynamical systems which are governed by non-linear equations of motion may become unpredictable in a rather different way. Suppose that, as some parameter which expresses the degree of non-linearity is increased, a system becomes overstable and starts to oscillate, in a way that may be represented in phase space by orbital motion round a closed trajectory with a well-defined period T. Then at some larger value of the parameter the trajectory may undergo a small but significant change, such that it closes on itself after two orbits rather than one; the period is then $2T$ rather than T. Further increases in non-linearity lead to further period doublings, until eventually, at a well-defined *accumulation point*, the period becomes infinite. By then the trajectory has split so often that it no longer closes on itself, and two points in phase space which both lie on the trajectory are liable to be separated by an infinite length of it, even though the distance between them, measured as the crow flies, is infinitesimal. In such circumstances the behaviour of the system is unpredictable because deviations which are infinitesimal to begin with, due perhaps to noise, grow exponentially, like infinitesimal displacements from a state of unstable equilibrium [§8.1]. There are normally *windows* of the non-linearity parameter above the accumulation point within which the motion of the system becomes periodic again – in the largest of them the period undergoes a second doubling sequence, from $3T$ to $6T$ etc. – but outside these windows the motion is said to be *chaotic*.

Computations based upon a variety of apparently different mathematical models have shown that this *route to chaos* is governed by laws of remarkable universality, and the laws have been verified by experiments on a variety of real systems too. Among the real systems which obey them are fluid systems undergoing convection. That being so, it is scarcely surprising that in the heady days of discovery a hope took root that the theory of chaos would lead in due course to a deeper understanding of fluid turbulence in general. That hope is still alive, but the gulf between 'chaotic' motion, which normally involves only one or two degrees of freedom, and genuinely 'turbulent' motion, which always involves a very large number of degrees of freedom indeed, is wide and may ultimately prove unbridgeable.

It should be added that there are several other routes to chaos besides the one described above. In particular, there may be ranges of the non-linearity parameter within which a system's motion is essentially periodic, and predictable, for many periods in succession but is interrupted by bursts of chaos of unpredictable duration; this phenomenon is called *intermittency*. There are aspects of universality in it, as there are in the period doubling phenomenon, and it too is displayed by real systems as well as model ones. Further investigation of intermittency in this technical sense may one day help to explain various intermittent effects which accompany the onset of turbulence in fluid systems which are more complicated than convecting layers. It seems to have little to do, however, with the best known and most easily demonstrated example: the essentially predictable intermittency which accompanies the transition to turbulence in pipe flow [§9.3].

(ii) *The structure of homogeneous turbulence*

When the motion of fluid through a pipe is laminar, its velocity profile becomes independent of distance x along the pipe once x is larger than the so-called inlet length [§1.13]. Averaged over time, the motion becomes independent of x for large x in the case of turbulent pipe flow also, and when this state has been achieved the turbulence is said to be *well-developed*. To describe its *structure* we need to specify more than the profile of $\bar{u}$, since $\bar{u}\{x\}$ contains no information whatever about the local eddying motions which are a feature of turbulence. Upon what other characteristics of the velocity field should we concentrate attention?

According to G. I. Taylor and his successors, a statistical approach is appropriate; we should concentrate on velocity correlation functions, in particular on the second-order autocorrelation functions $\langle u_i\{\boldsymbol{x},t\}u_i\{\boldsymbol{x}',t\}\rangle$ and $\langle u_i\{\boldsymbol{x},t\}u_i\{\boldsymbol{x},t'\}\rangle$ – the angle brackets here imply ensemble averages, and the suffix i (= 1, 2 or 3) refers to the direction of the velocity component involved – and on the characteristic distances $(\boldsymbol{x}' - \boldsymbol{x})$ and times $(t' - t)$ over which these decay. Equivalently, we may concentrate on the power spectrum of $\tilde{\boldsymbol{u}}\{\boldsymbol{k},\omega\}$, the Fourier transform of $\boldsymbol{u}\{\boldsymbol{x},t\}$. These are measurable quantities, since fluctuations of local velocity may be monitored with the aid of hot-wire or laser anemometers.[1]

The structure of the turbulence that characterises pipe flow is significantly complicated by the presence of the pipe, so statistical theories are normally applied in the first instance to the ideal case of *homogeneous turbulence*, a case which one may attempt to realise experimentally by passing a fluid stream with uniform velocity $\boldsymbol{U}$ through a uniform grid of wires set normal to the flow. The vortices which are shed by the wires combine to produce, some way downstream, a turbulent field which may be described as well-developed and whose structure does not vary with position, in a transverse direction at any rate. It must vary slightly with distance from the grid as dissipation takes its toll of the turbulent kinetic energy, but this longitudinal variation is usually slow enough to be negligible and in that case the turbulence may be not only homogeneous but also, if viewed in a frame of reference which moves at $\boldsymbol{U}$ relative to the grid and in which

[1] Hot-wire anemometers exploit the fact that heat losses due to forced convection are proportional to $\sqrt{u_\perp}$, where $u_\perp$ is the component of fluid velocity normal to the axis of the wire. Laser anemometers exploit the Doppler shift of light scattered by small solid particles suspended in the fluid and moving with it.

$\bar{u}$ is therefore zero, virtually *isotropic*.[2] It is to the ideal case of ***homogeneous isotropic turbulence*** that many theories refer. They lead to various predictions which may be compared with experiment, concerning not only power spectra and their rates of decay but also the diffusion constants which govern rates of transport of molecules, or perhaps of heat, within the turbulent field.

To predict the structure of homogeneous isotropic turbulence is not the modest goal it may appear to be. An ensemble average of the Navier–Stokes equation shows that $\langle \boldsymbol{u} \rangle$ or $\bar{\boldsymbol{u}}$ (ensemble averages and time averages amount to the same thing in contexts where neither is changing systematically with time) depends to some extent on an average which is of second order in the fluid velocity, namely on $\langle (\boldsymbol{u} \cdot \boldsymbol{\nabla}) \boldsymbol{u} \rangle$, and it follows that $\bar{\boldsymbol{u}}$ cannot be rigorously calculated without some knowledge of the second-order correlation functions mentioned above. In the same way, however, these second-order quantities cannot be rigorously calculated without some knowledge of third-order correlation functions, and so on. There is a fundamental problem here, known as the *closure problem*. It occurs in different guises in many branches of physics: in quantum field theory, for example, and in classical theories of the molecular structure of liquids. Techniques for papering over the closure problem exist, but they are of interest to specialists only and beyond the scope of this book.

(iii) *Quasi-steady turbulent shear flows*

Fully developed turbulent pipe flow shares two of the properties of turbulent flow behind a grid: it is quasi-steady in the sense that $\bar{\boldsymbol{u}}$ is independent of time, and the turbulence which characterises it is thought to be locally isotropic, or almost so, over much of the pipe's cross-section. It differs from grid flow, however, in that $\bar{\boldsymbol{u}}$ is far from uniform, and the turbulence is liable to be inhomogeneous in consequence. In this respect it is representative of a class of problems of considerable importance in engineering practice, all of them involving turbulent fluids undergoing shear. Turbulent jets, turbulent wakes, and turbulent boundary layers fall within this class.

It will be clear from what has been said above that attempts to calculate the profile of mean velocity in a pipe, jet, wake or boundary layer from first principles, i.e. from the Navier–Stokes equation, have so far been frustrated – and seem likely to be frustrated for the foreseeable future – by the closure problem. That being so we are obliged to rely on heuristic theories, of which two sorts are available. In theories of one sort reliance is placed on the concept of an *eddy viscosity*, supposed to supplement the normal shear viscosity when turbulent motion is superposed on the shearing motion of the main stream, and the eddy

[2] Homogeneity and isotropy are not the same thing: homogeneity requires, for example, that $\langle u_1(\boldsymbol{x},t) u_1(\boldsymbol{x},t) \rangle = \langle u_1^2 \rangle$ is independent of x_1, x_2 and x_3, whereas isotropy requires that $\langle u_1^2 \rangle = \langle u_2^2 \rangle = \langle u_3^2 \rangle$.

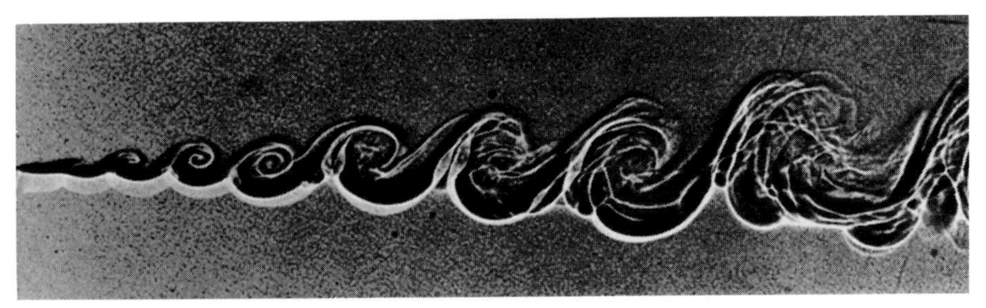

Figure 9.1 One of several published shadowgraphs [cf. fig. 8.24] of a turbulent mixing layer between streams of helium (above) and nitrogen (below) at a pressure of 4 atmospheres; the streams are flowing from left to right with different velocities (mean value about 7 m s^{-1}) such that ρU^2 is the same in both, and the point at which they first meet lies at the left-hand edge of the figure. The pattern of vortices is a continuously changing one and is not often as regular as this.
[From G. L. Brown & A. Roshko, *J. Fluid Mech.*, **64**, 775, 1974.]

viscosity is estimated in terms of the gradients of $\bar{u}$ and of what Prandtl called a *mixing length*. Theories of the other sort, collectively referred to as *scaling theories*, start from the hypothesis that $\bar{u}\{\boldsymbol{x}, t\}$ and other functions which describe the flow are in some respects *self-similar* or *self-preserving*, and in their subsequent development they rely extensively on dimensional analysis. The mixing length theory originally formulated by Prandtl gave results in rather good agreement with experimental data, but because these results were in some respects inconsistent with the results of later scaling theories, and because the whole idea of an eddy viscosity came to seem unconvincing, this theory went out of fashion round about 1950. The eddy viscosity concept can be made more plausible, however, and the most serious inconsistencies with scaling theory can be removed, by a modest reformulation of Prandtl's ideas. Thus reformulated, there seems rather little to choose between the mixing length approach and the scaling approach, at any rate in the context of quasi-steady flow. Both approaches involve hypotheses for which no *a priori* justification is available, and both depend on frequent appeals to experiment.

(iv) *Coherent structures in turbulence*

The photograph – or *shadowgraph* – in fig. 9.1 shows an instantanteous side view of two streams of gas which are moving from left to right past a thin plate which initially separates them, but which terminates at the left-hand edge of the field of view. Thereafter the streams start mixing with one another, and because they consist of different gases, with different refractivities, the refractivity within the so-called *mixing layer* is inhomogeneous; it is this inhomogeneity which is responsible for the shadow effect. Outside the mixing layer the velocities, U_1 in one gas and U_2 in the other, are uniform, but within the mixing layer the motion appears to be characterised by a spiral structure of remarkable regularity, such that the size of the spirals, and the separation between them, increases with distance from the point of origin of the mixing layer in a systematic (and self-preserving) manner. When the spirals are observed over a period of time they are found to be drifting from left to right at something like the mean velocity $\frac{1}{2}(U_1 + U_2)$ but also to be changing continuously, in such an unpredictable way that the motion has to be regarded as turbulent. The structure of the turbulence is evidently far from homogeneous, however; it is said to be *coherent* instead.

At first sight, the coherent structure in this case may be thought to be a relatively straightforward consequence of the Kelvin–Helmholtz instability [§8.11]; the spirals in fig. 9.1 are distinctly reminiscent of those sketched in fig. 8.23, which shows the expected appearance of an ideal vortex sheet between two streams of fluid moving at different speeds, a short time after this instability has begun to develop. However, the fact that the spacing between the spirals increases from left to right in fig. 9.1, whereas the drift velocity is uniform, implies that the mean frequency with which spirals pass a fixed point must decrease from left to right. That implies in turn that the small spirals on the left need to merge with their neighbours and reform in order to generate, a moment later, spirals resembling the large ones on the right. It seems that in this context the vorticity has a natural tendency to organise itself into rotating regions of steadily increasing size. The process is not understood, though the footnote on p. 335 may shed some light on it. The heuristic theories referred to above are quite unable to explain such subtleties, and the only aspect of fig. 9.1 which they can account for is the wedge-like shape of the mixing layer.

There are many other examples of turbulent flow in which transient large-scale *coherent structures* of this sort seem to arise spontaneously. The task of trying to explain them may seem to be of less fundamental importance than the task of trying to solve the closure problem, but it may well prove more rewarding.

The topics mentioned in this introduction are taken up in the sections that follow, at a level which is inevitably superficial. The single section devoted to the theory of homogeneous turbulence concerns only a few basic ideas associated with the name of Kolmogorov which can be justified by dimensional reasoning alone. These ideas are used in later sections devoted to turbulent shear flows with an uncritical freedom that is likely to leave some readers unsatisfied. No justification is offered, for example, for the tacit assumption that the turbulence remains more or less isotropic even in the presence of shear. Moreover, whereas homogeneous turbulence is described in terms of the Fourier components $\tilde{\boldsymbol{u}}\{\boldsymbol{k}\}$, the turbulence in shear flows is described more loosely, in terms of *eddies*. Anyone who has seen how the water in a fast river swirls as it moves over a rocky bottom will appreciate

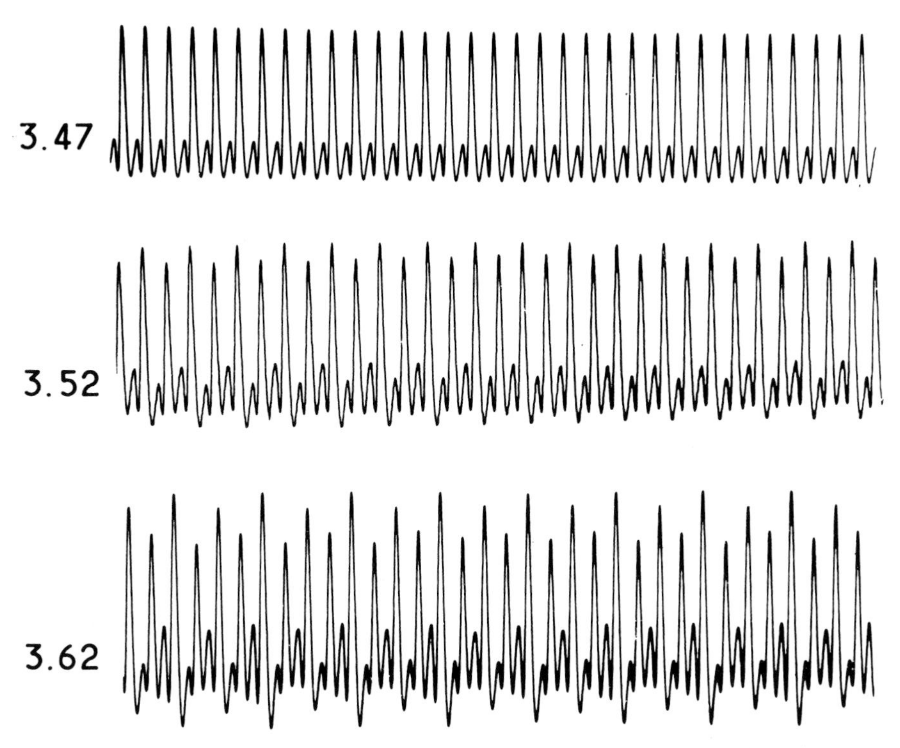

Figure 9.2 Variation of temperature (plotted vertically) with time in a rectangular Bénard cell containing mercury, for three different values (as marked on the diagram) of Ra/Ra_c; each trace spans a time interval of about 200 s.
[From A. Libchaber, C. Laroche & S. Fauve, *J. de Physique Lettres*, **43**, L-211, 1982.]

why it may well be appropriate to think in terms of eddies, and the general idea that large eddies correspond to small k and *vice versa* is surely acceptable. Nevertheless, the relation between the language of Fourier components and the language of eddies deserves more scrutiny than it receives below. As for the large eddies revealed in fig. 9.1, these and other examples of coherent structures are referred to in only three of the sections, and only in a glancing way.

Thus readers with a special interest in the subject matter of this chapter should pay particular attention to the suggestions for further reading which are to be found, as usual, at the end of it.

9.2 Period doubling and intermittency in Bénard convection

Figure 9.2 is reproduced from a paper describing experiments on Bénard convection in liquid mercury contained in a small rectangular cell with dimensions

$7 \times 7 \times 28$ mm, formed out of two thick horizontal plates of copper held apart by thermally insulating spacers. It shows how the temperature of the mercury at a point just above the middle of the lower copper plate varied with time, for three different values of the Rayleigh Number. Bénard convection begins at $Ra_c \approx 1700$ [§8.7], and the four convective rolls which can be accommodated in a cell of these dimensions are stable for a considerable range of Ra above that; within this range the temperature at a fixed point in the cell does not vary significantly with time. It first starts to oscillate when Ra/Ra_c reaches about 2, at which stage, in this fluid of very low Prandtl Number, the rolls become overstable and start themselves to oscillate with a well-defined period T of a few seconds. By the time $Ra/Ra_c = 3{\cdot}47$, corresponding to the top trace in fig. 9.2, the period has already doubled once. The other traces show it doubling twice more to become $8T$ at $Ra/Ra_c = 3{\cdot}62$, and yet one more doubling was detected. One of the standard results of chaos theory is that the intervals between successive period doublings should decrease as the accumulation point is approached according to the universal law

$$\frac{X_{n+1} - X_n}{X_{n+2} - X_{n+1}} \to 4{\cdot}6692016 \ldots \text{ (the \textit{Feigenbaum Number})} \qquad (9.1)$$

for large n, where X_n is the value of the appropriate non-linearity parameter at which the nth doubling occurs. The experiments of which fig. 9.2 is a partial record led to the result

$$\frac{Ra_3 - Ra_2}{Ra_4 - Ra_3} = 4{\cdot}4 \pm 0{\cdot}1.$$

In that particular experiment a horizontal magnetic field was used to damp the oscillations and enhance the effect, but the period-doubling sequence has been observed in similar experiments using gaseous and liquid helium and water, which are not influenced by magnetic fields because they are non-conducting. Small cells are needed because measurements on large ones tend to be confused by accidental disturbances of one kind or another in the otherwise regular convection pattern.

It needs to be stressed that the accumulation point (Ra_∞), which can readily be identified in these experiments, does not mark the onset of turbulence in the normal sense of that word. Not only are there windows of periodicity above the accumulation point, as described in §9.1(i); even outside these windows the motion of the fluid retains a great deal of the spatial order associated with stable convective rolls at lower Rayleigh Numbers. That can be shown by studying the correlations between the temperature fluctuations at two different points in a cell. Only when these correlations are no longer detectable is the motion fully turbulent. We shall return to the topic of turbulent Bénard convection in §9.10.

Figure 9.3 is reproduced from a paper on Bénard convection in a small cell containing a fluid of large Prandtl Number (silicone oil). Here the traces record the variation with time, at a point near the middle of the cell, of vertical velocity u_z

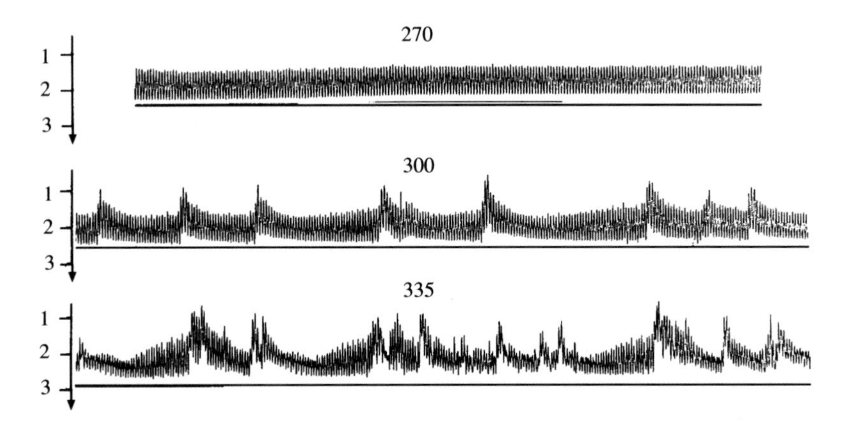

Figure 9.3 Variation of u_z (plotted vertically downwards, in units of mm s^{-1}) with time in a Bénard cell containing silicone oil, for three different values (as marked on the diagram) of Ra/Ra_c; in this case each trace spans a time interval of several hours.
[From P. Bergé, M. Dubois, P. Manneville & Y. Pomeau, *J. de Physique Lettres*, **41**, L-341, 1980.]

rather than temperature. At $Ra/Ra_c = 270$ u_z oscillates with uniform amplitude,[3] but as the Rayleigh Number is further increased the oscillations are increasingly disturbed by erratic and short-lived bursts of motion of a different kind. It is noticeable that at $Ra/Ra_c = 300$, though not perhaps at $Ra/Ra_c = 335$, the oscillations are not interrupted by the bursts; they continue through them with amplitude and phase essentially unchanged. That is not quite the behaviour to be expected of a non-linear dynamical system which is following the standard intermittency route to chaos. Nevertheless, the figure clearly demonstrates intermittency of a sort. Intermittency of another sort is described in the following section.

9.3 The transition to turbulence in pipe flow

According to Poiseuille's law [(1.33)] the discharge rate in pipe flow, which may be represented in dimensionless terms by the Reynolds Number ($Re = 2\rho Q/\pi a\eta$), is proportional to the magnitude of the pressure gradient ∇p (an essentially positive quantity, whereas $\nabla_3 p$ in (1.33) is negative). When turbulence is fully

[3] The increase in Pr shifts the onset of chaos, as explained in §8.5, to values of Ra/Ra_c which are much larger than those to which the traces in fig. 9.2 refer.

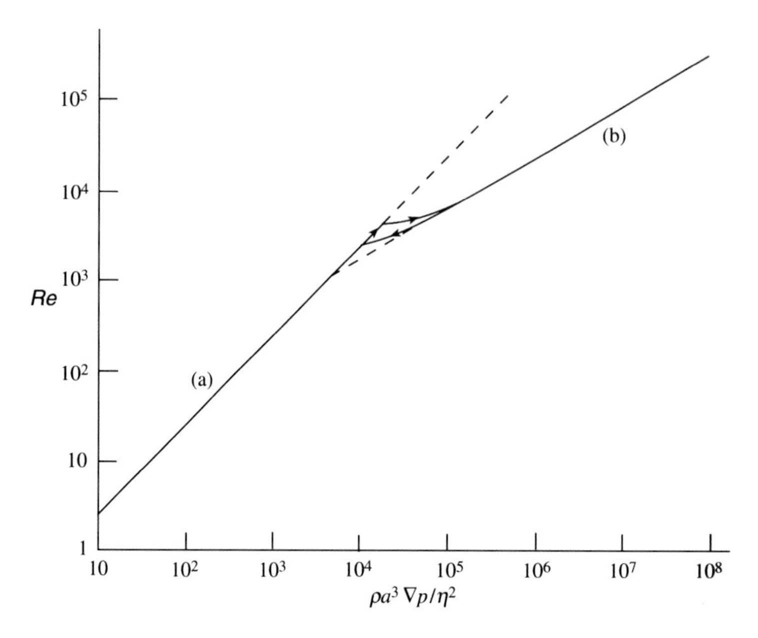

Figure 9.4 Log–log plot of Reynolds Number, which in this context is a dimensionless represention of discharge rate, *versus* a dimensionless representation of longitudinal pressure gradient, for flow through a smooth straight pipe of circular cross-section. The two straight lines represent (a) Poiseuille's law [(1.33)] and (b) an empirical expression due to Blasius [(9.36)].

developed, however, *Re* is found experimentally to vary like $(\nabla p)^n$, where the exponent n normally lies in the range 0·5–0·6, the precise value depending upon how smooth the walls of the pipe are [§9.8]. Thus on a logarithmic plot of *Re* *versus* ∇p experimental data tend to lie on two straight lines, which normally intersect at a value of *Re* which is about 1000. The transition between these lines occurs well above $Re = 1000$; it is gradual rather than abrupt, and it is normally hysteretic as fig. 9.4 suggests. When the discharge rate is increased slowly from zero the first departures from Poiseuille's law are observed at an upper *critical value* of *Re* which may be as high as 10^5 if the walls are very smooth, if the whole apparatus is vibration-free, and if the entry is flared, but which is more likely otherwise to be about 4000. When the discharge rate is reduced slowly from a high value, however, laminar flow is not completely restored until *Re* has fallen to a *lower critical value* which is close to 2300.

The question of whether there is some absolute limit to the upper critical value has been intensively studied. Linear stability analysis suggests not; it suggests that the parabolic profile associated with ideal Poiseuille flow is intrinsically stable for all values of *Re*. A complete non-linear analysis would presumably show, however, that above $Re \approx 10^5$ only the tiniest disturbance is needed to trigger the transition to fully-developed turbulence. If the entry to the pipe is bluff rather than flared, of course, the transition is triggered there. Whether it is triggered by a Tollmien–Schlichting instability in a boundary layer attached to the inner surface of the pipe [§8.13], or by a Kelvin–Helmholtz instability in a boundary layer which has separated and which exists as a cylindrical vortex sheet enclosing a submerged jet of vorticity-free fluid [§§7.7 and 8.12], is scarcely material.

It is within the hysteretic transition range between the two straight lines of fig. 9.4 that intermittency tends to be observed. It can readily be demonstrated by allowing water from a constant head device to flow through a long horizontal tube – the longer the better – which has an internal diameter of a few millimetres. The mean velocity of the water $\langle \overline{u} \rangle$ ($= Q/\pi a^2$)[4] can be judged from the curvature of the parabola in which it falls after it emerges as a jet from the pipe's open end. When the pressure head is increased from a small value the parabola becomes steadily flatter as the water picks up speed until the transition region is reached, whereupon the emergent jet starts to pulsate. Given the right pressure head, the right entry conditions, and no vibration it may pulsate with remarkable regularity, the period in a typical case being of the order of seconds.

The following explanation of this effect appears to be correct in principle though it may be oversimplified. The flow becomes turbulent near the pipe's entry when Re first reaches the upper critical value. From there the turbulence spreads steadily downstream, to create what is vividly called a *turbulent slug*. For continuity reasons, the mean velocity must be the same in the turbulent slug as it is in front of the slug, where the fluid is still moving along the pipe in a laminar fashion. That means that the pressure gradient, or to be more precise its modulus, is greater in the slug than in front of it and hence, if the total pressure drop along the pipe is held constant, that once a slug has formed the pressure gradient in the laminar fluid must be less than it was previously. The slug therefore reduces Q and $\langle u \rangle$, and these quantities continue to fall as long as the slug is increasing in length. When they have fallen to such an extent that Re is only about 2300, laminar conditions are restored near the entry zone. The slug may continue to increase in length if the velocity with which its front end travels along the pipe, U_F, exceeds the velocity, U_R, with which its rear end follows after. In due course, however, it is voided from the pipe's exit. Thereupon Q increases to its original value and the process repeats itself. The process seems to bear little relation, as was indicated in §9.1(i), to the scenarios involving intermittency which have been mapped out as possible routes to chaos.

For given Q, the parabolic profile of $u\{r\}$ which characterises laminar pipe flow [(1.32)] and the profile of $\overline{u}\{r\}$ which characterises turbulent flow are distinctly

[4] The angle brackets represent in this context a spatial average over the cross-section of the pipe, rather than an ensemble average as in §9.1(ii).

different [§9.8]. Complications arise at interfaces between laminar and turbulent flow because of the mismatch, and the shapes adopted by the front and rear ends of a turbulent slug are not known. Their velocities can be measured, however, and for $Re \approx 4000$ one finds $U_F \approx 1{\cdot}5\langle u\rangle$ and $U_R \approx 0{\cdot}7\langle u\rangle$. There is little to say about these values except that they are both intermediate between the minimum value of u in the laminar fluid (which is zero, of course, in contact with the walls of the pipe) and its maximum value (which is $2\langle u\rangle$, on the pipe's axis). The ratio $U_R/\langle u\rangle$ falls steadily towards zero as Re is increased above 4000, but $U_F/\langle u\rangle$ stays more or less constant. When Re is reduced below 4000, however, $U_R/\langle u\rangle$ rises and $U_F/\langle u\rangle$ falls, and the lower critical value of Re is where they come together, both of them then being equal to (or perhaps slightly less than) unity. Below this lower critical value the flow is of necessity laminar because turbulent slugs, if they did exist, would shrink rather than grow.

9.4 The energy cascade in homogeneous turbulence

A qualitative picture of what happens when a uniform stream of fluid passes through a grid with spacing d, at a Reynolds Number which is large enough for turbulence to arise behind the grid, is as follows. The fluid exerts a drag force on the grid and the grid exerts a corresponding reaction on the fluid. This means that in the frame of reference S′ in which the mean velocity of the fluid is zero the grid does work upon the fluid, supplying it with kinetic energy at a uniform rate. This energy is initially associated with a fairly regular structure of eddies shed into the fluid from the wires of the grid, and the Fourier components of the velocity field just behind the grid, $\tilde{\boldsymbol{u}}'\{\boldsymbol{k}\}$ in S′, are large in the neighbourhood of some wavevector which is presumably of the order of $1/d$ and relatively small elsewhere. As time passes, however, i.e. as distance from the grid increases, eddies get twisted around by their neighbours and instabilities of one sort or another develop. The lines of vorticity embedded in the fluid become increasingly convoluted in consequence, and as each line becomes longer so the density of lines increases – this is the effect of *intensification of vorticity by stretching* described in §7.1. However, the lines of vorticity are not permanently embedded; they can diffuse. Hence they can come together with other lines of opposite sense, and they ultimately disappear by forming closed loops which collapse to a point. During this evolutionary process the energy initially associated with wavevectors in the neighbourhood of $1/d$ is passed to smaller values of k until it is ultimately dissipated as heat.

Imagine now a rather different situation, in which a body of fluid whose mean velocity is everywhere zero is being continuously stirred in some way. The stirring feeds kinetic energy continuously into large-scale motions characterised by a wavevector k_o say, at a rate ε_o per unit mass of fluid, and these motions feed the energy in a continuous *cascade* to larger k, with continuous dissipation as heat as the end result. The steady-state spectrum of the resultant homogeneous and isotropic turbulence may be characterised by a function $E\{k\}$ ($\propto \tilde{\boldsymbol{u}}'\{k\}^2$), where $E\{k\}\mathrm{d}k$ is the mean kinetic energy per unit mass which is stored in the range of wavevectors betweeen k and $k + \mathrm{d}k$. The dimensions of ε_o are $[\mathrm{L}]^2[\mathrm{T}]^{-3}$ while those of E are $[\mathrm{L}]^3[\mathrm{T}]^{-2}$, so that ε_o^2/k^5E^3 is a dimensionless combination. If we may assume that E depends only on ε_o, k and perhaps ρ, i.e. if we may assume that k_o and η are irrelevant, we may at once conclude that

$$E \sim \varepsilon_o^{2/3}\, k^{-5/3}. \tag{9.2}$$

The sign $\sim$ in this pseudo-equation, and in similar ones below, implies rather more than proportionality of the two sides; it implies an expectation, based partly on hope and partly on experience, that the numerical coefficient of proportionality is of order unity. In fact, the suppressed coefficient on the right-hand side of (9.2) is thought to be about 1·5.

The assumption that k_o is irrelevant has been disputed, and there exist theories of homogeneous turbulence which predict corrections to (9.2) in which k_o appears; the evidence from experiment seems to be against them, however. As for the assumption that η is irrelevant, this is sound enough at small values of k but, as the following argument shows, it fails at large k where viscous dissipation becomes important. The dissipation within a range $\mathrm{d}k$ may be calculated from (6.26): per unit mass, it is of order $\nu k^2 E\mathrm{d}k$, where ν is the kinematic viscosity, η/ρ. On account of this dissipation the rate at which kinetic energy is transferred down the cascade, say ε per unit mass and unit time, must dwindle as k increases, with

$$\frac{\mathrm{d}\varepsilon}{\mathrm{d}k} \sim -\,\nu k^2 E. \tag{9.3}$$

But if (9.2) is satisfied at small values of k where dissipation is negligible and where ε is equal to ε_o, it is plausible to suppose that the related pseudo-equation

$$E \sim \varepsilon^{2/3} k^{-5/3} \tag{9.4}$$

is satisfied where dissipation, though not negligible, is still relatively small, and this hypothesis, taken in conjunction with (9.3), implies that

$$\frac{\mathrm{d}\ln E}{\mathrm{d}\ln k} \sim -\frac{5}{3} - \frac{2\nu}{3}\frac{k^{4/3}}{\varepsilon^{1/3}}. \tag{9.5}$$

We may conclude that viscosity is negligible where k is small compared with a wavevector k_K defined by

$$k_\mathrm{K} = \left(\frac{\varepsilon_o}{\nu^3}\right)^{1/4}, \tag{9.6}$$

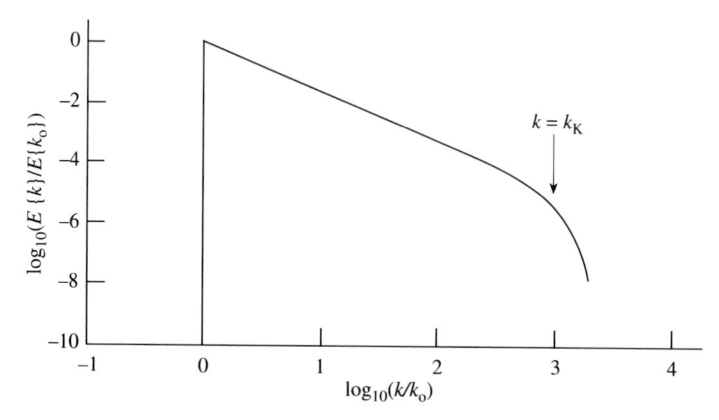

Figure 9.5 Sketch of the three-dimensional Kolmogorov spectrum for the case $k_K/k_o = 10^3$. The sharp rise at $k = k_o$ is idealised and would be difficult to realise in practice, while the shape of the curve near $k = k_K$ conforms only roughly to the guess represented by (9.5). The straight section of the curve between the two cut-offs has a slope of $-\frac{5}{3}$.

on the grounds that the second term on the right-hand side of (9.5) is there small compared with the first. Once k becomes comparable with k_K, however, viscosity can be ignored no longer; according to (9.5), it has the effect of cutting off the spectrum rather abruptly, in the manner suggested by fig. 9.5. Equation (9.2) is called, after the person who first drew attention to it, the *Kolmogorov* $-\frac{5}{3}$ *power law*, and the inverse of the high-k cut-off, i.e. $1/k_K$, is the *Kolmogorov scale length*.

Now once a Kolmogorov spectrum has been established in the manner described above it should persist when stirring ceases, for a time at any rate; even though the cascade is then turned off at its source, there is enough energy stored in the motions for which $k \approx k_o$ to maintain the rate of energy transfer at larger values of k. Thus (9.2) is applicable to the homogeneous turbulence behind a grid throughout what is called the *inertial subrange* of k, such that k_o (i.e. $1/d$) $\ll k \ll k_K$, though ε_o should in this context be regarded as the rate at which energy is dissipated per unit mass rather than the rate at which energy is supplied. In practice, the inertial subrange in grid turbulence tends to be rather short, and the experimental data which most strikingly confirm Kolmogorov's prediction are those obtained from studies of turbulence on a large scale, in the oceans or in the atmosphere. Figure 9.6 shows a spectrum measured in a channel off the coast of Vancouver Island, using a flowmeter towed by a ship. At its narrowest point the channel width (L_{min}) is only 3 km and the tide runs through it with a speed (u_{max}) of up to 15 knots, so the Reynolds Number defined by $Re = u_{max}L_{min}/\nu$ is 10^{10} or more; at such a large value, intense turbulence is inevitable at the point where the data were collected, downstream from the narrows where $\overline{u}$ was only about one

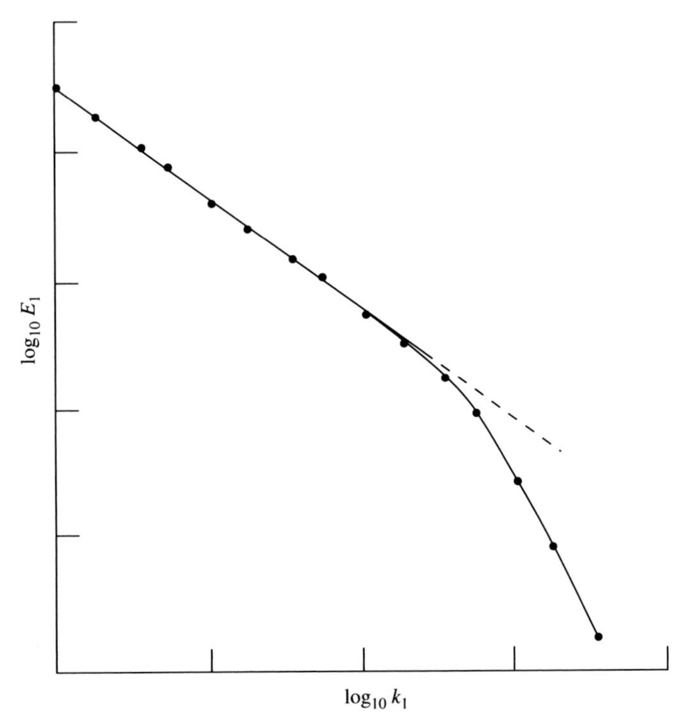

Figure 9.6 Section of a one-dimensional spectrum determined experimentally, in circumstances where k_K/k_o was probably at least 10^7. Between successive marks on the abscissa scale k_1 changes by a factor 10, while between successive marks on the ordinate scale E_1 changes by 100, as in fig. 9.5. The straight line fitted to the points on the left of the diagram has a slope of $-\frac{5}{3}$ exactly.
[Based on data plotted in H. L. Grant, R. W. Stewart and A. Moilliet, *J. Fluid Mech.*, **12**, 241, 1962.]

fifth of u_{max}. The spectrum is actually a one-dimensional projection of the three-dimensional spectrum we have so far discussed, the quantity plotted along the ordinate scale being $\log_{10}E_1$ rather than $\log_{10}E$, $E_1\{k_1\}dk_1$ being the mean kinetic energy per unit mass which is stored in the range of wavevectors whose components in the x_1 direction lie between k_1 and $k_1 + dk_1$. It can readily be shown, however, that where the three-dimensional spectrum obeys a $-\frac{5}{3}$ power law the one-dimensional spectrum should do the same, though elsewhere the two are bound to differ. The fit of the experimental points in the inertial subrange of fig. 9.6 to a straight line of slope $-\frac{5}{3}$ is impressive.

Incidentally, there are several different ways of defining Reynolds Numbers for turbulent flows such as that one. Once the turbulence is fully developed, the velocity of the stream in which it has originated and the width of the aperture through which that stream has emerged are arguably of less significance than the root mean square value of the fluctuations in fluid velocity, i.e. of

$$\boldsymbol{u}'\{\boldsymbol{x}\} = \boldsymbol{u}\{\boldsymbol{x}\} - \overline{\boldsymbol{u}}\{\boldsymbol{x}\}, \tag{9.7}$$

and the scale length $1/k_o$ which describes the long wavelength cut-off of the spectrum. From (9.2) with 1·5 as the coefficient on the right-hand side we have, when k_K is so much larger than k_o as to be irrelevant in this context,

$$u'_{\mathrm{rms}} = \left(2\int_{k_o}^{k_K} E\mathrm{d}k\right)^{1/2} \approx 2\,\frac{\varepsilon_o^{1/3}}{k_o^{1/3}}, \tag{9.8}$$

so an alternative definition of the Reynolds Number is

$$Re = \frac{u'_{\mathrm{rms}}}{k_o\nu} \approx 2\,\frac{\mathrm{e}_o^{1/3}}{\nu k_o^{4/3}} = 2\,\frac{k_K^{4/3}}{k_o^{4/3}}. \tag{9.9}$$

By the time $\overline{u}$ has fallen to only about one fifth of u_{max}, however, almost all the kinetic energy initially stored as $\frac{1}{2}u_{\mathrm{max}}^2$ per unit mass has been transferred to the fluctuations and is stored as $\frac{1}{2}u'^{\,2}_{\mathrm{rms}}$ instead, so u_{max} and u'_{rms} must have been virtually the same in the situation to which fig. 9.6 refers. Moreover, the scale length $1/k_o$ was probably much the same as L_{min}. Hence the Re of (9.9) may not have been very different from the Re of the previous paragraph, and when both of them were about 10^{10} the ratio k_K/k_o, according to (9.9), was over 10^7.

9.5 Eddy viscosity and the mixing length

Most of the problems to be discussed in succeeding sections involve fluids in a fully-developed (though not necessarily homogeneous) state of turbulence which are undergoing quasi-steady shear flow. By way of introduction to them, consider a turbulent fluid whose local velocity $\boldsymbol{u}$ has components $(\overline{u}_1 + u'_1, u'_2, u'_3)$, where $\overline{u}'_1, \overline{u}'_2$ and $\overline{u}'_3$ all vanish and where $\overline{u}_1$ varies with x_2 but does not depend upon x_1 or x_3. We may suppose the stream to have reached a state in which the fluctuations of velocity which $\boldsymbol{u}'$ describes are continuously deriving kinetic energy from the mean motion which $\overline{\boldsymbol{u}}$ describes, and therefore from whatever external agency is maintaining the mean motion, at just the right rate to balance the losses due to viscous dissipation. Hence we may take the non-vanishing averages $\overline{u'^2_1}$ etc. to be likewise independent of x_1 and x_3.

Across any fixed plane normal to the x_2 axis, momentum in the x_1 direction is transferred in two ways. Firstly, it is transferred, as in the corresponding laminar flow situation, by the viscous shear stress; the average of this is clearly

$$\overline{s}_3 = \eta\,\overline{\left(\frac{\partial u_1}{\partial x_2} + \frac{\partial u_2}{\partial x_1}\right)} = \eta\,\frac{\partial \overline{u}_1}{\partial x_2}$$

[(6.3)]. Secondly, it is transferred by particles of fluid which cross the plane, carrying momentum with them; per unit area, the average rate of transfer due to particle transport (in the same sense as the transfer due to $\bar{s}_3$) is[5]

$$- \rho\overline{u_1 u_2} = - \rho\overline{u_1' u_2'} = \bar{s}_{3,\mathrm{Re}} \tag{9.10}$$

The quantity $\bar{s}_{3,\mathrm{Re}}$ defined by (9.10) plays the role of a shear stress and is known as the *Reynolds stress*. It is normally obliged to vanish for symmetry reasons when $\mathrm{d}\bar{u}_1/\mathrm{d}x_2$ is zero, but when $\mathrm{d}\bar{u}_1/\partial x_2$ is non-zero the Reynolds stress is clearly liable to be non-zero too, because there is then a systematic difference in u_1 between particles which are crossing the plane with positive u_2 and those which are crossing it with negative u_2.

That qualitative point should need little explanation, for all physicists are familar with the concept of Reynolds stress in the context of the kinetic theory of gases. In a gas undergoing laminar shear flow with $u_1 = u_1\{x_2\}$, $u_2 = u_3 = 0$ and $\boldsymbol{u}' = 0$, the stress s_3 is wholly a Reynolds stress, attributable to momentum transferred by particles which are simply the molecules of the gas. In that context the shear stress is well known to be proportional to $\mathrm{d}u_1/\mathrm{d}x_2$, and the coefficient of proportionality which constitutes the shear viscosity of a gas is well known to be given in terms of the r.m.s. molecular velocity c_{rms} (as measured in S′) and the mean free path ℓ by the formula

$$\eta \sim \rho c_{\mathrm{rms}} \ell \tag{9.11}$$

[(A.35)]. Can we relate turbulent Reynolds stresses to local velocity gradients in the same sort of way? If we allow for them, as must always be formally possible, by adding to the normal shear viscosity η an *eddy viscosity* η_{e}, defined by the equation

$$\bar{s}_{3,\mathrm{Re}} = \eta_{\mathrm{e}} \frac{\mathrm{d}\bar{u}_1}{\partial x_2}, \tag{9.12}$$

will η_{e} turn out to be a predictable quantity? Can we hope to deduce a formula for it by direct analogy with (9.11)?

All attempts to develop that attractively simple idea involve assertions which cannot be fully justified, and the following argument is no exception. Let us suppose at the start that even though the turbulence is not necessarily homogeneous it is nevertheless described by the energy spectrum $E\{k\}$ of (9.2). The wavevector k_{o} at which this spectrum cuts off on the low-k side plays a large part in the argument, but the reader should for the time being suspend curiosity as to what determines this.

Granted that first step, it is not difficult to estimate the velocity which plays the role of c_{rms} in (9.11). Let us think of the particles of fluid which are to play the role

[5] A related result may be found in the discussion of acoustic streaming in §7.14. Note that the Reynolds stress is strictly speaking a second-rank tensor, of which $\bar{s}_{3,\mathrm{Re}}$ is, in the present context, the only significant component.

of gas molecules as being macroscopic spheres of fixed radius a, whose centres of mass have instantaneous velocities $\boldsymbol{U}'$ in the frame S′; we will choose the magnitude of a shortly. It can then be shown that

$$\overline{\boldsymbol{U}' \cdot \boldsymbol{U}'} = 2 \int_{k_o}^{k_K} E(f\{ka\})^2 \mathrm{d}k,$$

where

$$f\{ka\} = \frac{3\{\sin(ka) - (ka)\cos(ka)\}}{(ka)^3}.$$

Because the function $f\{ka\}$ tends to unity in the limit $ka \to 0$, the root mean square value of U' when $k_o a \ll 1$ is just the u_{rms} described by (9.8), and

$$U'_{\mathrm{rms}} \sim \frac{\varepsilon_o^{1/3}}{k_o^{1/3}} \tag{9.13}$$

remains an adequate approximation, reliable to within a factor 2 or thereabouts, over most of the range $0 < k_o a < 1$.

To estimate the length ℓ which plays the role of the molecular mean free path in a gas is not so easy, for the centre of mass of a spherical particle of fluid cannot be regarded as moving in straight lines between isolated collisions; it surely follows an erratically curved path under the influence of the fluid which surrounds it. If it is a small particle ($k_o a \ll 1$) it is swept along by its surroundings and its effective mean free path is almost wholly determined by them. If it is a large particle, however, such that $k_o a \geqslant 1$, it should be legitimate to treat the mean velocity of the surrounding fluid as uncorrelated with $\boldsymbol{U}'$. In that case, we may perhaps assert that it experiences, besides a succession of random buffets that are continually accelerating it in fresh directions, a systematic retarding force of magnitude $-6\pi\eta_e a\boldsymbol{U}'$.[6] Such a retarding force would cause the autocorrelation function $\overline{\boldsymbol{U}'\{t\} \cdot \boldsymbol{U}'\{t+\tau\}}$ to decay exponentially with τ, its lifetime being of order $\rho a^3/\eta_e a$. Hence the effective mean free path is perhaps given when $k_o a \geqslant 1$ by

$$\ell \sim U'_{\mathrm{rms}} \frac{\rho a^2}{\eta_e},$$

from which it follows that if we may indeed write

$$\eta_e \sim \rho U'_{\mathrm{rms}} \ell \tag{9.14}$$

[6] This assertion is based upon Stokes's law [(6.68)]. However, because the fluid particle is viewed as a sphere of fixed radius rather than as a lump whose shape is continuously changing, composed of a fixed set of smaller particles, the stresses primarily responsible for altering its momentum are Reynolds stresses. Hence the use of η_e in place of η; according to (9.16), η is negligible by comparison with η_e at large Reynolds Numbers. An appeal to Stokes's law is adequately justified in this context, since the result expressed by (9.15) suggests that $\rho a U'_{\mathrm{rms}}/\eta_e$ – a Reynolds Number, but a quite different Reynolds Number from the one just referred to – is of order unity.

[cf. (9.11)] then we may also write

$$\ell \sim a.$$

We now have an estimate for U'_{rms} which is adequate when $k_\text{o}a < 1$ and an estimate of sorts for ℓ which may be adequate when $k_\text{o}a > 1$. The best we can do is to choose $a \approx 1/k_\text{o}$, in which case (9.14) reduces, in view of (9.13), to

$$\nu_\text{e} = \frac{\eta_\text{e}}{\rho} \sim \frac{U'_{\text{rms}}}{k_\text{o}} \sim \frac{\varepsilon_\text{o}^{1/3}}{k_\text{o}^{4/3}}. \tag{9.15}$$

It then follows [(9.6) and (9.9)] that

$$\frac{\nu_\text{e}}{\nu} \sim \frac{k_\text{K}^{4/3}}{k_\text{o}^{4/3}} \sim Re. \tag{9.16}$$

Finally, we must consider what determines the rate ε_o per unit mass at which kinetic energy is supplied to the turbulence by the main stream and subsequently dissipated as heat. In laminar shear flow of the type we are considering the rate of dissipation per unit mass is $\nu(\mathrm{d}u_1/\mathrm{d}x_2)^2$. When the mean shear stress is $\eta_\text{e}(\mathrm{d}\bar{u}_1/\mathrm{d}x_2)$ rather than $\eta(\mathrm{d}u_1/\mathrm{d}x_2)$ the corresponding rate is surely $\nu_\text{e}(\mathrm{d}\bar{u}_1/\mathrm{d}x_2)^2$, and if this quantity is substituted for the ε_o in (9.15) one obtains

$$\nu_\text{e}^3 \sim \frac{\nu_\text{e}}{k_\text{o}^4}\left(\frac{\mathrm{d}\bar{u}_1}{\mathrm{d}x_2}\right)^2,$$

or

$$\nu_\text{e} \sim \frac{1}{k_\text{o}^2}\left|\frac{\mathrm{d}\bar{u}_1}{\mathrm{d}x_2}\right|. \tag{9.17}$$

This is Prandtl's formula for eddy viscosity, though the justification for it differs from his, and $1/k_\text{o}$ – i.e. ℓ – is Prandtl's *mixing length*.

The mixing length evidently corresponds to the size of the largest eddies present, and the fact that it features prominently in (9.17) is indirect evidence that these eddies play a dominant role in determining the eddy viscosity. Just how big are they likely to be? The diameter of the coherent eddies revealed by the shadowgraph in fig. 9.1 is clearly the width of the mixing layer in which they arise, and it is probably true of most situations in which the turbulent region is restricted in a direction normal to the flow that it is only this restricted width which limits the eddy size; hence the suggestion in §9.4 that in the case of turbulence in a stream emerging from a narrow aperture $1/k_\text{o}$ may correspond to the width of the aperture. It is not easy to be more precise than that, which means, of course, that the magnitude of the eddy viscosity is always in some doubt. It will not escape the reader, moreover, that if $1/k_\text{o}$ is comparable with the lateral dimensions of the turbulent region then the same is true of the lengths a and ℓ which appear in the above analysis. Our reliance on an analogy with (9.11) may well seem perilous on that account, since most of the results of the kinetic theory of gases apply only when the particle size and the mean free path are both very small compared with the dimensions of the 'apparatus' under consideration. It is largely for those reasons, no doubt, that Prandtl's mixing length theory of eddy viscosity has gone out of fashion.

A particular problem with Prandtl's formula, and the principal reason why results based upon it are sometimes inconsistent with results based on scaling theories instead [§9.1(iii)], is that it predicts an eddy viscosity which vanishes where $\mathrm{d}\overline{u}_1/\mathrm{d}x_2$ vanishes. It turns out because of this that when the formula is applied to turbulent flows (whether in submerged jets or between parallel plates) for which the profile of $\overline{u}_1$ is necessarily symmetrical about the plane $x_2 = 0$ it predicts a so-called *velocity defect*, $\overline{u}_1\{0\} - \overline{u}_1\{x_2\}$, which is proportional to $|x_2|^{3/2}$ for small x_2. Thus it predicts an infinite curvature in the profile at $x_2 = 0$, which is highly implausible. But if it is true that large eddies play a dominant role in determining the eddy viscosity they must also play a dominant role in extracting energy from the main stream. We should recognise that at any one time a large eddy interacts with the main stream over a large range of x_2, and that being so the quantity we should substitute for ε_o in (9.15) is presumably $\nu_e\langle(\mathrm{d}\overline{u}_1/\mathrm{d}x_2)^2\rangle$ rather than $\nu_e(\mathrm{d}\overline{u}_1/\mathrm{d}x_2)^2$, where the angle brackets imply a spatial average over distances around the point of interest of order $1/k_o$. Prandtl's formula then acquires a non-local form, namely

$$\nu_e \sim \frac{1}{k_o^2}\left\langle\left(\frac{\mathrm{d}\overline{u}_1}{\mathrm{d}x_2}\right)^2\right\rangle^{1/2}. \tag{9.18}$$

The predicted eddy viscosity no longer vanishes where $\mathrm{d}\overline{u}_1/\mathrm{d}x_2$ vanishes, and the predicted velocity defect near the centre of symmetric profiles varies more sensibly, like x_2^2 rather than $|x_2|^{3/2}$.

9.6 A simple illustration of the scaling approach

Let us now tackle a specific problem in turbulent shear flow without invoking, initially at any rate, (9.17) or (9.18). Let us suppose that the mean motion is restricted to a thin layer of infinite extent in the x_1 and x_3 directions, centred on the plane $x_2 = 0$ and symmetrical about this plane, within which $\overline{u}_2 = \overline{u}_3 = 0$ while $\overline{u}_1$ depends only on x_2 and t.[7] If there is stationary fluid outside the layer, how does the layer thickness change with time?

[7] Because the velocities under consideration in this section are changing systematically with time, the ensemble averages which are of interest are *not* the same as time averages. The overbar notation is objectionable on that account but is nevertheless retained, (a) in order to facilitate comparison with results obtained in other sections, and (b) because angle brackets are being used to indicate averages over space [cf. (9.18)], and to use them here, as in §9.1(ii), to denote ensemble averages instead could well give rise to worse confusion.

We start with an assumption that the Navier–Stokes equation possesses solutions, of which the initial state represents one, which are *self-similar* or *self-preserving* in the sense that

$$\overline{u}_1\{x_2, t\} = \overline{u}_1\{0, t\} f_u \left\{\frac{x_2}{\delta}\right\} = u_o f_u \left\{\frac{x_2}{\delta}\right\}, \tag{9.19}$$

where u_o and the scale length δ change with time but the function f_u stays the same. Here f_u is an even function of its argument, defined in such a way that $f_u\{0\} = 1$; the suffix u serves to distinguish it from another function which appears shortly. When the moving layer is turbulent there is a fluctuating velocity field $\boldsymbol{u}'\{\boldsymbol{x}, t\}$ in addition, of course, but we may suppose the spectrum of this to be determined by the mean motion and we do not need to treat it separately. We now note that although the layer must grow in mass as it spreads, and although its kinetic energy must diminish on account of dissipation, its linear momentum in the x_1 direction must be conserved. On this account the product $m = u_o\delta$ and the Reynolds Number defined by

$$Re = \rho u_o \delta/\eta = m/\nu$$

are constants. If we look forwards in time we see the layer thickness increasing, of course, but we may also look backwards and see it decreasing, and it is convenient to choose the origin for t in such a way that $\delta\{0\} = 0$. The magnitude of $\delta\{t\}$ at any later time must be implied by the conditions at $t = 0$, but it is inconceivable that $\delta\{0\}$ (which is zero) and $u_o\{0\}$ (which is infinite) are separately of significance; it can only be their finite product m which matters. Hence $\delta\{t\}$ and $u_o\{t\}$ can depend only on m, t, η and ρ, and the only form of dependence for δ which is dimensionally consistent is

$$\delta\{t\} \propto \sqrt{\nu t}\, f_\delta\{Re\}.$$

Equation (1.28), which represents the exact solution to the problem when Re is small enough for the fluid to support laminar flow, evidently satisfies this relation. When Re is large and the moving layer is turbulent instead, however, we expect η to be irrelevant, because then the only role of the viscosity is to determine the high-k cut-off in the turbulent spectrum, k_K, and the Reynolds stress is surely not affected by this. In that case $f_\delta\{Re\}$ must be proportional to $\sqrt{Re}$, and we have

$$\delta\{t\} = \sqrt{B\nu Re t} = \sqrt{Bmt}; \quad u_o = \sqrt{m/Bt}. \tag{9.20}$$

The undetermined dimensionless 'constant' B is variable to the extent that different choices are possible for the scale length δ.

Now the function f_u is determined, when Re is so large that $\overline{s}_3$ is negligible compared with $\overline{s}_{3,\mathrm{Re}}$, by the equation of (mean) motion

$$\frac{\partial \bar{s}_{3,\mathrm{Re}}}{\partial x_2} = \rho \frac{\partial \bar{u}_1}{\partial t}.$$

In view of (9.19) and (9.20), this means that

$$\begin{aligned}\frac{\partial \bar{s}_{3,\mathrm{Re}}}{\partial x_2} &= \rho \left(\frac{\mathrm{d}u_\mathrm{o}}{\mathrm{d}t} f_u - \frac{u_\mathrm{o} x_2}{\delta^2} \frac{\mathrm{d}\delta}{\mathrm{d}t} f'_u \right) \\ &= -\frac{\rho u_\mathrm{o}}{2t} \left(f_u + \frac{x_2}{\delta} f'_u \right) \\ &= -\frac{\rho u_\mathrm{o} \delta}{2t} \frac{\partial}{\partial x_2} \left(\frac{x_2}{\delta} f_u \right),\end{aligned}$$

and hence, since $\bar{s}_{3,\mathrm{Re}}$ must vanish on the plane $x_2 = 0$ for symmetry reasons, that it may be written in the form

$$\bar{s}_{3,\mathrm{Re}} = \rho u_\mathrm{o}^2 f_s \left\{ \frac{x_2}{\delta} \right\}, \tag{9.21}$$

with

$$f_s \left\{ \frac{x_2}{\delta} \right\} = -\frac{1}{2} B \frac{x_2}{\delta} f_u \left\{ \frac{x_2}{\delta} \right\}. \tag{9.22}$$

Thus our assumption that $\bar{u}_1$ is self-similar implies that the Reynolds stress is self-similar too, and that it scales – on dimensional grounds it could hardly do otherwise – like ρu_o^2.

There are some problems of this nature which cannot be tackled by the route which the above analysis illustrates because δ (or its equivalent) is liable to depend upon four parameters excluding η instead of three. It is therefore worth noting that the argument can be inverted if preferred. Suppose that we choose (9.21) as one of our postulates, together with (9.19). On substituting into the equation of motion we find

$$\frac{u_\mathrm{o}^2}{\delta} f'_s = \left(\frac{\mathrm{d}u_\mathrm{o}}{\mathrm{d}t} f_u - \frac{u_\mathrm{o} x_2}{\delta^2} \frac{\mathrm{d}\delta}{\mathrm{d}t} f'_\mathrm{u} \right),$$

which, with $u_\mathrm{o}\delta = m$ (constant), reduces to

$$\frac{m}{u_\mathrm{o}^3} \frac{\mathrm{d}u_\mathrm{o}}{\mathrm{d}t} = \frac{f'_s}{\left(f_u + \frac{x_2}{\delta} f'_u \right)}.$$

Since the left-hand side of this equation is a function of t only whereas the right-hand side is a function of x_2/δ only, both sides must be constant, and it immediately follows that u_o must be proportional to $\sqrt{m/t}$. Equations (9.20) become in this way the end result of the analysis and do not require independent justification. An example of a problem which can be successfully tackled by postulating that the Reynolds stress is self-similar and scales like ρu_o^2 is the problem, referred to in the following section but not analysed there in full, of how a turbulent wake spreads sideways behind an object which is moving steadily through otherwise stationary fluid. It is not possible to establish how the width of the wake varies with distance from the object using dimensional analysis alone, because the wake is influenced by a fourth parameter – the object's speed.

We can get no further than (9.21) (or than (9.20), if (9.21) is treated as a postulate) unless we are willing to adopt an additional hypothesis, that the eddy viscosity defined by (9.12) is independent of x_2. That is equivalent to the hypothesis that the function f_s must vary with x_2/δ in the same way as f'_u, or let us say that

$$f_s = \frac{1}{2} B' f'_{\mathrm{u}},$$

where B' is independent of x_2. Equation (9.22) then integrates to

$$f_u = \exp\left(-\frac{Bx_2^2}{2B'\delta^2}\right). \tag{9.23}$$

We may choose δ, not yet specified in detail, to represent the r.m.s. width of this Gaussian profile, in which case $B' = B$. Equation (9.21) then becomes

$$\overline{s}_{3,\mathrm{Re}} = \frac{1}{2} B\rho u_o^2 f'_u \left\{\frac{x_2}{\delta}\right\} = \frac{1}{2} B\rho u_o \delta \frac{\partial \overline{u}_1}{\partial x_2},$$

which tells us that the eddy viscosity defined by (9.12), being given by

$$\eta_e = \frac{1}{2} B\rho u_o \delta = \frac{1}{2} B\rho m,$$

is independent of t as well as of x_2 but depends upon the momentum in the turbulent layer. If $B \sim 1$, as seems probable, then

$$\nu_e = \frac{\eta_e}{\rho} \sim \frac{1}{2} m. \tag{9.24}$$

Since the r.m.s. gradient of $\overline{u}_1$ in the Gaussian profile of (9.23) is of order u_o/δ, while the mixing length is presumably of order δ, that answer seems reasonably consistent with (9.18), though it is clearly inconsistent with Prandtl's formula in its unmodified form, (9.17). Had we had full confidence in (9.18), in fact, we might

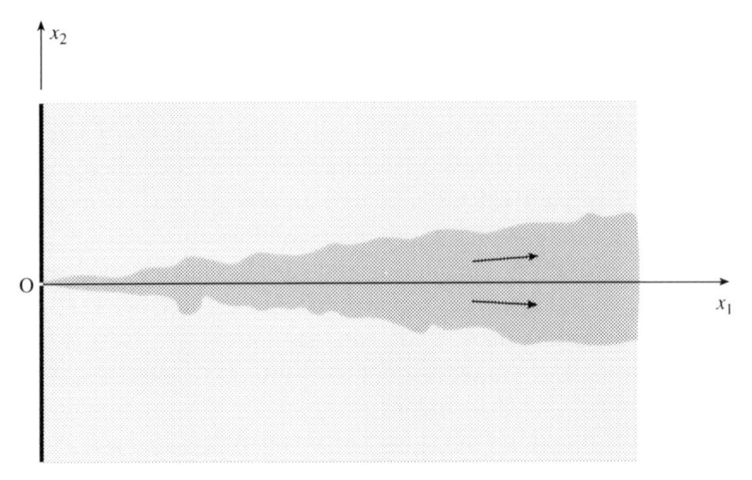

Figure 9.7 Cross-section through a planar turbulent jet (suggested by darker shading) emerging through a very thin slit at the origin O into a half-space containing otherwise stationary fluid.

have written something like (9.24) down at the start. It would then have been a trivial exercise to derive (9.23), by solving the equation of motion.

It seems likely that were experimental data available concerning the spreading of infinite plane turbulent sheets, which they are not, (9.18) would describe them rather well. Hearing that, the reader may well ask himself what the tortuous scaling argument has to offer; after all, it tells us nothing about the profile of $\overline{u}_1$ unless we add the assumption that the eddy viscosity is independent of x_2, and that assumption remains unsupported as long as we eschew the admittedly speculative argument by which (9.18) was justified in §9.5. Scaling arguments prove their worth in complex situations which cannot be analysed using (9.18) alone, but they always need to be supplemented by guesswork or by judicious appeal to experimental data.

9.7 Turbulent jets, wakes and mixing layers

Figure 9.7 shows a sketch of a submerged jet of fluid, forced out at high speed through a very small aperture at $x_1 = 0$ into a vessel filled with stationary fluid. The aperture is supposed to be a narrow slit extending in the x_3 direction, so the jet resembles the plane layer of moving fluid discussed in the last section. It differs, however, in that its thickness varies with x_1 rather than t. It also differs in that if a self-preserving solution of the form

$$\overline{u}_1\{x_2, x_1\} = u_o\{x_1\} f_u \left\{\frac{x_2}{\delta}\right\}$$

exists [cf. (9.19)] the quantity which must be independent of x_1 is not $m = u_o\delta$ but $M = u_o^2\delta$; M is proportional to the rate at which the jet transfers momentum across a fixed plane normal to the x_1 direction, and this does not vary with x_1 once a steady state has been reached. Hence the Reynolds Number, defined by $Re = u_o\delta/\nu = M/\nu u_o$, increases with x_1 as δ increases and u_o falls. At the slit itself and for a short distance in front of it the Reynolds Number may be small enough for the flow to be laminar. Submerged jets are notoriously unstable, however, for reasons touched on in §8.12, and Re does not need to be much greater than unity for them to develop turbulence. In what follows we shall assume the jet to be turbulent almost from the start and, pursuing the line of argument laid down in the previous section, we shall therefore ignore the effect of η on the rate at which it spreads.

If δ is completely determined by M, x_1 and ρ, it must for dimensional reasons be proportional to x_1, and the ratio δ/x_1 must be the same for all planar jets, independent of M and ρ. This conclusion is confirmed by experiments, which also show the semi-angle with which the jet spreads to be quite small. It is small enough for the lines of flow which describe the mean motion to be almost parallel, so the mean motion should be a solution of the following approximate equations, based on the equations used to describe boundary layers [(7.4) and (7.5)]:

$$\frac{\partial \overline{u}_1}{\partial x_1} + \frac{\partial \overline{u}_2}{\partial x_2} = 0,$$

$$\overline{u}_1 \frac{\partial \overline{u}_1}{\partial x_1} + \overline{u}_2 \frac{\partial \overline{u}_1}{\partial x_2} \approx \frac{\partial \overline{s}_{3,\mathrm{Re}}}{\partial x_2}.$$

As in the simpler example discussed in §9.6, the hypothesis that the eddy viscosity is independent of x_2, which amounts to the hypothesis that $\overline{s}_{3,\mathrm{Re}}$ varies with x_2 in the same way as $\partial\overline{u}_1/\partial x_1$, enables us to find a solution. It is

$$\overline{u}_1 = u_o \mathrm{sech}^2\left(\frac{x_2}{\sqrt{2}\delta}\right), \tag{9.25}$$

$$\overline{u}_2 = \overline{u}_1 \frac{x_2}{x_1} - u_o \frac{\delta}{2x_1} \tanh\left(\frac{x_2}{\sqrt{2}\delta}\right)$$

(the $\sqrt{2}$ in these expressions ensures that the sech^2 profile for $\overline{u}_1$ matches the Gaussian profile of (9.23) with $B' = B$ at small values of x_2/δ, and it has no significance otherwise), and ν_e turns out to be given by

$$\nu_e = \frac{1}{2}\left(\frac{\delta}{x_1}\right)^{3/2} (Mx_1)^{1/2} = \delta^2\left(\frac{u_o}{\delta}\right)\left(\frac{\delta}{2x_1}\right). \tag{9.26}$$

Thus ν_e is not constant in this context: since δ/x_1 and M are both constants, it increases with distance from the slit like $\sqrt{x_1}$. Its magnitude is smaller than naive

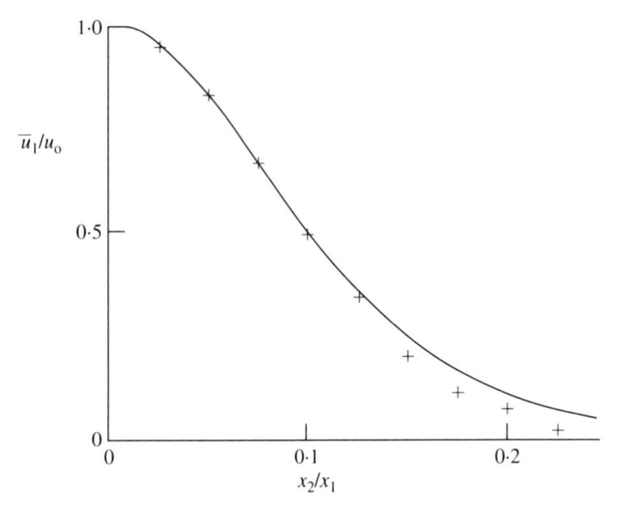

Figure 9.8 Velocity profile in a planar turbulent jet: experimental points fitted by a curve which represents (9.25) with $\delta = 0{\cdot}0803x_1$.
[Based on data plotted in A. A. Townsend, *The Structure of Turbulent Shear Flow*, Cambridge University Press, 2nd edition, 1976.]

application of (9.18) might lead one to expect, by a constant factor of perhaps $\delta/2x_1$ which is known from experimental evidence [fig. 9.8] to be about 0·04. The discrepancy may be partly attributable to deficiencies in the model on which (9.18) is based. It is also possible, however, that the structure of the turbulence in each part of the jet reflects the recent history of that part, i.e. that the local values of ε_o (the dissipation) and $1/k_o$ (the mixing length) are determined by the values of u_o and δ at some smaller value of x_1, where the jet was faster and thinner. Turbulent flows are not infrequently complicated by memory effects of this sort.

At the point where a two-dimensional jet first emerges from a slit its mean velocity profile is necessarily rather different from the self-preserving form of (9.25), but that equation is in adequate agreement with experimental data at values of x_1 which are large compared with the slit width. Figure 9.8 shows a typical fit. Discrepancies appear in the wings of the distribution, where $\overline{u}_1$ decays to zero more rapidly than the theory predicts, and it is mildly disturbing to discover that by fitting a sech^2 curve rather than a Gaussian curve we have made these discrepancies more noticeable rather than less noticeable. However, discrepancies are to be expected in the wings where turbulence gives way to vorticity-free motion. The rate at which mass is transferred across any fixed plane normal to x_1 is proportional to $u_o\delta$ and therefore increases with x_1 like $\sqrt{x_1}$, which means that vorticity-free fluid has to be drawn in from the surroundings to feed the turbulent jet; this is the phenomenon of *entrainment* briefly referred to in §2.17. The vorticity-free and turbulent regions are in principle quite distinct, but common sense suggests, and experiments confirm, that the dividing surface which separates them is rapidly changing and very irregular, as a result of the turbulent motion inside it. The convolutions in this surface have an appreciable depth, i.e. they cover an appreciable range of x_2 within which the local turbulence, if monitored experimentally with a small fixed probe, proves to be intermittent. Simple theories concerning the profile of $\overline{u}_1$ are inevitably liable to fail within this range.

Incidentally, the process of entrainment is thought to be akin to the process whereby a breaking wave traps bubbles of air beneath the surface of the ocean; small volumes of vorticity-free fluid are first engulfed by convolutions in the dividing surface, then ground up into smaller volumes, and ultimately infected with vorticity through the process of vorticity diffusion.

The arguments applied above to a two-dimensional submerged jet can be applied to three-dimensional jets which emerge through small circular holes rather than through slits, and the results are not very different. With some modification [see remarks in §9.6] they can also be applied to the wakes which accompany objects moving through otherwise stationary fluid. They suggest, for example, that behind a spherical object which is being drawn through fluid at a uniform speed such that the wake is turbulent the width of the wake should increase with distance x_1 measured from the sphere's centre like $x_1^{1/3}$, at any rate in the *far wake* where $\delta \ll x_1$. This prediction is confirmed by experiment. If the sphere is self-propelled the wake motion is more complicated, and δ may be proportional to $x_1^{1/5}$ instead.

As for plane turbulent mixing layers whose thickness $\delta(x_1)$ is effectively zero at $x_1 = 0$ [§9.1(iv) and fig. 9.1], the two quantities which must be independent of x_1 for these are the velocities U_1 and U_2 of the streams on either side. Thus δ is presumably determined by U_1, U_2, ρ and x_1, and it follows at once by dimensional reasoning that δ must be proportional to x_1. This prediction is obviously consistent with one aspect of the behaviour shown in fig. 9.1, and in fact the self-preserving profile for $\overline{u}_1$ which is observed experimentally in such mixing layers is in adequate agreement with calculations based upon the assumption that the eddy viscosity is independent of x_2 but proportional to x_1, which is what (9.18) implies in this context. However, no simple theory suffices to explain the existence of the coherent structures shown in fig. 9.1, nor of the coherent structures which feature in wakes and probably in jets as well.

9.8 Turbulent flow between parallel plates and in pipes

From the free turbulent shear flows discussed in §§9.6 and 9.7 we now move on to consider turbulent shear flow which is bounded by two parallel plates of infinite extent. The flow is to be thought of initially as driven by motion of one of the plates rather than by a longitudinal pressure gradient. If we picture a stationary

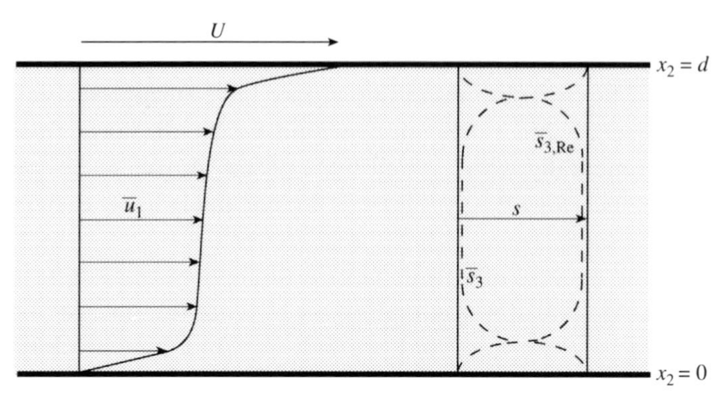

Figure 9.9 Turbulent shear flow between parallel plates, driven by motion of the upper one. The curve on the left suggests the profile of mean velocity $\overline{u}_1$. The broken curves on the right suggest how the uniform shear stress s is made up of two components which vary with x_2.

plate occupying the plane $x_2 = 0$ and a plate occupying the plane $x_2 = d$ which is moving in the x_1 direction with a constant velocity U, we have the turbulent analogue of the trivial problem of steady laminar shear flow discussed in §6.6(ii).

Such shear flow generates a mean shear stress which must be independent of x_2 in circumstances where U and $\overline{u}_1$ are no longer changing with time, and we may represent this constant quantity by the symbol s. It is of course the sum of two terms,

$$s = \overline{s}_3 + \overline{s}_{3,\mathrm{Re}} = (\eta + \eta_e)\frac{\mathrm{d}\overline{u}_1}{\mathrm{d}x_1}. \tag{9.27}$$

The first of these terms is negligible in free turbulent shear flows provided that the Reynolds Number is large enough, but it is never entirely negligible in wall-bounded shear flows. That is because, in the present context, the boundary conditions on the plane $x_2 = 0$ oblige not only $\overline{u}_1$ but also u_1', u_2' and u_3' to vanish there. It follows that $\partial u_1'/\partial x_1$ and $\partial u_3'/\partial x_3$ must vanish on this plane, and one may deduce from the continuity condition

$$\frac{\partial u_1'}{\partial x_1} + \frac{\partial u_2'}{\partial x_2} + \frac{\partial u_3'}{\partial x_3} = 0$$

that $\partial u_2'/\partial x_2$ does so too, and hence that u_1', u_2' and the Reynolds stress described by (9.10) tend to zero as x_2 approaches zero at least as fast as x_2, x_2^2, and x_2^3 respectively. Very close to the wall at $x_2 = 0$, therefore, it is the second term in (9.27) which is negligible, not the first [fig. 9.9]. In this region, known as the *viscous sublayer*, there must be a large gradient of mean velocity, such that

$$\frac{\mathrm{d}\bar{u}_1}{\mathrm{d}x_2} = \frac{s}{\eta}. \tag{9.28}$$

A similar viscous sublayer must exist, needless to say, near the wall at $x_2 = d$.

However, $\bar{s}_3$ is surely negligible compared with $\bar{s}_{3,\mathrm{Re}}$ in the middle of the fluid region, where $x_2 \sim \frac{1}{2}d$, and (9.18) for the eddy viscosity, for which our discussion of free shear flows has provided some justification, suggests that in this region we have

$$\frac{\mathrm{d}\bar{u}_1}{\mathrm{d}x_2} = \frac{s}{\eta_\mathrm{e}} \sim \frac{s}{\rho d^2|\mathrm{d}\bar{u}_1/\mathrm{d}x_1|},$$

or

$$\frac{\mathrm{d}\bar{u}_1}{\mathrm{d}x_2} \sim \frac{1}{d}\sqrt{\frac{s}{\rho}}.$$

Evidently this velocity gradient can be made as small as we wish by increasing d, while at the same time increasing U at whatever rate is necessary to maintain the same value of s. In the limit, the curve which, for given s, describes the variation of $\bar{u}_1$ with x_2 over the range $0 < x_2 < \frac{1}{2}d$ must become effectively independent of d; it must start off at $x_2 = 0$ with the large slope which (9.28) describes, and the slope must reduce smoothly to zero as x_2 increases to infinity.

Since $\bar{u}_1$ does not depend upon d or U in the limiting case, it must be completely determined by s, x_2, η (or ν) and ρ. Those parameters are such that $\sqrt{s/\rho}$ has the dimensions of velocity – and we may conveniently label this quantity by the symbol u_s – while $\sqrt{sx_2^2/\rho\nu^2} = x_2 u_s/\nu$ is dimensionless. It follows that the velocity profile sketched in the figure must be describable by a functional relationship of the form

$$\bar{u}_1 = u_s f\left\{\frac{x_2 u_s}{\nu}\right\}. \tag{9.29}$$

Hence for values of x_2 which lie well outside the viscous sublayer, such that the first term on the right-hand side of (9.27) is certainly negligible compared with the second, the eddy viscosity defined by (9.12) is such that

$$\frac{1}{\nu_\mathrm{e}} = \frac{\rho}{s}\frac{\mathrm{d}\bar{u}_1}{\mathrm{d}x_2} = \frac{1}{\nu} f'\left\{\frac{x_2 u_s}{\nu}\right\}. \tag{9.30}$$

Now the magnitude of the eddy viscosity in turbulent shear flow is influenced, according to the speculative argument used to justify (9.18), by the variation of $\bar{u}_1$ over a wide range of x_2. It seems reasonable to suppose, however, that the variation of $\bar{u}_1$ within the thin viscous sublayer has no noticeable effect on the eddy viscosity much further out, and that the eddy viscosity much further out is therefore independent of η. If so, the form of the function f for sufficiently large values of its argument must be such that

$$f'\left\{\frac{x_2 u_s}{\nu}\right\} \propto \frac{\nu}{x_2 u_s},$$

which integrates to

$$f\left\{\frac{x_2 u_s}{\nu}\right\} = \frac{1}{\mathsf{K}} \ln\left(\frac{x_2 u_s}{\nu}\right) + \mathsf{A}. \tag{9.31}$$

Here K (known as the *Kármán constant*) and A are numbers to be determined by experiment; one cannot expect to predict them accurately by matching (9.31) to the equation

$$f\left\{\frac{x_2 u_s}{\nu}\right) = \frac{x_2 u_s}{\nu} \tag{9.32}$$

which applies at very small values of x_2 [(9.28) and (9.29)] because the extent of the so-called *buffer zone* over which one form changes to the other is uncertain. The expression for the eddy viscosity which corresponds to (9.31) is of course

$$\eta_e = \mathsf{K}\rho x_2 u_s. \tag{9.33}$$

Attempts have been made to justify this result in other ways, but they carry rather little conviction.

Equations (9.29) and (9.31) together constitute the *logarithmic law of the wall*. Although it has been deduced here by reference to an idealised experiment requiring infinite plates an infinite distance apart, one of them moving with infinite velocity, it is expected to apply in other circumstances too. In particular, it is expected to apply, within an *inertial sublayer* such that x_2 is large compared with the thickness of the viscous sublayer but at the same time small compared with d, to flow between stationary plates a finite distance apart which is driven by a longitudinal pressure gradient. The condition that x_2 must be small compared with d is necessary on two grounds: when the shear stress is generated by a pressure gradient $\nabla_1 p$ it is related to its value s_o at $x_2 = 0$ by the equation

$$s = -\nabla_1 p\left(\frac{1}{2}d - x_2\right) = s_o\left(1 - 2\frac{x_2}{d}\right),$$

and to justify the logarithmic law of the wall in this context we need to assume that the term in x_2/d is immaterial; moreover, we need to ignore the effect on the flow near the plate at $x_2 = 0$ of the plate at $x_2 = d$, and this is presumably legitimate when $x_2/d \ll 1$ but not necessarily otherwise. Equation (9.31) has in fact been convincingly substantiated by experiments on pressure-driven flow between parallel plates. The experiments show that $\mathsf{K} \approx 0{\cdot}4$ and $\mathsf{A} \approx 5{\cdot}5$, and they also show that the logarithmic law of the wall starts to fail, as $x_2 u_s/\nu$ is reduced, when this quantity is about 30.

Whether or not the condition $x_2/d \ll 1$ is satisfied, the velocity profile for pressure-driven flow between parallel plates must satisfy a generalised version of (9.29), namely

$$\bar{u}_1 = u_{s_o} F_1 \left\{ \frac{x_2 u_{s_o} u_{s_o}}{\nu}, \frac{x_2}{d} \right\} = u_{s_o} F_2 \left\{ \frac{x_2}{d}, \frac{d u_{s_o} u_{s_o}}{\nu} \right\}.$$

However, the hypothesis (substantiated by the success of (9.31)) that the gradient of $\bar{u}_1$ is independent of ν except in the viscous sublayer and adjacent buffer zone places a constraint upon the function F_2, and it is not difficult to show that over the whole range $30\nu/u_{s_o} < x_2 < \frac{1}{2}d$ the so-called velocity defect, i.e. the difference between the local value of $\bar{u}_1$ and its value at the centre of the channel, must obey an equation of the form

$$\frac{\bar{u}_1\{\frac{1}{2}d\} - \bar{u}_1\{x_2\}}{u_{s_o}} = F\left\{\frac{x_2}{d} \text{ only}\right\}. \tag{9.34}$$

That too is confirmed experimentally. The observed dependence of the velocity defect on x_2 in the neighbourhood of $x_2 = \frac{1}{2}d$, where $\bar{u}_1$ passes through a maximum, is consistent with an eddy viscosity there of

$$\eta_e \approx 0{\cdot}5\rho d\bar{u}_{1,\max},$$

a conclusion which itself is adequately consistent with (9.18).[8] A curve which indicates how, for one particular value of the Reynolds Number, $\bar{u}_1$ varies between $x_2 = 0$ and $x_2 = d$ is sketched in fig. 9.10. The parabolic broken curve which is included for comparison is the profile associated with laminar flow under the same pressure gradient, i.e. for the same value of s_o $(= -\frac{1}{2}d\nabla_1 p)$, and the comparison makes clear that discharge rates are much reduced by turbulence.

The logarithmic law of the wall has also been verified in the form

$$\frac{\bar{u}\{r\}}{u_{s_o}} = \frac{1}{\mathsf{K}} \ln\left\{\frac{(a-r)u_{s_o}}{\nu}\right\} + \mathsf{A},$$

[8] This claim is based upon an examination of data plotted in figs. 95 and 97 of *Modern Developments in Fluid Dynamics, Vol. II*, ed. S. Goldstein, Clarendon Press, Oxford, 1938. Values of the dimensionless velocity defect are fitted there by functions which vary with $x_2' = \frac{1}{2}d - x_2$ like $x_2'^{3/2}$ for small x_2', and which therefore imply an eddy viscosity which goes to zero at the centre of the fluid layer like $x_2'^{1/2}$ [§§9.1(iii) and 9.5]. They can be fitted equally well, however, by a parabola corresponding to the constant value of η_e which is given here.

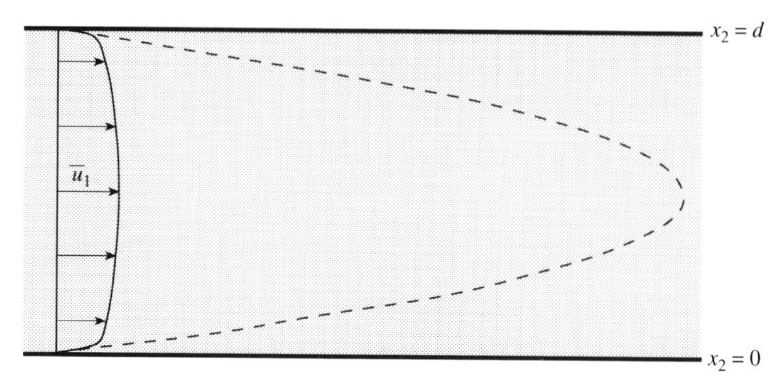

Figure 9.10 Profile of mean velocity for turbulent flow between stationary parallel plates, driven by a pressure gradient such that Re (= q/ν, where q is the discharge rate per unit width) is about $1{\cdot}4 \times 10^4$. The profile for laminar flow at the same pressure gradient is shown as a broken curve; turbulence reduces the value of $\overline{u}_1$ at the centre by a factor which in this case is 10. [Based on data obtained at a larger value of Re, plotted in A. A. Townsend, *The Structure of Turbulent Shear Flow*, Cambridge University Press, 2nd edition, 1976.]

where s_o $(= -\frac{1}{2}a\nabla_3 p = +\frac{1}{2}a\nabla p)$ is now the shear stress at $r = a$, for the case of turbulent flow along a smooth cylindrical pipe of radius a. Indeed, this law holds over a much larger range of r than we have any right to expect; it fails in the viscous sublayer and in the adjacent buffer zone, of course, but the value which it predicts for the mean velocity is correct to within a few parts per cent on the axis of the pipe, at $r = 0$.

For values of Re which are less than about 10^5, however, and larger than about 4×10^3 to ensure that turbulence is fully-developed and no longer intermittent [§9.3], the data may be fitted almost equally well over the whole range of r, including the viscous sublayer, by the purely empirical formula

$$\frac{\overline{u}\{r\}}{u_{s_\text{o}}} = 8{\cdot}6\left\{\frac{(a-r)u_{s_\text{o}}}{\nu}\right\}^{1/7}. \tag{9.35}$$

With

$$u_{s_\text{o}} = \sqrt{\frac{s_\text{o}}{\rho}} = \sqrt{\frac{a\nabla p}{2\rho}}$$

this formula corresponds to

$$Q = \int_0^a \overline{u}2\pi r\mathrm{d}r \approx 11{\cdot}2a^2 Re^{1/8}\sqrt{\frac{a\nabla p}{\rho}}, \tag{9.36}$$

from which one may deduce that the thickness of the viscous sublayer and buffer zone combined is about

$$30\frac{\nu}{u_{s_o}} \approx \frac{300}{Re^{7/8}}a. \tag{9.37}$$

Hence the viscous sublayer and the buffer zone occupy about 30% of the cross-sectional area of the pipe when $Re \approx 5 \times 10^3$, but they become less significant as Re increases.

The relation between discharge rate and pressure gradient expressed by (9.36) was first proposed by Blasius. It must be emphasised that only within the limited range $5 \times 10^3 < Re < 10^5$ does it provide a good fit to experimental data, and moreover that it works only for smooth pipes; if the inner surface of the pipe is rough, the discharge rate is significantly less.

9.9 Turbulent boundary layers

When fluid moves past a solid obstacle at high speed the boundary layer may become turbulent [§8.13]; the skin friction experienced by the obstacle then increases, though the form drag may decrease [§7.8]. The theory of turbulent boundary layers deserves discussion because of its relevance to the design of aircraft etc., but only the simplest case will be treated here: the case of a flat plate in the plane $x_2 = 0$ past which fluid is moving in the x_1 direction with uniform velocity U. In this case there is no longitudinal pressure gradient to complicate the problem [§7.2]. We will suppose the whole boundary layer to be turbulent, though this is only true in practice if turbulence is deliberately triggered at the plate's leading edge ($x_1 = 0$).

A turbulent layer which is bounded on one side by a solid plate and on the other side by vorticity-free fluid is a hybrid between the wall-bounded shear flows discussed in §9.8 and the free shear flows discussed in §9.7. Immediately adjacent to the plate there is a viscous sublayer, within which $\bar{u}_1$ is described in terms of the local shear stress exerted on the plate, $s_o\{x_1\}$, and the related velocity $u_{s_o} = \sqrt{s_o/\rho}$ by (9.29) and (9.32). Throughout the viscous sublayer the shear stress s is almost equal to s_o, and it is found experimentally to remain almost equal to s_o up to a value of x_2 which corresponds to about 20% of the total boundary layer thickness $\delta\{x_1\}$.[9] Thus over the range between $x_2 \approx 30\nu/u_{s_o}$ and $x_2 \approx 0{\cdot}2\delta$, which constitutes the inertial sublayer, the logarithmic law of the wall applies in its universal form, i.e. $\bar{u}_1$ is described by (9.29) and (9.31). At values of x_2 beyond $0{\cdot}2\delta$ the shear stress diminishes and the logarithmic law becomes unreliable. However, a velocity defect law of the form

$$U - \bar{u}_1 = u_{s_o}F\left\{\frac{x_2}{\delta}\right\} \tag{9.38}$$

[9] The symbol δ is used throughout this section to denote the layer thickness which contains virtually all the vorticity present in the fluid; it is an abbreviation for $\delta_{99\%}$, say, rather than $\delta_{90\%}$ [§7.2]

may be expected to apply wherever η is irrelevant to the motion, i.e. for all $x_2 > 30\nu/u_{s_o}$. Equation (9.38) is a version of (9.34), of course, differing only in so far as the boundary layer thickness here replaces the channel width d, and in so far as u_{s_o} and δ are not constants but vary with x_1; in principle the coefficient F in (9.38) could be a function of x_1/δ as well as of x_2/δ, but experimental data justify the hypothesis that it is not. Within the inertial sublayer, though not outside it, (9.38) must describe the same logarithmic variation of $\overline{u}_1$ with x_2 that (9.31) describes, so

$$F\left\{\frac{x_2}{\delta}\right\} = -\frac{1}{\kappa}\ln\left(\frac{x_2}{\delta}\right) + \text{constant} \tag{9.39}$$

within this range, and it follows from (9.29), (9.31), (9.38) and (9.39) that

$$\frac{U}{u_{s_o}} = \frac{1}{\kappa}\ln\left(\frac{\delta u_{s_o}}{\nu}\right) + C. \tag{9.40}$$

Here C is yet another 'constant' which is variable to the extent that it depends upon the choice of δ; experimental evidence suggests that with $\delta = \delta_{99\%}$ it is roughly 9.

We shall discover shortly that U/u_{s_o} varies rather slowly with Reynolds Number ($Re = Ux_1/\nu$) and that a typical value for this ratio is 25. It evidently corresponds to

$$\frac{\delta u_{s_o}}{\nu} \approx \exp\{0{\cdot}4(25 - 9)\} \approx 600.$$

But we know from the logarithmic law of the wall that at $x_2 \approx 30\nu/u_{s_o}$ the mean velocity is such that

$$\frac{\overline{u}_1}{u_{s_o}} \approx 2{\cdot}5 \ln 30 = + 5{\cdot}5 \approx 14.$$

Hence in this typical case [fig. 9.11] the thickness of the viscous sublayer and the adjacent buffer zone combined – a thickness denoted below by δ_{vsl} – is only about 5% (i.e. 30 ÷ 600) of δ, the full thickness of the boundary layer, yet $\overline{u}_1$ rises over this narrow range to almost 60% (i.e. 14 ÷ 25) of its ultimate value, U. The profiles of mean velocity in turbulent boundary layers are evidently very different from the profiles discussed in §7.2.

In §7.2 we were able to deduce how δ varies with x_1 by examining the momentum flux in the boundary layer. The argument is still useful when the boundary layer is turbulent, but it requires modification because $\overline{u}_1/U$ is no longer a unique function of x_2/δ. Because δ_{vsl} is so small we may use (9.38) to write the equation

$$s_o = \rho \frac{\mathrm{d}}{\mathrm{d}x_1}\int_0^{\infty} (U\overline{u}_1 - \overline{u}_1^2)\mathrm{d}x_2$$

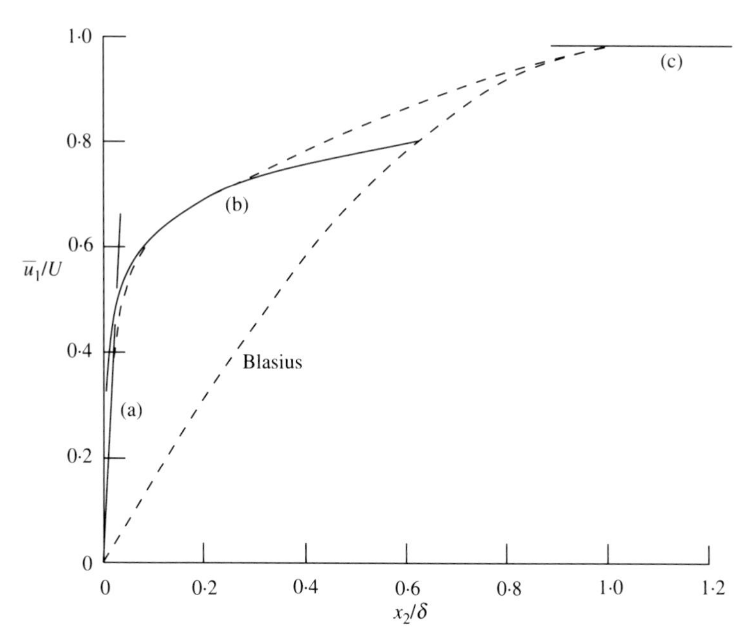

Figure 9.11 Velocity profile in a turbulent boundary layer, for the particular case discussed in the text: the layer is supposed to be attached to a plane surface, it is turbulent for all values of x_1 from the leading edge ($x_1 = 0$) onwards, the velocity U outside the boundary layer is uniform, and *Re* ($= Ux_1/\nu$) is about 2×10^6. The three full curves represent: (a) (9.32), which applies within the viscous sublayer; (b) the logarithmic law of the wall, i.e. (9.31), which in this case applies over the range $0{\cdot}05 < x_2/\delta < 0{\cdot}2$; and (c) $u_1 = U$, which applies for $x_2/\delta > 1$. The broken curves which link (a) to (b) and (b) to (c) are schematic.

A labelled broken curve shows Blasius's profile for a laminar boundary layer. Note that whereas the flow in a laminar boundary is self-similar, in the sense that the velocity profile has the same shape for all x_1, the flow in a turbulent boundary layer is not.

[(7.9)] in the following approximate forms:

$$u_{s_o}^2 \approx \frac{\mathrm{d}}{\mathrm{d}x_1}\int_{\delta_{\mathrm{vsl}}}^{\delta} (Uu_{s_o}F - u_{s_o}^2F^2)\mathrm{d}x_2$$

or, in view of the fact that U is significantly larger than the velocity defect $u_{s_o}F$ over most of the range covered by the integral,

$$u_{s_o}^2 \approx \frac{\mathrm{d}}{\mathrm{d}x_1}\int_{\delta_{\mathrm{vsl}}}^{\delta} Uu_{s_o}F\mathrm{d}x_2 \approx DU\frac{\mathrm{d}}{\mathrm{d}x_1}(u_{s_o}\delta), \tag{9.41}$$

where the 'constant' D – it appears to be about 3·6 – can be estimated only from experimental data concerning the form of the function F in regions where (9.39) is not necessarily valid. Another equation involving the rates of change of u_{s_0} and δ may be obtained by differentiating (9.40) with respect to x_1, and on combining this with (9.41) one obtains

$$\frac{\mathrm{d}\delta}{\mathrm{d}x_1} \approx \frac{1}{D}\left(\frac{u_{s_0}}{U} + \frac{1}{\mathrm{K}}\frac{u_{s_0}^2}{U^2}\right). \tag{9.42}$$

Hence $\mathrm{d}\delta/\mathrm{d}x_1$ should be about 0·012 when U/u_{s_0} is about 25. A related equation for $\mathrm{d}u_{s_0}/\mathrm{d}x_1$ is available, so in principle one should be able to predict, given D, how δ and u_{s_0} depend upon x_1 and hence estimate the skin friction which an edge-on plate experiences when it is covered by a turbulent boundary layer.

The solution is not, however, straightforward. Empirically, it is found that the drag force per unit breadth due to skin friction on one side of a plate of length L is adequately described, up to $Re = UL/\nu = 3 \times 10^6$ at any rate, by the formula

$$0{\cdot}037 Re^{-1/5} \rho U^2 L. \tag{9.43}$$

This result implies that

$$\frac{U}{u_{s_0}} \approx 5{\cdot}8 Re^{1/10}, \tag{9.44}$$

so the value of 25 for U/u_{s_0} which has been used for illustrative purposes above corresponds to $Re \approx 2 \times 10^6$.

Turbulent boundary layers contain coherent structures, which have been intensively investigated but are still not fully understood. They seem to be associated with intermittent eruptions of fluid from the viscous sublayer, where it is heavily laden with vorticity, into the boundary layer's outer regions.

9.10 Bénard convection at large Rayleigh Numbers

Investigations of the relation between the Nusselt Number Nu and the Rayleigh Number Ra in Bénard convection [§§8.7 and 9.2] have been extended in recent years to values of Ra as high as 10^{15}. Figure 9.12 shows some of the results obtained, in experiments where the convecting fluid was very cold helium gas contained in a cell with a height of 8·7 cm and a circular horizontal cross-section of diameter 8·7 cm. Gaseous helium is particularly suited for such work, for η and κ are both small and can be made smaller by lowering the temperature, while ρ can be significantly increased by increasing the applied pressure; the Prandtl number Pr is slightly dependent on temperature and pressure, but that is probably not material because the influence of Pr on Nu appears to be small.

The authors of the paper from which the figure is reproduced believe the convection to become genuinely turbulent (as opposed to chaotic) at $Ra \approx 5 \times 10^5$, but their results oblige them to distinguish two types of turbulence,

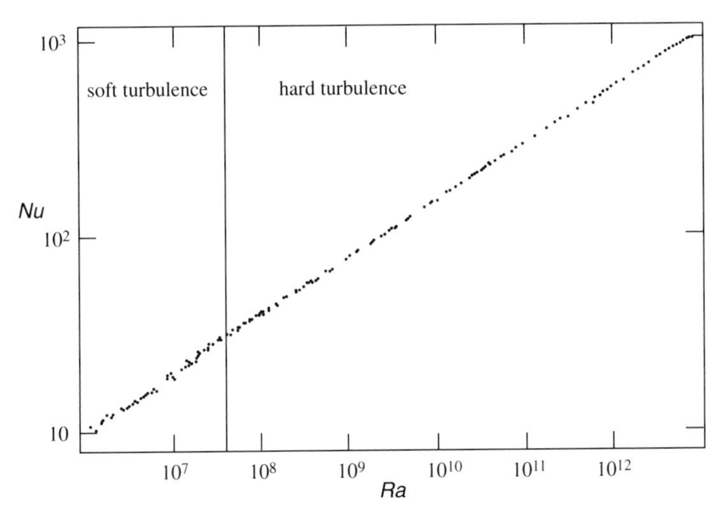

Figure 9.12 Dependence of Nusselt Number on Rayleigh Number in Bénard convection at large values of *Ra*.
[From B. Castaing, G. Gunaratne, F. Heslot, L. Kadanoff, A. Libchaber, S. Thomae, X.-Z. Wu, S. Zaleski & G. Zanetti, *J. Fluid Mech.*, **204**, 1, 1989.]

'soft' and 'hard'; soft turbulence persists up to $Ra \approx 4 \times 10^7$ and hard turbulence then takes over. The transition is marked by a change in the signals received from a bolometer which measures the temperature at a point half way up the cell: the r.m.s. excursions of temperature here become significantly smaller once the hard regime is reached. It is also marked, however, by a change in slope of the line mapped out by the experimental points in fig. 9.12. In both regimes the results can be adequately fitted by a power law relation of the form $Nu \propto Ra^n$, but whereas the exponent n takes the value $\frac{1}{3}$ in the soft regime, as observed by many previous experimenters, it is $0{\cdot}282 \pm 0{\cdot}006$, i.e. close to $\frac{2}{7}$, in the hard regime.

The Rayleigh Number is defined in this context as $\alpha g\theta_0 d^3/\chi\nu$, where θ_0 is the temperature difference between the top and bottom plates and where d is the distance between them [(8.15)]. Where the $\frac{1}{3}$ power law applies, therefore, the vertical heat flux H, being related to the Nusselt Number by $H = Nu\kappa\theta_0/d$, is independent of d for given θ_0. That can only mean that the heat flux is limited by relatively thin boundary layers adjacent to the plates, the thermal resistance of which is large compared with the thermal resistance of the much thicker layer of fluid which separates them.

The concept of an *eddy conductivity* κ_e helps one to understand and quantify that picture. When convection is occurring, the local temperature excess of the fluid on any horizontal plane is liable to fluctuate about its mean value θ by an amount τ [(8.37) and the footnote on p. 315], and the local vertical velocity is similarly liable to fluctuate about its mean value $\overline{u}_z$ by u_z'. These fluctuations are correlated, for the obvious reason that particles of fluid which are moving upwards with positive u_z' remember that they have come from regions where the mean temperature was higher and *vice versa*, and the contribution to H which arises from convection as opposed to conduction depends upon the correlation. To first order in τ it may be expressed as $\rho c_p \overline{\tau u_z'}$, which means that if the total heat flux is to be related to the gradient of mean temperature by the equation

$$H = -(\kappa + \kappa_e)\frac{d\theta}{dz} \tag{9.45}$$

the eddy conductivity must be defined so that

$$\kappa_e \approx \rho c_p \overline{\tau u_z'}\left(\frac{d\theta}{dz}\right)^{-1}. \tag{9.46}$$

Equation (9.46) is not accurate beyond the first order because, in the spirit of the Boussinesq approximations [§8.5], it does not take into account the variations of density $\alpha\rho\tau$ which accompany variations of temperature. These variations determine, however, the rate at which the gravitational field feeds energy into the turbulent motion. The total rate at which gravity does work on the fluid within a horizontal layer of infinitesimal thickness is given, to first order in τ, by

$$-\overline{(\rho + \alpha\rho\tau)g(\bar{u}_z + u_z')} \approx -\rho g\bar{u}_z + \alpha\rho g\,\overline{\tau u_z'}$$

per unit volume. This is of course zero, because the rate at which mass is transported across any horizontal plane is necessarily zero. However, the negative term on the right-hand side describes work done on, rather than by, the gravitational field; the energy for this comes from the heat source and heat sink which maintain the plates at temperatures $T \pm \frac{1}{2}\theta_o$ and the process by which it is turned into work involves thermal expansion of relatively cold fluid which has reached the bottom plate and is coming into equilibrium there, and thermal contraction of relatively hot fluid near the top plate. It is only the positive term which is relevant in the present argument, and this corresponds, if ε_o is the rate at which the turbulent motions are supplied with kinetic energy per unit mass, to

$$\varepsilon_o = \alpha g\overline{\tau u_z'} = -\frac{\alpha g\kappa_e}{\rho c_p}\frac{d\theta}{dz}. \tag{9.47}$$

Now when the concept of an eddy viscosity was explored, in §9.5, we used our knowledge of the kinetic theory of gases to justify $\nu_e \sim \varepsilon_o^{1/3}/k_o^{4/3}$ as an interim result and then estimated ε_o in terms of ν_e. The same approach is possible here. The thermal conductivity of a gas differs from its viscosity by a factor which is close to c_V [(8.16) with $Pr \approx 1$] and therefore to c_p, so we may estimate that

$$\kappa_e \sim \rho c_p \frac{\varepsilon_o^{1/3}}{k_o^{4/3}},$$

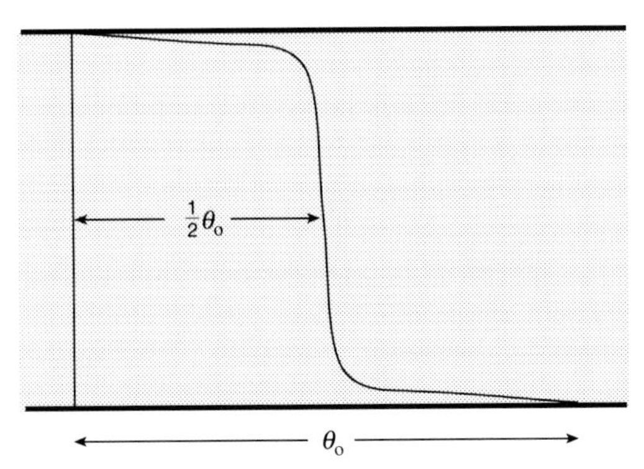

Figure 9.13 Hypothetical profile for the excess temperature θ in a fluid which is undergoing Bénard convection in the soft turbulence regime. Note the similarity between this and the profile sketched for $\bar{u}_1$ in fig. 9.9.

and when ε_o is described by (9.47) this corresponds to

$$\kappa_e \sim \frac{\rho c_p}{k_o^2}\left(-\alpha g\,\frac{d\theta}{dz}\right)^{1/2}. \tag{9.48}$$

This result is to be compared with Prandtl's formula for η_e, (9.17). It is open to a similar criticism, that κ_e is more likely to depend upon the temperature gradient averaged over a region of dimensions comparable with the mixing length $1/k_o$ than on the local value of this gradient.

Let us use (9.48) to estimate, for given H, the temperature gradient half way between the top and bottom plates of a convection cell; on the assumption that κ_e is very much larger than κ in this region, just as η_e is very much larger than η in the centre of a channel in which turbulent shear flow is occurring, it is such that

$$H \approx \kappa_e\,\frac{d\theta}{dz} \sim \frac{\rho c_p}{k_o^2}\,(\alpha g)^{1/2}\left(\frac{d\theta}{dz}\right)^{3/2}.$$

That suggests that if there were no boundary layers, and if $1/k_o \sim d$ as one expects, the temperature difference across the cell would be

$$d\,\frac{d\theta}{dz} \sim d\left(\frac{H^2}{\rho^2 c_p^2 \alpha g d^4}\right)^{1/3},$$

which is smaller than the actual temperature difference θ_o by a factor

$$\frac{d}{\theta_o}\frac{d\theta}{dz} \sim \left(\frac{Nu^2}{Ra Pr}\right)^{1/3}. \tag{9.49}$$

A glance at fig. 9.12 will show that at $Ra \approx 10^6$, where the soft domain begins, $Nu \approx 10$ for a fluid for which $Pr \approx 1$. Throughout this domain, therefore, the factor described by (9.49) should be less than 0·05. That is small enough to justify the assumption that it is the thermal resistance of the boundary layers which limits H, and hence to explain the fact that Nu is proportional to $Ra^{1/3}$.

The boundary layers, incidentally, may not correspond at all closely to the boundary layers discussed in the previous section, for the fluid which separates them is certainly not undergoing steady, vorticity-free, motion. Nevertheless they must both contain viscous sublayers, within which the convective heat flux $\kappa_e(\mathrm{d}\theta/\mathrm{d}z)$ must go to zero at the surface of each plate at the same rate as Reynolds stresses do and for much the same reason [§9.8], and within which we may therefore expect

$$\frac{\mathrm{d}\theta}{\mathrm{d}z} \approx -\frac{H}{\kappa}.$$

It can be argued that adjacent to the viscous sublayers there are inertial sublayers, within which κ_e, already significantly larger than κ, is proportional to distance from the surface in the same way that η_e is proportional to x_2 according to (9.33); a logarithmic variation of θ with distance from the surface would be expected in these sublayers, and the sketch in fig. 9.13 was drawn with that idea in mind. There is little to be gained by speculating on such points, however, for the fact that Nu ceases to be proportional to $Ra^{1/3}$ when Ra exceeds 4×10^7 shows that an important ingredient is missing from the model of the turbulent convection process on which the above analysis is tacitly based.

We have been assuming that the structure of the turbulence which exists between two plates is much the same whether it is driven by horizontal shear or by a vertical temperature gradient, and that it can be adequately characterised in both cases by a dissipation rate ε_o and a mixing length $1/k_o$. Flow visualisation experiments suggest, however, that distinctly different coherent structures arise in these two cases. The intermittent eruptions of vorticity from the viscous sublayer which are observed in turbulent boundary layers are replaced, in turbulent convection, by the intermittent appearance of what are called *thermals*. Thermals are columns of fluid which are rising from the boundary layer adjacent to the hot plate, or falling from the boundary layer adjacent to the cold one, through fluid which is moving relatively slowly. They penetrate the surrounding fluid to a height (or depth) which is a good deal greater than their diameter, but they normally terminate eventually with a mushroom cap, in which the rising (or falling) fluid rolls up into something resembling a vortex ring [§4.15]. Since thermals are often visible where smoke is rising from a cigarette, and since every reader will have seen grim photographs of much larger thermals – and their mushroom caps – associated with nuclear explosions in the atmosphere, they need rather little description. Thermals may become detached from the boundary layers in which they have originated if the supply of hot (or cold) fluid which is needed to feed them runs out, and in that case the cap may then continue to rise (or fall) on its own. A detached cap is sometimes called a *plume*, though in many accounts of this subject the words 'thermal' and 'plume' seem to be used interchangeably.

Flow visualisation experiments using a liquid with $Pr = 5{\cdot}6$ in a cubic Bénard cell of dimensions 18·5 × 18·5 × 18·5 cm have revealed a complicated pattern of events in the regime of hard turbulence, at $Ra \approx 10^9$. The fluid rises in one corner of the cell in intermittent thermals, which penetrate the entire height of the cell. When one of these thermals meets the opposing boundary layer it sets up a wave-like disturbance in this, which grows into what the experimenters have called a *swirl* and which thereby triggers the release, in the opposite corner, of a falling thermal. That falling thermal triggers the release of another rising thermal, and so on. A scaling theory has been put forward to explain the fact that Nu is proportional to $Ra^{2/7}$ in the hard regime, but for details of this the original paper should be consulted.

Further reading

Most of the material covered in this chapter is discussed at a similar level in two very different books: *An Introduction to Turbulence and its Measurement*, Pergamon Press, Oxford, 1971, by P. Bradshaw, and *Physical Fluid Dynamics*, Clarendon Press, Oxford, 2nd edn. 1988, by D. J. Tritton. Tritton's account is less detailed than this one in some respects but more detailed in others, and readers who wish to know more about recent work on coherent structures cannot do better than read Tritton on the subject and follow up his references to the original literature.

Of the many good books devoted entirely to chaos theory which are now available, *Universality in Chaos*, Adam Hilger, Bristol, 1984, a reprint volume selected and introduced by P. Cvitanović, is recommended, because it includes the papers on convection which are quoted in §9.2.

Modern developments in the fundamental theory of turbulence are discussed in *The Physics of Fluid Turbulence*, Clarendon Press, Oxford, 1990, by W. D. McComb, which includes comprehensive references to earlier work. Applications of scaling theories to turbulent shear flow are discussed in *The Structure of Turbulent Shear Flow*, Cambridge University Press, 2nd edn. 1976, by A. A. Townsend, and in *A First Course in Turbulence*, MIT Press, Cambridge (Mass.), 1987, by H. Tennekes & J. L. Lumley. The scaling theory which purports to explain the dependence of Nu on Ra at large values of Ra is outlined in the paper from which fig. 9.12 is copied; the reference is given in the caption. The flow visualisation experiments referred to in that connection are described by G. Zocchi, E. Moses & A. Libchaber, in *Physica A*, **166**, 387, 1990.

10

Non-Newtonian fluids

10.1 Introduction

Fluids were defined in §1.2 as materials which cannot withstand a shear stress, however small, without deforming, and it was suggested there that glaziers' putty should be classified as a plastic solid rather than a fluid because it appears to hold its shape indefinitely unless subjected to appreciable force. But does it really do so? If we were to watch it for a very long time (and to find some way of preventing it from drying out during the process) might we not see putty flow under its own weight? After all, lead pipes flow visibly under their own weight given a century or two in which to do so, as anyone in Cambridge may verify by inspection of some of the older buildings there. The process by which lead flows, known as *creep*, involves vacancy diffusion, and provided that a specimen of lead is polycrystalline on a fine scale its creep rate should be proportional to the shear stress acting upon it; if so, then according to the definition given in §1.2 it is a fluid – a liquid rather than a gas, of course – though its viscosity is certainly enormous, greater than the viscosity of water by many orders of magnitude. If apparently solid materials such as polycrystalline lead are really liquid, is putty really liquid too? And if putty is not, what about chewing gum, or toothpaste, or yoghurt, or mayonnaise, or a host of similar substances which do not appear to flow under their own weight but which flow readily enough when squeezed?

These questions are not easy to answer experimentally, and the answers are of rather little practical importance. For our purposes, the significant point is that the effective viscosity of such materials (defined as shear stress divided by rate of shear), which may or may not be infinite at infinitesimal rates of shear, becomes smaller as the rate of shear increases. In so far as they are liquid, therefore, they are *non-Newtonian liquids*, for which the constitutive relations connecting stress and rate of deformation are essentially non-linear.

There are very many liquids, or quasi-liquids, whose flow properties cannot be described by linear constitutive relations, though some of them, as we shall see, become more viscous rather than less viscous as the rate of shear increases. Examples to be found in every home include, apart from ones already mentioned, white of egg, condensed milk, shampoo and blood. Examples to be found in the processing plants of modern industry include molten polymers, polymers in solution, and a great variety of slurries, colloidal dispersions, fermentation products, foams, and suchlike. All of them owe their non-linearity to the fact that when they are subjected to shear their structure alters in some way: in a slurry which consists of anisotropic solid particles suspended in water, for example, the particles tend to align under shear; in a molten polymer the long-chain molecules are entangled with one another and tend to become stretched in the direction of shear as a result. As those examples suggest, non-Newtonian liquids generally become *anisotropic* when they are sheared, and this greatly complicates the task of describing them. Whereas the constitutive relations for an incompressible isotropic fluid involve only one viscosity coefficient, the constitutive relations for an incompressible fluid with uniaxial anisotropy, e.g. a suspension of aligned needle-shaped particles, involve *five* such coefficients, each of them independent of the others. If the fluid is biaxial rather than uniaxial, then more than five different viscosities need to be distinguished. Each of the viscosity coefficients is likely to depend upon the rate of shear, and since the dependence is rarely linear it cannot be adequately described with less than three parameters for each.

Another source of complication lies in the fact that changes of structure, from isotropic to anisotropic or *vice versa*, take a finite time. In general, a given velocity gradient such as $\partial u_1/\partial x_2$ is unlikely to produce a significant change of structure unless its inverse, which has the dimensions of time, is comparable with or less than the structural relaxation time. This rough rule of thumb explains why simple liquids such as water are Newtonian: the molecules in such liquids could in principle be forced by rapid shear into anisotropic configurations, but the relaxation time would be extremely short, probably not more than 10^{-12} s (this being the time required for a typical small molecule travelling at a speed characteristic of a temperature of a few hundred degrees Kelvin to move through a substantial fraction of its own diameter), and the velocity gradients required would therefore be enormously larger than those which can be realised in conventional viscometers. Conversely, non-Newtonian liquids possess long relaxation times; in some cases they are extremely long, to be measured in hours or days or even years. That being so, it is hardly surprising that the stresses present in a specimen of non-Newtonian liquid which is undergoing shear are not necessarily determined by the instantaneous velocity gradients alone; they may also depend upon on the past history of the specimen, e.g. on the time that has elapsed since these velocity gradients were first established. Not only do the constitutive relations have many terms, therefore; each of these terms may need to be expressed, using yet more parameters, as an integral over times past.

The word *rheology* was coined in the 1920's to describe the scientific study of flow properties in general. The subject may be said to embrace classical fluid dynamics, but the particular concern of rheologists lies with non-Newtonian liquids and plastic solids. The equipment of the experimental rheologist consists principally of viscometers, or *rheometers*, in which the material under test can be subjected to various types of deformation, and with which the associated stress components may be measured: the rate of deformation may be held constant and the stresses measured once a steady state has been established; it may be raised impulsively from zero to some predetermined level and the stresses measured as a function of time thereafter; or it may be made to oscillate in magnitude with uniform amplitude and frequency. The theoretical rheologist has traditionally regarded his first task as being to formulate constitutive relations which are consistent with all the experimental data thus obtained; ideally they should possess some degree of universality, in the sense that they can be successfully applied, with different parameters, to a range of different materials. His second task has been to use these constitutive relations to predict the behaviour of the materials in question when they are not confined in rheometers of relatively simple geometry but are flowing through orifices, say, or round obstacles of some sort. Their equations of motion may be too complex to be solved analytically, but the theoretical rheologist can nowadays hope to solve them numerically instead, using high-speed computers.

Progress along those lines has, alas, been slow: non-Newtonian materials display so many different patterns of behaviour that equations formulated to describe the behaviour of one have tended to prove useless for another. According to some experts it is time for a change in the method of attack, away from emphasis on the mathematical structure of the constitutive relations, and certainly away from empirical attempts to discover relations of universal validity, and towards emphasis on the physical structure of the particular material under discussion, on how it is affected by flow, and on how it determines the material's behaviour. Significant advances have been made in recent years by proponents of this philosophy, in the understanding of liquid polymers and polymer solutions in particular.

Most rheologists would restrict use of the label *non-Newtonian fluid* to substances which, because they possess long structural relaxation times, are non-linear and tend to remember their past treatment. In a wider sense, however, any fluid is non-Newtonian if the simple Newtonian equations (6.3) and (6.9) are insufficient to describe its behaviour, or if it is ruled by the laws of quantum rather than of Newtonian mechanics. Besides sections on polymers and suspensions, therefore, this chapter includes sections on *liquid crystals*, which are anisotropic even when at rest; on *plasmas*, which are anisotropic when threaded, as they usually are, by magnetic lines of force; and on *liquid helium*. It is not remotely possible to discuss so many different materials in a manner that is fully coherent, let alone complete, within the scope of a single chapter. What follows is frankly a ragbag, therefore, containing a limited number of items, only loosely connected with one another, which happen to have engaged the author's attention. Emphasis is placed throughout on phenomena which are conspicuously non-Newtonian or which lend themselves to demonstration.

10.2 Linear viscoelasticity

Here we take up one of the ideas hinted at above, that even in a liquid as simple as water the molecular structure has a tendency to become anisotropic during shear, though the degree of anisotropy may be minute and insufficient to affect the viscosity in any measurable way.

Imagine first a *solid* specimen of some isotropic elastic material, which at time t is suddenly sheared about the x_3 axis through some small angle γ ($0 < \gamma \ll 1$), in such a way that a molecule situated at the point (x_1, x_2, x_3) is carried virtually instantaneously to the point $(x_1 + \gamma x_2, x_2, x_3)$; the specimen is then slightly anisotropic (in this case in a biaxial sense), because all distances in the direction of $(\hat{x}_1 + \hat{x}_2)$ have been slightly stretched and all distances in the direction of $(\hat{x}_1 - \hat{x}_2)$ have been slightly diminished.[1] The deformation induces an instantaneous shear stress $s_3 = G\gamma$, where G is the shear modulus of the solid, but if the solid is able to creep then the shear stress will subsequently fall [fig. 10.1] as individual atoms adjust their positions in ways that remove anisotropy without altering the shape of the specimen as a whole; we may express its value at any later time $t + T$ by an equation of the form

$$s_3 = G\gamma f\{T\}, \tag{10.1}$$

where the T-dependent fraction f is such that $f\{0\} = 1$. In real solids there may well be a limit to the amount of creep which can occur. If, however, there is no such limit – if f decays to zero, at a rate which is fast enough for its integral over time to converge – then the our hypothetical 'solid' should in some circumstances display, as the following analysis shows, the behaviour we associate with Newtonian liquids.

Imagine the specimen to be deformed in a continuous fashion, so that in the interval between $t' = t - T$ ($t' < t, T > 0$) and $t' + dt' = t - T - dT$ ($dt' > 0, dT < 0$) the angle of shear increases by $d\gamma$. This increase generates a contribution to s_3 of which a fraction $f\{T\}$ persists at the later time t. Provided that the degree of anisotropy is at all times very small we are entitled to treat the response of the specimen as linear, and in that case, though not otherwise, we may add the residual stresses generated by a succession of earlier increments in γ to obtain the following result:

[1] $\hat{x}_1$ and $\hat{x}_2$ represent unit vectors in the x_1 and x_2 directions.

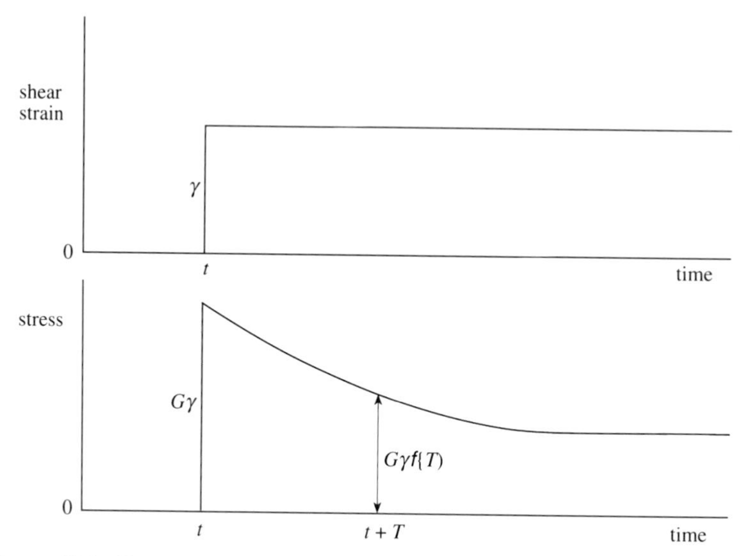

Figure 10.1 In a solid which is suddenly sheared about the x_3 axis through a small angle γ (upper diagram) a shear stress $s_3 = G\gamma$ develops. This stress is liable to decay if the solid can creep (lower diagram), and it may settle down to a steady value less than $G\gamma$ as the figure suggests. Liquids respond to sudden shear strain in a similar fashion, but the stress decays eventually to zero.

$$s_3\{t\} = G\int_{-\infty}^{t} \frac{\mathrm{d}\gamma}{\mathrm{d}t'}\{t'\}f\{t-t'\}\mathrm{d}t' = G\int_{0}^{\infty} \frac{\mathrm{d}\gamma}{\mathrm{d}t'}\{t-T\}f\{\mathrm{T}\}\mathrm{d}T. \qquad (10.2)$$

In two extreme cases its meaning is transparent. When the time taken to establish the shear strain, starting from $\gamma = 0$, is very short compared with a structural relaxation time τ defined by the equation

$$\tau = \int_{0}^{\infty} f\{T\}\mathrm{d}T, \qquad (10.3)$$

we may take f outside the integral in (10.2) and replace it by unity, thus obtaining the standard relation for a solid which does not creep, namely

$$s_3\{t\} = G\gamma\{t\}. \qquad (10.4)$$

When the relaxation time is very short, on the other hand, we may take $\mathrm{d}\gamma/\mathrm{d}t'$ outside the integral instead and write

$$s_3\{t\} = G\frac{\mathrm{d}\gamma}{\mathrm{d}t}\{t\}\int_{0}^{\infty} f\{T\}\mathrm{d}T = G\tau\frac{\mathrm{d}\gamma}{\mathrm{d}t}\{t\}.$$

This is the equation which holds for a Newtonian liquid subjected to planar shear of the sort we have considered, namely

$$s_3\{t\} = \eta \frac{\mathrm{d}u_1}{\mathrm{d}x_2}\{t\}, \tag{10.5}$$

but with a viscosity given by

$$\eta = G\tau. \tag{10.6}$$

The shear modulus of ice is something like 10^9 N m^{-2}, and the (unmeasurable) instantaneous shear modulus of water is perhaps of that order too. Thus if τ for water is about 10^{-12} s, as was suggested in §10.1, the viscosity of water should according to (10.6) be about 10^{-3} N m^{-2} s, and so indeed it is.

So how does a material respond to shear when its relaxation time is neither very long nor very short? An explicit answer requires knowledge of the form of $f\{T\}$. If we follow Maxwell, who was the first person to think along these lines, and set

$$f\{T\} = \mathrm{e}^{-T/\tau}, \tag{10.7}$$

then the shear stress associated with a particular shear strain which is oscillating with angular frequency ω and small amplitude is given by

$$s_3\{t\} = -\,\mathrm{i}\omega G\tau\{t\} \int_0^\infty \mathrm{e}^{\mathrm{i}\omega T}\,\mathrm{e}^{-T/t}\,\mathrm{d}T.$$

Depending upon one's point of view, this is equivalent *either* to the solid-like (10.4) with G replaced by a complex, frequency-dependent, shear modulus

$$G\{\omega\} = -\frac{\mathrm{i}\omega\tau}{1 - \mathrm{i}\omega\tau}\,G\{\infty\}$$

which is imaginary at frequencies which are small compared with τ^{-1}, *or* to the fluid-like (10.5) with η replaced by a complex, frequency-dependent, viscosity

$$\eta\{\omega\} = \frac{1}{1 - \mathrm{i}\omega\tau}\,\eta\{0\}$$

which is imaginary at high frequencies. Within the framework of a linear theory such as this we may deal with strains which are varying with time in a manner which is not simple harmonic by breaking them down into simple harmonic Fourier components and adding the stresses due to each. The general conclusions are, of course, that when the material is strained over times that are short compared with τ it responds like an elastic solid, that when it is strained over times that are long compared with τ it responds like a viscous liquid, and that in intermediate cases it responds hysteretically, with some dissipation of energy as heat. It is said to be *viscoelastic*.

Incidentally, it may be shown by differentiating both sides of (10.2) with respect to t that that equation may also be written in differential form, if the decay of $f\{t\}$ is exponential, as

$$\frac{\mathrm{d}s_3}{\mathrm{d}t} + \frac{s_3}{\tau} = G\frac{\mathrm{d}\gamma}{\mathrm{d}t}. \tag{10.8}$$

Some authors prefer to take this as their starting point in approaching the subject of linear viscoelasticity, and it is frequently justified by reference to a mechanical analogue suggested by Maxwell, consisting of a spring and dashpot in series; the force applied to them provides an analogue for s_3 and the total extension (the sum of separate extensions in the spring and dashpot) provides an analogue for γ. The spring-and-dashpot model adds rather little, however, to one's comprehension of the subject, especially as the decay of $f\{T\}$ is very rarely exponential for real viscoelastic substances.

A viscoelastic substance which most readers will have played with is available from toyshops under the name of *Silly Putty*; it is said to consist of clay particles dispersed, not in oil as in the case of plasticine, but in a silicone-based polymer. When a lump of this is put on a table it flows out into a sheet with a completely flat surface, betraying no sign that it is not a normal liquid albeit one with a large viscosity, but roll the lump into a ball and it will bounce as though it were made of rubber. When a lump of Silly Putty is flowing slowly, and perhaps when it is bouncing quickly, the extent to which its structure differs from that of Silly Putty at rest may be sufficiently small for the linear theory of this section to be applicable to it. When it is being drawn out rapidly into a thin filament, however, the structural changes may well be so profound that a non-linear theory would be needed to describe the process properly. In fact Silly Putty responds to rapid stretching by breaking, as though it were a brittle solid. We shall meet other examples of non-linear viscoelasticity in §10.6 below.

10.3 Viscosity in a uniaxial liquid

For the purposes of this section it will be best to forget about liquids which become anisotropic only when they are flowing, and to consider a liquid whose degree of anisotropy is unaffected by its motion and which is therefore anisotropic even when at rest. To simplify the discussion we shall assume this liquid to be uniaxial rather than biaxial. Nematic liquid crystals are uniaxial, and in them the axis about which there is rotational symmetry is known as the *director*. We may as well adopt that usage here, and it will also be helpful to borrow from liquid crystal physics the idea that the director may in principle be *anchored* in any desired orientation by application of a suitable field; magnetic fields of quite modest strength will anchor the director very effectively in a bulk specimen of nematic. Note that when the director is anchored by a field the argument used in §1.2 to show that s_{ij} and s_{ji} are always equal falls to the ground: the viscous stresses are free to exert a finite torque on any infinitesimal element of fluid because, following infinitesimal rotations of all the molecules or suspended particles within the element whose orientations collectively define the director, any such viscous torque is neutralised by a counter-torque exerted by the field.

In circumstances where the director of a uniaxial fluid is anchored in the x_3 direction, the linear equations for shear stress which replace (6.3) may be written as:

$$s_{12} = s_{21} = \eta_3 \left(\frac{\partial u_1}{\partial x_2} + \frac{\partial u_2}{\partial x_1} \right), \tag{10.9}$$

$$s_{23} = \eta_4 \frac{\partial u_2}{\partial x_3} + \eta_1 \frac{\partial u_3}{\partial x_2}, \quad s_{32} = \eta_2 \frac{\partial u_2}{\partial x_3} + \eta_4 \frac{\partial u_3}{\partial x_2}, \tag{10.10}$$

$$s_{31} = \eta_4 \frac{\partial u_3}{\partial x_1} + \eta_2 \frac{\partial u_1}{\partial x_3}, \quad s_{13} = \eta_1 \frac{\partial u_3}{\partial x_1} + \eta_4 \frac{\partial u_1}{\partial x_3}, \tag{10.11}$$

where η_1, η_2, η_3 and η_4 constitute (in the particular system of notation favoured here) four of the five coefficients of viscosity referred to in §10.1. Note that the coefficient of $\partial u_2/\partial x_3$ in the equation for s_{23} and the coefficient of $\partial u_3/\partial x_2$ in the equation for s_{32} are (like the corresponding coefficients in (10.11)) identical. Their identity, which constitutes the only surprise in equations (10.9)–(10.11), is an example of an *Onsager relation*, and readers who are unfamiliar with Onsager relations, and with the subject of irreversible thermodynamics in which they arise, should take it on trust.

From the above equations one may readily deduce, by permuting the suffices, the similar equations which apply when the director is anchored along x_1 or x_2 instead of x_3. The diagrams in fig. 10.2 represent steady planar shear flow between two flat plates which are normal to the x_2 axis, such that u_1 increases in a linear fashion with x_2 but is independent of x_1 and x_3, while u_2 and u_3 are zero. The viscosity coefficient which constitutes the ratio between the shear stress s_{12} which acts on the plates and the velocity gradient $\mathrm{d}u_1/\mathrm{d}x_2$, and which is therefore in principle measurable using a set-up of this nature, is η_1 if the director is anchored along x_1 [fig. 10.2(a)], η_2 if it is anchored along x_2 [fig. 10.2(b)], and η_3 if it is anchored along x_3 [fig. 10.2(c)].

The linear equations for normal stress which replace (6.9) when the director is anchored along x_3 may be written in terms of a fifth coefficient, η_5, as

$$\begin{aligned} p_1 &= p - \frac{2}{3}\left(2\eta_3 \frac{\partial u_1}{\partial x_1} - \eta_3 \frac{\partial u_2}{\partial x_2} - \eta_5 \frac{\partial u_3}{\partial x_3}\right) \\ p_2 &= p - \frac{2}{3}\left(-\eta_3 \frac{\partial u_1}{\partial x_1} + 2\eta_3 \frac{\partial u_2}{\partial x_2} - \eta_5 \frac{\partial u_3}{\partial x_3}\right) \\ p_3 &= p - \frac{2}{3}\left(-\eta_3 \frac{\partial u_1}{\partial x_1} - \eta_3 \frac{\partial u_2}{\partial x_2} + 2\eta_5 \frac{\partial u_3}{\partial x_3}\right). \end{aligned} \tag{10.12}$$

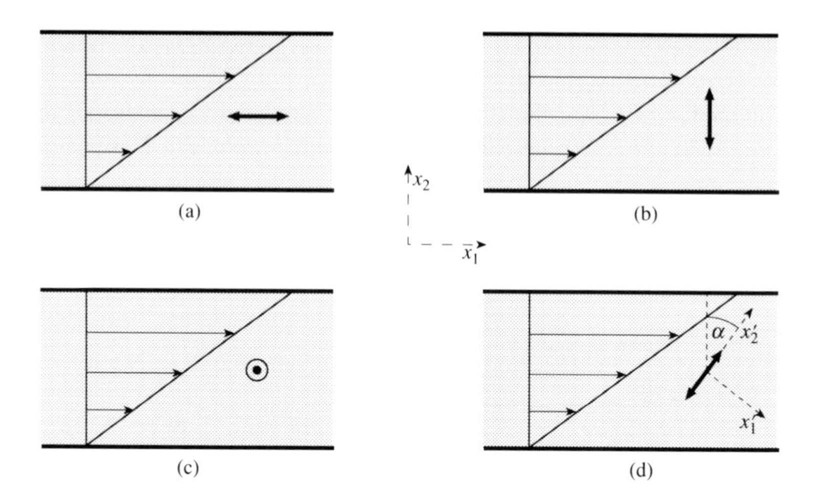

Figure 10.2 Four cases of planar shear in a uniaxial fluid; they are distinguished by the orientation of the director, which is indicated by a double-headed arrow in (a), (b) and (d), and by a point-and-circle in (c). The shear stress s_{12} is proportional to η_1 in case (a), to η_2 in case (b), and to η_3 in case (c). In the more general case (d), the relevant viscosity coefficient is described by (10.15).

Hence for incompressible flow, when $\nabla \cdot \boldsymbol{u} \approx 0$, we have

$$\frac{1}{2}(p_1 + p_2) - p_3 = \eta_{\text{ext}} \frac{\partial u_3}{\partial x_3}, \tag{10.13}$$

where

$$\eta_{\text{ext}} = \eta_3 + 2\eta_5. \tag{10.14}$$

The quantity on the left-hand side of (10.13), and hence η_{ext} and η_5, can in principle be determined experimentally, from the force needed to extend a filament of the fluid of known cross-sectional area in which the director is oriented longitudinally. When molten polymers or polymer solutions are drawn out into filaments they acquire uniaxial anisotropy, as we shall see below, and in that context rheologists refer to the quantity η_{ext} which is defined by (10.13) as the *extensional viscosity*. As (10.14) makes clear, the extensional viscosity for a Newtonian fluid in which there is no distinction between η_3 and η_5 is three times larger than the normal shear viscosity η.

What is the measurable viscosity

$$\eta^* = \frac{s_{12}}{\mathrm{d}u_1/\mathrm{d}x_2}$$

in the situation represented by fig. 10.2(d)? In this case the director is anchored along the x_2' axis of a reference frame S′ which differs from the original reference frame S by a left-handed rotation about the x_3 axis through an angle α, and the question can readily be answered by working out the velocity gradients in S′, by using equations (10.9)–(10.12) to find the associated stress components in S′, and by using the sort of arguments that were used to justify (1.1) and (1.2) to transform the stresses back into S. The answer turns out to be

$$\eta^*\{\alpha\} = \eta_2 \cos^4 \alpha + (\eta_{\text{ext}} + \eta_3 - 2\eta_4) \sin^2 \alpha \cos^2 \alpha + \eta_1 \sin^4 \alpha. \quad (10.15)$$

It shows, in so far as it involves η_4, that once η_{ext} has been measured η_4 is in principle measurable too.

Expressions for the normal stress components in S in the situation represented by fig. 10.2(d) may be obtained in the same manner. In particular, it may be shown that

$$\begin{aligned} p_3 = p_3' &= 3p - p_1' - p_2' \\ &= p + \frac{du_1}{dx_2} \sin \alpha \cos \alpha \left(\frac{1}{3}\eta_{\text{ext}} - \eta_3\right). \end{aligned}$$

In the form

$$p_3 = p + 2\zeta_{12} \sin \alpha \cos \alpha \left(\frac{1}{3}\eta_{\text{ext}} - \eta_3\right), \quad (10.16)$$

this last result holds even when the fluid undergoes shear about the x_3 axis which is not planar, i.e. when $\partial u_2/\partial x_1$ is non-zero as well as $\partial u_1/\partial x_2$. Here ζ_{12} is a component of the tensor ζ_{ij} defined by (6.18) which by itself describes shear which is vorticity-free, i.e.

$$\zeta_{12} = \frac{1}{2}\left(\frac{\partial u_1}{\partial x_2} + \frac{\partial u_2}{\partial x_1}\right). \quad (10.17a)$$

Unsurprisingly, the quantity

$$\omega_{12} = \frac{1}{2}\left(\frac{\partial u_1}{\partial x_2} - \frac{\partial u_2}{\partial x_1}\right), \quad (10.17b)$$

which describes a component of the fluid's angular velocity [(6.19)], plays no part in determining p_3.

When an isotropic Newtonian fluid undergoes planar shear flow the normal stresses remain isotropic; the rate of shear cannot affect p_1, p_2 or p_3 for obvious symmetry reasons. The fact that they become anisotropic when an anisotropic fluid is subjected to shear has some quite spectacular consequences, as we shall see in §10.6, and we shall return to (10.15) and (10.16) in that section. First, however, we need to discuss what determines the angle α when no field is present to anchor the director.

10.4 A theory of flow birefringence

It has been known since the time of Maxwell that simple liquids which are completely isotropic in their properties when at rest but which are composed of non-spherical molecules tend to become birefringent when subjected to shear. The phenomenon, called *flow birefringence*, is also exhibited by solutions in which the solute molecules are non-spherical, and by suspensions of macroscopic solid rods, fibres or platelets. In searching for an explanation we may start by considering a simple model, of solid prolate spheroids immersed in a Newtonian liquid of viscosity η. The spheroids are supposed to be far enough apart from one another for the interactions between them to be ignored, so let us focus on the behaviour of a single spheroid when the velocity of the surrounding fluid at large distances from it is described by the equations

$$u_1 = u_{o,1} + (\zeta_{12} + \omega_{12})x_2, \quad u_2 = u_{o,2} + (\zeta_{12} - \omega_{12})x_1$$

[(10.17a) and (10.17b)]. Its centre of mass will tend to move with the fluid so that it experiences no drag force, and no generality is lost if we choose the origin to coincide with the centre of mass and choose $u_{o,1}$ and $u_{o,2}$ to be zero. Its semi-major axis and semi-minor axes are denoted below by a and b respectively.

What torque does the spheroid experience when its major axis is situated in the (x_1, x_2) plane at an angle α to the $+x_2$ axis, as suggested by fig. 10.3, and is rotating about the x_3 axis in a left-handed sense (clockwise in the figure) with angular velocity $\dot{\alpha}$? An approximate answer to that question is available, for use in circumstances such that the Reynolds Number defined by $\rho\zeta_{12}a^2/\eta$ is small compared with unity.[2] It is

$$-G_3 \approx A\eta\{(a^2 - b^2)\zeta_{12}\cos 2\alpha + (a^2 + b^2)(\omega_{12} - \dot{\alpha})\}, \qquad (10.18)$$

where $-G_3$ acts clockwise in fig. 10.3, and where A is a positive coefficient of which the only feature of significance here is that it scales like a and b. Unsurprisingly, the component of torque due to ζ_{12} vanishes when $a = b$, and it also vanishes when $\alpha = \pm\pi/4$; when the shear is vorticity-free, therefore, the spheroid can rest in stable equilibrium at $\alpha = +\pi/4$, with its major axis along the axis of extension. Equally unsurprisingly, the component due to the liquid's clockwise angular velocity ω_{12} remains finite when $a = b$ but vanishes when the spheroid is rotating at the same rate. Now in circumstances where the creeping flow approximation is justified for the liquid the inertia of the solid spheroid is likely to be negligible, and we may assume it to rotate at such a rate that G_3

[2] G. B. Jeffery, *Proc. Roy. Soc. A*, **102**, 161, 1922.

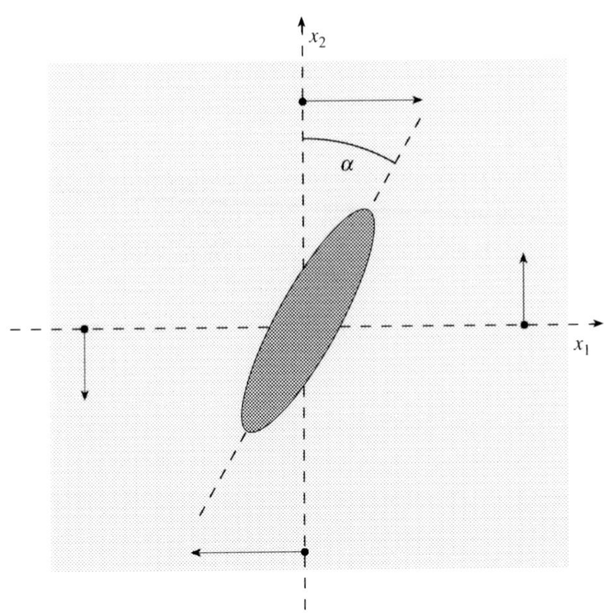

Figure 10.3 A solid prolate spheroid surrounded by isotropic fluid which, as arrows representing fluid velocity relative to the centre of the spheroid are intended to suggest, is undergoing vorticity-free shear and simultaneous rotation about the x_3 axis.

vanishes. This means that in the case of planar shear such that ω_{12} and ζ_{12} are equal we have

$$\dot{\alpha} \approx \frac{1}{2}\frac{\mathrm{d}u_1}{\mathrm{d}x_2}(1 + \lambda \cos 2\alpha), \left[\lambda = \frac{a^2 - b^2}{a^2 + b^2}\right]. \tag{10.19}$$

Equation (10.19), which can be extended to spheroids whose major axes do *not* lie in the (x_1, x_2) plane, suggests that small prolate spheroidal inclusions in an isotropic liquid undergoing this type of planar shear will precess continuously about the x_3 axis, with an angular velocity which is at a minimum when $\alpha = \pm\pi/2$, i.e. when their major axes lie in (x_1, x_3) planes.

At first sight it seems safe to conclude that the spheroids are more likely to be found in that particular orientation than in any other, but that conclusion is not in accordance with experimental observations on, say, dilute solutions of non-spherical molecules, to which the model should be applicable: the preferred orientation for the axis of alignment, i.e. for the director, is found to correspond to $\alpha = \pi/4$ when $\mathrm{d}u_1/\mathrm{d}x_2\,(>0)$ is small, though it does change towards $\alpha = \pi/2$ as $\mathrm{d}u_1/\mathrm{d}x_2$ increases. The explanation lies in the fact that the angular velocity described by (10.19) turns out to be, in a typical case, much smaller in magnitude than the random angular velocities which solute molecules are bound to possess as a result of thermal agitation. The effects of thermal agitation may be illustrated by reference to a two-dimensional model for the three-dimensional problem which really confronts us. Let us suppose that the spheroids are in some way constrained so that their major axes are confined to the (x_1, x_2) plane, and that their orientations within that plane are described by a distribution function $f\{\alpha\}$, such that $f\{\alpha\}\mathrm{d}\alpha$ is the fraction of spheroids whose major axes lie in the angular range between α and $\alpha + \mathrm{d}\alpha$ at any one time. Then f must satisfy the equation

$$\frac{\partial f}{\partial t} = -\frac{\partial(f\dot{\alpha})}{\partial \alpha} + D\frac{\partial^2 f}{\partial \alpha^2}, \tag{10.20}$$

where $\dot{\alpha}$ is given by (10.19) and where D is a rotational diffusion coefficient – here supposed for simplicity to be independent of the rate of shear – which recognises the dispersive effects of random thermal motion. Both sides of (10.20) must vanish in the steady state, in which case we may integrate the right-hand side to show that

$$\frac{\partial f}{\partial \alpha} = \mu(f - f_o + \lambda f \cos 2\alpha), \qquad \left[\mu = \frac{1}{2D}\frac{\mathrm{d}u_1}{\mathrm{d}x_2}\right].$$

Here f_o is an integration constant chosen to ensure that the gradient $\partial f/\partial\alpha$ averages to zero over one complete cycle, as it must do because f is single-valued; f_o is in fact the distribution function in the absence of shear, which is $(2\pi)^{-1}$ if one chooses to normalise f over the full angular range of 2π, or π^{-1} if one recognises that α and $\alpha + \pi$ are equivalent in this problem and normalises over π instead. Now $\partial f/\partial\alpha$ vanishes where f passes through a maximum, so the maxima occur at angles α_{max} such that

$$\cos 2\alpha_{max} = -\frac{f_{max} - f_o}{\lambda f_{max}}.$$

The ratio $(f_{max} - f_o)/f_{max}$ is very small when μ is very small, and in this limit we have $\alpha_{max} \approx \pi/4$ (or $5\pi/4$), as observed experimentally. But as μ increases so $(f_{max} - f_o)/f_{max}$ and α_{max} increase, until in the limit of very large μ, at which rotational diffusion becomes irrelevant, the distribution is described for all α by the equation

$$f = \frac{f_o}{1 + \lambda \cos 2\alpha} \quad [\propto \dot{\alpha}^{-1}], \tag{10.21}$$

and we have $\alpha_{max} = \pi/2$ (or $3\pi/2$).

A three-dimensional version of the above analysis has been applied, with fair success in the limit of weak alignment, not only to dilute solutions but also to pure liquids in which all the molecules are rod-shaped. It cannot be said, however, to constitute a satisfactory theory of flow birefringence in concentrated suspensions

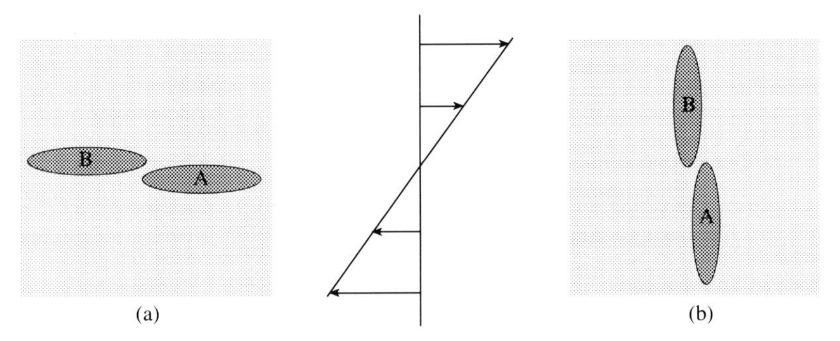

Figure 10.4 Part of a suspension of aligned prolate spheroids which is undergoing planar shear: (a) $\alpha = \pi/2$, (b) $\alpha = 0$.

of macroscopic rod-shaped particles. Rotational diffusion is bound to be less significant for macroscopic particles than for individual molecules, and the theory therefore implies for suspensions a distribution function which does not differ greatly from the limiting form described by (10.21), in which case

$$\frac{f_{\max}}{f_o} \approx \frac{1}{1-\lambda} = \frac{a^2+b^2}{2b^2}.$$

The alignment is therefore liable to be nearly complete if $a \gg b$, even at small rates of shear. In that case, and if the concentration of particles is quite large, we cannot legitimately calculate the torque exerted on each particle on the assumption that the fluid medium surrounding it is isotropic. We must surely allow in some way for interactions between neighbouring particles.

Minor modifications to (10.19) are unlikely to prove sufficient. It is intuitively clear that the continuous precession of individual rods which that equation purports to describe must be completely suppressed if their concentration is so great that every one of them is virtually in contact with several neighbours, and the following argument suggests there may be no shear-induced precession even at quite moderate concentrations, such that there is still plenty of room for rods to rotate if they really want to do so. Figure 10.4 shows close-up views of two suspensions of prolate spheroids in a Newtonian liquid, which are perfectly aligned by a strong field of some sort; the suspensions are undergoing planar shear, as suggested by the velocity profile in the centre, and the principal difference between (a) and (b) lies in the orientation of this field. Since the field only prevents the spheroids from rotating and does *not* prevent them from being swept along by the liquid which surrounds them, a collision between A and B appears in both cases to be imminent. As A and B come together the liquid between them will be squeezed out sideways, and because this liquid is viscous the pressure which acts on the surface of each spheroid will increase. The normal forces which act on the spheroids as a result may be sufficient to prevent them from actually touching, but whether they touch or not is immaterial: during their collision or near-collision they each experience an impulsive torque about their centres of mass, which acts in a clockwise sense in case (b) but in an anticlockwise sense in case (a). In case (b), the effect on a single spheroid of a succession of such events will surely be to increase the average torque otherwise described (with $\alpha = 0$) by (10.18). In case (a), however, interactions between ellipsoids are liable to decrease the torque described (with $\alpha = \pi/2$) by (10.18), and they *may* reverse it in sign. If they do so – if $-G_3$ is positive for $\alpha = 0$ but negative for $\alpha = \pi/2$ – then there is bound to be an intermediate value of α at which it vanishes. If the director adopts that orientation when the suspension is subjected to planar shear, then no field will be needed to keep the rods pointing more or less along it.

10.5 Flow alignment: a general approach

An alternative approach to the problem of flow birefringence in concentrated suspensions of rod-shaped particles – or of *flow alignment*, since the birefringence to which alignment leads is really a secondary aspect of the problem – is suggested by the remarks in the previous paragraph, taken together with the analysis in §10.3. If it is true that the rods become almost parallel to one another when the suspension is undergoing shear, then we have to deal with a fluid which is markedly anisotropic in a uniaxial sense and which needs to be described by five viscosity coefficients η_1, η_2 etc. If this fluid is being steadily sheared about the x_3 axis, and perhaps rotated about that axis too, then the torque per unit volume which it experiences when the director is *not* anchored but is precessing in the (x_1, x_2) plane with angular velocity $\dot{\alpha}$ is given in terms of these coefficients by

$$-g_3 = s_{12} - s_{21} = s'_{12} - s'_{21}$$

(where the primes indicate shear stresses in the frame S′ of fig. 10.2(d)) and hence by

$$-g_3 = (\eta_2 - \eta_1)\zeta_{12}\cos 2\alpha + (\eta_2 + \eta_1 - 2\eta_4)(\omega_{12} - \dot{\alpha}). \qquad (10.22)$$

This result is similar in form to (10.18) for the torque on an individual ellipsoid, but it differs in that the coefficient of $(\omega_{12} - \dot{\alpha})$ is not necessarily greater in magnitude than the coefficient of $\zeta_{12}\cos 2\alpha$. We may take it for granted that in a concentrated and well-aligned suspension of long thin rods (as opposed to the suspension of discs which is considered briefly below) η_2 is considerably greater than both η_1 and η_4; this has to be so if the mean torque exerted on an individual rod by planar shear is to be clockwise in the configuration of figs. 10.2(b) and 10.4(b), and if it is to be significantly greater, e.g. by the factor a^2/b^2 which (10.18) suggests, than the corresponding mean torque in figs. 10.2(a) and 10.4(a). Thus $|\eta_2 + \eta_1 - 2\eta_4|$ is less than $|\eta_2 - \eta_1|$ if it happens that η_4 is greater than η_1. In that case there exist orientations at which the director is able to remain in equilibrium in the absence of any field without the individual rods precessing, because they are such that $g_3 = 0$ when $\dot{\alpha} = 0$. When the fluid is undergoing vorticity-free shear with $\omega_{12} = 0$ they evidently correspond to $\alpha = \pm\pi/4$. When it is undergoing planar shear with $\omega_{12} = \zeta_{12}$, on the other hand, they correspond to solutions of the equation

$$\cos 2\alpha = -\frac{\eta_2 + \eta_1 - 2\eta_4}{\eta_2 - \eta_1}. \tag{10.23}$$

It is not difficult to see that in the first case, and when $\zeta_{12} > 0$, the solution which describes stable as opposed to unstable equilibrium is $\alpha = +\pi/4$; in that case the rods tend to line up in the direction in which the suspension is being stretched. In the second case, again when $\zeta_{12} > 0$, the stable solution is such, if corrections of order η_1/η_4 and η_4/η_2 may be disregarded, that

$$\alpha \approx \frac{\pi}{2} - \sqrt{\frac{\eta_4 - \eta_1}{\eta_2}}. \tag{10.24}$$

Experimental data for η_1, η_2 and η_4 are hard to come by for suspensions of well-aligned rods, no doubt because it is impossible to produce a field which will anchor the director effectively. Such data are available, however, for a number of nematic liquid crystals of the sort that are used in the display elements of digital watches and similar devices. These materials consist entirely of small rod-shaped molecules, rather than of rod-shaped particles in suspension in a Newtonian liquid, and the molecules are never perfectly aligned; the so-called *order parameter*, defined as the ensemble average $\langle \frac{3}{2}\cos^2\beta_i - \frac{1}{2} \rangle$ where β_i is the angle between the director and the long axis in the ith molecule, is typically about 0·5 rather than 1. Nevertheless the above theory, and (10.23) in particular, applies to them. Measurements on nematic MBBA (methoxy-benzylidene-butyl-aniline) at a temperature of about 25°C, for example, suggest $\eta_2 \approx 0{\cdot}104$, $\eta_4 \approx 0{\cdot}0255$, and $\eta_1 \approx 0{\cdot}0238$ kg m^{-1} s^{-1}.[3] Those figures, when inserted into (10.23), imply that when MBBA is subjected to planar shear such that $\omega_{12} = \zeta_{12} > 0$ the director sets, in the absence of any field to constrain it, at $\alpha \approx +82°$. That result is fairly typical of many nematic liquid crystals, and we shall explore some of the consequences of flow alignment in these materials in §10.8 below.

A similar argument may be applied to suspensions of particles which are disc-shaped rather than rod-shaped, and which are uniaxial because, as a result of shear, the planes of all the discs are almost parallel to one another. The director is then normal to the discs, and the largest of the three relevant viscosity components is η_1 rather than η_2. Stable equilibrium in planar shear for which $\omega_{12} = \zeta_{12}$ turns out to be possible if $\eta_4 > \eta_2$, and it is characterised when $\zeta_{12} > 0$ by a value of

[3] C. Gähwiller, *Phys. Lett. A*, **36**, 311, 1971.

α which is slightly less than zero rather than slightly less than $\pi/2$. Both rods and discs, therefore, are likely to align in such a way that they lie almost, though not quite, in planes on which shear is taking place. In both cases flow alignment tends to reduce the apparent viscosity of a suspension subjected to planar shear; the measurable viscosity coefficient η^*, should according to (10.15) be close to η_1 in the case for which $\eta_2 \gg \eta_1$, $\alpha \approx \pi/2$, and close to η_2 in the case for which $\eta_1 \gg \eta_2$, $\alpha \approx 0$.

In principle, the theory outlined above should be applicable to non-Newtonian liquids in general, provided that the anisotropy induced in them by flow is uniaxial; a similar approach may be applied to biaxial fluids, but the complications are inevitably greater. Suppose that the long-chain molecules in a molten polymer, for example, are stretched during steady planar shear flow in such a way that they tend to run parallel to one another, and suppose that the polymer becomes locally uniaxial in consequence. If we knew the viscosity coefficients η_1, η_2 etc. we would be able, using the theory, to predict the orientation of the director. However, the degree of anisotropy is not normally independent of the shear rate in this example, as it is in the case of nematic liquid crystals and as we have assumed it to be in the case of suspensions once the rate of shear is sufficient to cause almost perfect alignment. The viscosity coefficients therefore depend upon the rate of shear – hence the non-linearity of the constitutive relations – and they may also depend upon the orientation of the director which we hope to predict. To find a *self-consistent* description of the response of a molten polymer to steady planar shear may be far from easy.

To describe the polymer's response in situations where the rates of shear to which each fluid element is subject are changing with time is naturally harder still, because then its viscoelastic tendencies need to be taken into account.

10.6 Non-Newtonian effects in polymeric liquids

The flow properties of a polymeric liquid depend to a considerable extent upon its molecular weight, upon the degree of cross-linking if any, and, where the polymer is dispersed in an inert solvent, upon its concentration. The remarks in this section apply chiefly to long-chain polymers which are not cross-linked, and which are either molten, with no solvent present, or else dispersed at concentrations which are not small enough for the polymer molecules to be treated as virtually independent of one another. They apply, that is, to situations in which each polymer molecule is constrained by many entanglements with other molecules, from which it can extricate itself only by a process known as *reptation*.[4] An aqueous solution of high-molecular-weight polyacrylamide at a concentration in

[4] The word reptation is derived from a Greek root and implies snake-like motion.

the range 1%–2% by weight may be used to demonstrate most of the effects described.

One of the striking features of such polymeric liquids, for which reptation models provide an adequate explanation, is that they are *tension-thickening*: that is to say, their extensional viscosity increases as the rate of extension increases. It increases in such a way that at large rates of extension *Trouton's ratio*, which is the ratio of η_{ext} to the apparent shear viscosity η^*, as measured under conditions that approximate to the ideal of steady planar shear flow at a rate of shear which is comparable with the rate of extension at which η_{ext} is measured, invariably exceeds the value 3 that applies for Newtonian fluids and may exceed it by several orders of magnitude. It is a fair assumption that in such cases η_{ext} is much larger than η_1, η_2, η_3 and η_4 individually.

The tension-thickening property of many polymeric liquids explains their *spinnability*, i.e. the fact that they may readily be drawn out into fine threads. Water can exist as a thin cylindrical jet if it is forced under pressure through an orifice, of course, but water jets are subject to the Rayleigh–Plateau instability and soon break up [§8.3]. If one dips one's finger into golden syrup – a liquid which is just as Newtonian as water but very much more viscous – one may draw a thread of sorts out of it and, because the rate at which the Rayleigh–Plateau instability develops is inversely proportional to viscosity, this thread does not immediately break up into drops. While it is being extended, however, it is vulnerable in another way. Under a tensional force F which is bound to be uniform along the length of the thread if its inertia is negligible, the rates of longitudinal extension and of lateral contraction (du_1/dx_1 and $-a^{-1}da/dt$ respectively, where the x_1 direction is taken to lie along the axis of the thread and where a is the thread's radius) are proportional to $F/\eta a^2$, or to $F/\eta_{ext}a^2$ in circumstances where η and η_{ext} need to be distinguished. Hence $-da/dt$ is proportional to a^{-1}, which implies that small differences of radius along the length of a thread get larger as time passes and that *necks* tend to develop in consequence. Because of tension-thickening, polymeric threads show this tendency to a much reduced extent; a local decrease in a and increase in F/a^2 must increase du_1/dx_1 and $-a^{-1}da/dt$ slightly, but the consequent increase in η_{ext} does much to stabilise the system.

Some polymer solutions may be persuaded with care to display what is called the *open syphon effect*. It is initiated by tilting the solution from a full beaker (A) to an empty one (B) in a thin stream. The stream is sufficiently stable, for the reason outlined above, to remain intact while beaker A is restored to the vertical and placed on a stand above beaker B, and, even though it is only by climbing up the side of A that the liquid can now get to B, the flow continues until most of A's contents have been transferred.

However, such polymer solutions not only have a large extensional viscosity, they are also viscoelastic. Thus if a falling stream is suddenly broken, for example by cutting it with a pair of scissors, the two parts of it contract in the way that a

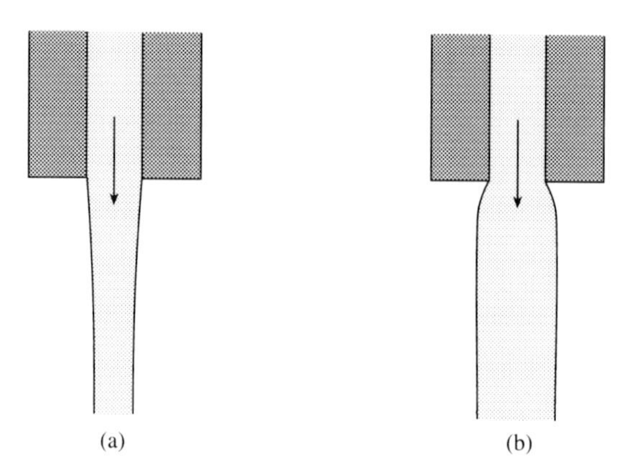

Figure 10.5 Liquid emerging from a capillary tube: (a) Newtonian liquid, (b) polymeric liquid, showing the die-swell effect.

stretched rubber band contracts when it too is cut. This makes a good demonstration, and with a little practice it is possible for the demonstrator to persuade all the liquid above the cut to snap suddenly back into the beaker or bottle from which it is flowing.

A less spectacular manifestation of non-Newtonian behaviour in polymeric liquids is the *die swell effect*; it is seen where a jet of liquid emerges from the open end of a capillary tube through which it has been moving in a laminar fashion, under the influence of a longitudinal pressure gradient. Except at very small values of Re a jet of say golden syrup would narrow in these circumstances [fig. 10.5(a)], but jets of polymeric liquids normally increase markedly in radius [fig. 10.5(b)]. Where it is inside the capillary the liquid is undergoing what is virtually planar shear, in the sense that its longitudinal component of velocity u_1 depends upon radius r. Equation (10.16) and the argument in §10.5 suggest that in these circumstances the so-called *hoop stress*, which acts at right angles to both the radius vector and the axis of the tube and which tends to increase the length of any ring-shaped element of fluid which encircles the axis at constant r, is given in cylindrical coordinates by

$$p - p_\theta = -\left|\frac{du_1}{dr} \sin\alpha \cos\alpha\right| \left(\frac{1}{3}\eta_{\text{ext}} - \eta_3\right),$$

where $\cos\alpha$ may be small but is not quite zero. In so far as $\frac{1}{3}\eta_{\text{ext}}$ exceeds η_3 this hoop stress is compressive; it tends to shorten ring-shaped elements of fluid rather than to stretch them. When the liquid emerges from the capillary the rate of shear drops suddenly to zero and the hoop stress disappears, but because the liquid is viscoelastic it carries with it a memory of its experience inside the capillary. It therefore responds to the removal of the compressive hoop stress by expanding.

Figure 10.6 The Weissenberg effect in an aqueous solution of polyacrylamide. When the elevation of the surface is as marked as this the flow pattern is not steady, for liquid adjacent to the rod climbs upwards in a continuous fashion while liquid in the outer parts of the meniscus runs down again. At slightly higher speeds of rotation, liquid accumulating at the top of the meniscus parts company with the rod and is flung radially outwards. [Photograph by Keith Papworth, Cavendish Laboratory, Cambridge.]

A related effect, illustrated by a photograph in fig. 10.6, is the *Weissenberg effect*. The photograph shows a dish containing a 2% aqueous solution of high-molecular-weight polyacrylamide, with a vertical rod of about 9 mm diameter dipping into it. The rod is rotating about its axis at about 5 revolutions per second, and the liquid is therefore rotating too, with a velocity u_θ which decreases with increasing radius r. The effect of such rotation on a dishful of golden syrup would be to displace the liquid away from the rod and to lower the level of its surface there [§2.7], but the effect on a polyacrylamide solution is evidently quite the reverse.[5]

[5] Professor Weissenberg demonstrated this effect at a lecture in Cambridge, shortly after he had announced its discovery in *Nature* in 1947. Having demonstrated both its occurrence and its non-occurrence, using two otherwise similar viscous liquids, he was pressed by a member of the audience to reveal what these liquids were. He appeared strangely reluctant to do so, but after some persuasion he was heard to say: 'This one is mouse poison; that one is rat poison'. I owe this anecdote to Brian Pippard.

Were the liquid Newtonian, the vertical displacement of the surface would at low rates of rotation be everywhere small compared with the diameter of the rod and the depth to which it is submerged, and moreover the shear viscosity which occurs in (6.54) for the torque per unit length g_3 would of course be independent of r. In those circumstances the circulation of the liquid, in a dish whose radius was much larger than that of the rod, would reach a vorticity-free steady state characterised by the equation $u_\theta \propto r^{-1}$. The Weissenberg effect may most easily be analysed on the assumption that for polymeric liquids also the circulation reaches a vorticity-free steady state, though the fact that the effective shear viscosity of any non-Newtonian liquid is liable to vary with the rate of deformation (and therefore with r in the present context) would make this assumption a dubious one, even at small rates of rotation; where the vertical displacement of the surface is as large as it is in fig. 10.6 the assumption is wholly unjustified, for reasons indicated in the figure caption. If it be granted nevertheless, we may deduce the normal stress in the vertical direction, p_z, from (10.16), using either $+\pi/4$ or $-\pi/4$ for α depending upon whether $\zeta_{\theta r}$ ($=\partial u_\theta/\partial r$, or $-u_\theta/r$) is positive or negative [§10.5]; at radius r it is given by

$$p_z = \frac{1}{3}(p_z + p_r + p_\theta) + \frac{|u_\theta|}{r}\left(\frac{1}{3}\eta_{\text{ext}} - \eta_3\right),$$

or, since it is obvious on symmetry grounds that p_r and p_θ are equal when the shear is vorticity-free and $\alpha = \pm\pi/4$, by

$$p_z = p_r + \frac{3|u_\theta|}{2r}\left(\frac{1}{3}\eta_{\text{ext}} - \eta_3\right).$$

But p_z is related to the atmospheric pressure p_{A} and to depth below the surface, which may be written in the notation of fig. 2.2 as $z_\text{o} + \zeta\{r\}$, where $\zeta\{r\}$ – not to be confused with any component of ζ_{ij} – represents the elevation of the surface at radius r, by the equation

$$p_z = p_{\text{A}} + \rho g(z_\text{o} + \zeta).$$

Hence

$$p_r = p_{\text{A}} + \rho g(z_\text{o} + \zeta) - \frac{3|u_\theta|}{r^2}\left(\frac{1}{3}\eta_{\text{ext}} - \eta_3\right),$$

and it follows by equation of the inwards force per unit mass, $\rho^{-1}\text{d}p_r/\text{d}r$, to the centripetal acceleration of the liquid that

$$g\frac{\text{d}\zeta}{\text{d}r} = \frac{u_\theta^2}{r}\left(1 - \frac{\eta_{\text{ext}} - 3\eta_3}{\rho|u_\theta|r}\right) \tag{10.25}$$

in the steady state. In the experiment illustrated by fig. 10.6, $(\eta_{ext} - 3\eta_3)/\rho$ may well have been 10 $m^2\ s^{-1}$ or more. It is no surprise, therefore, that the non-Newtonian correction term in (10.25) was sufficient to reverse the sign of $d\zeta/dr$.

Perhaps the best way to measure η^* for a non-Newtonian fluid is to use a *Couette viscometer* [§6.9(ii)] in which the gap between the two cylinders is very narrow compared with their radius, though the much-used *cone-and-plate viscometer* provides an acceptable alternative. Both of these devices provide a reasonable approximation to the ideal of steady planar shear, and when the rate of shear is large enough to ensure that $\eta_2 \gg (\eta_4 - \eta_1)$ in a liquid which becomes uniaxial under shear, so that the alignment angle α is almost $\pi/2$ and described by (10.24), an adequate approximation to (10.15) should be

$$\eta^* \approx \eta_1 + (\eta_{ext} + \eta_3 - 2\eta_4 - 2\eta_1)\,\frac{\eta_4 - \eta_1}{\eta_2},$$

or say

$$\eta^* \approx \eta_1 + \eta_{ext}\,\frac{\eta_4 - \eta_1}{\eta_2} \qquad (10.26)$$

if η_{ext} is relatively large. In so far as the first term on the right-hand side of (10.26) dominates the second, the liquid is likely to be *shear-thinning*, in the sense that η^* falls as the rate of shear increases: η_1 is surely likely to be less than the viscosity of the isotropic liquid at very low rates of shear and to fall further as the degree of anisotropy increases. However, the second term could dominate the first, and in that case a liquid which is tension-thickening could be *shear-thickening* as well. In practice, the sort of polymeric liquids we have discussed in this section are more often shear-thinning than shear-thickening at modest rates of shear, though shear-thickening does seem to be the rule at very high rates. It is because these liquids so often combine tension-thickening with shear-thinning that their Trouton ratios tend to be so high.

A striking manifestation of shear-thinning in polymeric liquids is a phenomenon referred to as the *spurt effect*, not infrequently observed when such liquids are extruded through cylindrical capillary tubes: plotted as a function of the magnitude of the pressure gradient ∇p the discharge rate Q displays a hysteretic transition between two quite different regimes [fig. 10.7]. It is thought that above the transition the velocity profile in the capillary has an abrupt change of slope in it, at some radius r_c which decreases towards zero as the pressure gradient increases further [fig. 10.8]. The shear stress $-\eta^*(du_1/dr)$ is given by $\frac{1}{2}r\nabla p$ and increases with radius in a linear fashion between $r = 0$ and $r = a$, but at $r = r_c$ there is a sudden increase in $|du/dr|$ and a sudden fall in η^*.

The rate of flow of water through a capillary may be affected in a quite different but equally dramatic manner by the addition of long-chain polymers in much

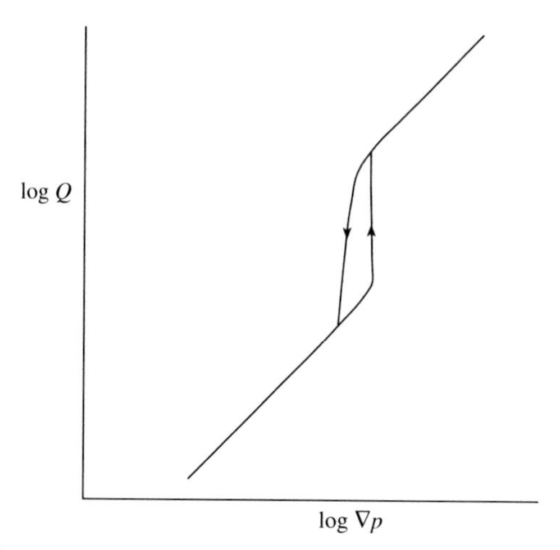

Figure 10.7 The spurt effect: the discharge rate through a capillary suddenly changes, by a factor of perhaps 100.

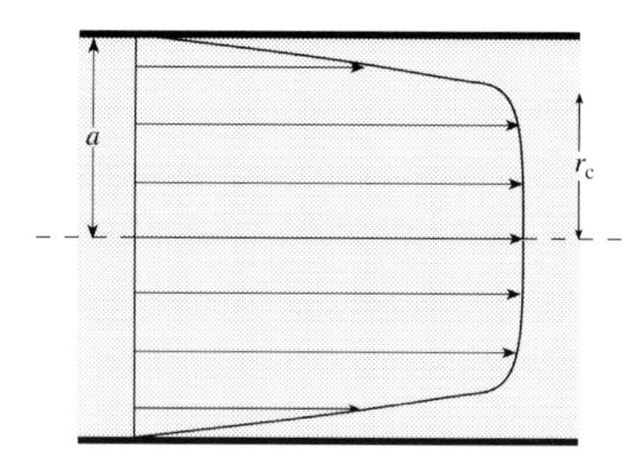

Figure 10.8 Hypothetical velocity profile above the spurt effect transition.

smaller concentrations than those at which the spurt effect manifests itself. In the case illustrated by fig. 10.9, the addition of 0·05% of polyethyleneoxide decreases the flow rate in the laminar region by roughly 20%, a decrease which can safely be attributed to a 20% increase in viscosity. In the turbulent region, however, the flow rate of the solution is *greater* than that of pure water at the same pressure gradient, not less, and by up to 60%. Polymer additives not only enable water at large values of the Reynolds Number to flow more freely through tubes, they also reduce the drag force which it exerts on obstacles in its path. These effects, undoubtedly non-Newtonian, are still not fully understood.

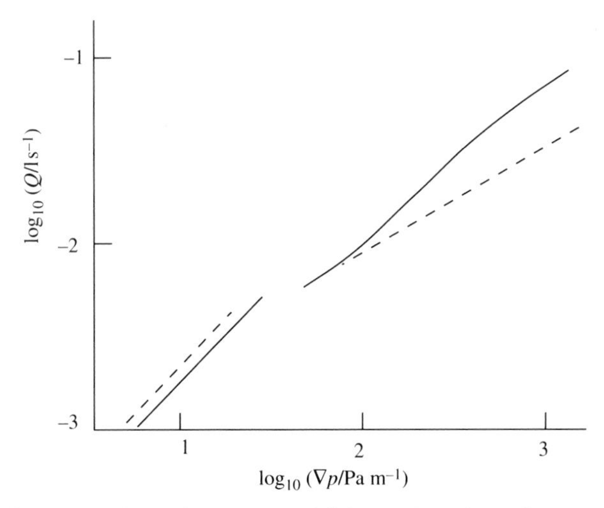

Figure 10.9 The effect of a polymer additive on flow through a long pipe with an internal diameter of about 3 mm; the broken line represents the variation of discharge rate Q with pressure gradient ∇p for pure water [cf. fig. 9.4] and the full line represents the same variation for water containing polyethyleneoxide at a concentration of 500 p.p.m. by weight.
[Based on results quoted in P. S. Virk and E. W. Merrill, *Viscous Drag Reduction*, ed. C. S. Wells, Plenum Press, New York, 1969.]

10.7 Non-Newtonian effects in suspensions

Put a spoonful or two of powdered cornflour in a wineglass and mix it into a smooth paste with about one third of the volume of cold water. When the glass is gently tilted, the paste appears to have the consistency of whipping cream and to flow as easily. Try to stir it with a spoon, however, and you will find this to be hard work. Try to jerk it out of the glass and it may well refuse to move. Here is an extreme example of shear-thickening behaviour, in a concentrated suspension of particles which are virtually spherical. It seems that the particles repel one another, and that they therefore tend to adopt, when left to themselves, a rather regular structure which enables them to stay out of contact. Shear motion forces them into contact, however, and it forces them into a less regular structure such that they, together with the interstices between them, occupy a larger volume; the term *dilatancy* is used to describe this effect. The water present in the mixture is no longer sufficient to fill all the interstices, so the paste becomes noticeably drier as well as stiffer on shearing, just as the surface of sand on a damp beach becomes drier when one stands on it.

Dilute clay slurries behave quite differently; they are shear-thinning rather than shear-thickening, again to an extreme degree. Take some fuller's earth, or *bentonite*, one form of which is useful as a catalyst and is therefore stocked by chemical suppliers; it is a white powder consisting of very small silicate platelets derived from volcanic ash. Shake some of this up with a good deal of water and centrifuge the mixture to precipitate the larger platelets. Then concentrate the supernatant liquid by evaporation, and you will be left with a translucent *gel* which is sufficiently rigid not to flow out of a test tube when the test tube is turned gently upside down. Tap the test tube, however, and its contents miraculously liquefy, turning from a *gel* into a *sol* which is almost as labile as the water of which it so largely consists. Bentonite gels owe their rigidity to the fact that each silicate platelet is in contact with one or more neighbours and is held in contact by weak forces which are electrical in origin. These forces are thought to hold an edge of one platelet to a face of another; the structure of the gel probably resembles in some ways the structure of a house of cards, though it must be a lot more random. Like a house of cards it is very sensitive to disturbance, and once the links between adjacent platelets have been broken the suspension is able to flow. Rapid shear-thinning occurs, presumably because the effect of shear is to orient the platelets so that they are almost parallel [§10.5], and once oriented they no longer interfere with one another.

Most suspensions of non-spherical particles which are not concentrated enough for dilatancy to be a problem are shear-thinning, like bentonite gels, at modest rates of shear. No doubt flow alignment is always partly responsible, but there is evidence that in some cases another factor is at work; it seems that planar shear may *translate* suspended particles as well as rotate them, and may thereby persuade them to adopt a layered arrangement which favours easy shear. The existence of layers which lie in planes of shear, i.e. which are normal to the x_2 axis in the geometry of fig. 10.2, and the surprising regularity of their spacing, is revealed in visible light by flow-induced interference effects. Presumably the layers break up eventually, because shear-thinning behaviour is almost invariably succeeded, in suspensions as well as in non-Newtonian polymeric liquids, by shear-thickening at high rates of shear, and the onset of shear-thickening is often quite abrupt. It is possible that in some uniaxial suspensions of aligned particles there is a critical rate of shear at which the factor $(\eta_4 - \eta_1)$ which occurs inside the square root in (10.24) – or $(\eta_4 - \eta_2)$ if the particles are disc-shaped rather than rod-shaped – changes from positive to negative; following such a change of sign the director could not remain fixed in stable equilibrium, whatever its orientation, and break up of the layers would seem inevitable. Other instabilities may, however, be involved.

Bentonite gels and cornflour paste are not viscoelastic in the normal sense, but they nevertheless remember their previous treatment in the way that viscoelastic fluids do: when suddenly brought to rest, it takes them an appreciable time to recover the rigidity (in one case) and lability (in the other) which characterise them once they have reached equilibrium. The setting process in a bentonite gel involves rotation of individual platelets as a result of Brownian motion and it can be remarkably slow; the setting time may be several hours. The combination of properties displayed by bentonite suspensions, namely pronounced shear-thinning and an appreciable recovery time, is called *thixotropy*, while the combination displayed by cornflour paste is known as *anti-thixotropy* or *rheopexy* – though the latter word seems to have been coined for a different purpose originally. Neither phenomenon is restricted to suspensions, for polymeric liquids which are thixotropic or anti-thixotropic can also be found. Thixotropy in particular is of considerable industrial importance. It makes paints easier to apply, for example, and it is a crucial factor in the technology of drilling though rock; the right sort of thixotropic mud will both support the drill when it is stationary and lubricate it when it is moving.

10.8 Flow phenomena in nematic liquid crystals

Literally hundreds of organic substances are now known which exist as crystalline solids at moderately low temperatures and as isotropic liquids at moderately high temperatures, but which display a variety of stable *mesophases* in between. The common feature of mesophases which are classed as *liquid crystals* is that their structure is locally anisotropic, normally in a uniaxial sense. The so-called *smectic* (i.e. 'soapy' or 'waxy') liquid crystals have structures in which the molecules, besides being aligned with respect to one another, possess some degree of long-range translational order, and many of these are not really liquid at all. However, the *nematic* (i.e. 'threaded') mesophases of compounds such as MBBA or the cyano-biphenyl materials which commonly feature in liquid crystal displays [§10.5] are as labile as water. Other organic compounds whose molecules are almost equally compact display labile mesophases which are *cholesteric* or *discotic* rather than nematic, while polymeric substances and two-phase (*lyotropic*) systems display labile mesophases with their own special characteristics for which yet more names have been coined. However, the experimental work which has so far been done on the fluid dynamics of liquid crystals has been done almost exclusively on uniaxial nematics, and it is only to uniaxial nematics that the rest of this section refers.

The orientation of the director in a flowing nematic is determined by four influences which tend to compete with, and in the steady state balance, one another. Firstly, there is the influence of flow alignment; in the case of simple planar shear this tends to rotate the director until it lies almost, though not quite, in the direction in which the fluid is moving (i.e. the x_1 direction in fig. 10.2) [§10.5]. Secondly, there is the influence exerted by applied fields, if any. Thirdly, there is the influence exerted by the solid surfaces which contain the liquid; these may not matter much in the interior of a specimen of large volume, but the experimental observations to be discussed relate to layers whose thickness was only a few millimetres. Techniques exist – and are exploited in the manufacture of

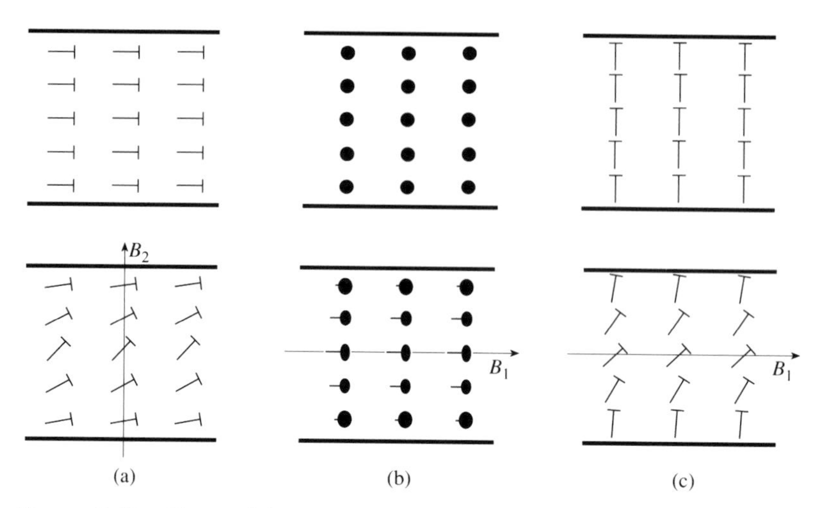

Figure 10.10 Six possible states for a layer of nematic contained between parallel plates; imaginary 'nails', supposed to be pointing along the director, are used to indicate its local orientation. The three sketches in the upper row represent 'single crystal' states of minimum free energy, which can be achieved in practice by appropriate treatment of the plate surfaces. The three sketches below these are intended to suggest the distortions which result when magnetic fields are applied in the directions indicated. Columns (a), (b) and (c) illustrate *splay*, *twist* and *bend* distortion respectively.

display devices – for treating solid surfaces so as to anchor the director in the nematic liquid immediately adjacent to them; it may be anchored so that it is normal to the surface, or it may be anchored so that it is in the plane of the surface and lies in a specified direction, and in both cases the anchoring is strong. The fourth influence is exerted by the *curvature elasticity* (to call it *stiffness* would really be better) of the nematic itself.

The effects of curvature elasticity may be understood in general terms by reference to the diagrams in fig. 10.10, which indicate how the orientation of the director might vary with x_2 in a nematic layer contained between two plates which are normal to x_2, and which have been treated so as to anchor the director in the same direction. The diagrams in the top row show (from left to right) the equilibrium states associated with surface anchoring in the x_1, x_3 and x_2 directions respectively; in each of these there is no variation, and the nematic may be thought of as a 'single crystal'. The diagrams below them suggest states in which the director has been rotated away from its orientation in the single crystal state by application of a magnetic field. In (a) rotation about x_3 through an angle α which varies with x_2 but not with x_1 or x_3 has introduced a type of continuous distortion into the director field known as *splay*, while (b) and (c) illustrate *twist* and *bend* distortion respectively. In the first case the nematic liquid experiences a torque per unit volume about the x_3 axis of magnitude

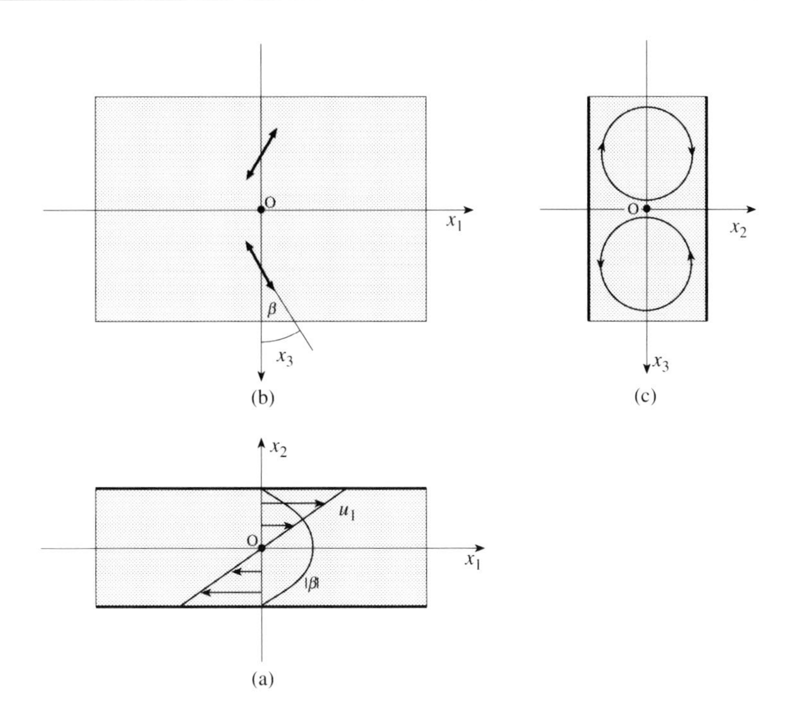

Figure 10.11 Three sections through a layer of nematic liquid contained between parallel plates, which have been treated so as to favour orientation of the director in the x_3 direction. Motion of one plate in the $+x_1$ direction and of the other in the $-x_1$ direction establishes a velocity profile in the liquid which, initially at any rate, is linear, as shown in diagram (a). On account of this shear the director tends to twist away from the x_3 direction, as shown in diagram (b), where the director is represented by a double-headed arrow; diagram (a) suggests how the angle of twist, $|\beta|$, varies with x_2. A secondary circulation of the sort suggested by diagram (c) results; the sense of the circulation shown in (c) is correct only if $\eta_3 > \eta_1$ as is normally the case.

$$g_3 = K_1 \frac{\partial^2 \alpha}{\partial x_2^2},$$

where K_1 is the *splay coefficient* of the nematic, which restores equilibrium by rotating the director back again when the field is switched off. Similar torques act in the other two cases, determined by the *twist* and *bend coefficients* respectively, K_2 and K_3.

Consider [fig. 10.11] a layer of nematic which is undergoing planar shear flow in the x_1 direction, with no fields present, between two plates in the planes $x_2 = \pm\frac{1}{2}d$ (moving in opposite directions) which have both been treated in such a way that in contact with them the director is anchored along x_3. Once the rate of shear exceeds some critical value the 'single-crystal' state in which the director is everywhere parallel to x_3 becomes unstable with respect to a perturbation which twists the director about the x_2 axis through an angle $\beta\{x_2\}$, which may be positive or negative, towards the orientation almost parallel to x_1 which is favoured by flow alignment, i.e. towards $\beta = \pm\pi/2$. When the critical value is exceeded by a small margin only, the steady state is such that β is approximately proportional to $\cos(\pi x_2/d)$, its amplitude being small compared to $\pi/2$; in this state the torque which is responsible for flow alignment is balanced everywhere by a counter-torque due to curvature elasticity of magnitude $K_2(\partial^2\beta/\partial x_2^2)$ [cf. fig. 10.10(b); flow alignment plays the role which in that case is played by the field B_1]. But the fact that the fluid is anisotropic means that where β is neither zero nor $\pm\pi/2$ the shear stress component s_{13} is non-zero as well as s_{12}; in fact

$$s_{13} \approx -(\eta_3 - \eta_1)\sin\beta\cos\beta\,\frac{du_1}{dx_2}.$$

Hence the fluid experiences a transverse force per unit volume of magnitude

$$f_3 = \frac{ds_{13}}{dx_2} \approx -(\eta_3 - \eta_1)\cos 2\beta\,\frac{du_1}{dx_2}\frac{d\beta}{dx_2}.$$

Because $d\beta/dx_2$ changes sign on the plane $x_2 = 0$, this force acts in different directions in the two halves of the specimen, where $x_2 > 0$ and where $x_2 < 0$, and it therefore imposes on the fluid a secondary circulation. In practice the specimen is found, once the critical rate of shear has been exceeded, to form *rolls* parallel to the x_1 axis whose thickness in the x_3 direction is comparable with d, and the sense of circulation within each roll, together with the sign of β, alternates from one roll to the next in the manner suggested by the figure.[6]

This roll phenomenon constitutes one member of a substantial class of flow instabilities displayed by nematic liquid crystals which owe their occurrence to the anisotropy of these materials. Another member is the instability which leads to Bénard-like convection rolls in a layer of nematic which is heated from *above* rather than below [§8.7]. The latter effect is explained in outline in the caption to fig. 10.12.

The phenomena to be observed in nematic layers which are driven well beyond some initial instability, e.g. by increasing the rate of shear or by increasing the temperature difference between their top and bottom surfaces, are even more varied and complex than the phenomena referred to briefly in §8.7 and §9.10. Rapid circulation of a nematic, however it is caused, invariably leads to the production of *disclinations*, a disclination being a line singularity round which there is marked distortion – splay, twist, bend or some combination of these – in the director field. Thus if, when looking through a polarising microscope at a drop of nematic enclosed between a slide and a cover slip, one slightly disturbs the

[6] E. Guyon & P. Pieranski, *J. Phys.* (*Paris*), **36**, C1-203, 1975.

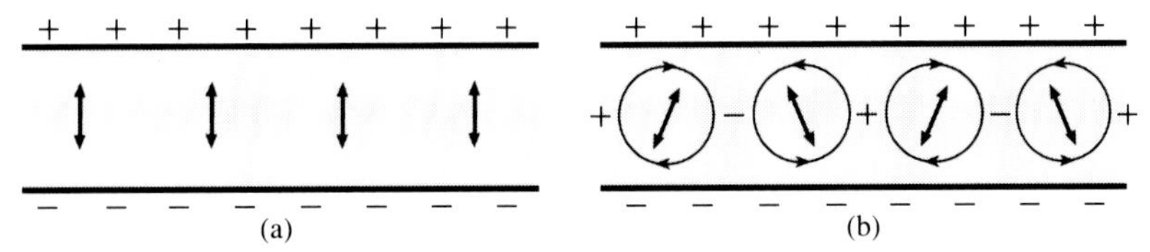

Figure 10.12 Bénard convection in a layer of nematic liquid heated from above: the + and − signs imply 'relatively hot' and 'relatively cold' respectively. The plates which confine the liquid are supposed to have been treated so as to favour orientation of the director in a vertical direction, so below the onset of convection the director is uniformly vertical (diagram (a)); the heat flux $\boldsymbol{H}$ is then parallel to the temperature gradient but in the opposite direction, i.e. it is vertically downwards. Just above the onset of convection (diagram (b)) the director in the interior of the layer is subject to a rotation about the y axis, say, which varies with x in a periodic fashion. Because the thermal conductivity of nematics is always greater along the director than at right angles to it, $\boldsymbol{H}$ is rotated in the same sense; it acquires a periodic horizontal component H_x, which carries heat towards the regions on the central plane which are labelled + in (b) and away from the regions labelled −. In these regions the liquid is hotter and less dense, or colder and more dense respectively, than it would be otherwise. The less dense liquid rises and the more dense liquid falls, and it is the resulting rolling motion which maintains the periodic distortion in the director field against the neutralising effects of curvature elasticity.

cover slip, a mass of disclinations will immediately appear; these are the 'threads' which give the nematic phase its name. They appear in closed loops which slowly collapse to a point and disappear if the fluid is left undisturbed but which may grow and branch if the disturbance continues. If fluid dynamicists ever cease to regard the subject of turbulence in isotropic fluids as a challenge, let them turn to turbulence in bulk specimens of nematic which are permeated by disclinations. As yet, this area of the subject has scarcely been touched.

10.9 Plasmas in magnetic fields

The plasmas which occur naturally in the Earth's ionosphere and in the further reaches of space, which can be created under laboratory conditions in discharge tubes, and which may one day be created routinely inside fusion reactors, are gases under such conditions of temperature and pressure that they are almost completely ionised. Like any gas they can support longitudinal sound waves, in which the positively charged ions and the negatively charged free electrons move together, but they can also support longitudinal *plasma oscillations* in which the ions and electrons move in opposite directions. If we regard the plasma as a superposition of an ionic fluid and a free-electron fluid, having densities ρ_i and ρ_e ($\ll \rho_i$) and velocities $\boldsymbol{u}_i$ and $\boldsymbol{u}_e$ respectively, then in the case of a plane sound wave of angular frequency ω and wavelength λ which is propagating in say the x_1 direction we have

$$u_{i,1} \approx u_{e,1} \propto e^{i(kx_1 - \omega t)},$$

whereas for the corresponding plasma oscillation we have

$$u_{i,1} \approx -\frac{\rho_e}{\rho_i} u_{e,1} \propto e^{i(kx_1 - \omega t)}.$$

The dispersion relation for sound waves in plasmas is the standard one for a normal gas, namely (3.10), though there is room for argument about the appropriate value to use for γ. The dispersion relation for plasma oscillations is

$$\omega^2 \approx \frac{\rho_e e^2}{\varepsilon_0 m^2}(1 + 3\lambda_D^2 k^2), \tag{10.27}$$

where e and m are the electronic charge and mass respectively, and where

$$\lambda_D = \sqrt{\frac{\varepsilon_0 m k_B T}{\rho_e e^2}}$$

is the *Debye screening length*.[7] In a sound wave the mass density of the plasma fluctuates but the charge density does not; the plasma remains virtually neutral throughout. In a plasma oscillation it is the charge density which oscillates and the mass density which is virtually unaffected.

The fact that a plasma can support sound waves is of course irrelevant to its flow properties as long as it is moving at speeds which are small compared with the velocity of sound [§1.8], and the fact that it can support plasma oscillations is even more irrelevant; plasma oscillations arise only when the plasma is disturbed by an electromagnetic field of some sort with a frequency spectrum extending into the range for which (10.27) has real solutions. In that case, how does a plasma differ from the same gas in an un-ionised state, in which charge density fluctuations are impossible because the individual molecules are uncharged? The answer is that its electrical conductivity is great enough to ensure that if a magnetic field happens to be present its lines of force behave as though they were *embedded* in the plasma, just as lines of vorticity are embedded in a fluid whose viscosity is small [§7.1].

Consider the rectangular element of plasma in a magnetic field which is sketched in fig. 10.13(a). When this is deformed by planar shear in the x_2 direction, as in fig. 10.13(b), the conductivity of the plasma is irrelevant because ($\boldsymbol{B} \wedge \boldsymbol{u}$ being zero everywhere) it experiences no electric field; the plasma responds to this type of shear in the same way as any other fluid. When it is deformed by shear in the x_1 direction, however, it carries the magnetic lines of

[7] In the ionosphere, λ_D is typically about 2 mm, and the angular freqency of plasma oscillations whose wavelength is large compared with that is about $4 \times 10^7\ \mathrm{s}^{-1}$.

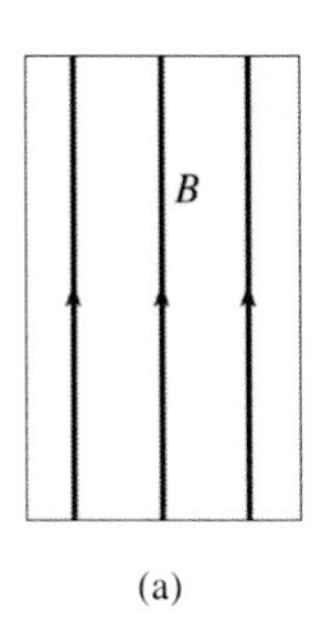

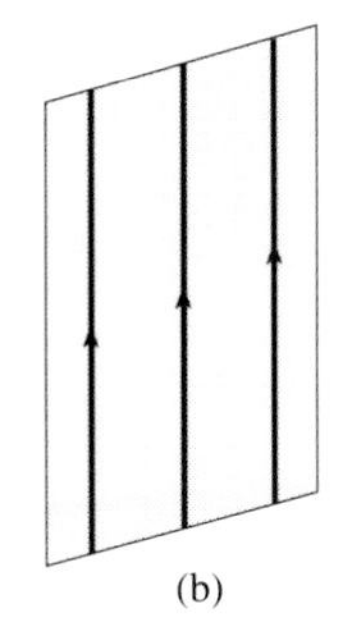

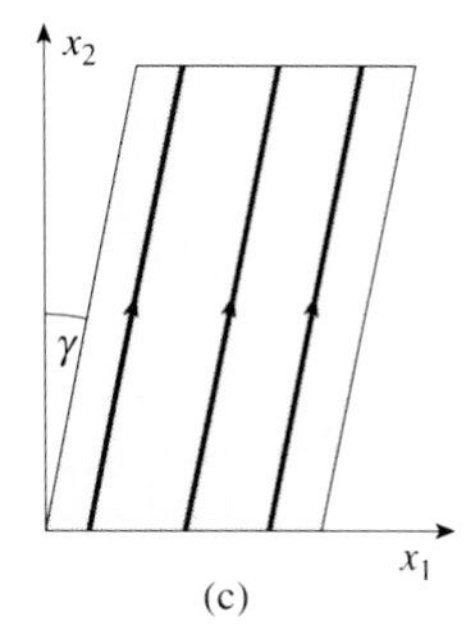

Figure 10.13 A rectangular block of plasma (diagram (a)) in which a uniform magnetic field $\boldsymbol{B}$ is embedded. The spacing between the lines of force remains unchanged when the plasma is sheared in the x_2 direction (diagram (b)) but decreases by a factor cos γ when it is sheared in the x_1 direction instead (diagram (c)).

force with it, as in fig. 10.13(c). Consequently, the field strength rises from B to $B \sec \gamma$, where γ is the angle of shear, and the energy stored in the field per unit volume rises from $B^2/2\mu_o$ to

$$\frac{B^2 \sec^2 \gamma}{2\mu_o} \approx \frac{B^2}{2\mu_o}(1 + \gamma^2) \quad [\gamma \ll 1].$$

The increase implies the existence of a shear stress which is independent of the rate of deformation; it implies that the plasma responds to this type of shear like a solid rather than a fluid, and that its shear modulus G is B^2/μ_o. Consequently, it is possible to propagate transverse sound waves (i.e. shear waves) through a plasma in a magnetic field in a way that is not possible for a normal fluid, as long as the direction of propagation coincides with that of the magnetic field, and their velocity is approximately[8]

$$c_A = \sqrt{\frac{G}{\rho}} = \sqrt{\frac{B^2}{\mu_o \rho}}. \qquad (10.28)$$

The existence of these so-called *Alfvén waves* and the validity of (10.28) for the *Alfvén velocity* c_A have been confirmed in laboratory experiments.

The flow properties of plasmas in magnetic fields, which cannot be examined further here, form the subject matter of *magnetohydrodynamics* or *MHD*.

10.10 Liquid helium

The equation of state of an ideal gas which obeys quantum mechanics rather than classical mechanics depends upon whether the molecules of the gas are fermions

[8] The approximation is valid provided that c_A is very much less than the velocity of light and provided that the angular frequency of the wave is much less than the ionic cyclotron frequency, $\Omega_c = |e|B/m_{mol}$.

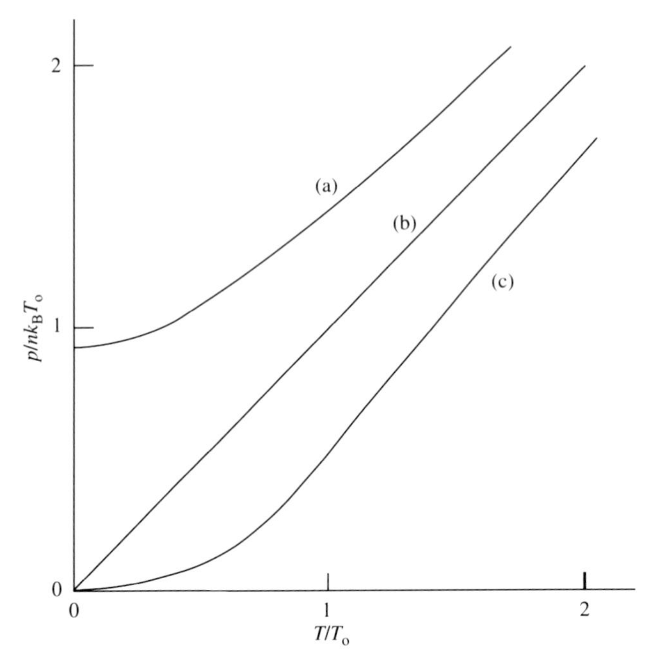

Figure 10.14 The equation of state of an ideal gas composed of (a) fermions and (c) bosons, compared with (b) the classical equation $p = nk_BT$.

or bosons; if they are fermions the pressure p is greater than its classical value of nk_BT, while if they are bosons the reverse is true. In each case the equation of state may be represented by a single curve on a plot of p/n against T, if the axes are scaled as in fig. 10.14 by a density-dependent temperature T_o, defined by the equation

$$k_BT_o = \frac{2\pi\hbar^2}{m_{mol}}\left(\frac{n}{2{\cdot}162g}\right)^{2/3}, \tag{10.29}$$

where g is the spin degeneracy factor for the molecule (i.e. 1 for a boson with spin quantum number zero, 2 for a fermion with spin quantum number $\frac{1}{2}$, and so on). The temperature thus defined has no special physical significance for a fermion gas, but for a boson gas it is the temperature at which, on cooling, the process known as *Bose–Einstein condensation* begins. Below this temperature bosons start to accumulate in the lowest energy state available to them, the number density associated with this state being

$$n\left\{1 - \left(\frac{T}{T_o}\right)^{3/2}\right\}.$$

Technically, an ideal boson gas undergoes a third-order phase transition at T_o, between a 'normal' phase and a 'condensed' phase. There is a discontinuity here in the derivative of the specific heat, $(\partial c_V/\partial T)_n$, and also, though this is not detectable in the figure, in $(\partial^2 p/\partial T^2)_n$.

In real as opposed to ideal gases, the ratio T/T_o is normally so large compared with unity that quantum mechanical corrections to the classical equation of state are completely negligible; in air at standard temperature and pressure, for example, T/T_o is about 6×10^4. The corrections are not quite negligible for gaseous helium which has been cooled until it is on the brink of liquefaction, partly because the temperature at which helium liquefies is exceptionally low (at standard pressure it is 4·2 K) and partly because m_{mol}, to which T_o is inversely proportional [(10.29)], is small. Thus in the saturated vapour phase of ^{4}He, the common isotope of helium, T/T_o is only about 5, and n should on that account be about 4% greater – greater rather than smaller because ^{4}He is a boson – than the ideal classical value of $p/k_B T$. This quantum-mechanical effect is masked in practice, however, by a larger one which is essentially classical in origin: a density shift for which intermolecular interactions are responsible.

When helium vapour liquefies at standard pressure its density increases by a factor of about 8. That increase is exceptionally small, of course, a more typical value for the factor by which density increases on liquefaction being 10^3. The anomaly is attributable to the influence on liquid helium of its zero point energy, and zero point energy is also held responsible for the fact that helium is the only known substance which cannot be made to solidify by cooling alone; it remains liquid at all accessible temperatures – down to the absolute zero as far as we know – unless subjected to a pressure in excess of 2·5 MPa (25 atm.). Now if it were legitimate to treat the liquid phase of ^{4}He as an ideal gas its condensation temperature T_o would be about $8^{2/3}$ times greater than that of the saturated vapour; to be precise, it would be 3·13 K. No phase transition is observed at that temperature, but there *is* a transition at 2·2 K, where the liquid phase known as Helium 1 changes to another liquid phase called Helium 2. The transition is more nearly second-order than third-order; it is marked by a λ-type anomaly in the specific heat which is much more pronounced than the anomaly associated with Bose–Einstein condensation in an ideal boson gas. Nevertheless, the transition between Helium 1 and Helium 2 may be seen as a sort of Bose–Einstein condensation, rather drastically modified by the effects of intermolecular interactions.

The flow properties of Helium 1 are in no way peculiar, but the flow properties of Helium 2 are very unusual indeed. Judged by its ability to flow through fine capillaries without any pressure gradient to drive it its viscosity appears to be zero, though resistance is liable to appear suddenly if a critical flow rate is exceeded. Thus a glass beaker containing Helium 2 will rapidly empty itself by the open syphon effect, the open syphon in this case [cf. §10.6] consisting of a very thin film of liquid covering the surface of the glass; its thickness at the top of the beaker is rarely more than 100 times the atomic diameter and may be a good deal less than that. If one sets out to measure the viscosity of Helium 2 in a different way, however, by measuring the rate of decay of torsional oscillations of a solid disc suspended in the liquid, the result obtained is comparable with the viscosity of Helium 1. One may reconcile two such different results by saying that Helium 2 behaves as though it were a mixture of two components, one of them *superfluid* and the other *normal*; it is the superfluid component which flows through capillary tubes and thin films without resistance, and the normal component which damps an oscillating disc. In a classic experiment performed by Andronikashvili in 1946 the effective density of the normal component, ρ_n, was determined as a function of temperature by measuring the period of torsional oscillations, and hence the moment of inertia, of a stack of discs with spacers in between them; the spacers were thin enough to ensure that all the normal fluid in the interstices between the discs moved with them, while the superfluid remained stationary. Andronikashvili demonstrated that ρ_n/ρ decreases from unity at $T = 2{\cdot}2$ K towards zero in the limit $T \to 0$.

The entropy of the superfluid component appears to be zero at all temperatures, and superfluid can change into normal fluid only by absorbing heat. Now when waves of ordinary sound propagate through Helium 2 the velocities of the two components may presumably be thought of as oscillating in phase with one another, like the velocities $\boldsymbol{u}_i$ and $\boldsymbol{u}_e$ of a plasma through which a sound wave is propagating. In that case, is it possible to excite longitudinal waves in which the two velocities oscillate in antiphase, analogous to the plasma oscillations mentioned in §10.9? If so, the quantities which vary sinusoidally from place to place should be the entropy density and temperature of the liquid rather than its mass density. Experimental observation of this predicted phenomenon – of the propagation of *second sound* as opposed to *first sound* – confirmed the so-called *two fluid model* for Helium 2 in a convincing fashion.

When the two fluid model was first proposed, the superfluid component was thought to correspond to the $n\{1 - (T/T_o)^{3/2}\}$ molecules per unit volume which, in an ideal boson gas at a temperature below T_o, are expected to occupy the single-particle state of lowest energy, and the normal component was identified with the $n(T/T_o)^{3/2}$ molecules per unit volume which remain in excited states; intermolecular interactions were held to blame for the fact that $\rho_n\{T\}/\rho$ and $(T/T_o)^{3/2}$ are in practice distinctly different. However, the idea that the collective ground state of a specimen of Helium 2 – the state to which it would be confined if T could be reduced to zero – bears any resemblance to the ground state of an ideal boson gas is extremely implausible; as in any liquid, the positions of the individual molecules must be very strongly correlated. Nowadays, therefore, we take the correlated collective ground state as our starting point and examine the quantised excitations which it can sustain. Most of the possible modes of excitation

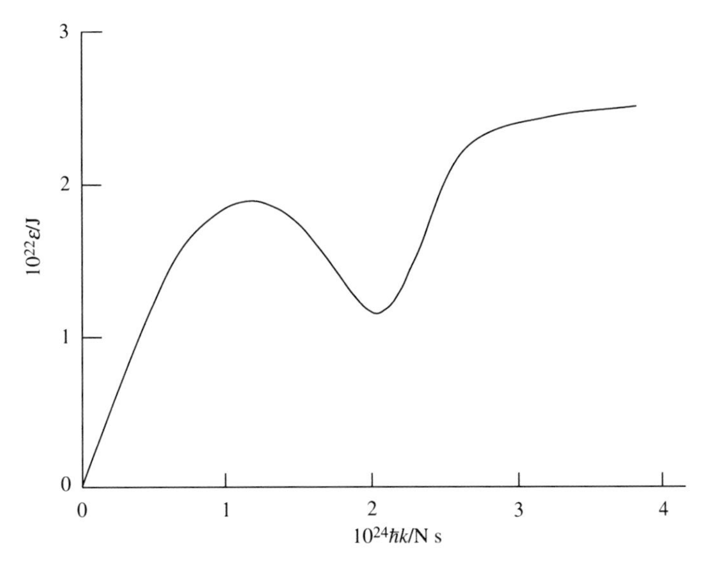

Figure 10.15 The quasi-particle excitation spectrum for Helium 2. [From R. J. Donnelly, J. A. Donnelly & R. N. Hills, *J. Low Temp. Phys.*, **44**, 471, 1981.]

correspond to almost independent *quasi-particles*, and the normal component of Helium 2 is thought of as a gas-like assembly of such quasi-particles, created by thermal excitation of a non-gas-like collective ground state.

The dispersion curve which describes how the energy $\varepsilon\{k\}$ of a quasi-particle in Helium 2 varies with its quasi-momentum $\hbar\boldsymbol{k}$ ($\boldsymbol{k}$ being a wavevector)[9] has been determined experimentally, using the technique of coherent neutron scattering, and is shown in fig. 10.15. The slope of this curve where it passes through the origin is just the velocity of first sound, about 240 m s^{-1}, and in this region the quasi-particles are clearly *phonons*. Near the minimum in the curve, where $\varepsilon/\hbar k$ falls to about 60 m s^{-1}, they are referred to as *rotons*, though that name is probably a bit misleading; rotons are no longer believed to involve localised rotation of the fluid, as once upon a time they were. It has been suggested that they are microscopic vortex rings, but they are probably just modified phonons; to scatter a neutron coherently one needs to disturb the density of the scattering matrix, and while the density of Helium 2 is certainly disturbed when waves of first sound are present it is not obviously disturbed when vortex rings are present instead.

Whatever the origin of the roton minimum in fig. 10.15, it explains the observed temperature dependence of ρ_n and of the specific heat of Helium 2 rather well. As T rises from zero, more and more quasi-particles – rotons in particular – are

[9] In so far as quasi-particles of small k correspond to *phonons*, it is clear from the discussion in §3.7 that the momentum carried by a quasi-particle may differ from $\hbar\boldsymbol{k}$; hence the reference to *quasi-momentum*.

thermally excited, until ρ_n eventually becomes equal to the total density ρ; at that stage, presumably, the last trace of the original collective ground state vanishes and thereafter we have normal Helium 1. It has to be admitted, however, that the phase transition itself is not easy to discuss from the quasi-particle point of view; in that respect, and in that respect only, the older model of Bose–Einstein condensation has the edge.

Let us now try to understand the superfluidity of the collective ground state of Helium 2, i.e. its apparent lack of viscosity, in the context of a hypothetical experiment in which a solid sphere submerged in Helium 2 at $T = 0$ is drawn through it with uniform velocity U. In this context the quasi-particles are almost wholly irrelevant. They are irrelevant because, just as a boat which is moving through calm water cannot generate surface waves unless it moves faster than the minimum wave velocity [§§5.9 and 5.11], so no quasi-particles can be generated by the moving sphere unless U exceeds the minimum value of $\varepsilon/\hbar k$, i.e. 60 m s^{-1} or thereabouts. To judge by what is observed in cases of flow through capillaries, the critical value of U at which, in this hypothetical experiment, Helium 2 would cease to behave as a superfluid would be significantly smaller than that, not more than 1 m s^{-1} at the most.

To describe the ground state of a system composed of N indistinguishable molecules whose positions are strongly correlated with one another we need a many-body wavefunction $\Psi_o\{\boldsymbol{x}_1, \boldsymbol{x}_2, \ldots \boldsymbol{x}_n, \ldots \boldsymbol{x}_N\}$, such that $|\Psi_o|^2 \mathrm{d}\boldsymbol{x}_1 \mathrm{d}\boldsymbol{x}_2 \ldots \mathrm{d}\boldsymbol{x}_n \ldots \mathrm{d}\boldsymbol{x}_N$ is the probability that the small volumes $\mathrm{d}\boldsymbol{x}_1$, $\mathrm{d}\boldsymbol{x}_2$ etc. are simultaneously occupied by the centres of molecules. To describe the same system in motion at $T = 0$ with velocity $\boldsymbol{u}\{\boldsymbol{x}\}$ we presumably need, on the assumption that the motion has no effect on the intermolecular correlations which characterise the ground state, a wavefunction Ψ such that $|\Psi|^2$ and $|\Psi_o|^2$ are identical for all configurations, i.e. for all sets of values of $\boldsymbol{x}_1, \boldsymbol{x}_2$ etc.; that is to say, we need a wavefunction which differs from Ψ_o in phase but not in amplitude, thus

$$\Psi = \Psi_o e^{i\Phi\{\boldsymbol{x}_1, \boldsymbol{x}_2, \ldots \boldsymbol{x}_n, \ldots \boldsymbol{x}_N\}}. \tag{10.30}$$

Moreover, the energy E corresponding to the moving state presumably needs to exceed the energy E_o of the ground state by

$$E - E_o = \frac{1}{2} m_{\mathrm{mol}} \sum_n \boldsymbol{u}\{\boldsymbol{x}_n\} \cdot \boldsymbol{u}\{\boldsymbol{x}_n\}. \tag{10.31}$$

It may readily by shown by substitution of (10.30) into the relevant many-body Schrödinger equation, however, that (10.31) is not satisfied unless

$$\left(\frac{\hbar}{m_{\mathrm{mol}}}\right)^2 \sum_n \left(\boldsymbol{\nabla}_n\Phi \cdot \boldsymbol{\nabla}_n\Phi - 2i\boldsymbol{\nabla}_n\Phi \cdot \frac{\boldsymbol{\nabla}_n\Psi_o}{\Psi_o} + i\nabla_n^2\,\Phi\right) = \sum_n \boldsymbol{u}\{\boldsymbol{x}_n\} \cdot \boldsymbol{u}\{\boldsymbol{x}_n\},$$

(10.32)

where $\boldsymbol{\nabla}_n$ represents the gradient operator which has components $(\partial/\partial x_{n1}, \partial/\partial x_{n2}, \partial/\partial x_{n3})$. But since $\boldsymbol{\nabla}_n \Psi_o$ is equally likely to be positive or negative whereas Ψ_o has the same sign throughout in a boson system, the term on the left-hand side of (10.32) which involves $\boldsymbol{\nabla}_n \Psi_o$ should be virtually zero when summed over n – and is exactly zero when averaged over all configurations. Hence (10.32) implies that we may identify $(\hbar/m_{\text{mol}})\boldsymbol{\nabla}_n \Phi$ with $\boldsymbol{u}\{\boldsymbol{x}_n\}$; we need not bother about the term involving $\nabla_n^2 \Phi$ since this too should be zero if the density is uniform and $\boldsymbol{\nabla} \cdot \boldsymbol{u} = 0$.[10]

Now imagine a configuration in which one may discern a closed loop on which lie, like beads on a necklace, M occupied sites, and suppose this configuration to be varied in a manner that corresponds to displacing the beads around the necklace. If each bead is displaced to such an extent that it exactly replaces the one that was initially in front of it, the process restores the original configuration, which means that it changes Φ by zero or else by some integral multiple of 2π. But this change of phase is evidently given by

$$\sum_{m=1}^{\mathrm{M}} \boldsymbol{\nabla}_m \Phi \cdot (\boldsymbol{x}_{m+1} - \mathbf{x}_m) = \frac{m_{\text{mol}}}{\hbar} \oint \boldsymbol{u} \cdot \mathrm{d}\boldsymbol{x} = \frac{m_{\text{mol}} K}{\hbar},$$

where K is the circulation round the loop. If it is zero for all loops then K is zero for all loops and the motion of the superfluid is everywhere vorticity-free. If it is not zero for all loops then vortex lines must be present, the strength K of which must be an integral multiple of $2\pi\hbar/m_{\text{mol}}$.

That argument, due to Onsager, Feynmann and others, is conjectural in some respects, but the conclusions to which it leads have been accepted as correct since the mid-1950's, following experiments in Cambridge by Hall and Vinen and many similar experiments elsewhere. These showed that when a specimen of Helium 1 which is rotating like a solid body inside a rotating can, and which is therefore permeated throughout by lines of vorticity, is suddenly transformed by cooling into superfluid Helium 2, the lines of vorticity become concentrated in an array of vortex lines, leaving the fluid otherwise vorticity-free. They confirmed, moreover, that the vortex lines are quantised as predicted. Much work has been done since then on quantised vortex lines in Helium 2, but little is known for certain even now about the structure of their cores. According to one very simple model the core is a tube filled only with the vapour phase of helium, at a pressure which is almost zero. To maintain the cylindrical surface of the tube in equilibrium requires a pressure difference of $-\sigma/a$, where σ is the surface tension of Helium 2 and a is the tube's radius [(2.36)], whereas according to Bernoulli's theorem the pressure in the liquid just outside the tube is

[10] Note that for a system which contains only one particle, governed by periodic boundary conditions, the idea that $(\hbar/m)\boldsymbol{\nabla}\Phi$ can be identified with the particle's velocity is obvious.

$$p = -\frac{1}{2}\rho u^2 = -\frac{\rho K^2}{8\pi a^2}.$$

For a line such that $K = 2\pi\hbar/m_{\mathrm{mol}}$, therefore, the model is self-consistent only if

$$a = \frac{\rho K^2}{8\pi\sigma} = \frac{\pi\rho\hbar^2}{2\sigma m_{\mathrm{mol}}} \approx 0{\cdot}5\ \text{Å}.$$

That estimate for the core radius is too small – it is less than the radius of a single ^{4}He atom – for the hollow-core model to be taken seriously, but in one respect at least it is not seriously misleading: the cores are undoubtedly extremely thin.

So why does superfluid Helium 2 at $T = 0$ offer no resistance to a submerged solid sphere which is moved slowly through it? Because it adopts the vorticity-free pattern of dipolar backflow that an Euler fluid would adopt in the same situation and because this pattern resists change; as long as it does so, a drag force is out of the question [see remarks on the subject of d'Alembert's paradox in §§4,7 and 7.8]. Vorticity cannot enter superfluid Helium 2 from the surface of the sphere, as it would enter a classical viscous fluid in the same situation, unless vortex lines for which K is at least $2\pi\hbar/m_{\mathrm{mol}}$ are able to lift off the surface, and nucleation of the lift-off process is effectively inhibited, when U is small, by a potential barrier. What happens once U exceeds the critical velocity (normally less than 1 m s^{-1}, as indicated above) is still not understood in detail.

Because the phase transition between Helium 1 and Helium 2 has always been seen as a Bose–Einstein condensation of a sort, it was taken for granted for many years that the isotope ^{3}He, which consists of fermions, would exhibit only one liquid phase, and that its flow properties would be unremarkable. Far from it! During the 1970's, when ^{3}He derived from nuclear reactions became available in bulk quantities, it was found to display a normal phase plus two superfluid phases which are stable at very low temperatures, and a third superfluid phase can be stabilised by application of a magnetic field. In all three superfluid phases the ^{3}He atoms – or to be more accurate the quasi-particles which play the role of independent atoms at densities such that the effects of interatomic interactions are strong – are paired with one another, and because they are paired with parallel spins ($S = 1$ rather than $S = 0$) all three phases are anisotropic.

Further reading

An Introduction to Rheology, Elsevier, Amsterdam, 1989, by H. A. Barnes, J. F. Hutton & K. Walters, forms a useful supplement to this chapter. Readers wishing to learn more about polymeric liquids than that introductory text can tell them may refer to *Dynamics of Polymeric Fluids, Vol. I, Fluid Mechanics*, John Wiley, New York, 1977, by R. B. Bird, R. C. Armstrong & O. Hassager, and to *The*

Theory of Polymer Dynamics, Clarendon Press, Oxford, 1986, by M. Doi & S. F. Edwards.

A comprehensive account in book form of the continuum properties of liquid crystals is provided by P. G. de Gennes in *The Physics of Liquid Crystals*, Clarendon Press, Oxford, 1974.

For readers wishing to explore the large subject of MHD, *A Textbook of Magnetohydrodynamics*, Pergamon Press, Oxford, 1965, by J. A. Shercliff, may provide a suitable starting point.

An Introduction to Liquid Helium, Clarendon Press, Oxford, 2nd edition 1987, by J. Wilks & D. S. Betts, introduces the reader to the principal experimental facts and theoretical ideas concerning ^{3}He as well as ^{4}He; it includes references to earlier and more elementary monographs on ^{4}He, of which the most compendious is *The Properties of Liquid and Solid Helium*, Clarendon Press, Oxford, 1967, by J. Wilks. Two useful books of later date devoted to liquid helium are *The Superfluid Phases of Helium 3*, Taylor & Francis, London, 1990, by D. Vollhardt & P. Wölfle, and *Quantised Vortices in Helium 2*, Clarendon Press, Oxford, 1991, by R. J. Donelly.

Appendix

One-dimensional sound waves in gases

A.1 Small-amplitude theory excluding attenuation

Sound waves in solids may be longitudinal or transverse, but in fluids – in normal Newtonian fluids at any rate – they are always longitudinal; transverse (i.e. shear) waves are attenuated so rapidly that their amplitude falls over a single wavelength by a factor $e^{-2\pi}$ [(1.30)]. When a plane sound wave is propagating in the x_1 direction through otherwise stationary fluid, therefore, the transverse components of velocity u_2 and u_3 are zero, and u_1 does not depend upon x_2 or x_3. The motion is essentially one-dimensional, and suffices to distinguish one direction from another are not needed. The relevant equations of motion, when viscous stresses are negligible, are

$$\frac{\partial(\rho u)}{\partial x} = -\frac{\partial \rho}{\partial t}, \tag{A.1}$$

which is the continuity condition in one dimension [(2.1)], and

$$-\frac{1}{\rho}\frac{\partial p}{\partial x} = \frac{\partial u}{\partial t} + u\,\frac{\partial u}{\partial x}, \tag{A.2}$$

which is Euler's equation [(2.3) and (2.7)].

Since the fluid is supposed to be stationary in the absence of any sound wave, u is proportional to the wave amplitude. For the time being we are concerned only with waves whose amplitude is infinitesimal, so we may greatly simplify the above equations by linearising them, i.e. by discarding terms which are of second order in u. One such term is obviously $u(\partial u/\partial x)$. Less obviously, in

$$\frac{\partial(\rho u)}{\partial x} = \rho_o\,\frac{\partial\{(1+s)u\}}{\partial x} = \rho_o\left(\frac{\partial u}{\partial x} + s\,\frac{\partial u}{\partial x} + u\,\frac{\partial s}{\partial x}\right),$$

where ρ_o is mean density and s is condensation, i.e. $(\rho - \rho_o)/\rho_o$, we may ignore all but the first of the final three terms. Similarly, ρ may be replaced by ρ_o in $\rho^{-1}(\partial p/\partial x)$. The equations then simplify to:

$$\frac{\partial u}{\partial x} \approx -\frac{\partial s}{\partial t}, \tag{A.3}$$

$$\frac{\partial u}{\partial t} \approx -\frac{1}{\rho_o}\frac{\partial p}{\partial x}. \tag{A.4}$$

In terms of a compressibility defined by

$$\beta_o = \frac{1}{\rho_o}\left(\frac{\partial \rho}{\partial p}\right)_o = \left(\frac{\partial s}{\partial p}\right)_o,$$

where the suffices denote – as in ρ_o – values taken at $s = 0$, (A.4) may be expressed as

$$\frac{\partial u}{\partial t} \approx -\frac{1}{\beta_o \rho_o}\frac{\partial s}{\partial x},$$

and by elimination of u between this and (A.3) we may arrive at the one-dimensional wave equation

$$\frac{\partial^2 s}{\partial x^2} \approx \beta_o \rho_o \frac{\partial^2 s}{\partial t^2}. \tag{A.5}$$

This has travelling-wave solutions of the form

$$s \propto e^{i(kx-\omega t)}, \tag{A.6}$$

for which the wave velocity is

$$c_{s,o} = \frac{\omega}{k} = \sqrt{\frac{1}{\beta_o \rho_o}}. \tag{A.7}$$

A.2 Riemann's treatment for arbitrary amplitudes

Let us now define a *local* velocity of sound c_s in terms of the local compressibility $\beta = \rho^{-1}(\partial\rho/\partial p)$ by the equation

$$c_s = \sqrt{\frac{1}{\beta\rho}}. \tag{A.8}$$

Equations (A.1) and (A.2) may then be recast in forms that slightly resemble one another:

$$\frac{\partial \rho}{\partial t} + u\frac{\partial \rho}{\partial x} + \rho\frac{\partial u}{\partial x} = 0, \tag{A.9}$$

$$\frac{\partial u}{\partial t} + u\,\frac{\partial u}{\partial x} + \frac{c_\mathrm{s}^2}{\rho}\frac{\partial \rho}{\partial x} = 0. \tag{A.10}$$

But if we can find a function $F\{x, t\}$ such that

$$\begin{aligned} \frac{\partial F}{\partial t} &= \frac{c_\mathrm{s}}{\rho}\frac{\partial \rho}{\partial t} \\ \frac{\partial F}{\partial x} &= \frac{c_\mathrm{s}}{\rho}\frac{\partial \rho}{\partial x}, \end{aligned} \tag{A.11}$$

we can make (A.9) and (A.10) look even more similar:

$$\frac{\partial F}{\partial t} + u\,\frac{\partial F}{\partial x} + c_\mathrm{s}\,\frac{\partial u}{\partial x} = 0, \tag{A.12}$$

$$\frac{\partial u}{\partial t} + u\,\frac{\partial u}{\partial x} + c_\mathrm{s}\,\frac{\partial F}{\partial x} = 0. \tag{A.13}$$

Equations (A.12) and (A.13) can then be added and subtracted to yield

$$\frac{\partial(F + u)}{\partial t} + (c_\mathrm{s} + u)\,\frac{\partial(F + u)}{\partial x} = 0 \tag{A.14}$$

$$\frac{\partial(F - u)}{\partial t} + (-c_\mathrm{s} + u)\,\frac{\partial(F - u)}{\partial x} = 0. \tag{A.15}$$

In this form they tell us that on any plane perpendicular to the x axis which moves with velocity $c_\mathrm{s} + u$ the quantity $(F + u)$ remains constant, while $(F - u)$ remains constant on planes which move at $-c_\mathrm{s} + u$. We may infer that if $(F - u)$ happens to be the same everywhere at one instant of time it remains uniform and constant thereafter. We then have a *simple wave* disturbance, associated with propagation in the $+x$ direction only, in which the variation of u with x and t mirrors the variation of F. We can evidently arrange, by adjusting the zero for F, that for such a simple wave $u = F$ everywhere, and in that case the velocity with which planes of constant $(F + u)$ propagate is $c_\mathrm{s} + F$. Simple waves travelling in the opposite direction, for which $u = -F$, are also possible of course.

The conditions expressed by (A.11) are evidently satisfied by

$$F = \int_{\rho_\mathrm{o}}^{\rho} \frac{c_\mathrm{s}}{\rho}\,\mathrm{d}\rho, \tag{A.16}$$

where ρ_o is the density of undisturbed fluid outside the region affected by the sound wave. If the fluid is an ideal gas, and if the compressions and rarefactions which it undergoes when sound passes through it are virtually isentropic [§3.2], c_s is proportional to $\rho^{(\gamma-1)/2}$ and this choice for F corresponds to

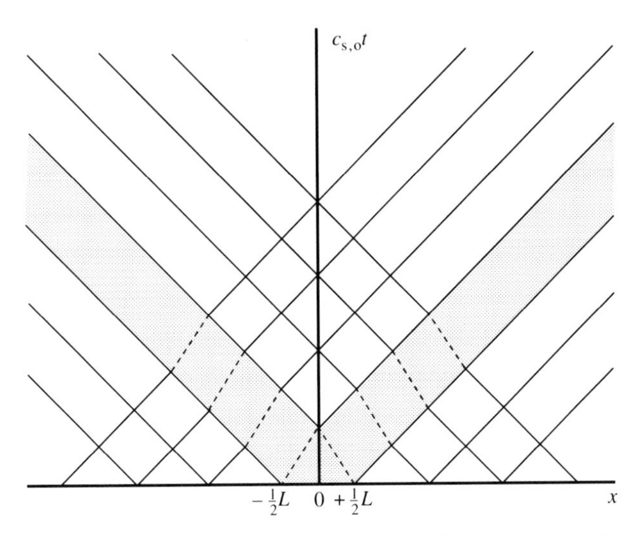

Figure A.1 Two simple waves separating from a (symmetrical) disturbance confined at $t = 0$ to the region $-\frac{1}{2}L < x < +\frac{1}{2}L$. The inclined lines are contours on which $F \pm u = 0$.

$$F = \frac{2}{\gamma - 1}(c_s - c_{s,o}); \tag{A.17}$$

when the refererence frame is such that the velocity u_o of the undisturbed gas is zero, no further adjustment is required to the zero for F to ensure that $u = F$ everywhere in a simple wave which is travelling in the $+x$ direction. Hence the velocity with which planes of constant $(F + u)$ propagate in such a simple wave is

$$c_s + u = c_s + F = \frac{\gamma + 1}{\gamma - 1} c_s - \frac{2}{\gamma - 1} c_{s,o}. \tag{A.18}$$

This corresponds exactly to (3.42), and indeed all the conclusions reached in §3.6 are substantiated by the above analysis. When $u = F$, planes of constant $(F + u)$ are planes of constant F, of course, and since (A.17) defines F uniquely in terms of ρ (for given ρ_o) they are also planes of constant ρ. It is because the velocity with which these planes propagate increases with ρ that excess density profiles of simple waves distort in the way they do.

Figure A.1 is a space–time diagram to illustrate the constraints which apply to the development of an arbitrary one-dimensional density disturbance – not necessarily a simple wave – in a body of otherwise stationary gas of uniform density ρ_o. At time $t = 0$ the disturbance is supposed to be localised within a length L centred on $x = 0$. Since F and u are both zero outside this region at $t = 0$, it follows that along all the lines drawn in the diagram which start from points where

$x > +\frac{1}{2}L$ or $x < -\frac{1}{2}L$ at an inclination to the t axis of $+45°$, and which represent the trajectories of planes propagating with velocity $c_s + u$, the quantity $(F + u)$ must be zero; note that these lines are straight except where they cross the regions which are shaded in the diagram, within which c_s may differ from $c_{s,o}$ and u may differ from zero. Similarly, $(F - u)$ must be zero along all the lines inclined in the opposite sense. It follows that, except in the two shaded bands, F must be zero everywhere – which implies that ρ must equal ρ_o – and so must u. Once the two shaded bands have separated from one another, in fact, the disturbance is bound to consist, whatever its nature at $t = 0$, of two simple wave trains propagating in opposite directions with length not exceeding L, and behind each of these simple waves the gas is stationary and has the same density – and hence the same pressure and temperature – as it had originally.

The above account of Riemann's elegant transformation and of the quite subtle argument which he based upon it is a condensed one, which the reader may well have difficulty in digesting at a first reading. The argument applies, *mutatis mutandis*, to fluids other than ideal gases. It does not reveal the whole story once a shock front has formed, however, largely because the constant which relates c_s to $\rho^{(\gamma-1)/2}$ is then different on the two sides of the front.

A.3 Attenuation due to thermal conduction

We now return to the linearised theory of §A.1 and consider the effect of dissipative processes which have so far been left out of account. We may start with the general equation analogous to (2.1) which expresses conservation of energy rather than conservation of mass, namely

$$\nabla \cdot (\rho e' \boldsymbol{u} + \boldsymbol{H} + p\boldsymbol{u}) = -\frac{\partial(\rho e')}{\partial t}, \qquad \text{(A.19)}$$

with

$$e' = e + \frac{1}{2}u^2 + gz. \qquad \text{(A.20)}$$

Here e is the intrinsic internal energy of the fluid per unit mass, while e' is internal energy per unit mass in a larger sense, with kinetic and gravitational potential energies included. Each of the three terms inside the brackets on the left of (A.19) represents an energy flux. The first term, $\rho e' \boldsymbol{u}$, is associated with the transport of energy-laden mass. The second, $\boldsymbol{H}$, is the heat flux associated with thermal conduction, which is given in terms of the thermal conductivity κ by $-\kappa\nabla T$. The third, $p\boldsymbol{u}$, represents the rate at which energy is transferred per unit area across any fixed plane by the work which the fluid on one side does on the fluid on the other. Equation (A.19) may be simplified with the aid of (2.1) and (2.5) to

$$\nabla \cdot (\boldsymbol{H} + p\boldsymbol{u}) = -\rho \frac{\partial e'}{\partial t} - \rho(\boldsymbol{u} \cdot \nabla)e' = -\rho \frac{\mathrm{D}e'}{\mathrm{D}t}, \tag{A.21}$$

but in the context of one-dimensional sound waves of infinitesimal amplitude it can be simplified much further, by ruthless elimination of all second-order terms and also of the gravitational term in e', which cannot be relevant when the wavelength is very much less than the scale height of the atmosphere (about 20 km). For an ideal gas in which $e = c_V T$ it then becomes

$$-\kappa_0 \frac{\partial^2 T}{\partial x^2} + p_0 \frac{\partial u}{\partial x} \approx -\rho_0 c_V \frac{\partial T}{\partial t}. \tag{A.22}$$

We now seek to eliminate T and u from (A.22) in favour of the condensation s, using (A.3), (A.4) and the equation of state for an ideal gas. The task is simplified if we first differentiate both sides of (A.22) twice with respect to x, thereby transforming it to

$$\left(\kappa_0 \frac{\partial^2}{\partial x^2} - \rho_0 c_V \frac{\partial}{\partial t}\right) \frac{\partial^2 T}{\partial x^2} = p_0 \frac{\partial^3 u}{\partial x^3}. \tag{A.23}$$

The equation of state is

$$T = \frac{1}{c_p - c_V} \frac{p}{\rho} = \frac{1}{c_p - c_V} \frac{p}{\rho_0(1 + s)} \tag{A.24}$$

[(3.26)], and to first order in s it differentiates thus:

$$\frac{\partial T}{\partial x} \approx \frac{1}{c_p - c_V}\left(-\frac{\partial u}{\partial t} - \frac{p_0 \partial s}{\rho_0 \partial x}\right) \quad \text{[using (A.4)]},$$

$$\frac{\partial^2 T}{\partial x^2} \approx \frac{1}{c_p - c_V}\left(\frac{\partial^2 s}{\partial t^2} - \frac{p_0}{\rho_0} \frac{\partial^2 s}{\partial x^2}\right) \quad \text{[using (A.3)]}.$$

From (A.3) alone we have

$$\frac{\partial^3 u}{\partial x^3} \approx \frac{\partial^3 s}{\partial x^2 \partial t}.$$

Hence (A.23) becomes, after a little manipulation,

$$\frac{\kappa_0}{\rho_0 c_V} \frac{\partial^2}{\partial x^2}\left(\frac{p_0}{\rho_0} \frac{\partial^2 s}{\partial x^2} - \frac{\partial^2 s}{\partial t^2}\right) \approx \frac{\partial}{\partial t}\left(\frac{\gamma p_0}{\rho_0} \frac{\partial^2 s}{\partial x^2} - \frac{\partial^2 s}{\partial t^2}\right). \tag{A.25}$$

To describe completely the effects of thermal conduction on a small-amplitude sound wave we need this fourth-order differential equation, which combines aspects of the one-dimensional diffusion equation with aspects of the one-dimensional wave equation in an intriguing way. The factor $\kappa_o/\rho_o c_V$ which appears on the left-hand side is, of course, just the thermal diffusivity χ_o.

For (A.6) to be a solution to (A.25), k and ω must satisfy the dispersion relation

$$\chi_o k^2 \left(\frac{p_o}{\rho_o} k^2 - \omega^2\right) = \mathrm{i}\omega \left(\frac{\gamma p_o}{\rho_o} k^2 - \omega^2\right),$$

or

$$(\gamma + \mathrm{i}\varepsilon) \frac{p_o}{\rho_o} k^2 \approx (1 + \mathrm{i}\varepsilon)\omega^2, \tag{A.26}$$

with

$$\varepsilon = \frac{\chi_o k^2}{\omega} = \frac{\chi_o \omega}{c_s^2}. \tag{A.27}$$

At very low frequencies where ε is negligible this corresponds to (A.7), with β_o taking its isentropic value, $1/\gamma p_o$. At frequencies so high that $1/\varepsilon$ is negligible it again corresponds to (A.7), but with β_o taking its isothermal value, $1/p_o$. At audible frequencies, where in practice ε is much less than unity [§3.2] but not completely negligible, its roots are complex.

Let us choose ω to be real and write

$$k = k' + \mathrm{i}k'' = k' \left(1 + \mathrm{i}\frac{\alpha}{4\pi}\right). \tag{A.28}$$

With k', k'' and α all real, this implies a wave with velocity ω/k' whose energy density attenuates like

$$\mathrm{e}^{-2k''x} = \mathrm{e}^{-\alpha k' x/2\pi} = \mathrm{e}^{-\alpha x/\lambda} \approx 1 - \alpha x/\lambda.$$

Thus α is the fractional power loss per wavelength, a quantity often referred to as the *attenuation coefficient* (though other meanings are sometimes attached to this term). On inserting (A.28) into (A.26) and equating real and imaginary parts one obtains, to first order in ε,

$$\frac{\omega}{k'} = c_s \approx \sqrt{\frac{1}{\gamma p_o}}$$

(it is only in the second order that thermal conduction affects the velocity of sound) and, for the attenuation coefficient due to thermal conduction alone,

$$\alpha = \alpha_\kappa \approx 2\pi\varepsilon \frac{\gamma - 1}{\gamma} = 4\pi^2 \frac{\gamma - 1}{\gamma} \frac{\chi_o}{c_s \lambda}. \tag{A.29}$$

For a gas such as air in which $\gamma \approx 1{\cdot}4$ this corresponds, in view of (3.17), (3.10) and (3.11), to

$$\alpha_\kappa \approx 20\,\frac{\ell}{\lambda},$$

where ℓ is the molecular mean free path.

A.4 Attenuation due to viscosity

In a fluid which has viscosity the pressure is anisotropic and Euler's equation is insufficient; we should replace $\partial p/\partial x$ where it occurs on the right-hand side of (A.4) by

$$\frac{\partial}{\partial x}\left\{p - \left(\frac{4}{3}\eta + \eta_\mathrm{b}\right)\frac{\partial u}{\partial x}\right\} \tag{A.30}$$

[(6.22)], where η is the shear viscosity and η_b the bulk viscosity of the fluid. Within the framework of a linearised theory the effect of the substitution is straight-forward. In view of (A.3), which is unaffected by viscosity, (A.4) becomes

$$\frac{\partial u}{\partial t} \approx -\frac{1}{\beta_\mathrm{o}\rho_\mathrm{o}}\frac{\partial s}{\partial x} - \frac{1}{\rho_\mathrm{o}}\left(\frac{4}{3}\eta + \eta_\mathrm{b}\right)\frac{\partial^2 s}{\partial x \partial t}, \tag{A.31}$$

and elimination of u between (A.3) and (A.31) leads to

$$\left\{1 + \beta_\mathrm{o}\left(\frac{4}{3}\eta + \eta_\mathrm{b}\right)\frac{\partial}{\partial t}\right\}\frac{\partial^2 s}{\partial x^2} = \beta_\mathrm{o}\rho_\mathrm{o}\frac{\partial^2 s}{\partial t^2}. \tag{A.32}$$

The equivalent dispersion relation is

$$\left\{1 - \mathrm{i}\omega\beta_\mathrm{o}\left(\frac{4}{3}\eta + \eta_\mathrm{b}\right)\right\}k^2 = \beta_\mathrm{o}\rho_\mathrm{o}\omega^2. \tag{A.33}$$

Here again the roots are complex. On substituting $(k' + \mathrm{i}k'')$ for k and equating imaginary parts one obtains, for the attenuation coefficient to first order in $\omega\beta_\mathrm{o}\eta$ due to viscosity alone,

$$\alpha = \alpha_\eta \approx \frac{8\pi\omega\beta_\mathrm{o}\eta}{3} = \frac{16\pi^2\eta}{3\rho_\mathrm{o}c_\mathrm{s}\lambda}. \tag{A.34}$$

The bulk viscosity η_b has here been ignored, but we return to it in §A.5 below. Now the kinetic theory of gases tells us that[1]

$$\eta \approx \rho\ell\bar{c}_\mathrm{mol}. \tag{A.35}$$

[1] This formula is often quoted with a factor $\frac{1}{3}$ on the right-hand side, but a more reliable value for the numerical coefficient is $5\pi/32$.

Using this result in conjunction with (A.34) we may estimate that for a gas such as air

$$\alpha_\eta \approx 30\,\frac{\ell}{\lambda}.$$

The suffix here indicates a contribution due to shear viscosity alone.

Contributions to attenuation from different sources are expected to be additive, and the 'classical' expression for the total attenuation coefficient in a gas is

$$\alpha = \alpha_\kappa + \alpha_\eta, \tag{A.36}$$

with α_κ and α_η given by (A.29) and (A.34) respectively. It provides a satisfactory fit to experimental data for monatomic gases, but for polyatomic gases, including air, it is seriously deficient. It implies that a plane sound wave with a frequency of 1 kHz and a wavelength of 0·3 m, travelling through air at normal temperature and pressure, loses about 5% of its power per kilometre. In the Earth's atmosphere, the loss rate is normally a good deal greater than that.

A.5 Additional attenuation mechanisms

A monatomic ideal gas stores energy in only one way, namely as translational kinetic energy of its molecules, but a polyatomic one can store it in molecular rotations and vibrations as well. At low frequencies the thermal capacity and the specific heat ratio (say γ_o) reflect all these possibilities: for the linear CO_2 molecule at room temperature, for example, the thermal capacity per molecule at constant volume is about $\frac{7}{2}k_B$, of which $\frac{3}{2}k_B$ is attributable to the translational degrees of freedom, k_B to the rotational ones, and a further k_B to the not-fully-excited vibrational ones; γ_o is therefore about $\frac{9}{7}$ for carbon dioxide. At high frequencies, however, the vibrational degrees of freedom become irrelevant because they do not have time to respond to the oscillations of temperature which accompany a sound wave, and the effective specific heat ratio (say γ_∞) is then greater than γ_o: in carbon dioxide at room temperature γ_∞ is about $\frac{7}{5}$. Over a range of intermediate frequencies there is both dispersion – as the velocity of sound increases from $\sqrt{\gamma_o p/\rho}$ to $\sqrt{\gamma_\infty p/\rho}$ – and additional attenuation. The additional attenuation predicted by theory turns out to be described by

$$\alpha_{\text{vib}} = 2\pi\,\frac{\gamma_\infty - \gamma_o}{\sqrt{\gamma_\infty \gamma_o}}\,\frac{\omega\tau_{\text{vib}}}{1 + (\omega\tau_{\text{vib}})^2}, \tag{A.37}$$

where τ_{vib} is a relaxation time determined by the strength of the coupling between the translational and vibrational degrees of freedom, and in many polyatomic gases this is large enough to make the classical attenuation insignificant by comparison. Carbon dioxide is a case in point. Experimental results for α in this

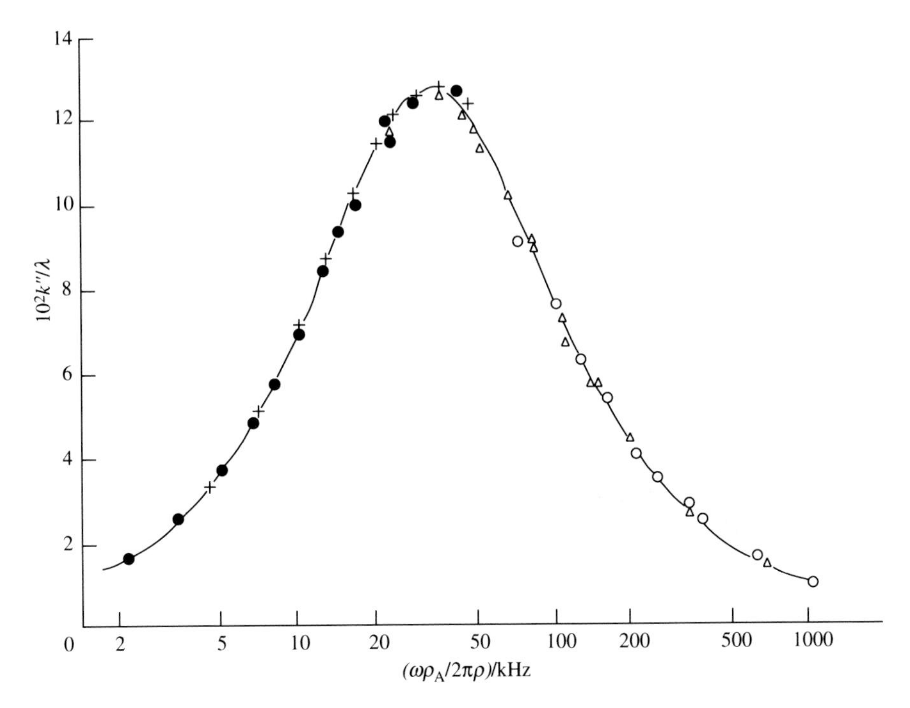

Figure A.2 Sound absorption in carbon dioxide at 23 °C for various values of density ρ (ρ_A being density at normal atmospheric pressure) and for four different frequencies between 1·4 kHz and 3·0 MHz. Since τ_{vib} is inversely proportional to density at constant temperature, the quantity plotted on a logarithmic scale along the abscissa axis is proportional to $\omega\tau_{vib}$.
[Based on results quoted by H. O. Kneser, in *Physical Acoustics, Principles and Methods*, Vol. II Part A, ed. W. P. Mason, Academic Press, New York, 1965.]

gas, plotted in fig. A.2, exhibit a peak which can be closely fitted by (A.37) with $\tau_{vib} \approx 4 \times 10^{-6}$ s at normal temperature and pressure.

The attenuation of sound in the Earth's atmosphere is due almost entirely to the oxygen present in it, which contributes to α an absorbtion peak which is similar in form to the peak shown in fig. A.2. It can likewise be described by (A.37), though at normal temperatures the factor $(\gamma_\infty - \gamma_o)/\sqrt{\gamma_\infty\gamma_o}$ is smaller for oxygen than it is for carbon dioxide by a factor of about 100. The frequency at which the peak has its maximum is probably about 50 Hz in pure oxygen, corresponding to a value of about 3 ms for the relaxation time τ_{vib}. Considering that the time between collisions for a single oxygen molecule is of order 3×10^{-10} s, this value for τ_{vib} is astonishingly long. One has to remember, however, that the quantum of energy associated with the vibration of an O_2 molecule is considerably bigger than $k_B T$ at normal temperatures; that is why the vibrational modes are so little excited, and

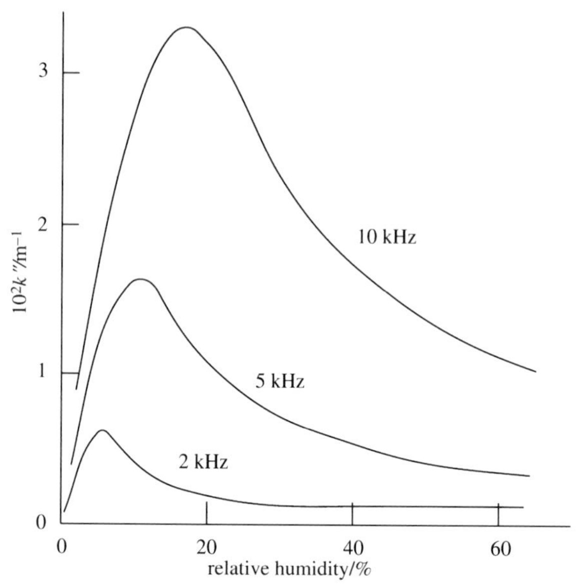

Figure A.3 Inverse of attenuation length versus relative humidity for air at 20 °C.
[Based on results quoted by H. O. Kneser, in *Physical Acoustics, Principles and Methods*, Vol. II Part A, ed. W. P. Mason, Academic Press, New York, 1965.]

why the difference between γ_∞ and γ_o for oxygen is so small. On this account, collisions in which a vibrational quantum is created at the expense of translational or rotational energy, or else destroyed, are extremely rare.

In the atmosphere, however, there is always water vapour present, and it has a catalytic effect in greatly reducing τ_{vib} for the oxygen. The frequency at which the absorption peak occurs is thereby shifted into the kHz range, and the more humid the atmosphere the higher this frequency is. Some results are shown in fig. A.3 for $\alpha/2\lambda$ (i.e. for k'', the inverse of the attenuation length) in the atmosphere at 20 °C, plotted against relative humidity for three fixed frequencies. One may estimate by extrapolation of the data plotted there that when the relative humidity is only about 5% the 1 kHz sound wave of §A.4 may lose 5% of its power in under 10 m rather than in 1 km.

If the vibrational degrees of freedom in a polyatomic gas take a finite time to reach equilibrium with the translational degrees of freedom, is the same not true of the rotational degrees of freedom? The answer is yes, and in principle rotational relaxation contributes another absorbtion peak, also described by (A.37) but with γ_∞ in place of γ_o, $\frac{5}{3}$ (the value of γ appropriate for a monatomic gas) in place of γ_∞, and a different relaxation time τ_{rot} in place of τ_{vib}. However, τ_{rot} is very much shorter than τ_{vib}, and the rotational relaxation peak is centred on a frequency which is normally beyond the range of measurement. All we normally see of it, therefore, is the low frequency tail for which

$$\alpha_{\rm rot} = 2\pi \frac{\frac{5}{3} - \gamma_\infty}{\sqrt{\frac{5}{3}\gamma_\infty}} \omega \tau_{\rm rot}, \qquad \text{(A.38)}$$

and in so far as $\alpha_{\rm rot}$ is then proportional to ω it is indistinguishable in form from the α_η which (A.34) describes. This additional absorption may in fact be regarded as essentially viscous in origin but attributable to the *bulk viscosity* $\eta_{\rm b}$, which we left on one side in passing from (A.33) to (A.34). To make the $\alpha_{\rm rot}$ of (A.38) correspond to the additional term in α_η which arises when $\frac{4}{3}\eta$ is replaced by $\frac{4}{3}\eta + \eta_{\rm b}$, we need to choose

$$\eta_{\rm b} = \frac{\frac{5}{3} - \gamma_\infty}{\sqrt{\frac{5}{3}\gamma_\infty}} \rho c_{\rm s}^2 \tau_{\rm rot}. \qquad \text{(A.39)}$$

Where the bulk viscosity of gases is referred to in the literature, it is generally stated to be zero for a monatomic gas but comparable with the shear viscosity for a polyatomic one and, on the assumption that $\tau_{\rm rot}$ is of the same order as the collision time $\ell/\bar{c}_{\rm mol}$, such statements are consistent with (A.39) and (A.35). However, the statements are based upon measurements of the attenuation of sound at very high frequencies, that being the only context in which effects attributable to bulk viscosity are readily discernible. Were we able to detect such effects in flow experiments involving angular frequencies much less than $\tau_{\rm vib}^{-1}$, the apparent bulk viscosity would be [compare the low-frequency limit of (A.37) with (A.38)]

$$\eta_{\rm b} = \frac{\gamma_\infty - \gamma_{\rm o}}{\sqrt{\gamma_\infty \gamma_{\rm o}}} \rho c_{\rm s}^2 \tau_{\rm vib}. \qquad \text{(A.40)}$$

For a polyatomic gas with large $\tau_{\rm vib}$ it would be much greater than the shear viscosity.

In polyatomic liquids, there may be several relaxation processes analogous to the processes in gases whereby the internal degrees of freedom of the molecules come into equilibrium with the translational degrees of freedom. Each of these processes is liable to contribute a peak to the curve of sound attenuation versus frequency, and each of them contributes also to the effective bulk viscosity at frequencies well below the peak.

Further reading

An exposition of Riemann's treatment is to be found in *Waves in Fluids*, Cambridge University Press, 1978, by J. Lighthill. For a thorough discussion of the classical theory of sound attenuation, and of additional attenuation mechanisms in both gases and liquids, refer to *Ultrasonic Absorption*, Clarendon Press, Oxford, 1963, by A. B. Bhatia.

INDEX

For topics which arise frequently in the text, references are given only to pages on which these topics are introduced or explained.

It is over 350 years since Toricelli discovered the law obeyed by fountains, yet fluid dynamics remains an active and important branch of physics. Modern examples of the extraordinarily varied and intriguing phenomena with which it deals include solitons, chaotic behaviour in convecting systems, and non-Newtonian effects in polymer solutions. This book provides an accessible and comprehensive account of the subject, emphasising throughout the fundamental physical principals, and stressing the connections with other branches of physics.

Beginning with a gentle introduction, the book goes on to cover: Bernoulli's theorem and its implications; compressible flow and shock waves; potential flow; surface waves and ship waves; viscosity and highly viscous flow; vorticity dynamics and boundary layers; thermal convection and instabilities; turbulence and turbulent shear flow; the behaviour of non-Newtonian fluids, including nematic liquid crystals and liquid helium; and the propagation and attenuation of sound in gases.

Undergraduate or graduate students in physics or engineering who are taking courses in fluid dynamics will find this book invaluable, but it will also be of great interest to anyone who wants to find out more about this fascinating subject.

Front cover illustration shows flow past a sphere, made visible with coloured dyes. In this case the conditions are such that the boundary layer attached to the obstacle is laminar. In the case illustrated on the back cover of flow past a hemisphere, part of it is clearly turbulent. (Photographs by Henri Werlé, reproduced by courtesy of ONERA, France.)

Cover design: Colin Ward

CAMBRIDGE
UNIVERSITY PRESS

ISBN 0-521-42969-2